AF572568

Benchmark Papers in Geology

Series Editor: Rhodes W. Fairbridge
Columbia University

VOLUME

30 HOLOCENE TIDAL SEDIMENTATION/*George deVries Klein*
31 PALEOBIOGEOGRAPHY/*Charles A. Ross*
32 MECHANICS OF THRUST FAULTS AND DÉCOLLEMENT/*Barry Voight*
33 WEST INDIES ISLAND ARCS/*Peter H. Mattson*
34 CRYSTAL FORM AND STRUCTURE/*Cecil J. Schneer*
35 OCEANOGRAPHY: Concepts and History/*Margaret B. Deacon*
36 METEORITE CRATERS/*G. J. H. McCall*
37 STATISTICAL ANALYSIS IN GEOLOGY/*John M. Cubitt and Stephen Henley*
38 AIR PHOTOGRAPHY AND COASTAL PROBLEMS/*Mohamed T. El-Ashry*
39 BEACH PROCESSES AND COASTAL HYDRODYNAMICS/*John S. Fisher and Robert Dolan*
40 DIAGENESIS OF DEEP-SEA BIOGENIC SEDIMENTS/*Gerrit J. van der Lingen*
41 DRAINAGE BASIN MORPHOLOGY/*Stanley A. Schumm*
42 COASTAL SEDIMENTATION/*Donald J. P. Swift and Harold D. Palmer*
43 ANCIENT CONTINENTAL DEPOSITS/*Franklyn B. Van Houten*
44 MINERAL DEPOSITS, CONTINENTAL DRIFT AND PLATE TECTONICS/*J. B. Wright*
45 SEA WATER: Cycles of the Major Elements/*James I. Drever*
46 PALYNOLOGY, PART I: Spores and Pollen/*Marjorie D. Muir and William A. S. Sarjeant*
47 PALYNOLOGY, PART II: Dinoflagellates, Acritarchs, and Other Microfossils/*Marjorie D. Muir and William A. S. Sarjeant*
48 GEOLOGY OF THE PLANET MARS/*Vivien Gornitz*
49 GEOCHEMISTRY OF BISMUTH/*Ernest E. Angino and David T. Long*
50 ASTROBLEMES–CRYPTOEXPLOSION STRUCTURES/*G. J. H. McCall*
51 NORTH AMERICAN GEOLOGY: Early Writings/*Robert Hazen*
52 GEOCHEMISTRY OF ORGANIC MOLECULES/*Keith A. Kvenvolden*
53 TETHYS: The Ancestral Mediterranean/*Peter Sonnenfeld*
54 MAGNETIC STRATIGRAPHY OF SEDIMENTS/*James P. Kennett*
55 CATASTROPHIC FLOODING: The Origin of the Channeled Scabland/*Victor R. Baker*
56 SEAFLOOR SPREADING CENTERS: Hydrothermal Systems/*Peter A. Rona and Robert P. Lowell*
57 MEGACYCLES: Long-Term Episodicity in Earth and Planetary History/*G. E. Williams*
58 OVERWASH PROCESSES/*Stephen P. Leatherman*
59 KARST GEOMORPHOLOGY/*M. M. Sweeting*
60 RIFT VALLEYS: Afro-Arabian/*A. M. Quennell*
61 MODERN CONCEPTS OF OCEANOGRAPHY/*G. E. R. Deacon and Margaret B. Deacon*
62 OROGENY/*John G. Dennis*

A BENCHMARK® Books Series

SEAFLOOR SPREADING CENTERS
Hydrothermal Systems

Edited by

PETER A. RONA
Atlantic Oceanographic and
Meteorological Laboratories-
National Oceanic
and Atmospheric Administration

and

ROBERT P. LOWELL
Georgia Institute of Technology

STROUDSBURG, PENNSYLVANIA

Benchmark Papers in Geology, Volume 56
Library of Congress Catalog Card Number: 79-18265
ISBN: 0-87933-363-4

82 81 80 1 2 3 4 5
Manufactured in the United States of America

LIBRARY OF CONGRESS CATALOGING IN PUBLICATION DATA
Main entry under title:
Seafloor spreading centers.
(Benchmark papers in geology; 56)
Bibliography: p.
Includes indexes.
1. Sea-floor spreading—Addresses, essays, lectures. 2. Geothermal resources—Addresses, essays, lectures. 3. Heat budget (Geophysics)—Addresses, essays, lectures. 4. Hydrothermal deposits—Addresses, essays, lectures. I. Rona, Peter A. II. Lowell, Robert P.
QE511.7.S4 551.1 79-18265
ISBN 0-87933-363-4

Distributed world wide by Academic Press,
a subsidiary of Harcourt Brace Jovanovich,
Publishers.

SERIES EDITOR'S FOREWORD

The philosophy behind the Benchmark Papers in Geology is one of collection, sifting and rediffusion. Scientific literature today is so vast, so dispersed, and, in the case of old papers, so inaccessible for readers not in the immediate neighborhood of major libraries that much valuable information has been ignored by default. It has become just so difficult, or so time consuming, to search out the key papers in any basic area of research that one can hardly blame a busy person for skimping on some of his or her "homework."

This series of volumes has been devised, therefore, as a practical solution to this critical problem. The geologist, perhaps even more than any other scientist, often suffers from twin difficulties—isolation from central library resources and immensely diffused sources of material. New colleges and industrial libraries simply cannot afford to purchase complete runs of all the world's earth science literature. Specialists simply cannot locate reprints or copies of all their principal reference materials. So it is that we are now making a concerted effort to gather into single volumes the critical materials needed to reconstruct the background of any and every major topic of our discipline.

We are interpreting "geology" in its broadest sense: the fundamental science of the planet Earth, its materials, its history, and its dynamics. Because of training in "earthy" materials, we also take in astrogeology, the corresponding aspect of the planetary sciences. Besides the classical core disciplines such as mineralogy, petrology, structure, geomorphology, paleontology, and stratigraphy, we embrace the newer fields of geophysics and geochemistry, applied also to oceanography, geochronology, and paleoecology. We recognize the work of the mining geologists, the petroleum geologists, the hydrologists, and the engineering and environmental geologists. Each specialist needs a working library. We are endeavoring to make the task of compiling such a library a little easier.

Each volume in the series contains an introduction prepared by a specialist (the volume editor)—a "state of the art" opening or a summary of the object and content of the volume. The articles, usually some twenty to fifty reproduced either in their entirety or in significant extracts, are selected in an attempt to cover the field, from key papers of the last century to fairly recent work. Where the original works are in foreign languages, we have endeavored to locate or commission translations. Geologists, because of their global subject, are often acutely aware of the oneness of our world. The selections cannot therefore be restricted to any one country, and whenever possible an attempt is made to scan the world literature.

To each article, or group of kindred articles, some sort of "highlight commentary" is usually supplied by the volume editor. This commentary should serve to bring that article into historical perspective and to emphasize its particular role in the growth of the field. References, or citations, wherever possible, will be reproduced in their entirety—for by this means the observant reader can assess the background material available to that particular author, or, if desired, he or she too can double check the earlier sources.

A "benchmark," in surveyor's terminology, is an established point on the ground that is recorded on our maps. It is usually anything that is a vantage point, from a modest hill to a mountain peak. From the historical viewpoint, these benchmarks are the bricks of our scientific edifice.

RHODES W. FAIRBRIDGE

PREFACE

The papers selected for this volume constitute definitive steps in the realization that hydrothermal systems exist at seafloor spreading centers and that the existence of the systems has major implications for man's understanding of earth processes and his utilization of earth resources. With regard to earth processes, the significance of the role of hydrothermal systems at seafloor spreading centers in the global heat budget and in geochemical mass balances is only now emerging. With regard to earth resources, recognition of the geothermal energy potential of such systems and their activity in concentrating metallic mineral deposits in oceanic crust is exerting an increasing influence on exploration for future energy and mineral resources.

The subject is interdisciplinary, building upon geophysical, geochemical, and geological studies relevant to the physics and chemistry of hydrothermal systems and to axial processes of seafloor spreading centers comprising the Earth-encircling oceanic ridge-rift system. The papers are grouped into three categories: theory, dealing with flow of hot aqueous solutions through permeable media at seafloor spreading centers; experiment, treating laboratory studies of basalt-seawater interaction at elevated temperatures and pressures; and field, dealing with actual investigations. The relative number of pages selected for reprinting in each group is approximately proportional to the number of papers published in that group, with the majority of papers devoted to field investigations. Supplementary basic background references are cited, and relevant papers that pursue the various lines of research are identified in the introductions in order to place the reprinted papers in context of the entire developing subject.

PETER A. RONA
ROBERT P. LOWELL

CONTENTS

CONTENTS BY AUTHOR

SEAFLOOR SPREADING CENTERS

INTRODUCTION

The study of hydrothermal systems at seafloor spreading centers has only recently attained critical mass as a viable field of research. That hydrothermal systems might exist at seafloor spreading centers was inferred from the presence of the components of such systems, comprised of a volcanogenic heat source to drive convection, seawater as the fluid medium, and fractured volcanic rocks as the permeable solid medium (Elder 1965; Deffeyes 1970). The actual existence of hydrothermal systems at seafloor spreading centers and the characteristics of these systems have been determined within the past fifteen years in conjunction with the development of the theory of plate tectonics, which has focused attention on seafloor spreading centers as divergent plate boundaries where oceanic lithosphere is generated.

The theme for this volume originated as a symposium organized by the editors at the 1977 Annual Meeting of the Geological Society of America (Rona and Lowell 1978). The various studies reprinted and referenced in this volume provide an overview of the magnitude and a perspective on the nature of hydrothermal systems at seafloor spreading centers. The estimated heat loss attributed to hydrothermal convection at seafloor spreading centers is 2.0×10^{12} cal/s, which equals 20 percent of the Earth's total heat loss (Paper 7), which invalidates our prior concept of the equivalency of heat flow through ocean basins and continents. The estimated seawater flux through the rocks at seafloor spreading centers is $1.3\text{-}9 \times 10^{17}$ g/yr, which is sufficient to cycle the entire mass of the world ocean every 5-11 m.y. (Paper 30) and to effectively control the chemistry of certain elements in seawater. The oceanic crustal generation rate at seafloor spreading centers is 5×10^{16} g/yr (Deffeyes 1970), 2-20 times smaller than the seawater flux; this difference has profound implications for basalt-seawater interaction (Paper 4). Prior to realization of the magnitude of the flux of seawater and its role in convective heat transfer through oceanic ridges, the oceans were considered a passive sink for elements weathered from the continents and transported by rivers. With the emerging realization of the magnitude of hydrothermal processes involving seawater as a reactive hydrothermal medium at temperatures ranging up to about 400°C, hydrothermal systems at seafloor spreading centers are recognized as active sources for the exchange of elements with fluxes comparable to those derived from rivers. Our understanding of the geochemical cycles and mass budgets of elements is undergoing radical revision as a consequence of this realization.

The estimates of magnitude of hydrothermal convection pertain to the global seafloor spreading center system. A distinction must be made regarding the intensity of hydrothermal activity (Paper 31) to achieve perspective. Low-intensity hydrothermal activity, characterized by solution temperatures less than 100°C and generally close to the ambient temperature of seawater, temperature gradients less than 100°C/km, and the production of low-temperature hydration reactions (zeolite metamorphic facies), is ubiquitous at seafloor spreading centers (Hall and Robinson, 1979). High-intensity hydrothermal activity, characterized by solution temperatures between 100° and 400°C, temperature gradients between 100° and 1300°C/km, and the concentration of hydrothermal mineral deposits (not necessarily high-temperature minerals), is extremely localized at seafloor spreading centers. The various hydrothermal mineral deposits described in this volume are considered products of extremely localized high-intensity hydrothermal systems (Paper 31) and are not characteristic of the major portion of the 52,000 km-long worldwide seafloor spreading center system. The degree of intensity of hydrothermal activity, the variation in products releated to intensity, and the controlling seafloor structural and thermal factors are subjects of ongoing research.

The future viability of research on hydrothermal systems at seafloor spreading centers is assured because the research is making major contributions to man's understanding of the earth on both basic and applied levels. On the basic level the research is filling large gaps in our knowledge of global geochemical mass balances and the earth's thermal regime. On the applied level the developing knowledge is providing the background to deal with contemporary environmental and resource concerns. Knowledge of the hydrothermal exchange of elements between seawater and basalt at seafloor spreading centers is basic to understanding the impact of natural and man-made pollution on the ocean environment. Knowledge of hydrothermal processes of metal concentration at seafloor spreading centers is instrumental in opening oceanic crust as a prospective new exploration province for critical metals at sea and on land (White 1968; Dmitriev et al. 1970; Sillitoe 1972, 1973; Hutchinson 1973; Duke and Hutchinson 1974; Bonatti 1975; Strong 1976; Rona 1977; Paper 31).

Part I

THEORETICAL STUDIES

Editors' Comments on Papers 1 and 2

1 **LISTER**
Qualitative Models of Spreading-Center Processes, Including Hydrothermal Penetration

2 **LOWELL**
Circulation in Fractures, Hot Springs, and Convective Heat Transport on Mid-Ocean Ridge Crests

Theoretical models of hydrothermal circulation in the oceanic crust are founded on the fundamental model developed in the classic paper of Lapwood (1948). Lapwood determined the critical condition for the onset of thermal instability in a homogeneous, fluid-saturated, horizontal slab of porous material, heated from below. This basic model has been used to interpret heat flow data obtained at seafloor spreading centers (Palmason 1967; Talwani et al. 1971; Paper 19; Paper 6). However, because the physical conditions at seafloor spreading centers are, in many respects, much different from the physical situation represented by Lapwood's simple model, several deviant models have been developed. The more recent models fall mainly into two classes: a) porous medium models in which the permeability is assumed to be due to fine-scale interconnected fractures such that Darcy's Law governs the flow rate through the material; (b) fractured rock models in which the permeability is principally due to discrete, widely spaced fractures and/or fault zones. A third model is the water-penetration model of Lister (1974; Paper 1; 1979). The porous medium models are, for the most part, essentially extrapolation of the basic Lapwood model to the finite-amplitude regime by using finite-difference analogues of the heat and fluid flow equations. For example, Ribando et al. (1976) consider two-dimensional finite amplitude convection in an infinite porous slab in an attempt to model the heat flow data from the Galapagos of Williams et al. (Paper 19). They examine the effects of different types of thermal and fluid bondary conditions as well as the effect of a decreasing permeability with depth. They conclude that convection in a crustal layer with a constant permeability of 4.5×10^{-12} cm^2 can account for the heat flow variations of the Galapagos. Moreover, they conclude that conductive heat flux measurements yield the entire heat flux through the ocean floor and that convective heat transport is important only if extensive hot springs exist.

Somewhat different numerical models have been developed by Fehn and Cathles (1978, 1979) and Patterson and Lowell (1979). These models take into account the effect of intrusive activity at seafloor spreading centers. A particular advantage of the Fehn and Cathles model is that it can incorporate lateral variations in permeability. However, these models also assume an isothermal upper boundary and have no convective heat transfer across the ocean floor. The bulk permeabilities assumed are on the order of 10^{-11} cm^2 or less.

The second class of theoretical models is that assuming a permeability controlled by discrete, widely spaced fractures. These models have been developed by Bodvarsson and Lowell (1972), Lowell (Paper 2), and Sleep and Wolery (1978). In contrast to the numerical treatments of porous medium convection, the fracture models are mathematically much simpler, and they do have convective heat losses. These models also fit the observed heat flow data and may be more useful in describing localized submarine geothermal phenomena, such as hot springs and near-bottom water temperature anomalies, than porous medium models. Typical calculations consider vertical fractures of a few millimeters width, separated distances of the order of kilometers, thus yielding permeabilities of roughly 10^{-8}–10^{-9} cm^2.

Finally, the water penetration model presented in a series of papers by Lister (1974; Paper 1; 1979) deserves special mention. In this model permeability in the oceanic crust is generated by thermal contraction cooling, the downward penetration of the fractured/unfractured boundary being of the order of several meters per year. During this growth phase of the permeable region or "active phase" of circulation, convection is very vigorous and heat is rapidly extracted from the rock. The lifetime of the active phase is expected to be of the order of hundreds of years, corresponding to a total depth of penetration of several kilometers. After the "active" phase is completed, convection continues in the permeable rock so long as sufficient temperature differences and permeability exist. The unique feature of this model is that it gives a theoretical estimate for the initial permeability of the oceanic crust (i.e., the permeability is not an ad hoc assumption as in other models). The value obtained is approximately 10^{-7} cm^2, which is several orders of magnitude greater than that assumed in porous medium models or discrete fracture models. The principal difficulty of the model, however, is that it is a one-dimensional model and does not take into account the presence of preexisting large scale fractures and/or faults. These may lead to preferential penetration of fluid, with the result that the permeability may be highly nonuniform.

To summarize, we can say that models have been developed that can account for the scatter in and, in part, for the low mean values of conductive heat flow measured near seafloor spreading centers. At present, the heat flow data alone is not sufficient to discriminate among the models. In particular, the heat flow data does not seem able to determine the scale of permeability in the oceanic crust (i.e., whether convection can best be described by porous, medium/fine scale fracture models or by large scale, widely spaced discrete fractures and/or fault zones. It is probable that geochemical data indicating the volume of the crust that interacts chemically with the circulating seawater may help discriminate among the models.

We also point out that the models developed to date are at best just order-of-magnitude models. Several important factors that are pertinent to the physical situation in the oceanic crust have been treated only in a very approximate manner. For example, the effect of sediment cover on convection has received very limited treatment. Skilbeck and Anderson (1979) derive a relationship that indicates that a sediment layer seals convection if $K_s/ahK_b \ll 1$ where K_s, K_b are the permeabilities of the sediment and basalt, respectively, a is the wavenumber of the convection cell, and h is the sediment thickness.

It is not entirely known, however, under what conditions convection in the underlying basalt may be manifested in the heat flow distribution, even if there is negligible water movement through the sediments. Recent measurements indicate the occurrence of nonlinear temperature gradients (Anderson et al. 1978). These have been interpreted as resulting from vertical water movement across the sediment–ocean bottom interface, but other possibilities exist.

Another problem involves the effect of rough seafloor topography on convection. Lister (Paper 5) has suggested the importance of topography on the location of upwelling and downwelling sites of circulating seawater: high heat flux (ascending fluid) correlated with topographic highs, low heat flux (descending fluid) correlated with topographic despressions. Heat flux thus correlates with topography in a manner opposite to the correlation anticipated from refraction of conductive heat flux. Laboratory and theoretical work on the effects of topography on hydrothermal circulation is currently being carried out by Hartline et al. (1975, 1978). Theoretical calculations are also being carried out by Lowell (1979, 1980). Little work has been done on the problem of convection driven simultaneously by topography and heating from below.

There is also no adequate interpretation of the near-bottom water temperature anomalies that have been measured by towed thermistor packages (Paper 19; Rona et al. 1975; Lowell and Rona 1976; Rona 1978; Crane and Normark 1977; Paper 20). Interpretations have been made using models either of thermal plumes or of turbulent diffusion of heat. The data, however, do not lie in the range over which either of these models is strictly applicable. Consequently, better experiments, as well as more appropriate models, are required in order to determine accurately what amount of heat transfer is indicated in near-bottom water temperature anomalies. It should be pointed out that the plume equations in Williams et al. (Paper 19) and Crane and Normark (1977) are dimensionally incorrect. The correct form is given in Lowell and Rona (1976).

Finally, there are no models that examine closely the evolutionary aspects of marine hydrothermal circulation.

Much theoretical work remains, and we anticipate that it will be carried forward in coordination with the continued collection and analysis of heat flow and geochemical data.

1

Reprinted from *Tectonophysics* **37**:203–218 (1977)

QUALITATIVE MODELS OF SPREADING-CENTER PROCESSES, INCLUDING HYDROTHERMAL PENETRATION

C.R.B. LISTER

Departments of Geophysics and Oceanography, University of Washington, Seattle, Washington 98195 (U.S.A.)

(Revised version received May 7, 1976)

ABSTRACT

Lister, C.R.B., 1976. Qualitative models of spreading-center processes, including hydrothermal penetration. In: S. Uyeda (editor), Subduction Zones, Mid-Ocean Ridges, Oceanic Trenches and Geodynamics. Tectonophysics, 37 (1—3): 203—218.

Observations on active geothermal areas suggest that water can penetrate rapidly into hot rock, and this is confirmed by semi-quantitative theory. Upwelling of magma material at an oceanic spreading center places hot rock close to an ample supply of cold water. Clues to the nature of the interaction can be found in the structure of ophiolite suites, and in simple physical induction on the nature of the upwelling. The subsurface dike injection zone seems to be dominated by conductive cooling and to provide a barrier against water penetration as long as it is heated by liquid magma in a chamber below. As soon as the magma has crystallized into a cumulate, water penetration proceeds rapidly to the base of the crust. It is probably stopped by meeting an ultramafic layer of olivine phenocrysts separated from the upwelling melt. The olivine crystals may be oriented by a fluidized-bed process near the spreading axis, forming a thin acoustically anisotropic layer. Below, mantle material depleted by partial melting grades back to primitive composition at depth. The mantle material should remain viscous enough to upwell in a broad zone, and the coupling of distributed upwelling to a thick rigid crust formed close to the spreading axis is responsible for the central valley and block tectonics characteristic of slow-spreading ridges. At fast spreading rates, the larger magma chamber decouples a thinner rigid surface layer and smooth topography results.

INTRODUCTION

Evidence has been accumulating that water penetration into young oceanic crust is a major factor in determining the thermal and physical state of the rocks. Heat-flow measurements have shown conclusively that the thermal flux through sediment ponds near the ridge crest is far too low to be compatible with new lithosphere formed at reasonable temperature (Lister, 1972). Nevertheless, careful analysis of the decay of topographic height with age has shown that this *is* compatible with a lithosphere of reasonable rock parameters formed at the melting point (Davis and Lister, 1974). The *distribution* of heat-flow values over medium-scale topography was the vital evidence that confirmed the existence of hydrothermal circulation in the basement rocks (Lister, 1972). Detailed investigations since then have suggested that the lateral scale of the convection is about 8 km between peaks (Williams et al., 1974) and that the fluctuations are too smooth for the water flow to be confined to widely spaced fractures as suggested by Bodvarsson and Lowell (1972). If permeability is distributed through the rocks, then the convection cells should have a similar aspect ratio to the 1.6 found for λ/d in laboratory experiments (Elder, 1965). This would imply active circulation to a depth of 5 km, so that the early thermal history of the entire oceanic crust may be dominated by geothermal systems.

Highly metalliferous sediments are found near active ocean ridges (Bostrom and Petersen, 1969; Dymond et al., 1973; Piper, 1973) and the geochemical flux needed to generate them should be associated with the circulation of hydrothermal waters. Diffusion of even the more mobile chemical species is slow in the silicate rock matrix, and therefore complex processes of chemical change, such as those discussed by Wolery and Sleep (1976), require that the water and rock be in intimate contact.This means that the permeability should be distributed through the rock as many cracks a short distance apart. A mechanism for the formation of this kind of permeability has been given by Lister (1974). The boundary between the cracked, permeable rock and relatively undisturbed hot rock can be treated as a propagating thermal front. Rapid cooling and thermal contraction of the rock permit the development of horizontal tension even under high overburden pressure, and the cracks propagate. The theory predicts water penetration rates of the order of 60 m year^{-1}, based on very approximate rock parameters and a cracking temperature of 800°K. An independent estimate based on the thermal output, spacing and apparent lifetimes of New Zealand geothermal areas suggests that a rate of 2 m year^{-1} may be more appropriate (Lister, 1976).

It seems as if a plausible general mechanism for water penetration has been established, and that natural rate of the process has been estimated to within one or two orders of magnitude. The most salient feature of the results is this rapid rate of penetration: meters per year. Geothermal areas should have a highly active phase lasting only a few thousand years, during which time the thermal output should be large. After active penetration has ceased, circulation may continue, but the thermal flux is then merely the conductive flux from below. It is as if the conduction-cooling boundary of the new lithosphere were moved down to the base of the crust by the penetration of water. This paper is an attempt to divine the consequences of such a sudden cooling process for the mechanism of generation of new sea floor.

GENERALIZED OCEAN CRUST STRUCTURE

Discussion of the mechanisms operating at a spreading center requires at least a working model for the rock types and textures characteristic of each depth range in the crust and lithosphere. It is clear even from a cursory examination of any ridge crest bathymetric profile that reality is not represented well by any scheme of simple layering. Secondary volcanism and faulting are largely responsible for the rugged topography commonly observed, but the existence of a few smooth sections of the East Pacific Rise suggests that these processes are not primary or essential to the generation of a basic crust and mantle structure. Therefore a layered model can be used as the starting point for a generalized discussion so long as its layers are clearly understood to be quite variable at best, and possibly dissected by other rock types.

The relationship of ophiolite suites to oceanic crust has been discussed by Coleman (1971), and the actual structures by Moores and Vine (1971). There is general agreement between the seismic velocity structure observed in the deep ocean and the pressure-corrected velocities measured on ophiolite samples (Christensen and Salisbury, 1975). Though considerable reconstruction is needed to generate a simple suite of layers, and some idealization of the original crust may be inevitable, four distinct units can be discerned. At the top there are the expected sediments and pillow basalts, often in a melange. Under this is usually a "sheeted complex" consisting of essentially 100% dikes, and where the individual dikes are discernible throughout the 1—2 km thick layer. Below are gabbro masses showing the cumulate structures usually associated with the orderly cooling of large magma bodies. The basal layer is ultramafic rock showing little sign of more than partial melting, and varying from highly silica-depleted dunite to less depleted harzburgite. These layers have been depicted in Fig. 1 with two additions. The first is the sequence of geothermal convective systems that penetrate at least the crustal layers in young ocean floor. The second is the grading associated with separation and settling of the components in magma generated by the partial melting of "primitive" mantle material as it upwells under a spreading center.

The crustal layer of basaltic composition must be offset by depleted ultramafic rocks that grade back toward the original mantle composition in the layer labeled as peridotite. However, it is unlikely that primitive material could experience enough partial melting from pressure release alone to leave a residue of pure olivine dunite. Such materials are more likely to be made by the settling of olivine crystals from the fluid in a magma chamber. A possible mechanism considers the escape of a small fraction of partial melt from its host matrix. Whatever the initial method of separation on a microscopic scale, the liquid must eventually upwell through relatively narrow channels or fissures. The channeling is assisted by the energy release associated with the upwelling of a fluid of lower density than the host rock. If the density contrast were 0.6 g cm^{-3}, for example, upwelling from a depth of 30 km would result in the liberation of 40 cal. cm^{-3} of frictional

Fig. 1. Idealized ocean crust structure. The differentiates are materials that separate from upwelling magma containing phenocrysts. Mantle material depleted by partial melting is arbitrarily labeled 'peridotite'. Hydrothermal circulation is indicated by the blue arrows.

heating, enough to melt out the channels. The relatively high velocity flow in these channels would be capable of carrying dense phenocrysts up with the liquid. As soon as the fluid is dispersed in a magma chamber these crystals would settle to form a layer of dunite, and this would happen before crystals from the cooling of the melt could be deposited. Thus, the "differentiation boundary" refers to the redistribution of components from the fluid mush that separates from mantle material after pressure release, and not to the graded depletion due to partial melting.

THERMAL CONSTRAINTS AND THE SHEETED COMPLEX

Having generated a magma of basaltic composition by the pressure-release melting of asthenosphere material, one must begin to consider its interaction with the cold ocean boundary above. The mechanisms associated with the extrusion of pillow lavas are well enough understood not to need discussion here. The sheeted complex represents the real quasi-steady-state boundary between magma and cold water, and is a thick enough layer of uniformly solid material to dominate the mechanics of new oceanic plate. It is obvious how continuous but not steady extension, and continuous upwelling of magma, can form a structure composed of nothing but dikes. The thermal regime associated with this process is, however, highly dependent on whether or not water percolation is important.

Consider first the case of pure conductive cooling. Material is supplied intermittently at the melting point, and deposits its latent heat and specific heat of cooling into the walls. Mathematical modeling of this process is tricky but has been discussed extensively by Sleep (1975). The model assumes constant horizontal velocity for the material as soon as emplaced, and takes an average of the temperature at the emplacement axis through the cycle of dike intrusion and cooling. It cannot model the existence of a large magma chamber, because vertical heat transport by convection maintains a nearly constant heat flux at the top of the chamber while the crystals settle at the bottom. The zone of dike intrusion is represented fairly well, and the point at which the magma just remains molten between dike injections is

Fig. 3. Completed picture of the main processes active at a fast spreading center. High temperatures are indicated by orange, red and pink, except in the ultramafic layer. Generalized flow directions are indicated by arrows in the mantle and in the water-saturated region. Because the magma chamber isolates the thin rigid crust from the upwelling trough in mantle material, the system can spread smoothly without significant surface tectonics. Orientation of the olivine crystals in the ultramafic layer occurs due to stretching of the central region while it is fluidized by upwelling magma.

Pillows
Sheeted Complex
Cumulates
Ultramafics
Peridotit
Differentiates

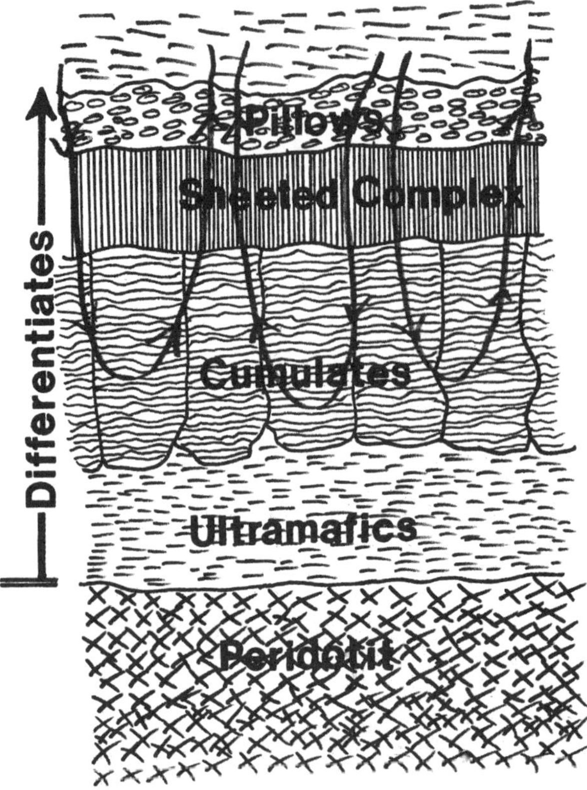

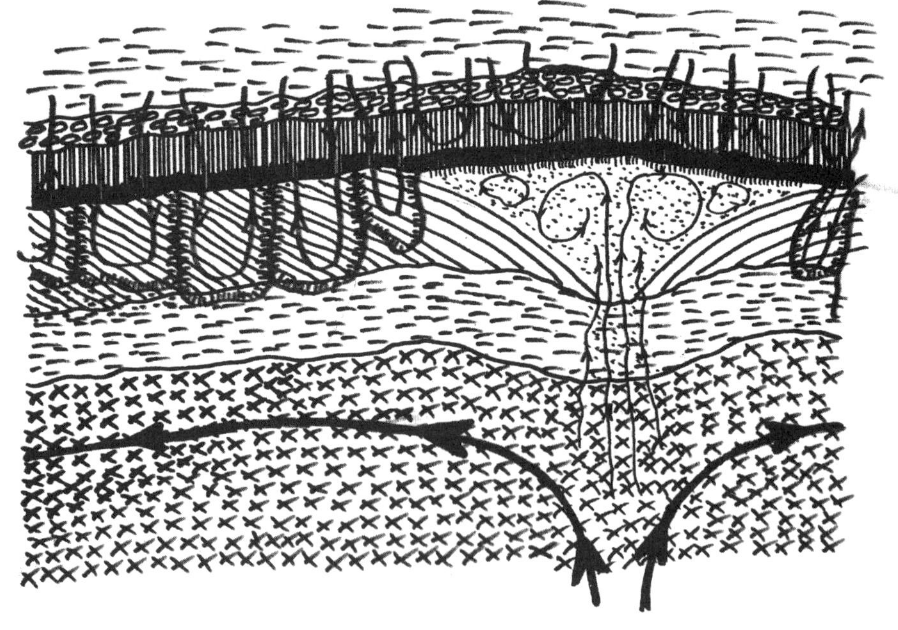

14

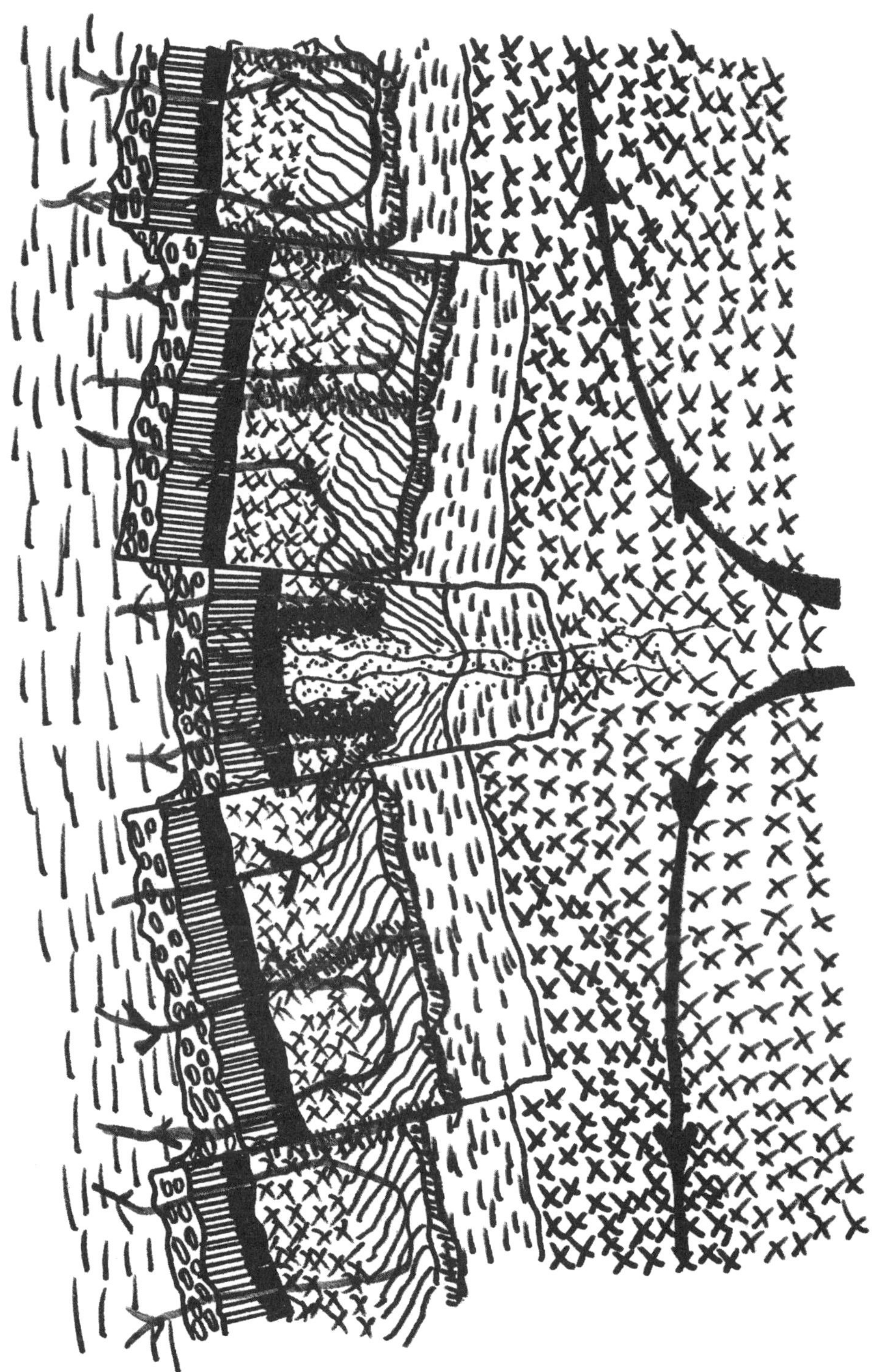

15

given accurately by the intersection of the solidus isotherm with the axis of spreading. This is a lower limit for the extent of the sheeted complex because no dike "chilled margins" can be generated when accretion of solid rock by freezing is continuous. The smooth transition between solidified dike and permanent magma is an important feature of the mechanism because it can account for the re-splitting of the most recent dike rather than quasi-random dike injection in tensioned crust above a magma chamber. The maximum thickness of the sheeted complex is a function only of rock/magma properties and spreading rate: it is of the order of 700 m at 5 cm $year^{-1}$, 1100 m at 3 cm $year^{-1}$ and over 3 km at 1 cm $year^{-1}$ (Sleep, 1975, fig. 1). The point at which chilled margins become visible on the dikes depends greatly on the rate-of-freezing differences needed to produce noticeable grain size changes, and on the actual thickness of the dikes. It is unlikely to require that the geologically indentifiable sheeted complex be less than 70% of the above quoted thicknesses.

The conductive model is capable of producing a sheeted complex, but one of thickness limited by spreading rate. If water percolation is an important contributor to heat transfer, the sheeted complex could be considerably thicker, in fact, almost indefinitely thick. Comparison between the thicknesses observed in ophiolite suites and those expected is not practical, since it is unlikely that reliable spreading rates will ever be estimated for the slivers of oceanic crust now found on land. Although there is good reason to suppose that the sheeted complex is the magnetic recording layer that generates the marine anomalies observed at the ocean surface, determination of the thickness of the main magnetized layer is difficult. The surface pillow lavas should be sufficiently randomized by the mechanics of the extrusion process not to have a significant magnetic expression at a distance: attempts at modeling a magnetization to fit the field observed near the ocean floor tend to confirm this (Larson et al., 1974). The gabbros below the sheeted complex are cooled by the downward propagation of nearly horizontal isotherms, and reversals in their magnetization should also have little surface expression. Thus the sheeted complex should comprise the magnetic layer, but the thickness of the layer affects only the amount of "overshoot" in the anomaly amplitude that would be observed above a sharp reversal transition. A thicker magnetized layer should generate smoother anomalies, and this tendency should be reinforced by the narrower magnetized blocks associated with the slower spreading rate. Unfortunately, there are variations in the sharpness of the magnetization boundary generated at a real ridge crest, since

Fig. 4. Simplified and somewhat speculative picture of spreading at a slow rate. Colors as in Fig. 3. A narrow magma chamber fails to isolate the rapidly developed thick rigid crust from the axial trough inherent in broad mantle upwelling, and block tectonics result. A broader zone of partial melting than of crustal magma retention will cause secondary intrusion and volcanism near the spreading center. This process is not shown because of the pictorial complexity it would introduce.

the dike-injection center may move about. There are enough variations in the "sharpness" of Pacific anomalies generated at roughly similar spreading rates (Pitman et al., 1968) to make quantitative estimates of magnetic layer thickness impractical. All one can conclude is that water percolation, though necessary to account for low axial valley heat-flow measurements, does not seem to play a major rôle in the formation of the sheeted complex if the thicknesses observed in ophiolite suites are representative of crust generated at moderate spreading rates.

The lack of a dramatic cooling of the sheeted complex by penetrating hydrothermal convection may appear strange when one considers the high penetration speeds calculated even for very low cracking temperatures (Lister, 1974). However, the stress situation in the sheeted complex departs radically from the simple one-dimensional model, and, in fact, represents an extreme case of the "pressure-case" dominated penetration discussed qualitatively by Lister (1976). The one-dimensional model assumes that each cooled layer is of infinite extent and thus cannot contract at all without cracking. The ends of a ridge segment are unlikely to offer any resistance to large-scale shrinkage of the crustal layer (e.g. Turcotte, 1974). Any tension developed in surface layers must be balanced by compression below, and, if magma can upwell in major shrinkage cracks, the compressive layer must support the full difference between the lithostatic and hydrostatic pressures, integrated down to the depth of penetration. A schematic diagram of the stress situation is given in Fig. 2. The thin layer cool enough to sustain high compressive stress tends to stabilize the system against geometric instability as long as a magma heat source is maintained below. Any convective cell that

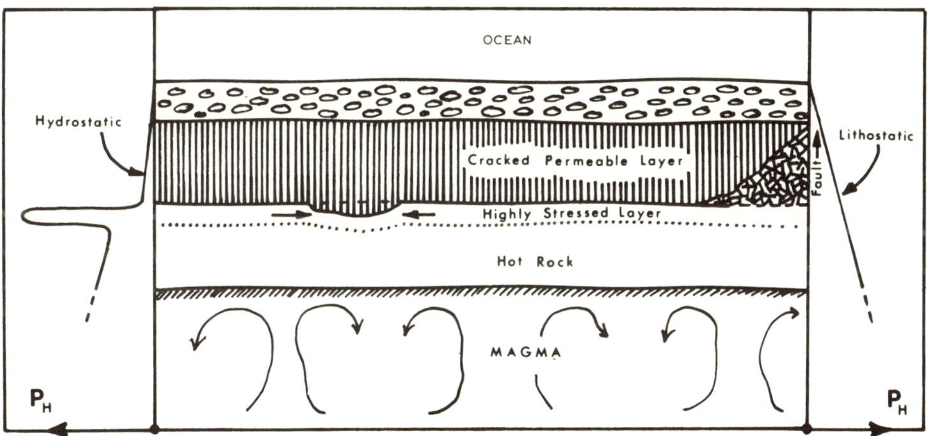

Fig. 2. Diagram of the stress system in a surface layer partially penetrated by water, when faulted regions are under normal lithostatic pressure (graph of horizontal pressure at right). The lower *hydrostatic* pressure in the permeable region must be compensated by a highly stressed layer below so that horizontal forces balance. Incipient instability in the cracking front boundary (shown at left center) is inhibited by stress concentration in the thinner, warmer, and distorted stressed layer.

penetrates deeper than average causes increased local stress, and therefore local compressive creep, and also higher local thermal flux (Fig. 2). The presence of the magma chamber prevents any large-scale thickening of the compressed layer and so permits a kind of equilibrium to be established with only partial water penetration.

MAGMA CHAMBER COOLING

The presence of an equilibrium impermeable layer between the circulating seawater and the magma chamber permits a crude estimate to be made of the crystallizing time of the cumulates. The hydrothermal systems maintain the top of the impermeable layer at near seawater temperature, while the base is at the solidus of basaltic magma (say 1200°C). The conductivity of basalt is about $4.5 \cdot 10^{-3}$ cal. cm^{-1} s^{-1} $°C^{-1}$ (Clark, 1966), so that a 1 km thick layer would pass a heat flow of 54 HFU or 1700 cal. $year^{-1}$. Assuming a latent heat of about 300 cal. cm^{-3}, 3 km of magma would be frozen in 50,000 years by heat losses through the top alone. On fast spreading ridges, say 5 cm $year^{-1}$, the magma chamber would extend about 2.5 km on each side of the crest, since lateral heat losses are small. On slow-spreading ridges, lateral losses to nearby penetrative hydrothermal systems should dominate, and, in spite of a thicker impermeable layer, only a small volume of magma may be present. If the rate is slow enough, conductive cooling alone would prevent the existence of any magma chamber at all, and spreading would have to become highly episodic (Sleep, 1975).

The overall effect of partial water penetration into the sheeted complex should be to somewhat thicken the layer at high spreading rates, and reduce the width of the magma chamber underneath it. The effects at slow spreading ridges should be expressed mainly by higher lateral heat losses at depth, since the narrow magma chamber top would prevent even quasi-equilibrium conditions from developing.

PENETRATIVE CONVECTION

As soon as the cumulate has crystallized, there is no longer a magma-convection heat source at the base of the sheeted complex. Cooling begins to propagate downwards, and the stabilizing influence of the heat boundary is lost. Under these conditions, three-dimensional "cold-finger" convective structures should develop, and propagate downward slightly more slowly than the one-dimensional model prediction for their cracking temperature (Lister, 1976). Rates similar to those estimated for the New Zealand geothermal areas should prevail: 0.2—20 m $year^{-1}$. Even the slowest of these rates is substantially faster than sea-floor spreading, and cold water should penetrate to the depth of the Mohorovicic discontinuity within a negligible increment of distance from the ridge crest. The various mechanisms that may terminate water penetration have been discussed at length (Lister,

1974). The only point that need be added is the probable influence of the change in rock compositions at the base of the differentiated crustal layer. As mentioned above, a transition from gabbro to ultramafic material should occur at the base of the magma chamber. Olivine has the property of being readily attacked by water and metamorphosed into serpentine and brucite, with a substantial volume increase (Coleman and Keith, 1971). It is most likely that this process, with or without a contribution from static fatigue of the rock, causes closing of the cracks and cessation of penetration.

PLATE THICKNESS AND UPWELLING

The thermal constraints of the sheeted complex, and a great deal of circumstantial evidence, suggest that the divergence in the velocity field of the top surface of the lithosphere is confined to a very narrow region. This requires that all the material flux needed to supply new crust upwells in the same narrow zone. Since the bulk of the oceanic crust is made of cooled magma products (Fig. 1) there is no difficulty in the requirement: the viscosity of a basaltic magma is many orders of magnitude lower than that of lithosphere or asthenosphere. A different situation applies below the crust: upwelling asthenosphere material will have a lowered viscosity because of the partial-melting process, but not sufficiently so to upwell in a very narrow column or plume. Moreover, the supply of asthenosphere material must be derived at a depth below the partial-melting zone, so that the upwelling must begin by being relatively broad.

How much mantle material flows horizontally away from the ridge crest with the lithosphere is difficult to estimate, since it depends on the distribution of viscosity in the upwelling. All realistic thermal models of the lithosphere itself imply that it thickens with age (e.g. Oxburgh and Turcotte, 1968; Lister, 1972; Parker and Oldenburg, 1973), and begins by being quite thin. Viscous drag on the underside of it entrains a considerable additional volumetric flow with the moving surface, and the existence of a large overall volumetric upwelling is confirmed by the considerable volume of pressure-release partial melt. The details of the flow pattern are well beyond this discussion, but the surficial mechanism is clearly affected by the fact that the mantle upwelling is broader than the accretion zone of a rigid crust.

If a rigid skin develops the full spreading velocity at the axis, but the full material supply is not upwelled there, then conservation of mass can be satisfied only by the presence of an axial valley (Deffeyes, 1970). Segments of the skin are uplifted where they reside on the flank of the upwelling, and, since there is a lateral gradient of upwelling, they must also be tilted outward. The existence of such tilting has been verified by acoustic reflection profiling of turbidite sediments on the Juan de Fuca ridge (Davis and Lister, in press). It is well known that median valleys are normal on slow-spreading ridges and absent on fast-spreading ridges, with the Juan de Fuca ridge at 3 cm year^{-1} showing transitional behavior. Two aspects of the mechanism of spreading can account for this: the decoupling effects of a magma chamber, and the reduction of entrained mantle flow due to lower viscosity beneath fast-spreading ridges.

The Juan de Fuca ridge is particularly interesting because the rate of sedimentation seems to influence the crestal structure. Over most of the ridge, a central valley exists only as a small depression of 200 m or so. At the north end, where sediments flood onto the spreading center from the nearby continental margin, there is an enormous central valley reaching a depth of almost 5 km beneath the sediment pile (Davis and Lister, in press). An intriguing possibility, suggested by the thermal and water-percolation analysis above, is that the sediment blanket prevents the formation of the usual quasi-stable impermeable sheeted complex as there is too little surficial conductive cooling. The paradoxical result is that cooling extends deeper than usual near the injection zone because of penetrative hydrothermal convection. This removes the buffering magma chamber, and increases the thickness of the mantle layer flowing with the plate, so that the transitional ridge shows the tectonic features of a fully developed slow-spreading ridge. There is some evidence that cooling has penetrated to such a depth that the spreading of that central valley has slowed almost to a standstill, and the center of activity has jumped to the west (Davis and Lister, in press).

COMPLETED QUALITATIVE MODELS

Now that the constituent mechanisms of spreading have been developed to a sufficient qualitative level, it is possible to assemble diagrams showing the whole system in operation. The fast-spreading ridge is the most straightforward and is shown in Fig. 3. A weak "central valley" exists in the mantle layer, but is decoupled from the surface by the magma chamber. Water penetration is limited to the upper part of the sheeted complex by thermal stabilization until crystallization of the cumulates is complete. At this point, rapid penetrative convection of water cools the whole crustal layer, and may extend a short distance into the ultramafic material. Once the depth of water circulation in the crust has stabilized, conductive cooling of the upper mantle begins at that reference depth, and the lithosphere thickens according to the square-root law of thermal cooling (Parker and Oldenburg, 1973; Davis and Lister, 1974).

The case of the slow-spreading ridge is more speculative because of its complexity, and the probable importance of some of the non-uniform processes that are hard to incorporate into a simple model. An attempt to picture the mechanism has been made in Fig. 4. Here a small magma chamber is cooled both laterally and at the top by nearby water circulation. The lateral closeness of the water penetration, and the width of the magma chamber, are both exaggerated in the figure relative to what they should be at a really slow spreading ridge (1 cm year^{-1}). The basic difference between Fig. 4 and Fig. 3 is just the spatially more rapid cooling of the new material because of the slower spreading rate. The dramatic difference in tectonic structure is due to the coupling of differential upwelling in the mantle material to the thick rigid crustal blocks. Instead of a smooth surface trending almost uniformly down from the ridge crest, there is large-scale block faulting that develops the ridge-and-valley topography so characteristic of the Mid-Atlantic ridge. The fact that upwelling and partial melting should occur in a mantle region much wider than the accretion zone of the crust suggests that secondary volcanism and/or intrusion may be an important process. It was not practical to include this in the diagram due to the graphic difficulties even in a multicolor presentation. However, one minor point that was included is the difference in the cumulate structure between material laterally accreted by cooling of the walls, and material deposited by crystal settling. The former is essentially a dikeless sheeted complex that should exist as a narrow transition region even at a fast spreading center. At a slow spreading ridge it would be a major feature of the gabbroic layer.

TEST OF THE THERMAL PREDICTION

The principal prediction of any model of spreading that incorporates water percolation is the concentration of heat output near the ridge crest. Not only is the high thermal flux from the incipient cooling of the mantle transferred to the ocean, but the entire heat content of the crust is removed by the penetrative convection of water. This prediction cannot be tested readily in the open ocean because of the rapid dispersal of any output of hot water. Only the largest thermal plumes that are generated during the highly active phase of hydrothermal penetration can be detected (Lister, 1972; Scott et al., 1974). However, the water temperatures around one active ridge in a closed basin have been investigated by Detrick et al. (1974), and data from their fig. 4 can be reinterpreted to demonstrate the ridge crest output. Four near-bottom potential temperature profiles from a region just south of the ridge crest are replotted in Fig. 5. The variations between profiles are largely due to lack of synopticity: current meter data in the area (Detrick et al., 1974) show that water can move up to 50 km in 23 days, while the profiles are up to 25 km apart and taken over a comparable period of time.

The average history of water in the basin can be modeled by supply at a given source potential temperature over a topographic sill, rapid spreading out over the floor of the basin, and then a gradual rise toward the outflow level at an approximately constant flow rate. As the deep basin is stratified by upward increasing potential temperature, the horizontal eddy diffusion rate should be much larger than the vertical eddy diffusion rate. Then each horizontal layer in a restricted basin, such as that south of the ridge crest, can be considered laterally mixed, while the vertical mixing should be quite restricted. This is a crude approximation, but can be improved upon only if synoptic temperature and current meter data are available for the whole

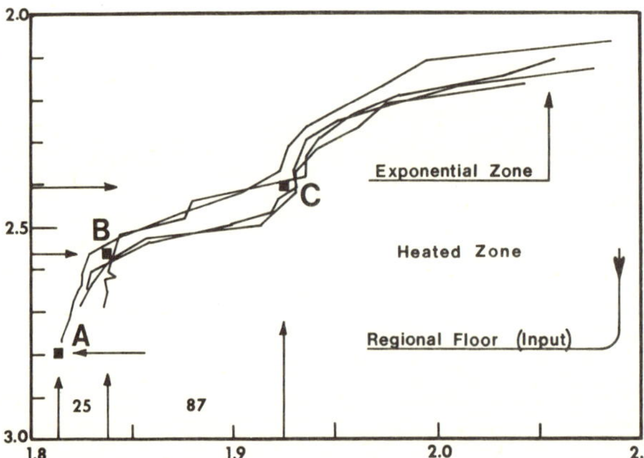

Fig. 5. Potential temperature profiles (°C) against depth (km) from near the Galapagos Rift Zone in the Panama basin, replotted from Detrick et al. (1974). The mean ridge crest height is at the level of point *C*, while point *A* is approximately the basin-floor level south of the spreading center. Point *B* is placed arbitrarily at the inflexion in the profiles. Temperature differences *AB* and *BC* are shown in millidegrees C.

basin. Where the horizontal cross-section of the basin is relatively constant, the upward flow velocity should be fairly constant also. In the absence of significant bottom heating, the temperature profile would be exponential downwards if the effective diffusivity were constant (Carslaw and Jaeger, 1959). There is such a region in Fig. 5 for depths above the mean ridge crest at 2400 m (point *C*). High gradients in the mid-water that can flow through the basin above several sills decay downward into a nearly isothermal zone.

Below the depth of the mean ridge crest, the temperature profiles show two quasi-linear regions: steeply changing temperature just below the ridge crest (*BC*), and then a more gradual decrease toward the floor of the basin (*AB*). The striking linearity of two of the profiles could be a random artifact, but the crude flow model would actually show this if the heat output of thermally shrinking topography were added to the water. Each incremental drop in the height of cooling lithosphere requires the release of an approximately constant amount of heat, since the shrinkage at any place is $\alpha \Delta T$ while the heat loss is $\rho c \Delta T$: they are linearly proportional to the local ΔT. Here α is expansion coefficient (corrected for isostatic compenstion), ρ is density, c is specific heat and ΔT is a temperature change. If the heat release from the change in topographic height is added to the water layer at that height its temperature will rise a constant amount per unit time (spreading rate and upward advection approximately constant).

If there were no hydrothermal heat release from the crustal layer, the linear profile would join the base of the exponential temperature profile due to diffusion from above. Instead there is a strong thermocline in the depth range of the ridge crest, indicating a large localized heat output. In semi-quantitative terms, a topographic height drop of 230 m produces a rise of 25 millidegrees on the flank, but a crestal height drop of only 150 m produces a heating of 87 millidegrees (Fig. 5). The height drop due to rapid cooling of the crustal layer near the spreading axis is not observed in the topography. The heat-loss equivalent in topographic height can be estimated by matching the total heat content of the basaltic layer with simple cooling of mantle material. If 6 km of mantle with an α of $5.6 \cdot 10^{-5}$ (Lister, in press) cool through 1500°C (the approximate equivalent heat content of molten basalt), the topography would drop by 500 m. Thus the heating of the crestal region is really equivalent to a total topographic drop of 650 m, bringing the ratio of the topographic drops, 2.8, much closer to the ratio of temperature rises, 3.5.

The above analysis is very crudely approximate, but it does predict the generation of a thermocline at the ridge crest depth when a circulating closed basin contains a spreading center. The only alternative explanations for such an observed thermocline are that it is the result of highly discontinuous feed of cold water over the basin sill, or that it is produced by a large contemporary lava flow at the ridge crest. Both these alternatives could be ruled out if a reinvestigation of the area demostrated a similar pattern after several years.

CONCLUSIONS

The first-order model of a spreading center incorporating the effects of rapid water penetration into hot rock appears to be consistent with the observed data. Moderate-sized magma chambers are permitted near the axis at fast spreading rates because water does not penetrate close to the hot isothermal top boundary of magma. Nevertheless, the entire crustal section is fully cooled within a few km of the ridge crest even on the East Pacific Rise, and so is consistent with the normal crustal layering found by seismic refraction measurements near the crest (Snydsman et al., 1975).

Another hitherto puzzling aspect of the seismic data may become more comprehensible. Strong velocity anisotropy is observed in the suboceanic mantle near the ridge crest, but diminishes on older litosphere (Snydsman et al., 1975). An important feature of the models in Figs. 3 and 4 is the layer of olivine crystals that develops at the base of the magma chamber. This layer is fluidized near the axis by the upwelling of magma through it, and it is also being stretched in the direction of spreading. These conditions are ideal for the alignment of anisotropic crystals, and the long axis of the crystals becomes parallel to the stretching. Since the long axis of olivine crystals is also the direction of the highest seismic velocity (N.H. Christensen, personal communication, 1974), the highest velocities should be observed transverse to the ridge. This is the pattern that is observed, and the anisotropy decreases with age by an increase in the velocity observed parallel to the spreading axis, while the transverse velocity remains approximately constant

(Snydsman et al., 1975). The model is consistent with this also, since the anisotropic layer is a thin one, underlain by relatively isotropic material. Near the ridge crest, the temperature of the underlying layer is still high and the velocity is reduced substantially by increased temperature (N.H. Christensen, personal communication, 1976). The fastest velocity observed is the anisotropically low velocity of the surface mantle layer. As cooling penetrates into the mantle below the anisotropic layer, the velocity there rises and is observed as soon as it exceeds that in the layer above. In the fast direction, the anisotropic layer remains faster than even fully cooled mantle material below, so that the observed velocity does not change.

Well-known laws of heat conduction (Sleep, 1975) and the qualitative theory of water penetration into hot rock have been applied to the surficial mechanisms at an oceanic spreading center. Together with simple ideas about the derivation of magma from upwelling asthenosphere material, they have produced a physically workable ridge crest mechanism. The first-order predictions of the model are consistent with marine geophysical data, at least for the case of fast spreading ridges. Further testing can be accomplished by unravelling some of the complexities at slow spreading centers, such as the Mid-Atlantic Ridge, and careful comparison of model predictions with the structures observed in ophiolite suites.

ACKNOWLEDGEMENTS

This work was supported by National Science Foundation grant DES 73-06593-A01.

I wish to thank S. Uyeda for the moral encouragement that caused this work to be undertaken.

REFERENCES

Bodvarsson, G. and Lowell, R.P., 1972. Ocean-floor heat flow and the circulation of interstitial waters. J. Geophys. Res., 77: 4472—4475.

Bostrom, K. and Petersen, M.N.A., 1969. The origin of aluminum ferro-manganoan sediments in areas of high heat flow on the East Pacific Rise. Mar. Geol., 7: 427—447.

Carslaw, H.S. and Jaeger, J.C., 1959. Conduction of Heat in Solids. Oxford Univ. Press, 2nd ed., 388 pp.

Christensen, N.I. and Salisbury, M.H., 1975. Structure and constitution of the lower oceanic crust. Rev. Geophys. Space Phys., 13: 57—86.

Clark, S.P. Jr., 1966. Thermal conductivity. In: S.P. Clark Jr. (editor), Handbook of Physical Constants. Geol. Soc. Am. Mem., 97: 461—482.

Coleman, R.G., 1971. Plate-tectonic emplacement of upper-mantle peridotites along continental edges. J. Geophys. Res., 76: 1212—1222.

Coleman, R.G. and Keith, T.E., 1971. A chemical study of serpentinization — Burro Mountain, California. J. Petrol., 12: 311—328.

Davis, E.E. and Lister, C.R.B., 1974. Fundamentals of ridge crest topography. Earth. Planet. Sci. Lett., 21: 405—413.

Davis, E.E. and Lister, C.R.B., in press. Tectonic structures on the Juan de Fuca ridge. Geol. Soc. Am. Bull.

Deffeyes, K.S., 1970. The axial valley: a steady-state feature of the terrain. In: H. Johnson and B.L. Smith (editors), Megatectonics of Continents and Oceans. Rutgers Univ. Press., Camden, N.J., pp. 194—222.

Detrick, R.S., Williams, D.L., Mudie, J.D. and Sclater, J.G., 1974. The Galapagos spreading center: bottom water temperatures and the significance of geothermal heating. Geophys. J.R. Astron. Soc., 38: 627—637.

Dymond, J., Corliss, J.B., Heath, G.R., Field, G.W., Dasch, E.J. and Veeh, H.H., 1973. Origin of metalliferous sediments from the Pacific Ocean. Geol. Soc. Am. Bull., 84: 3355—3372.

Elder, J.W., 1965. Physical processes in geothermal areas. In: Terrestrial Heat Flow. Am. Geophys. Union Monogr., 8: 211—239.

Larson, R.L., Larson, P.A., Mudie, J.D. and Spiess, F.N., 1974. Models of near-bottom magnetic anomalies on the East Pacific Rise crest at 21°N. J. Geophys. Res., 79: 2686—2689.

Lister, C.R.B., 1972. On the thermal balance of a mid-ocean ridge. Geophys. J.R. Astron. Soc., 26: 515—535.

Lister, C.R.B., 1974. On the penetration of water into hot rock. Geophys. J.R. Astron. Soc., 39: 465—509.

Lister, C.R.B., 1976. Qualitative theory on the deep end of geothermal systems. 2nd U.N. Geothermal Symp. Vol.: in press.

Lister, C.R.B., in press. Estimators for heat flow and deep rock properties based on boundary layer theory. Tectonophysics.

Moores, E.M. and Vine, F.J., 1971. The Troodos massif, Cyprus, and other ophiolites as oceanic crust, evaluation and implications. Philos. Trans. R. Soc. London, Ser. A, 268: 443—466.

Oxburgh, E.R. and Turcotte, D.C., 1968. Mid-ocean ridges and geotherm distribution during mantle convection. J. Geophys. Res., 73: 2643—2661.

Parker, R.L. and Oldenburg, D.W., 1973. Thermal model of ocean ridges. Nature, 242: 137—139.

Piper, D.Z., 1973. Origin of metalliferous sediments from the East Pacific Rise. Earth Planet. Sci. Lett., 19: 75—82.

Pitman, W.C., III, Herron, E.M. and Heirtzler, J.R., 1968. Magnetic anomalies in the Pacific and sea-floor spreading. J. Geophys. Res., 73: 2069—2085.

Scott, R.B., Rona, P.A., McGregor, B.A. and Scott, M.R., 1974. The TAG hydrothermal field. Nature, 251: 301—302.

Sleep, N.H., 1975. Formation of oceanic crust: some thermal constraints. J. Geophys. Res., 80: 4037—4042.

Snydsman, W.E., Lewis, B.T.R. and McClain, J., 1975. Upper-mantle velocities on the northern Cocos plate. Earth Planet. Sci. Lett., 28: 46—50.

Turcotte, D.L., 1974. Are transform faults thermal contraction cracks? J. Geophys. Res., 79: 2573—2577.

Williams, D.L., Von Herzen, R.P., Sclater, J.G. and Anderson, R.H., 1974. The Galapagos spreading center: lithospheric cooling and hydrothermal circulation. Geophys. J. R. Astron. Soc., 38: 587—608.

Wolery, T.J. and Sleep, N.H., 1976. Hydrothermal circulation and geochemical flux at mid-ocean ridges. J. Geol: in press.

2

Reprinted from *Royal Astron. Soc. Geophys. Jour.* **40**:351-365 (1975)

Circulation in Fractures, Hot Springs, and Convective Heat Transport on Mid-ocean Ridge Crests

Robert P. Lowell

(Received 1974 October 15)*

Summary

Hot springs in continental geothermal areas, thermal springs which are not associated with known high-temperature areas, and hot springs associated with hydrothermal circulation on ocean ridge crests are described by models based on fluid flow in fractures. A steady state model shows that fractures of the order of a few millimetres wide can carry a substantial convective flow and that the convective flow depends upon the third power of the fracture width. The steady state model also furnishes estimates of conductive temperature losses in springs and gives estimates of the depths of circulation for thermal springs in the south-eastern United States which are in good agreement with available field data.

In many cases the temperature and flow rate of springs is non-stationary. This is particularly true of hot springs in high temperature geothermal areas and it is expected to be true of springs on mid-ocean ridge crests. Time-dependent models for springs show that the main effect of the circulation is to lower the regional geothermal gradient. Non-stationary convection controlled by fractures can explain the variability of heat flow data obtained near ocean ridge crests. A numerical example shows that convection in a block 3 km wide, containing fractures 3 mm wide and 5 km deep, and circulating for 10^4 years gives rise to a hot spring with a temperature of 125 °C and a flow rate of 0·14 kg $m^{-1} s^{-1}$. Such a spring discharging at the sea floor would give rise to an unmeasurably small temperature anomaly in the sea water. The convective heat transfer due to such a circulation system is roughly 200 times greater than the heat transfer that would have been achieved by conduction alone.

Introduction

Hot springs, often with temperatures near 100 °C, are a common surface expression of high-temperature geothermal systems. Chemical analysis generally indicates that the water of such springs is meteoric in origin, implying that hot springs result from circulating groundwaters which extract heat from high-temperature subsurface rocks. The driving mechanism of hot springs in geothermal areas is a combination of pure hydrostatic pressure and differential pressures due to the thermal expansion of water.

* Received in original form 1974 August 2.

In addition to the hot springs associated with active, high-temperature geothermal areas, ' warm ' or ' thermal ' springs with temperatures a few tens of degrees above the mean annual air temperature occur in various parts of the world, without association with known high-temperature areas. Examples of these are the thermal springs in the south-eastern United States, in Georgia, Virginia and North Carolina. The temperatures of thermal springs are generally believed to be caused by downward circulation and heating by the normal geothermal gradient (Waring, Blankenship & Bentall 1965). A suitable geologic structure allows the water to return to the surface as a spring.

Hot and warm springs are also expected to exist on the sea floor. A large thermal spring has been located off the west coast of Florida (Pyle 1974, private communication). Recent heat-flow measurements (Talwani, Windish & Langseth 1971; Lister 1972; Williams *et al.* 1974) indicate the existence of hydrothermal circulation on mid-ocean ridge crests; and there is evidence which suggests hydrothermal discharge on the Mid-Atlantic Ridge crest (Scott *et al.* 1974) as well as on the Galapagos spreading centre (Williams *et al.* 1974). Since discrete zones of high permeability (faults, dikes, and deep fractures) almost certainly exist on ocean ridge crests, and since there is little sediment cover, it seems likely that hot and warm springs are not unusual phenomena there.

It is worthwhile, therefore, to develop theoretical thermal models for the circulation systems which give rise to hot and warm springs, both continental and submarine. In particular, the effectiveness of thermal convection as a driving mechanism will be estimated. A steady state model will be used to estimate the convective flow rate of a thermal spring and to relate the flow rate to the conductive temperature loss of the fluid as it rises to the surface. The results will be applied to thermal springs in the south-eastern United States. A time dependent model will be derived to show the temporal behaviour of the spring temperature for young springs. The time dependence of the temperature and flow rate will be given for a non-stationary convecting spring, and these results will be related to processes on ocean ridge crests.

Springs on land are usually found in fault and fracture zones (Waring *et al.* 1965) and deep fractures form the main vertical permeability in Icelandic geothermal areas (Bodvarsson 1964a). On ocean ridge crests, the vertical permeability is probably due mainly to faults, dikes, and fractures. Consequently, the spring models will be based on fluid flow in narrow, vertical fractures.

A steady state model

A steady state model will be derived for a transverse fracture system (Fig. 1) as described by Bodvarsson & Lowell (1972). The results, however, may be applied to a longitudinal system (in which the fluid is confined to a single fracture) with minor

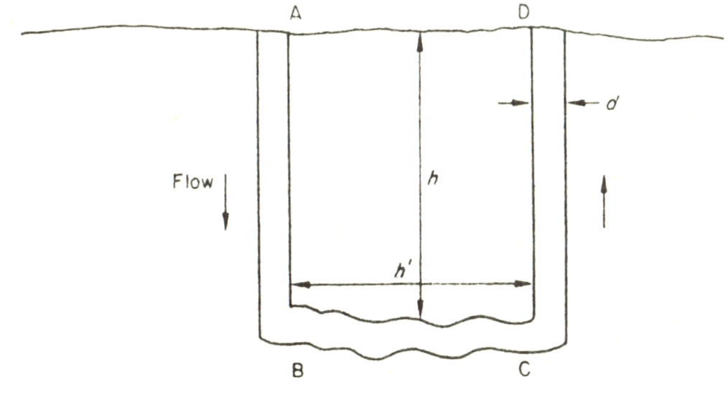

Fig. 1. Sketch of transverse flow system.

modifications. A transverse system consisting of two vertical fractures of width d and depth h and a horizontal contact zone of length h' is shown in Fig. 1. The fluid enters the fracture on the left at A, descends to B, flows along the horizontal contact to C, and returns to the surface at D. Assume that the system has been flowing long enough to attain a steady state and that heat is transferred only to the fluid in the vertical segments. Let descending fluid enter at A at temperature $T = 0$ and flow slowly so that at depth x (positive downwards), the temperatures of the fluid and rock will be equal. At the bottom of the fracture at a depth $x = h$, the fluid temperature will be $T(h) = \beta h$, where β is the geothermal gradient.

In the ascending fluid of temperature $T(x)$, assume that the heat transfer between the fluid and rock can be expressed in terms of a simple contact resistance. Thus at depth x, the heat transfer across the rock-fluid interface is,

$$\psi.(T(x) - \beta x). \tag{1}$$

The heat transfer coefficient ψ is of the form,

$$\psi = K/L \tag{2}$$

where K is the thermal conductivity of the rock and L represents the thickness of the resistance. Let the specific heat of the fluid be s and the mass flow per unit length of the fracture be q. The heat transfer in the ascending fluid is described by the differential equation,

$$sq(dT/dx) = \psi(T - \beta x). \tag{3}$$

The temperature distribution in the ascending fluid is found to be,

$$T(x) = (T_0/\mu h)[\mu x + 1 - \exp(\mu(x-h))], \tag{4}$$

where $\mu = \psi/sq$ and T_0 is the temperature at the base of the fracture $T(h)$.

The surface temperature of the spring is of primary interest. Setting $x = 0$ in (4) gives,

$$T(0) = (T_0/\mu h)(1 - \exp(-\mu h)). \tag{5}$$

Upon substituting for μ and ψ, (5) becomes,

$$T(0) = (sqLT_0/Kh)(1 - \exp(-Kh/sqL)). \tag{6}$$

For $q \ll Kh/sL$, the exponential term in (6) may be neglected, so that,

$$T(0) \simeq sqLT_0/Kh. \tag{7}$$

For small flows, the temperature of the spring increases linearly with the flow rate. This result is in agreement with the observations on springs in Iceland (Bodvarsson 1950). On the other hand, for large flows such that $Kh/sqL \ll 1$, a Maclaurin expansion of (6) yields,

$$T(0) \simeq T_0(1 - Kh/sqL).$$

The fractional temperature loss due to thermal conduction is,

$$\Delta T/T_0 \simeq Kh/2sqL. \tag{8}$$

The parameter L is a measure of the lateral distance to which the temperature field is perturbed as a result of the fracture circulation. On physical grounds, one would expect that L should increase as the depth of circulation h increases. If we simply choose $L = h$ and assume $K = 2{\cdot}5\ \mathrm{W\,m^{-1}\,{}^\circ C^{-1}}$ and $s = 4{\cdot}2 \times 10^3\ \mathrm{j\,kg^{-1}\,{}^\circ C^{-1}}$, the conductive temperature loss in a fracture flowing at $0{\cdot}1\ \mathrm{kg\,m^{-1}\,s^{-1}}$ is 0·6 per cent, whereas for a flow of $10^{-3}\ \mathrm{kg\,m^{-1}\,s^{-1}}$, the conduction loss is of the order of 25 per cent. It should be pointed out that the conductive cooling depends principally on the flow rate q and the heat transfer coefficient $\psi = K/L$. The choice $L \simeq h$ is substantiated by the heat flow calculation carried out in the Appendix. The flow rate used in calculating the conduction loss is the total flow (hydrostatic plus convective). Consequently, the estimated losses are valid regardless of the driving mechanism.

To estimate the convective flow rate in the fracture system, we assume laminar flow in the fractures. For convenience, it is assumed that there is no flow resistance in the horizontal section BC (Fig. 1). The mass flow is then given by the well-known relation (Lamb 1932, p. 528) for flow of a viscous fluid in a flat channel of length $2h$ and width d,

$$q = d^3 H/24\nu h \tag{9}$$

where ν is the kinematic viscosity of the fluid and H is the total head driving the flow. If we assume that the head H arises only from the thermal expansion of the fluid then it is given by,

$$H = \alpha\gamma \int_0^h T(x)\,dx - \alpha\gamma T_0\, h/2 \tag{10}$$

where α is the thermal expansion coefficient of the fluid and γ is its specific weight.

Inserting the temperature distribution (4) into (10) and using (9), we obtain an equation for the flow,

$$(1/\mu h)(1 - \exp(-\mu h)) = 1 - 24/R \tag{11}$$

where

$$R = \alpha\gamma\beta s d^3\, L/K$$

is a dimensionless number. The convective flow for a few representative fracture widths was calculated from (11) assuming typical water and rock parameters, a geothermal gradient of 0·03 °C m^{-1}, and a fracture depth of 10^3 m. It was also assumed that $L = h$. The results are given in Table 1. Table 1 shows that a fracture 5 mm wide and 100 m long can carry a convective flow of approximately 8 kg s^{-1}, which is a substantial flow.

These results show the extreme sensitivity of the flow to the fracture width. In fact, if $d > 3 \times 10^{-3}$ m, the flow can be found by expanding (11) in a Maclaurin's series for $\mu h \ll 1$. This gives,

$$q \simeq \alpha\beta\gamma h d^3/48\nu. \tag{12}$$

The parameters in (12) are fairly well fixed for typical geological situations with the exception of d and h. The fact that the flow varies as the cube of the fracture width suggests that the flow estimate is rather insensitive to the many simplifying assumptions which have been made regarding other variables in the equation. It should be pointed out, however, that the results given by (12) will be somewhat inaccurate for fractures of 10^{-2} m width or greater. The Reynold's number for flow in a fracture 10^{-2}m wide is,

$$\mathrm{Re} = q/\rho\nu \simeq 600$$

Table 1

Steady state convection flow rate for a few typical fracture widths

Based on: $s = 4{\cdot}2 \times 10^3$ j kg^{-1} °C^{-1}, $\gamma = 10^4$ kg m^{-2} s^{-2}, $\alpha = 10^{-4}$ °C^{-1}, $K = 2{\cdot}5$ W m^{-1} °C^{-1}, $\nu = 10^{-6}$ m^2 s^{-1}

$d = 10^{-3}$ m	$q \simeq 4 \times 10^{-4}$ kg m^{-1} s^{-1}.
$d = 5 \times 10^{-3}$ m	$q \simeq 8 \times 10^{-2}$ kg m^{-1} s^{-1}.
$d = 10^{-2}$ m	$q \simeq 0{\cdot}6$ kg m^{-1} s^{-1}.

where the fluid density is $\rho = 10^3\ \mathrm{kg\,m^{-3}}$. Since the fracture walls are no doubt quite rough, a Reynold's number of 600 suggests that the flow may be turbulent. Hence, for wide fractures (9) would have to be replaced by an expression of the form,

$$q^2 \propto H$$

which is suitable for turbulent flow (Lamb 1932, p. 665).

The main results of this model are (1) fractures a few millimetres wide or greater can transport a significant convective flow, (2) the flow rate is very sensitive to the fracture width, and (3) the temperature loss due to thermal conduction in the ascending flow decreases with increasing flow rate.

Application to thermal springs in the south-east

There are several areas of thermal springs in the south-east, the most notable being near Warm Springs, Georgia, the Warm Springs Valley, Virginia and Hot Springs, North Carolina. The characteristics of the springs in these areas are described in Watson (1924), Hewitt & Crickmay (1937), Stose & Stose (1947), Reeves (1932), and Waring *et al.* (1965). Though the data on these springs are somewhat incomplete, it appears that their circulation is controlled by faults and fracture zones and that their temperatures can be explained by heating by a normal geothermal gradient. The pertinent data for a few of the springs are given in Table 2.

The temperature and discharge data which have been obtained over the past century or so indicate that these springs have achieved a steady state. It is useful, however, to construct an energy balance based on the steady state assumption in order to (1) estimate the required contact area to see whether the steady state is energetically feasible, and (2) estimate the fracture length and hence the flow rate q from the observed spring discharge Q measured in kilograms per second. The power output of the springs is given by $P = Qs\Delta T$, where ΔT is the difference between the average spring temperature and the mean annual air temperature. The regional heat flow

Table 2

Data on thermal springs in the south-eastern United States

Spring name and location	Average temperature (°C)	Discharge ($\mathrm{kg\,s^{-1}}$)
Warm Springs, Georgia	30·5	41
Lifsey Spring, Georgia	25	5·6
Warm Sulpher Springs, Virginia	35	82
Hot Springs, North Carolina	40	2·0

Table 3

Calculations on thermal springs in south-eastern United States based on fracture model

Spring name and location	Power output (W)	Estimated minimum contact area ($\mathrm{km^2}$)	Estimated flow rate q ($\mathrm{kg\,m^{-1}\,s^{-1}}$)	Estimated conduction temperature loss (%)	Theoretical depth of circulation (m)
Warm Springs, Georgia	$3{\cdot}5\times10^6$	70	$4{\cdot}1\times10^{-3}$	7	1100
Lifsey Spring, Georgia	$3{\cdot}5\times10^5$	7	$5{\cdot}6\times10^{-3}$	6	800
Warm Sulphur Springs, Virginia	$8{\cdot}6\times10^6$	170	$8{\cdot}2\times10^{-3}$	4	1300
Hot Springs, North Carolina	$2{\cdot}5\times10^2$	5	2×10^{-3}	13	1700

in the south-east is roughly 0·05 W m^{-2} (Reiter & Costain 1973; Diment, Urban & Revetta 1972). Assuming that the power output of the springs is derived from the geothermal heat flux, the water–rock contact area can be estimated (see Table 3). The contact areas given in Table 3 represent the minimum required. Since the water is unable to absorb all the geothermal heat, the actual contact area may be several times larger than that calculated. These results show that it is energetically feasible for the small springs to be in the steady state. The energy balance calculation is inconclusive with regard to Warm Springs and Warm Sulphur Springs, however. Assuming a steady state for all the springs considered the fracture length required for the spring can be estimated. Crude estimates based on the contact areas in Table 3, and assuming a subsurface path length of the order of 10–20 km, suggest a 10-km fracture length for the high discharge springs and a 1-km fracture length for the smaller springs. The flow q is then given by the observed spring discharge rate divided by the fracture length. From the values of q, the percentage of conductive temperature loss in the ascending flow can be calculated. These results are given in Table 3. For the springs considered, the percentage temperature loss is small according to the fracture model. Although these calculations are based on rather incomplete data (for example, the subsurface path length is not well known), the assumed values are geologically reasonable, and better data are not expected to seriously alter the results.

If we consider thermal conduction as the only temperature loss mechanism, and a geothermal gradient of 20 °C km^{-1}, we can estimate the depth of circulation of the groundwater. The depths (see Table 3) based on the fracture model are in good agreement with the depths inferred from geological data. For example, geological data indicates that the water emerging at Warm Springs, Georgia, circulates to a depth of approximately 1200 m (Hewitt & Crickmay 1937); and at Hot Springs, North Carolina, the water circulates down to 1500 m (Stose & Stose 1947). The thermal springs in the south-east are apparently driven by hydrostatic pressure. Geological evidence indicates that elevation at which water enters the permeable beds is substantially greater than the elevation of the springs. The hydrostatic heads resulting from such elevation differences are sufficient to yield the observed flow rates of the springs, provided the fractures are of the order of 0·5 mm wide. In such narrow fractures, the convective flow based on equation (11) is negligible.

It should be emphasized, however, that the fracture model is generally applicable, and that the model does give useful results with regard to conductive losses and circulation depths of thermal springs.

Time-dependent models

The steady state model discussed above furnishes useful results regrading thermal spring circulation. In many cases, however, energy requirements indicate that the heat supply of a spring is non-stationary. This is particularly true of springs with high temperature and high flow rate which are often found in geothermal areas and may be true for a thermal spring such as Warm Sulphur Springs, Virginia. For example, a hot spring flowing at 100 kg s^{-1} at 80 °C has a power output of $3{\cdot}2 \times 10^7$ W. To derive this power output from a regional geothermal heat flux of 0·1 W m^{-2} would require that the water-rock contact area be at least 320 km^2. Since the water would not absorb all the heat flow from below, the real contact area would probably need to be 3 or 4 times this amount. A realistic estimate of the contact area for such a spring to be in a steady state is 1000 km^2. Based on the fracture model in Fig. 1, if the subsurface path length ABCD of the spring is 10 km, the fracture length must be 100 km. This value is unrealistically high for typical geothermal areas, and many of the larger hot springs in geothermal areas are expected to be non-stationary.

Moreover, since the water flowing in fractures is expected to alter the rock temperature a substantial lateral distance from the fracture, the steady state situation will take a considerable time to set up. The time for the rock a given distance from the fracture to cool can be estimated by the Fourier number $N = at/S^2$, where a is the thermal diffusivity of the rock, S is the horizontal distance from the fracture-rock interface and t is the time (Carslaw & Jaeger 1959, p. 59). For appreciable cooling at a distance S, we may take $N = 1$. Then for $S = 10^3$ m, $a = 10^{-6}\ m^2\,s^{-1}$, we find $t = 30\,000$ years. This is a rather long time, and in geothermal areas, precipitation of minerals may clog flow channels (Bodvarsson 1964b) or tectonic activity may alter them. Consequently, in geothermal areas, the springs may not have been flowing long enough to attain a steady state. It is also reasonable to expect that springs on ocean ridge crests are non-stationary.

(a) *Time-dependent equations*

Bodvarsson (1969) derived the time-dependent equations governing the heat transfer between water flowing in a fracture and the adjacent rock. For the present, we need only to state these equations for the temperature of the ascending fluid and the adjacent rock. We set up a Cartesian co-ordinate system at D (Fig. 1) with the x axis pointing vertically downward, the y axis into the rock, and the z axis along the strike of the fracture. We will assume that the temperature in the rock $T(x, y, t)$ is independent of z. Furthermore, we will let the fracture width d be so small that the temperature of the fluid is constant across the fracture and hence equal to T at $y = 0$. Thus, the heat transfer problem is one of pure conduction in the rock and of pure convection along the x axis in the fracture. The rock temperature is found from,

$$a(\partial^2 T/\partial x^2 + \partial^2 T/\partial y^2) = \partial T/\partial t. \tag{16}$$

The convective transport term appears as a boundary condition at the fracture wall. Since the temperature field is symmetric about the plane $y = 0$, this condition has the form,

$$\rho_f sd(\partial T/\partial t) - sq(t)(\partial T/\partial x) = 2K(\partial T/\partial y); \quad \text{at} \quad y = 0. \tag{17}$$

Equation (17) states that the heat transported by the fluid equals the rate of heat transfer through the walls of the fracture. This expression has been given by Carslaw & Jaeger (1959, p. 396). The negative sign occurs since the ascending fluid is flowing in the negative x direction. In general, the flow q is a function of time.

Since the convective heat transfer in the fluid is much greater than conductive heat transfer in the rock and since the ratio $d/h \ll 1$, we have,

$$|\partial^2 T/\partial x^2| \ll |\partial^2 T/\partial y^2| \tag{18a}$$

$$|d\partial T/\partial y| \ll |q\partial T/\partial x|. \tag{18b}$$

Equations (16) and (17) then become, approximately,

$$a\partial^2 T/\partial y^2 = \partial T/\partial t \tag{19}$$

$$\partial T/\partial x = -\Phi(t)\,\partial T/\partial y; \quad \text{at} \quad y = 0 \tag{20}$$

where $\Phi(t) = 2K/sq(t)$.
Together with equations (19) and (20), we have the initial condition,

$$T(x, y, 0) = \beta x \tag{21}$$

where β is the geothermal gradient, and a boundary condition at the base of the fracture. We will consider for the moment only the early phase of the spring flow (which is consistent with the assumption of neglecting vertical heat conduction in the rock). In this case, we assume for convenience that at time $t = 0$ the fracture system opens, the water enters at A at $T = 0$ and circulates to C without absorbing heat from the rock. The water enters the base of the ascending fracture at temperature $T = 0$. Thus,

$$T(h, 0, t) = 0. \tag{22}$$

If the flow q is constant, equations (19) through (22) can be solved using the method of Laplace transforms (Lowell 1972). The result is

$$\begin{aligned} T(x, y, t) = \beta/\Phi\Big\{&2\surd(at/\pi)\Big[\exp(-y^2/4at) - \exp\Big(-(\Phi(h-x)+y)^2/4at\Big)\Big] \\ &-y\Big[erfc(y/2\surd(at)) - erfc\Big((\Phi(h-x)+y)/2\surd(at)\Big)\Big] \\ &\times \Phi x erf\Big((\Phi(h-x)+y)/2\surd(at)\Big)\Big\}. \end{aligned} \tag{23}$$

The temperature of the water is given by (23) with $y = 0$.

To find the surface temperature of the spring, we set $x = y = 0$ thus,

$$T(0, 0, t) = 2\beta/\Phi\{\surd(at/\pi)(1 - \exp(-\Phi^2 h^2/4at))\}. \tag{24}$$

On this model, the first water to flow from the spring has a temperature of 0 °C. The temperature then increases with time as the water absorbs heat from the highest temperature rocks at the base of the fracture. After a certain period of time, depending mainly upon the flow and the depth h, the lower portion of the rock is substantially cooled and the spring temperature begins to decrease. For $q = 10^{-2}\ \mathrm{kg\,m^{-1}\,s^{-1}}$, $h = 10^3$ m, $K = 2{\cdot}5\ \mathrm{W\,m^{-1}\,°C^{-1}}$, $a = 10^{-6}\ \mathrm{m^{-2}\,s^{-1}}$, the maximum spring temperature occurs after about 100 years. For the case in which $\Phi h/2\surd(at) \ll 1$ we have approximately,

$$T(0, 0, t) = \beta\Phi h^2/2\surd(\pi at). \tag{25}$$

The important feature of this model is the non-stationary nature of the heat supply. Since the heat supply is being depleted in time, it is clear that the non-stationary model will require a far smaller water-rock contact area than the steady state model. For example, consider a hot spring flowing at $100\ \mathrm{kg\,s^{-1}}$ at 80 °C. From equation (24), the total water transport after 1000 years can be shown to be approximately 10^5 kg per square meter of the water-rock interface. (Since on this model the water is assumed to take heat from the rock only during its ascent, a circulation depth of 3 km must be assumed in order to obtain the given spring temperature.) Since the total discharge of the spring after 1000 years is 3×10^{12} kg, the contact area is found to be approximately $30\ \mathrm{km^2}$. This is roughly one tenth of the contact area required for the same spring assuming the steady state conditions.

(b) *Non-stationary convection*

In the non-stationary model discussed above, the case of constant flow was considered. This model is satisfactory for some hydrostatically driven springs. In many cases, however, the flow is time dependent. For example, in convective springs, as the heat supply is gradually depleted, the convective flow must also decrease.

To treat non-stationary convection, again consider the transverse fracture system in Fig. 1. Water will be assumed to enter at A at temperature $T = 0$ and circulate through the fracture system, exchanging heat with the rock at all points along the flow path. For this model, the assumption (18a) of negligible vertical heat conduction will be dropped; however, (18b) will be retained. A Cartesian co-ordinate system is set up at A with the x axis downward and the y axis directed to the right. The convective transport terms now take the form,

$$\left.\begin{aligned} \pm sq(t)\,\partial T/\partial x &= 2K\,\partial T/\partial y; \quad \begin{matrix} y = 0 \\ y = h' \end{matrix} \\ sq(t)\,\partial T/\partial y &= 2K\,\partial T/\partial x; \quad x = h \end{aligned}\right\} \tag{26}$$

where the positive and negative signs refer to the vertical fractures at $y = 0$ and $y = h'$, respectively. The convective flow is given by equations similar to (9) and (10) which were used for steady convection. In the non-stationary case, we write for the flow parameter $\Phi(t)$,

$$1/\Phi(t) = \frac{\alpha\gamma s d^3}{48\nu K h} \int_0^h \left(T(x, h', t) - T(x, 0, t)\right) dx. \tag{27}$$

Equations (16), (21), (26), and (27) along with the boundary condition,

$$T(0, 0, t) = 0 \tag{28}$$

represent a set of non-linear integro-differential equations. The form of these equations suggests that they can be solved by an iteration procedure. We first calculate the initial flow from (27) and substitute it into equation (26). An approximation to the temperature field in the time interval $0 < t < \Delta t$ is then obtained using the initial flow. A new flow value at the time Δt is then obtained from (27). Successive iterations may be made for successive time intervals, the temperature field at the end of each interval being calculated on the basis of the flow at the beginning of the interval. The result will be a reasonable representation of the change of flow and temperature with time provided the time step is not too large.

To obtain the initial flow, we let the fracture system open at time $t = 0$ and assume that the initial temperature in the descending fluid is,

$$T(x, 0, 0) = 0.$$

Since there is initially a linear gradient in the rock, the initial flow will be given by,

$$1/\Phi_i = \alpha\gamma s \beta h d^3/96\nu K. \tag{29}$$

The above iteration procedure is ideally suited to the method of finite differences. Consequently, equation (16) was written in finite difference from using central differences for the space derivatives and a forward difference in time. The derivatives in (26) were represented by uni-directional differences, and the integration in (27) was carried out using the trapezoidal rule. The finite difference model was tested by treating the case of constant flow, neglecting vertical heat conduction. Using a grid with 26 points in the horizontal direction, 21 points in the vertical, and a time step of 8·8 years, the finite difference results were found to agree with the analytical result (23) to within 5 per cent. This agreement could have been improved by decreasing the time step, but the increased accuracy was not considered necessary in view of the expenses incurred for increased computation time and the large possible variations in the model.

The finite difference model was then used to treat non-stationary convection. It was assumed that the depth of the fractures h was 1 km and the separation h' was 1 km. The convective flow rate and spring temperature as functions of time are given in Figs 2 and 3 for a few representative fracture widths. Fig. 2 shows that the greater the initial flow (that is, the greater the fracture width), the greater the initial rate of decrease of the convective flow. The slight initial increase in the flow rate in Fig. 2 for the 2-mm fracture is a result of the initial condition which was assumed. This effect is similar to the initial temperature increase which was calculated in equation (24) for the constant flow rate spring. After a short time, depending on the fracture width, the results are quite insensitive to the initial condition. After the first few hundred years, the flow decreases monotonically in time by roughly a factor of three over 2500 years no matter what value of d is used. Fig. 3 shows that the spring temperature decreases with increasing fracture width. The cooling is so pronounced, that after 2000 years, in fractures 5 mm wide or greater, the conductive heat flow over

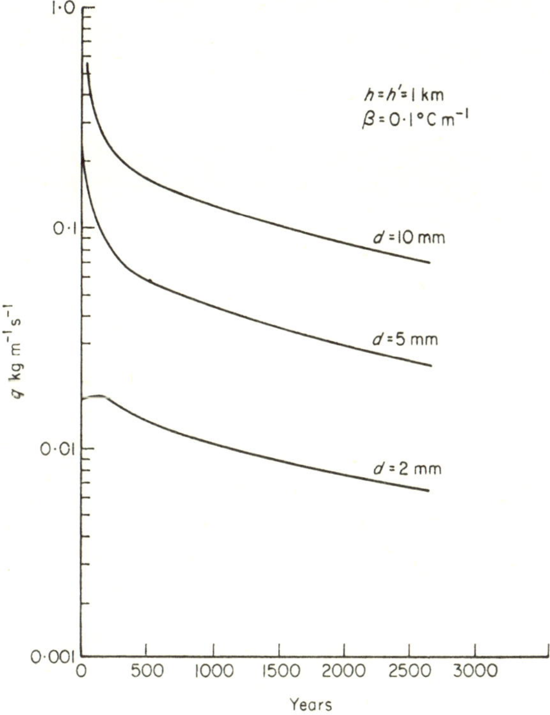

Fig. 2. Convective flow as function of time for various fracture widths.

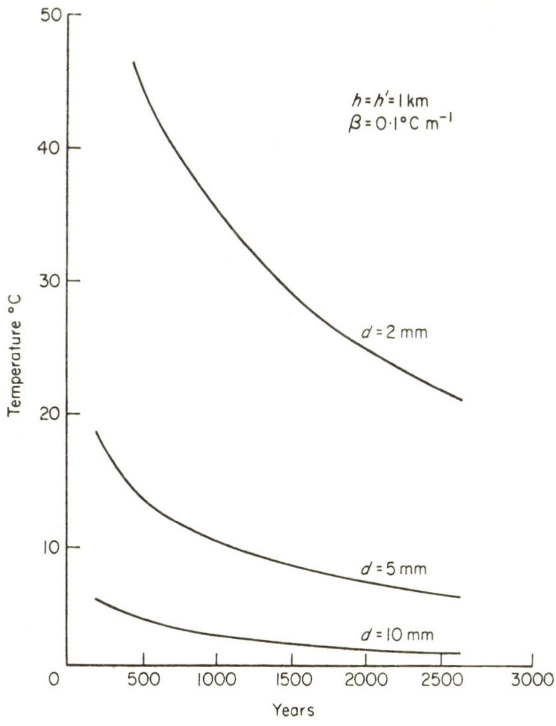

Fig. 3. Spring temperature as function of time for various fracture widths.

the whole block of width h' is found to be less than the initial conductive heat flow. It should be mentioned that, because the upward heat flux from below the fracture system has been neglected, the time dependent solutions derived here are valid only for the early phase of cooling. In their present form, these solutions do not reduce to the steady state solutions in the limit $t \rightarrow \infty$. However, the time dependent solutions given here show that the principal effect of the circulation is to cool the rock down to the depth h of the fracture system, thus lowering the regional geothermal gradient.

(c) *Hydrothermal circulation on mid-ocean ridges*

The heat flow data (Talwani *et al.* 1971; Lister 1972; Williams *et al.* 1974) suggests the existence of hydrothermal circulation on mid-ocean ridge crests. In fact, Williams *et al.* (1974) estimates that possibly 80 per cent of the heat transfer within 35 km of the Galapagos spreading centre axis takes place through hydrothermal vents. Consequently, hydrothermal circulation must form an integral part of ocean ridge models regarding heat flow. Such models are just beginning to be developed; and because of the complexity of the heat transfer processes on ridges, these models are quite simplistic (for example, Lister 1972; Bodvarsson & Lowell 1972; Sclater & Klitgord 1973). The models developed in this paper, while also quite simple, give additional insight into the problem of hydrothermal circulation at ridge crests. This model is particularly useful very near the ridge axis, where the absence of sediment cover and the high degree of fracturing and faulting is expected to give rise to circulation characterized by free discharge and recharge at the sea floor.

The circulation on ridge crests would tend to be controlled by discrete zones of high permeability. The vertical permeability is expected to be furnished mainly by dikes, the cracks between dikes and the adjacent rock, faults, and, possibly deep tensional fractures. The openings at the contacts between individual lava flows are expected to give rise to the main horizontal permeability. Although this is not an unrealistic model, it may be that topography partially controls the circulation (Lister 1972). There is also evidence that the crust on ridge crests may have a finer scale of permeability than is suggested above (Williams *et al.* 1974). In any event, there are almost certainly discrete zones of descending flow and of ascending flow which yield hot springs. It is worthwhile, therefore, to examine the implications of fracture-controlled circulation.

Since the fracturing and faulting is expected to be irregular and since flow channels may become clogged by precipitation of minerals, or tectonically altered, one would expect the crestal zone to be non-stationary and to contain several fracture-circulation systems of differing fracture widths, circulation depths and ages. Consequently, the conductive heat flow would be very irregular. Assuming a circulation system open to the sea floor, high conductive heat flow would be measured near zones of ascending flow provided (a) the fractures are narrow, (b) the circulation is deep, and (c) the system is young. Measurements near zones of descending flow would give rise to low values of conductive heat flow. Moreover, circulation in wide (1 cm) fractures of fairly shallow depth (1 km) separated by distances of the order of 1 km, would give rise to low conductive heat flow everywhere in the block containing the fracture system.

The overall effect of the hydrothermal circulation, however, is to cool the upper several kilometres of the crust (Sclater & Klitgord 1973). To apply the non-stationary fracture-convection model to the large scale cooling of the crestal zone, let us assume a circulation depth of $h = 5$ km and a horizontal separation between the fractures of $h' = 3$ km. Assuming a thermal conductivity of $K = 2{\cdot}5\ \mathrm{W\,m^{-1}\,{}^{\circ}C^{-1}}$ and a regional gradient $\beta = 100\ {}^{\circ}\mathrm{C\,km^{-1}}$, the conductive heat flow in the axial region is $0{\cdot}25\ \mathrm{W\,m^{-2}}$. This value is in reasonable agreement with measurements on the

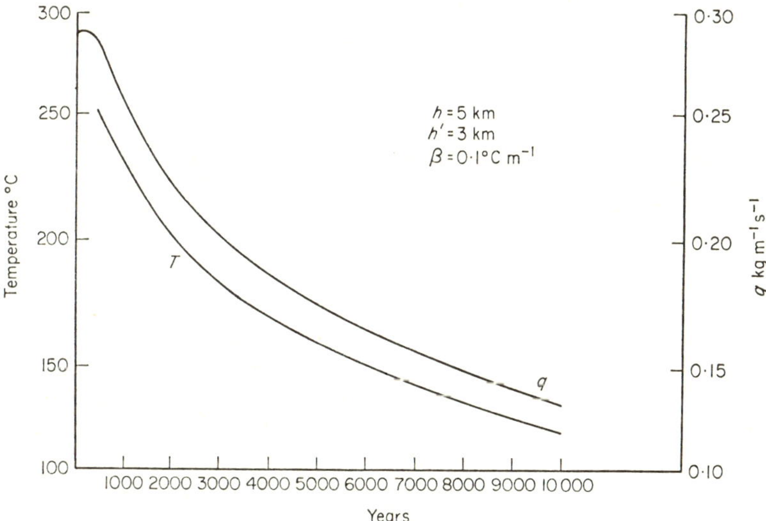

Fig. 4. Convective flow and temperature as function of time—example for ocean ridge crest circulation.

Galapagos spreading centre, and the assumed conductivity value is near that estimated for an 80 km thick lithosphere of pyrolite composition (see Sclater & Franceteau 1970). Although it is expected that the vertical permeability would decrease with increasing depth, the functional dependence is not well known. For simplicity, therefore, the finite difference calculations of $q(t)$ and $T(0, h', t)$ with the above values of h, h', K, β were carried out assuming a constant fracture width of 3 mm. The results are shown in Fig. 4.

After a circulation time of 10^4 years, the flow is found to decrease by roughly a factor of two from the initial value. Assuming a fracture length of 1 km the 10 000-year-old spring would discharge 135 kg s^{-1} at a temperature of about 125 °C. This is a large hot spring, but by no means unrealistic. The heat output of the spring is $7{\cdot}1 \times 10^4$ W m^{-1} of ridge axis. Following Lister (1972) we assume this spring discharges into a 100 m thick layer of bottom water flowing at 1 cm s^{-1} across the ridge. The spring discharge would raise the temperature of the bottom layer about 0·017 °C. Such a temperature increase would not be observable with present instrumentation.

To obtain an estimate of the effectiveness of the convective heat transfer, the total thermal output of the ascending fluid over a period of 10^4 years was calculated. The finite difference result yielded a total output of $4{\cdot}2 \times 10^{16}$ joules per metre length of ridge axis. On the other hand, if the block 3 km wide had transferred heat by conduction only, the total conductive heat transfer to the sea floor per metre length of ridge axis after 10^4 years would have been,

$$E = 0{\cdot}25 \text{ W m}^{-2} \times 3 \times 10^3 \text{ m } 3 \times 10^{11} \text{ s}$$
$$= 2{\cdot}25 \times 10^{14} \text{ j m}^{-1}.$$

Thus, the heat transferred by convection alone is nearly 200 times that which would have been transferred by conduction alone.

The conductive heat flow over the crustal block, after a circulation time of 10^4 years is shown in Fig. 5. Fig. 5 shows that the conductive heat flow over the block is less than normal over roughly one half of the block. Near the zone of ascending flow,

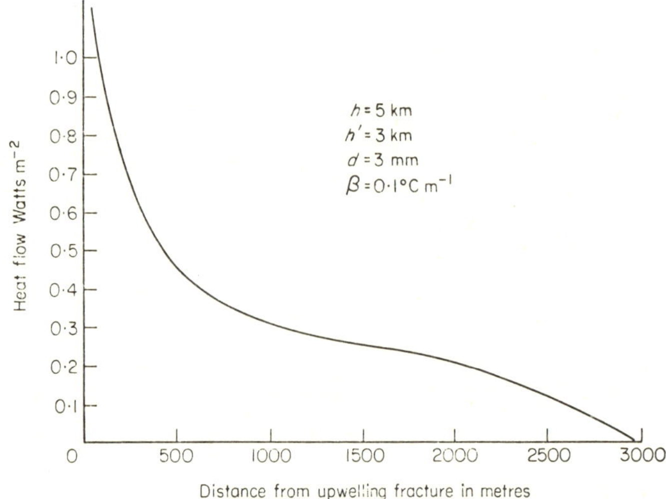

FIG. 5. Conductive heat flow versus distance from ascending flow after circulation time of 10^4 years.

the conductive heat flow becomes about 4 times the normal, which is not uncommonly observed on ridge crests. A non-stationary model for ridge crest circulation controlled by fractures a few millimetres wide, extending roughly 5 km deep, and separated by a few kilometres gives results in reasonable agreement with the data of Williams *et al.* (1974) for the Galapagos spreading centre. According to the fracture model, the observed variability of crestal heat flow data would be attributed to relatively shallow circulation systems of various depths, ages and fracture widths superimposed upon a deep circulation system.

Conclusions

In this paper, steady state and time-dependent models for hot springs have been derived on the basis of flow in a transverse fracture system. The steady state model showed that fractures a few millimetres wide or greater can carry a substantial convective flow and that the flow depends roughly on the third power of the fracture width. The steady state model gave useful results with regard to conductive temperature losses in springs. The depth of circulation for thermal springs in the south-eastern United States derived by this model were in good agreement with available field data.

The model for time-dependent convection in a transverse fracture system showed that the main effect of the circulation is to lower the geothermal gradient in the neighbourhood of the fracture system. By applying the time-dependent model to hydrothermal circulation on mid-ocean ridge crests, it was found that fracture-controlled circulation might account for the variability in observed heat flow data in the axial zone. Estimates of the amount of lithospheric cooling due to convection showed that the convective heat transfer after 10^4 years, due to circulation in a block 3 km wide and containing fractures 3 mm thick and 5 km deep, was 200 times the amount of heat which would have been transferred by conduction alone. The hot spring which would have resulted from this circulation would have given rise to an unmeasurable water temperature anomaly.

On the basis of the available heat flow data, the permeability structure of the crustal formations on ocean ridges cannot be uniquely determined. However, the fracture models give useful results with regard to both stationary and non-stationary convecting systems. Moreover, the models may be applicable to systems with scales of permeability of the order of a few tens of metres. Fracture-controlled circulation models may give useful results for many continental and marine geothermal systems.

Acknowledgments

I would like to thank G. Bodvarsson for suggesting the problem of convection in fractures and for the many informative discussions on various geothermal matters. I also thank C. O. Pollard and L. T. Long for their many helpful comments and criticisms on the original draft of this paper. This work was partially supported by the Oceanographic Section, National Science Foundation, NSF grant GA 41195.

School of Geophysical Sciences,
Georgia Institute of Technology,
Atlanta, Georgia 30332.

References

Bodvarsson, G., 1950. Geophysical methods in prospecting for hot water in Iceland, *Timarit Verkfraedingafelags Islands*, **35**, 49–59. (Translated from the Danish by Mr. Grainger, Dept. of Scientific and Industrial Research, Wellington, N.Z.)

Bodvarsson, G., 1964a. Physical characteristics of natural heat resources in Iceland, *Proc. U.N. Conf. New Sources Energy, Solar Energy, Wind Power, Geothermal Energy*, **2**, 82–89.

Bodvarsson, G., 1964b. Utilization of geothermal energy for heating purposes and combined schemes involving power generation, heating and/or by-products, *Proc. U.N. Conf. New Sources Energy, Solar Energy, Wind Power, Geothermal Energy*, **3**, 429–436.

Bodvarsson, G., 1969. On the temperature of water flowing in fractures, *J. geophys. Res.*, **74**, 1987–1992.

Bodvarsson, G. & Lowell, R. P., 1972. Ocean-floor heat flow and the circulation of interstitial waters, *J. geophys. Res.*, **77**, 4472–4475.

Carslaw, H. S. & Jaeger, J. C., 1959. *Conduction of heat in solids*, 2nd ed., Clarendon Press, Oxford, 510 pp.

Diment, N. H., Urban, T. C. & Revetta, F. A., 1972. Some geophysical anomalies in the eastern United States, in *Nature of the solid earth*, p. 544–572, ed. by E. C. Robertson, McGraw-Hill, New York.

Hewitt, D. F. & Crickmay, G. W., 1937. The warm springs of Georgia, their geologic relations and origin—a summary report, *U.S. Geol. Surv. Water Supply Pap.* 819, 40 pp.

Lamb, H., 1932. *Hydrodynamics*, 6th ed., Dover, N.Y., 738 pp.

Lister, C. R. B., 1972. On the thermal balance of a mid-ocean ridge, *Geophys. J. R. astr. Soc.*, **26**, 515–535.

Lowell, R. P., 1972. *An approach to thermal convection problems in geophysics with applications to the earth's mantle and ground water systems*, Ph.D. thesis, Oregon State University, 116 pp.

Reeves, F., 1932. Thermal springs of Virginia, *Virginia Geol. Survey Bull.*, **36**, 56 pp.

Reiter, M. A. & Costain, J. K., 1973. Heat flow in southwestern Virginia, *J. geophys. Res.*, **78**, 1323–1333.

Sclater, J. G. & Francheteau, J., 1970. The implications of terrestrial heat flow observations on current tectonic and geochemical models of the crust and upper mantle of the Earth, *Geophys. J. R. astr. Soc.*, **20**, 509–542.

Sclater, J. G. & Klitgord, K. D., 1973. A detailed heat flow topographic, and magnetic survey across the Galapagos spreading center at 86° W, *J. geophys. Res.*, **78**, 6951–6976.

Scott, R. B., Scott, M. R., Swanson, S. B., Rona, P. A. & McGregor, B. A., 1974. The TAG hydrothermal field, *EOS*, **55**, 293.

Stose, G. W. & Stose, A. J., 1947. Origin of the hot springs at Hot Springs, N.C., *Am. J. Sci.*, **245**, 624–644.

Talwani, M., Windisch, C. C. & Langseth, M. G., 1971. Reykjanes ridge crest: a detailed geophysical study, *J. geophys. Res.*, **76**, 473–517.

Waring, G. A., revised by R. R. Blankenship & R. Bentall, 1965. Thermal springs of the United States and other countries of the world—a summary, *U.S. Geol. Surv. Prof. Pap.* 492, 383 pp.

Watson, T. L., 1924. The temperatures of hot springs and the sources of their heat and water supply III. Thermal springs of the southeast Atlantic states, *J. Geol.*, **32**, 373–384.

Williams, D. L., Von Herzen, R. P., Sclater, J. G. & Anderson, R. N., 1974. Galapagos Spreading Centre: lithospheric cooling and hydrothermal circulation, *Geophys. J. R. astr. Soc.*, **38**, 587–608.

Appendix

Steady state heat flow near a spring flowing in a vertical fracture

To calculate the surface heat flow perturbation due to a flowing fracture, we represent the surface of the Earth by an infinite horizontal plane at temperature $T = 0$. The fracture acts as a heat source of strength $sq\,T(x)$, but since the conductive losses take place mainly near the Earth's surface, we can approximate the flowing fracture by a vertical plane of height h and temperature $T(h) = T_0$. We set up a Cartesian co-ordinate system at the intersection of the two planes with the x axis directed vertically downward and the y axis horizontal and perpendicular to the strike of the fracture. The problem which must be solved is that of steady state heat conduction in the region $x > 0$, $y > 0$, given by,

$$\left.\begin{aligned} &\partial^2 T/\partial x^2 + \partial^2 T/\partial y^2 = 0 \\ &T(0, y) = 0 \\ &T(x, 0) = T_0; \quad 0 \leqslant x \leqslant h. \end{aligned}\right\} \tag{13}$$

This is a standard problem in potential theory and the result is obtained by the use of Green's functions (Carslaw & Jaeger 1959, p. 361).

$$T(x, y) = T_0/\pi[2\tan^{-1} x/y - \tan^{-1}((x-h)/y) - \tan^{-1}((x+h)/y)]. \tag{14}$$

The surface heat flow due to the fracture is given by,

$$\text{H.F.} = (2T_0\,K/\pi)(1/y - y/(y^2+h^2)). \tag{15}$$

The total surface heat flow is given by adding (15) to the regional heat flux $K\beta$. It is useful to estimate the distance to which the flowing fracture perturbs the regional heat flux. Assuming a spring temperature of $T_0 = 20\,°\text{C}$ and a depth $h = 10^3$ m the distance y at which the heat flow perturbation reduces to 10 per cent of the regional heat flux is,

$$y \simeq 1{\cdot}9\ \text{km}.$$

Thus, the flowing fracture significantly perturbs the heat flux to a lateral distance of the same order as the depth of the fracture. This result tends to substantiate the choice $L = h$ which was made earlier. In the case of very deep fractures we take the limit as $h \to \infty$. The heat flow perturbation is then inversely proportional to the distance y. This model is inapplicable very near the fracture. Since the T_0 plane actually represents convected rather than conducted heat flux, the conducted heat flux at the fracture should be zero.

Part II

EXPERIMENT

Editors' Comments on Papers 3 and 4

3 HAJASH
Hydrothermal Processes along Mid-Ocean Ridges: An Experimental Investigation

4 MOTTL and SEYFRIED
Sub-Seafloor Hydrothermal Systems: Rock- vs. Seawater-Dominated

Experiments reacting seawater with basalt under the range of hydrothermal conditions supposed to prevail at seafloor spreading centers are providing insights and raising questions with regard to actual hydrothermal systems (See also Drever 1977, Benchmark volume 45). Initial experiments influenced by conditions documented at the Reykjanes geothermal area of Iceland (Paper 10) obtained promising results with regard to metal exchange and formation of alteration minerals (Bischoff and Dickson 1975; Paper 3).

M. J. Mottl and W. E. Seyfried (Paper 4) review the work of the most active "school" of basalt-seawater experimentalists centered at Stanford University and the U.S. Geological Survey in California (Dickson et al. 1963; Bischoff and Dickson 1975; Seyfried and Mottl 1977; Mottl and Holland 1978; Bischoff and Seyfried 1978). Recognizing that Reykjanes represents only one specific type of seawater-basalt interaction, this "school" has defined the water/rock mass ratio as a key parameter in determination of the chemistry of the solutions and the mineralogy of the solids (Paper 4). Another parameter undergoing experimental testing in conjunction with the water/rock mass ratio is fluid temperature gradient (Hajash 1977).

Questions raised by basalt-seawater interaction experiments pertain to the relation between experimental results and actual hydrothermal systems at seafloor spreading centers. One question concerning the relation between experimental and field products asks why anhydrite, which is produced in experiments, is generally absent in rocks recovered from oceanic ridges (Bischoff and Seyfried 1978). Anhydrite may simply be too metastable with respect to oxygenated seawater to be preserved at oceanic ridges (Bischoff, 1980). Another question is to what extent basalt-seawater alteration experiments performed in closed systems are representative of open hydrothermal systems at seafloor spreading centers. In particular, is the addition of normal seawater at shallow crustal levels sufficient to predominate over the water/rock mass ratio in determining solution chemistry and solid mineralogy? Limited available data suggest that such addition of seawater may be significant (Paper 27; Papers 24 and 25). Interaction between experimental and field observations continues to advance the formulation of fundamental generalizations in experimentally characterizing hydrothermal systems at seafloor spreading centers.

3

Reprinted from *Contrib. Mineralogy and Petrology* 53:205-226 (1975)

Hydrothermal Processes along Mid-Ocean Ridges: An Experimental Investigation

Andrew Hajash

Department of Geology, Texas A & M University, College Station, Texas 77843

Abstract. An experimental investigation of high-temperature seawater/basalt interactions has been conducted in order to better evaluate the geochemical and economic implications of hydrothermal circulation of seawater in the oceanic crust along active mid-ocean ridges. The results indicate that, as seawater reacts with basalt between 200° C and 500° C at 500–800 bars, the fluid tends to change from an oxygenated, slightly alkaline Na^+, Mg^{++}, $SO_4^=$, Cl^- solution to a reducing, acidic, Na^+, Ca^{++}, Cl^-, solution with Fe, Mn and Cu concentrations up to 1500, 190 and 0.3 ppm respectively. Silica concentrations in the fluid reach concentrations of 200–600 ppm; however, Al abundances remain very low ($\sim$0.5 ppm). Gray and green smectites, anhydrite, albite, tremolite-actinolite, chalcopyrite, pyrrhotite and hematite were the dominant alteration products formed.

These data imply that large-scale circulation of seawater in the oceanic crust could account for the Al-deficient metalliferous sediments associated with mid-ocean ridges and could be important in the genesis of certain Fe-Cu sulfide ore deposists. The process could also affect the geochemical budgets of certain elements and exert substantial control of the steady-state composition of seawater by removing excess Na and Mg and adding Ca, Si, and H to the oceans.

Introduction

Combined geological, geophysical, geochemical and isotopic data indicate that seawater penetrates the oceanic lithosphere to considerable depths and circulates through the hot basalt along active mid-ocean ridges. Such circulation would cool the oceanic crust by convection [1–7], cause intense metamorphism and substantial chemical exchange between seawater and basalt [8–12], and subsequently transport dissolved ions and complexes to upper crustal levels and into the overlying seawater. This process could, in turn, be instrumental in the formation of certain sediments [13–20] and in the generation of certain ore fluids and ore deposits [21–24]. In addition, the process could influence the geochemical mass balances of various elements in both seawater and the oceanic crust [9, 24–26].

The evidence for hydrothermal circulation of seawater in the oceanic crust and the historical development of the concept have recently been reviewed [24,

27]; for the sake of brevity, this information will not be discussed here. However a generalized model of submarine metamorphism and hydrothermal activity which is consistent with existing data is presented. In this model, seawater penetrates deep into the oceanic lithosphere in the vicinity of active mid-ocean ridges or into any region with adequate permeability [4, 9] and thermal energy. The seawater subsequently heats up and reacts with the surrounding basaltic rocks in such a way that the concentrations of Na^+, Mg^{++}, and $SO_4^=$ decrease and Ca^{++}, K^+ and SiO_2 increase in the seawater during predominantly greenschist facies metamorphism [26, 28–30]. In addition, it is probable that Fe, Mn, Cu, Ni, Co and other metals are leached from the rock [26, 29, 30]. The "evolved" hot seawater then rises through the oceanic lithosphere along unspecified temperature, pressure, oxygen fugacity and sulfur fugacity gradients precipitating various mineral species (including Fe–Cu sulfides) as suitable ambient conditions are encountered. As the metal-laden, NaCl solution discharges into the overlying seawater, the high pH and oxgen fugacity conditions cause the formation of colloidal ferric hydroxide and manganese oxide compounds that are enriched in trace elements adsorbed from the hydrothermal solution and seawater. Metalliferous sediments are formed as these compounds are incorporated into the sedimentary record. In addition, metal components, especially Mn, may precipitate as oxides directly from the hot solution near the discharge area resulting in the formation of hydrothermal deposits [31, 32].

In this model, seawater penetrates the oceanic crust and participates in various metamorphic processes resulting in significant chemical exchange with the basaltic rocks. In addition, seawater may undergo elemental exchange with basalts and gabbros during initial cooling [10]. However, the detailed nature of these processes is not known, but such information is critical in order to test the model and accurately evaluate this implications. For this reason, an experimental investigation of seawater/basalt interactions was undertaken. The experiments were conducted between 200° C and 500° C at 500–800 bars since these conditions reasonably approximate the temperature-pressure regime beneath mid-ocean ridges. The main objectives of this investigation were to determine:

1. the nature of the chemical changes that occur in the seawater during high-temperature reactions with tholeiitic basalt,
2. the chemical and mineralogical changes that take place in the basaltic glass during the alteration process, and
3. the relative rates of chemical exchange for some of the major elements under the conditions stated.

These basic experimental data will then allow more accurate evaluation of the geochemical and economic implications of large-scale hydrothermal circulation in the oceanic crust.

Methods

Experimental Methods. Most of the experiments were conducted in cold-seal pressure vessels using distilled water as the pressure medium, but a 70 ml "Bridgeman seal" vessel was used for some of the 200° C and 300° C runs. The vessels were pressurized before heating with an air-operated hydraulic pump. As the vessel was heated, water was gradually released to maintain constant pressure. Pressure was monitored by a bourdon tube gauge accurate to ±5 bars at 500 bars. Vertical resistance furnaces with automatic proportional temperature controllers provided a relatively constant temperature: typical variations for a 30 day experiment was ±15 degrees. A twelve-point chart recorder (0–1000° C, accuracy of ±0.5% of full scale) was used to record the temperature 2 or 3 times each day.

The starting material in all of the experiments consisted of oceanic tholeiite glass from six pillow basalt fragments in one dredge haul (TAG 72-16, Trans-Atlantic Geotraverse, 26°N, Mid-Atlantic Ridge, median valley). The glass was black and lustrous with no detectible signs of alteration. The outer centimeter of glass was carefully chipped off of each pillow, washed repeatedly with distilled water, dried, and ground to a powder (< 125 μm) in an agate mortar and pestal. The glass from all of the pillow fragments was then thoroughly mixed to obtain about 500 grams of homogeneous starting material.

Accurately measured portions of natural seawater and basaltic glass were sealed in cylindrical gold or platinum capsules (1 cm O.D. × 15 cm × 0.1 mm wall thickness) and allowed to react for 14 to 30 days at 200–500° C and 500–800 bars pressure. No external buffers were used to control the oxygen fugacity (f_{O_2}) since calculations indicate that, for 30 day runs at 200–500° C, insufficient H_2 will diffuse through the capsule walls for adequate buffering. Therefore, the f_{O_2} in the experiments is probably controlled entirely by phase relations in the capsule.

On completion of each run, the pressure vessel was quenched in air and then in water. The quench took 20–40 min depending on the temperature of the run. The capsule was removed from the pressure vessel and weighed prior to opening to check for possible mass exchange during the experiment. If the capsule weight before and after the run differed, it was rejected. The fluid from the capsule was then filtered through a 1.2 μm millipore filter and the pH measured. The solution was subsequently acidified with one drop of concentrated HCl to stabilize the dissolved species. All necessary dilutions were performed immediately after opening the capsules and the solutions were then stored in 200 ml polyethylene bottles. About one hour elapsed between the beginning of the quench and the final dilution. The solid material was removed from the capsule, washed three times with distilled water and dried at 105° C for one hour.

Analytical Procedures. The concentrations of Fe, Mn, Cu, Na, K, Ca, and Mg in the "seawater" solutions were determined on a Perkin-Elmer 306 atomic absorption spectrophotometer using an air-acetylene flame; Si and Al abundances were similarly obtained with nitrous oxide as the oxidant. Artificial seawater standards were used for these atomic absorption measurements to compensate for viscosity and light scattering effects. Analytical precision values for the various elements measured in seawater are expressed as standard deviations about the mean: Ca ±0.3%, Mg ±2.9%, Na ±2.3%, K±1%, Fe ±8%, Mn ±1.7%, Cu ±6%, and Si ±2%.

The fresh and experimentally altered solid phases were analyzed for major elements by atomic absorption spectrophotometry. The mineralogy was studied by standard petrographic, X-ray diffraction (XRD) and scanning electron microscopy (SEM) techniques. The material from each run was analyzed by X-ray diffraction after being dried for one hour at 105° C. Subsequent diffractograms were made after ethylene glycolation and after heating to 600° C for 2 hrs. These were conducted for clay mineral identification and to distinguish smectites from chlorites [33]. For the smectites, the (060) peak position was used to distinguish trioctahedral from dioctahedral structures. Whenever possible, individual mineral phases were separated for X-ray powder camera and chemical studies.

Presentation of Results

Fluid Chemistry. The fluid chemistry data from the natural seawater/basalt experiments are given in Table 1. Representative data from this table, identified by asteriks, are plotted as bar graphs in Figs. 1 and 2 to depict the general chemical trends observed.

The data show profound changes in the chemistry of the seawater during the experiments. Calcium increases substantially in seawater at all temperatures studied; the increases range from 200% to 400% of the initial value with most

Table 1. Fluid compositions from natural seawater/basalt experiments

T = temperature in °C; *P* = pressure in bars; *t* = duration of experiment in days; elemental concentrations given in parts per million (ppm); * refers to data plotted in Figs. 1–2. NSW = natural seawater

I.D.	*T*	*P*	*t*	pH	Water/Rock	Na	K	Ca	Mg	Fe	Mn	Cu	Al	Si
NSW	25	1	–	7.9	–	12200	418	430	1496	<0.1	<0.1	<0.1		3
#20*	200	500	14	6.3	5:1	11350	365	1400	3.6	0.4	0.26	0.15	<0.5	268
#26	200	500	14	–	5:1	insufficient fluid				0.62	0.28	–	–	–
#17	300	500	33	4.0	5:1	11075	423	1533	2.9	0.4	0.2	0.123	1.25	193
#18*	300	500	14	5.8	5:1	11725	405	1357	2.4	0.4	0.27	0.19	<0.5	205
#22	300	500	14	3.8	1:1	14000	445	1100	1.8	0.11	0.35	0.21	<0.5	220
#24	300	500	14	4.0	1:1	14500	519	1150	1.6	0.12	<0.1	0.35	<0.5	160
#19*	400	700	14	4.8	5:1	10950	457	1707	11.4	3.6	4.1	0.19	<0.5	340
#23	400	700	14	5.0	1:1	11900	600	2190	16.2	0.96	2.6	0.33	<0.5	200
#29	400/500	800	14/14	3.7	1:1	11800	400	1625	16.0	61.5	60.0	0.07	<0.5	593
#21*	500	800	14	3.3	5:1	10725	435	933	16.7	1057	90.0	0.33	0.5	625
#25	500	800	14	3.5	1:1	10475	573	925	47.0	1320	190	0.33	<0.5	643
#27	500	800	14	3.2	1:1	capsule ruptured during quench								
#28	500	800	14	3.2	1:1	10700	548	800	45	1100	131	0.07	<0.5	480
#30	500	800	25	3.1	3:1	12030	460	900	30	1500	63	0.06	<0.5	670

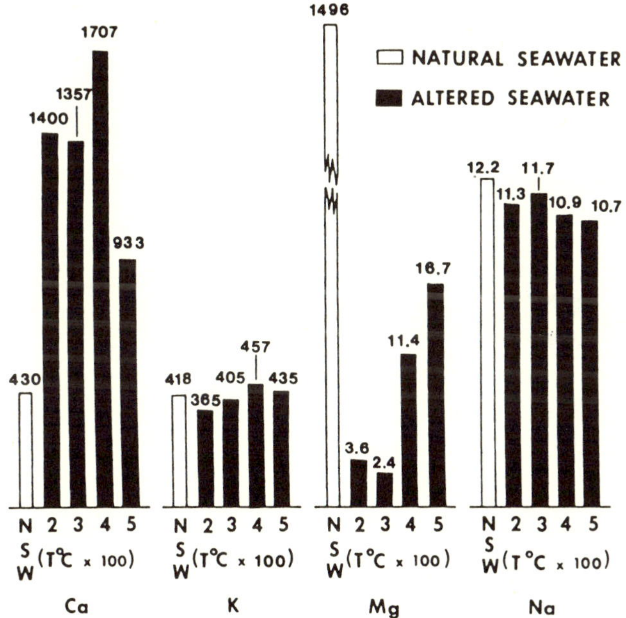

Fig. 1. Bar graphs of representative Ca, K, Mg and Na concentrations in the altered natural seawater solutions as a function of the temperature of the experiment. Concentrations of Ca, K and Mg are in parts per million; Na values are in parts per thousand

Fig. 2. Bar graphs of representative Fe, Mn and Si concentrations in the altered natural seawater solutions as a function of the temperature of the experiment. Concentrations are in parts per million

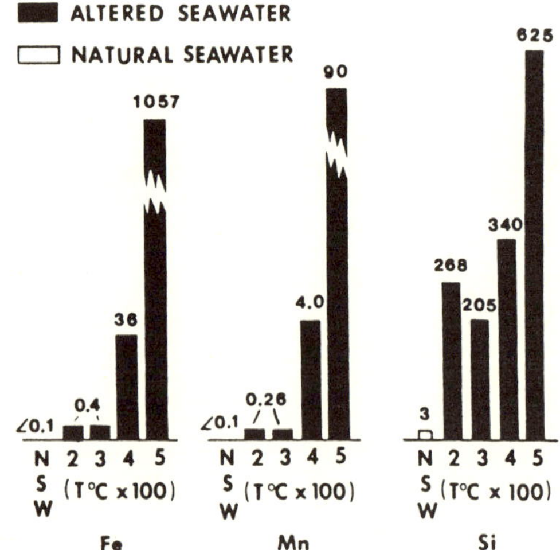

pronounced changes occurring in the 200–400° C range. The 500° C experiments resulted in increases of only 200% of the initial value.

Potassium abundances in the altered seawater are quite varied. The 200° C runs typically show decreases to about 80% of the initial value whereas the 300° C runs show no change or slight increases. Experiments conducted at 400–500° C shows K abundances that are 5–80% greater than the original concentration. For 14 day runs at 200–500° C, K tends to decrease slightly at lower temperatures and increase slightly at higher temperatures.

All of the experiments resulted in extreme depletion of Mg from the seawater. Of the 1 500 ppm Mg available in the initial seawater, 98–99% is removed and concentrated in the solid phase. The removal is slightly more effective at lower temperatures (Fig. 1).

Sodium abundances in the seawater tend to decrease upon reaction with the basaltic glass. The effect is accentuated at higher temperatures with typical decreases of approximately 12% at 500° C.

Fe and Mn concentrations in the seawater show exceedingly large increases as a function of increasing temperature. Even though both elements are below the detection limit of atomic absorption spectrophotometry (<0.1 ppm) in the initial seawater, Fe and Mn concentrations reach values of about 0.4 ppm and 0.26 ppm respectively in the 200° C and 300° C runs. The 400–500° C experiments show much higher abundances. Fe and Mn values typically reach more than 1 000 ppm and 100 ppm, respectively, in the 500° C runs.

The concentration of copper increased from 0.05 ppm in the initial seawater to 0.07–0.35 ppm in the altered waters. No reliable Ni abundances could be determined in the fluids.

It appears from the data that Al, in contrast to Fe, Mn and Cu, is essentially immobile in the seawater/basalt system investigated. Aluminum abundances are near or below the detection limit of atomic absorption spectrophotometry (~ 0.5 ppm) in all of the runs.

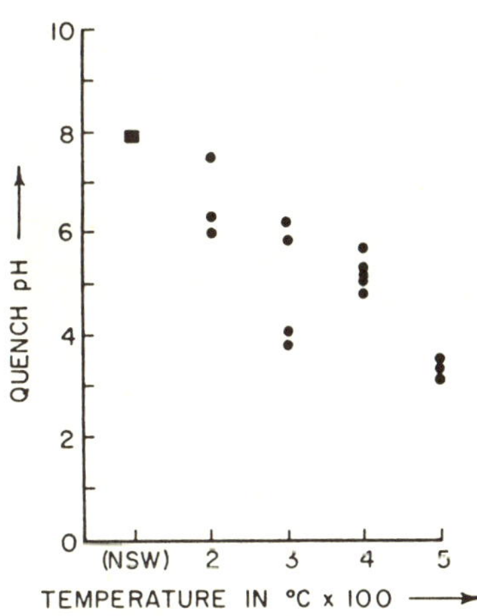

Fig. 3. Graph of the quench pH data as a function of the temperature of the seawater/basalt experiment.
The pH of the initial seawater (NSW) is designated by a solid square

Fig. 3 shows the variation of quench pH as a function of the temperature of the experiment. A strong trend of decreasing pH with increasing temperature is demonstrated.

It should be mentioned that, on opening, a small amount of gas escaped from the capsules. The gas from the 200° C runs was essentially odorless whereas the 300° C runs gave a moderate H_2S smell; the 400–500° C runs produced very strong H_2S odors. The smell of this gas, although not quantitatively significant, indicates that reduced sulfur species become more abundant at higher temperatures.

The following summarizes the changes in elemental abundances in the altered seawater with respect to the initial seawater:

1. Fe, Mn and Cu increase, especially at higher temperatures (400–500° C),
2. Ca increases rapidly,
3. Mg decreases rapidly to near zero,
4. K decreases initially at low temperatures (200° C) and increases slowly at higher temperatures,
5. Na decreases slightly,
6. Al remains very low, usually less than 0.5 ppm, and
7. the quench pH decreases with increasing temperature.

Chemical Analyses of the Altered Material. Bulk chemical analyses of the altered material produced by reactions with seawater are given in Table 2. For comparison, the composition of the basaltic glass starting material is also shown. For most elements, these data are less sensitive than the fluid chemistry data for determining elemental exchange trends; however, the two sets of information are generally consistent. Copper abundances in the solids show the most pronounced changes with substantial amounts being leached from the rock at all temperatures. A small amount of the leached copper remains in solution (Table 1); however, most of it appears to go into the alteration products (e.g., chalcopyrite) or is absorbed by the capsule walls if Au is used.

Table 2. Bulk chemical analyses of the basaltic glass starting material (SM) and the experimentally altered material

T = temperature of experiment in °C; *P* = pressure in bars; *t* = duration of experiment in days

I.D.	*T*	*P*	*t*	Na	K[a]	Ca	Mg	Fe	Mn	Al	Cu[a]
SM				2.20	820	7.60	4.84	7.15	0.129	7.8	82
#20	200	500	14	1.97	904	6.65	4.67	7.15	0.123	7.15	75
#22	300	500	14	2.06	769	6.48	4.15	6.80	0.115	6.11	61
#18	300	500	14	1.80	842	6.26	4.16	6.90	0.114	5.89	31
#17	300	500	33	2.03	730	6.04	4.41	6.72	0.115	6.02	36
#24	300	500	14	2.00	817	6.25	3.99	6.62	0.112	6.19	–
#23	400	700	14	2.14	605	6.62	4.21	6.90	0.118	6.38	23
#25	500	800	14	2.15	662	6.87	4.27	6.75	0.110	6.75	20

[a] Concentration in ppm; all others in percent.

Table 3. Mineralogy of the alteration products as a function of the temperature of the experiment

Starting material	Glass, a few phenocrysts of plagioclase, a few microphenocrysts of plagioclase and olivine
200° C experiments	Anhydrite; gray smectite (15.2 Å, air dried; 16.8 Å, glycolated; 060 = 1.53 Å)
300° C experiments	Anhydrite; gray smectite; analcite; prehnite(?)
400° C experiments	Gray smectite; green, Fe-rich smectite; tremolite-actinolite; albite; pyrrhotite
500° C experiments	Gray smectite; green, Fe-rich smectite; tremolite-actinolite; albite; epidote(?); cordierite(?); chalcopyrite; hematite; magnetite(?)

Mineralogy. The mineralogy of the altered material as a function of the temperature of the experiment is shown in Table 3. X-ray diffraction data for all of the minerals identified in the experiments are summarized in Table 4. A very fine-grained smectite, which showed the usual behavior of (001) upon glycolation and heating, was abundant in the 200° C and 300° C runs. Although chemical data are unavailable, saponite is suspected since almost all of the magnesium was removed from the seawater during the experiments. In addition, a strong X-ray diffraction peak at 1.53 Å (060) indicates a trioctahedral structure [33].

A light-green smectite (refractive index = 1.512 ± 0.008) became the dominant clay mineral at temperatures greater than 300° C. It commonly grew on the capsule walls as large, interconnecting flakes producing a "honeycomb" morphology (Fig. 4). Petrographic examination revealed abundant, small hematite plates imbedded in the smectite. Physical separation of the two phases for chemical analysis was not possible. Structural spacings determined by XRD (Table 4) are similar to both saponite and nontronite; however, accurate identification is not possible lacking reliable chemical data.

Table 4. X-ray diffraction data for minerals formed in seawater/basalt experiments. In all cases except magnetite, estimated and ideal relative intensities were in close agreement

Mineral	Peaks used for identification (Å)								
Albite	3.21	4.01	3.76	3.03	2.95	2.93	2.52		
intensity	100	80	80	80	70	50	70		
ideal peaks	3.20	4.03	3.75	3.03	2.94	2.92	2.52		
Analcite	3.42	5.57	2.93						
intensity	100	80	70						
ideal peaks	3.43	5.61	2.93						
Anhydrite	3.49	2.85	2.34						
intensity	100	33	22						
ideal peaks	3.49	2.85	2.34						
Chalcopyrite	8.51	3.38	3.04						
intensity	100	90	90						
ideal peaks	8.58	3.38	3.04						
Hematite	2.71	2.52	1.85	1.67	1.49	1.44	1.32	1.26	1.16
intensity	100	60	40	60	35	35	20	8	10
ideal peaks	2.69	2.51	1.84	1.69	1.48	1.45	1.31	1.26	1.16
Magnetite(?)	2.94	2.52	–	1.70	1.61	1.49	1.11	0.88	
intensity	70	100	70	60	85	85	60	40	
ideal peaks	2.97	2.53	2.09	1.71	1.61	1.48	1.09	0.88	
Pyrrhotite	2.98	2.64	2.07	1.72	1.32	1.16	1.05	0.996	
intensity	33	33	100	33	13	13	7	1	
ideal peaks	2.97	2.65	2.06	1.71	1.32	1.11	1.07	1.00	
Tremolite	8.84	4.82	4.50	3.90	3.15	2.98	2.72	2.54	2.07
intensity	100	10	20	16	100	40	90	40	45
ideal peaks	8.38	4.87	4.51	3.87	3.12	2.94	2.72	·2.53	2.04
Smectite (green)	15.2	4.54	2.63	2.52	1.74	1.53	1.33	1.31	
intensity	100	100	80	80	40	100	20	20	
Nontronite	15.4	4.56	2.64	2.56	1.72	1.52	1.32	1.30	
Saponite	15.6	4.56	2.63			1.53			
	100	40	30			50			

Smectite gray 1.52 Å air dried, 16.8 Å glycolated, 9.67 Å after heating (060) = 1.53 Å.

Euhedral crystals of anhydrite were abundant in the 200–300° C runs; they were usually found on the capsule walls and at the rock/water interface. A trace of anhydrite was found in the 400° C and 500° C runs. Analcite was restricted to the 300° C experiments.

The mineralogy of the 400° C and 500° C runs was very similar. Tremolite-actinolite grew as white fibrous mats between the capsule wall and the starting material (Fig. 5) and at the rock/water interface. Albite was found both as white, botryoidal masses (Fig. 6) and as clear euhedral prisms. Fig. 7 shows some of the large crystals of albite adjacent to small albite crystals that make up the botryoidal masses. The fibrous material in the upper, right-hand corner of Fig. 7 is tremolite. Fig. 8 shows some of the large albite crystals that formed at 500° C. The green smectite, albite and tremolite-actinolite were much more abundant

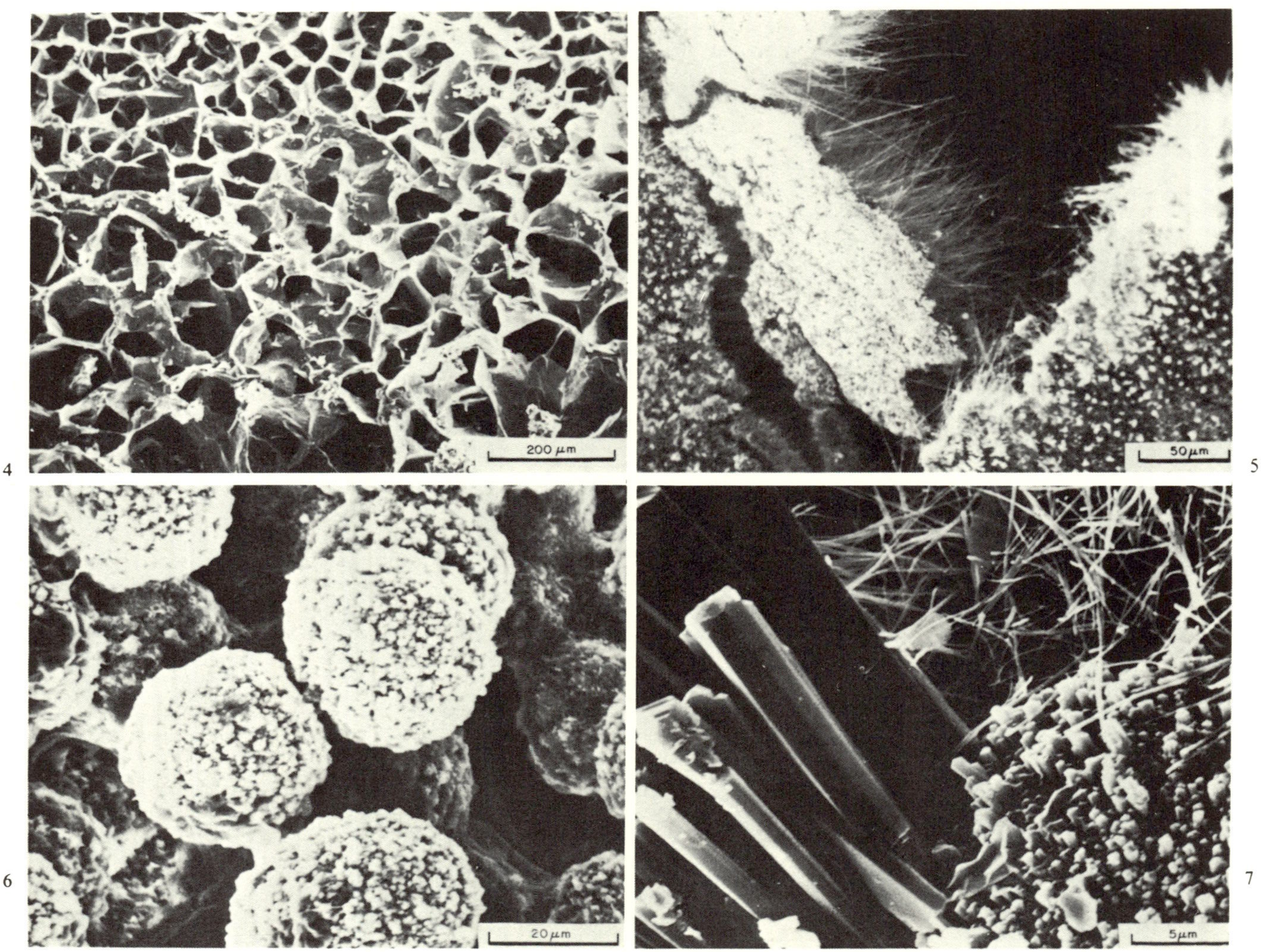

Fig. 4 Green smectite displaying "honeycomb" morphology

Fig. 5 White fibrous mats of tremolite-actinolite

Fig. 6 White botryoidal masses of albite

Fig. 7 Large euhedral prisms of albite adjacent to small albite crystals that make up the botryoidal masses

Fig. 8
Albite crystals under high magnification

Fig. 9
Pyrrhotite platelets

Fig. 10
Stack of hematite plates surrounding an albite(?) crystal

Fig. 11
Hematite plates under high magnification

at 500° C than at 400° C. In addition, cordierite was found by XRD in some of the 400° C and 500° C runs (Table 4).

Small but significant quantities of sulfide minerals were found in all of the 500° C runs and in one 400° C run. The sulfide phases were identified by powder-camera XRD studies of "hand-picked" mineral separates. Euhedral, hexagonal platelets of pyrrhotite (Fig. 9) were abundant in run #29; this experiment consisted of holding the capsule at 500° C for 14 days followed by 14 days at 400° C.

Chalcopyrite and hematite were coexisting phases at 500° C. Chalcopyrite formed as relatively large, euhedral crystals in one 500° C run but was found more typically as a fine-grained powder. Hematite crystals formed as small plates primarily at the rock/water interface or at the capsule wall/rock interface. Fig. 10 shows a stack of hematite plates surrounding an albite crystal; Fig. 11 shows some of the smaller hematite crystals.

Power-camera X-ray diffraction data from material containing predominantly hematite plates indicate that magnetite may also be present. Seven magnetite lines are present but two of the strongest lines (2.53 Å and 1.483 Å) correspond closely to major hematite lines (2.51 Å and 1.484 Å). Although positive identication cannot be made, the presence of lines unobscured by hematite indicate that magnetite may be present in small amounts with hematite and chalcopyrite.

Discussion

Experimental Results. The lack of quantitative mineral abundances and continuous sampling of the fluid limit understanding of the processes that control the mobilization and precipitation of the chemical entities. In addition, the concentrations of certain elements could have changed somewhat during the relatively long quench (20–40 min); notably, some Fe, Mn and Si could have precipitated at this time. However, the *trends* of chemical exchange are of primary importance and, in conjunction with mineralogical information, allow some generalizations to be made about relative elemental exchange rates and probable controlling reactions.

Calcium and magnesium concentrations in the seawater show the largest and most rapid changes of the major elements. Calcium increases in the seawater at all temperatures investigated, even though some Ca is removed by precipitation of anhydrite in response to its decreasing solubility with increasing temperature [34, 35]. As Ca is leached from the basalt, anhydrite precipitates as long as sufficient $SO_4^=$ is available. Since anhydrite is abundant at 200–300° C but absent or present only in trace amounts at 400–500° C, it appears that little $SO_4^=$ is present at the higher temperatures. Although $SO_4^=$ abundances in the altered fluid have not been measured, the absence of anhydrite and the strong H_2S smell emitted from the 400–500° C runs suggest that much of the initial $SO_4^=$ has been reduced to H_2S by interaction with basalt. Tremolite appears as a major alteration product at 400° C and 500° C and probably controls to a great extent the Ca abundance in the fluid. The appearance of large quantities of tremolite in the 500° C runs may explain why these experiments resulted in Ca increases of only 200% of the initial value, whereas the 200° C, 300° C and 400° C runs showed much greater increases (Fig. 1).

Magnesium is almost totally removed from the seawater regardless of the initial Mg concentration in the fluid, temperature of the experiment, grain size of the starting material or duration of the run. Mg removed from the fluid is probably incorporated into the octahedral layers of the smectite, a ubiquitous alteration product, along with Fe and Al. The high Mg content of the initial seawater is apparently a crucial factor in the formation of smectites instead of other clay minerals [36]. As mentioned earlier the removal of Mg from the seawater was slightly more effective at the lower range of experimental temperatures. The cause of this weak trend is unclear due to the complexity of the alteration products. Perhaps the appearance of the green smectite as an abundant alteration product in the 400° C and 500° C runs results from substitution of Fe^{+2} for Mg^{+2} in the smectite; this would presumably result in slightly more Mg in equilibrium with the alteration products at higher temperatures.

K appears to be less mobile than the other major elements in these relatively short experiments of 14–30 days duration. Even though K tends to increase in the seawater at high temperatures, less than 25% of the K available in the rock has been removed. In contrast, the seawater/basalt experiment of Mottl *et al.* [30], which lasted up to two years, show that K increases in the seawater up to concentrations of about 1 000 ppm. Bischoff and Dickson [26] also find K increasing in the seawater at 200° C.

The decrease in the Na abundances in the seawater on reaction with basaltic glass (Fig. 1) is probably due to the formation of analcite (300° C runs) and albite (400° C and 500° C runs).

Although Si concentrations in the initial seawater were very low ($\sim$ 3 ppm), the altered seawater solutions invariably contain a few hundred ppm of Si. The measured abundances are probably minimum values, however, since some Si may precipitate during the quench. Much of the silica removed from the rock is also precipitated as various alteration products.

The quench pH data of Fig. 3 indicate that the pH of the altered seawater is lowered in all of the experiments with the effect being more pronounced at higher temperatures. These quench pH data do not accurately represent the pH at the temperature of the run because the hydrogen ion complexes in the solution (H_2CO_3, HCO_3^-, H_2S, HS^-, HSO_4^-, and HCl) dissociate more at lower temperatures than at higher temperatures [34–40]. Therefore, the quench pH data are probably minimum values and are insufficient to accurately determine the pH at the temperature of the experiment.

Information about the nature of pH control during seawater/basalt interactions is very limited; a number of potential contributing reactions are listed below:

1. The formation of smectites, prehnite, tremolite-actinolite or any phase incorporating hydroxyls from the fluid could tend to increase the $[H^+]/[OH^-]$ radio and decrease the pH.

2. The oxidation of Fe^{++} and/or Mn^{++} in solution and precipitation of oxides would tend to lower the pH. For example, in the following equation,

six moles of H^+ are released for every two moles of Fe^{++} precipitated as hematite:

$$2Fe^{++}+3H_2O \leftrightarrows Fe_2O_3+6H^++2e^- .$$

3. If SO_2 is present at temperatures greater than 400° C, it could break down as follows [35],

$$4SO_2+4H_2O \leftrightarrows H_2S+3H_2SO_4 .$$

The subsequent dissociation of H_2S and H_2SO_4 could lower the pH.

4. Much of the pH lowering, however, would be offset by hydrolysis reactions (H^+ metasomatism) with the existing crystalline phases [26].

It is unclear, which reactions control the oxygen fugacity (f_{O_2}) during high-temperature seawater/basalt interactions. One possibility is equilibrium between magnetite and hematite as shown below:

$$\tfrac{1}{2}O_2+2Fe_3O_4 \leftrightarrows 3Fe_2O_3 .$$

Bischoff and Dickson [26] tested this possibility for seawater/basalt reactions at 200° C assuming that dissolved ferrous iron would be controlled by the same reaction. They concluded that the 200° C system was more reducing than that imposed by magnetite-hematite equilibrium and than an oxidized phase, such as nontronite, may be present. However, the presence of hematite and probably magnetite in the 500° C runs of this investigation suggests that equilibrium between these minerals may exert *some* control on the f_{O_2} at high temperatures. If this assumption is made, then the oxygen fugacity at 500° C would be approximately $10^{-17.3}$ atmospheres. This value of f_{O_2} and the 500° C mineral assemblage can be used to estimate the sulfur fugacity (f_{S_2}), which is believed to be quite high because of the strong H_2S smell generated and the Fe-sulfide phases formed. Fig. 12, calculated from thermodynamic data presented by Helgeson [40], is probably a reasonable representation of the phase boundaries in the Fe–S–O system at the experimental conditions. It is apparent from this figure that if chalcopyrite (similar stability field as FeS_2), hematite, and probably magnetite are coexisting phases as indicated by the 500° C mineralogy, then the sulfur fugacity must be about 10^{-2} atmospheres. If only chalcopyrite and hematite are present, then the f_{S_2} would be even higher. The system is surely more complicated than that presented here. For instance, if the green smectite which is abundant at 500° C is nontronite (Fe-rich), then the system may be even *more* reducing similar to the 200° C situation discussed by Bischoff and Dickson [26]. In any case, the sulfur fugacity is likely quite high. Such conditions could be produced by reduction of seawater sulfate or by release of primary sulfur from the rock, or both. Insufficient data are available to evaluate these two processes.

Although substantial amounts of heavy metals were leached from the basalt at all temperatures, the process was much more effective at higher temperatures. For instance, Fe, Mn and Cu reached concentrations up to 1 500, 190 and 0.3 ppm respectively, at 500° C. These exceedingly high abundances are probably due to the increased solubility of the metals under reducing, acidic conditions [41]. Under such conditions, both Fe and Mn favor the liquid phase. However, with increasing f_{O_2} or pH, Fe-bearing phases become stable before similar Mn-bearing phases. For example, pyrite becomes stable at total sulfur concentrations as

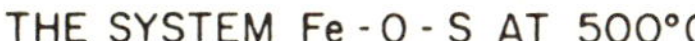

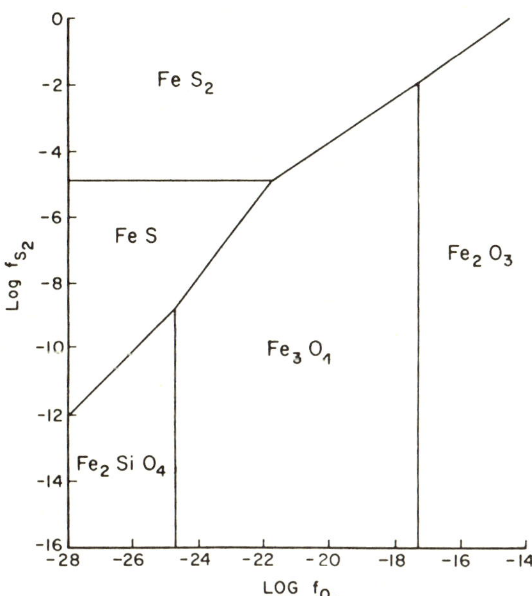

Fig. 12. Phase relations in the system Fe – S – O as a function of f_{S_2} and f_{O_2} at 500° C

low as 10^{-6} M even though no corresponding Mn phase can form [41]. Such selective removal of Fe from the fluid, which would lower the Fe – Mn ratio, seems to have occurred in the 400° C and 500° C experiments by the precipitation of pyrrhotite, chalcopyrite and hematite. No Mn-rich phases have been identified. This process may partially account for the marked contrast between the Fe/Mn ratios of the altered seawater (0.4 to 24) and the starting material (~ 55).

Within the range of experimental conditions investigated, the direction of elemental exchange between seawater and basalt is independent of temperature, grain size, rock/water ratio and duration of the run; however the magnitude and rate of exchange for some elements was greatly affected by temperature. For example Fe, Mn and Si exchange (Fig. 2) and Cu exchange (Table 2) was substantially enhanced by elevated temperatures; Na and K were moderately affected and Ca and Mg exchange showed little or no variation with temperature changes (Fig. 1).

The effect of pressure on the direction and extent of elemental exchange was not studied because previous investigations indicate that it is of little significance. Hawkins and Roy [36] performed over forty hydrothermal runs at temperatures ranging from 250° C to 425° C and pressures from 5000 to 25000 psi. They observed that these pressure variations had no effect on the rate of rock alteration or on the type of alteration products formed.

Variations in rock/water weight ratios from 1 : 1 to 1 : 5, with other conditions held constant, result in small, non-systematic differences in fluid compositions. As a matter of fact, the differences are smaller than the total experimental and analytical precision involved in reproducing a given experiment. This point is

Table 5. Fluid compositions from three 500° C experiments. All of the experimental conditions are identical except for the water/rock ratios. Symbols are defined in Table 1

I.D.	*T*	*P*	*t*	pH	Water/Rock	Na	K	Ca	Mg	Fe	Mn
NSW	25	1	–	7.9	–	12200	418	430	1496	<0.1	<0.1
#21	500	800	14	3.3	5:1	10725	435	933	16.7	1057	90
#25	500	800	14	3.5	1:1	10475	573	925	47.0	1320	190
#28	500	800	14	3.2	1:1	10700	548	800	45.0	1100	131

Table 6. Fluid compositions from two 300° C experiments and two 500° C experiments. Duration of the runs were significantly different; other conditions were the same. Symbols defined in Table 1

I.D.	*T*	*P*	*t*	pH	Water/Rock	Na	K	Ca	Mg	Fe	Mn
#22	300	500	*14*	3.8	1:1	14000	445	1100	3	0.11	0.35
#1C[a]	300	600	*236*	6.0	1:1	9814	1026	3263	1.8	0.23	0.35
#25	500	800	*14*	3.5	1:1	10700	580	900	51.8	1320	190
#3B[a]	500	1000	*268*	3.3	1:1	10318	1165	1211	29.2	1602	287

[a] Data from Mottl *et al.* [30].

illustrated in Table 5 where fluid compositions from three 500° C runs are compared. Note that #21 and #25 have different rock/water ratios but show smaller chemical differences than #25 and #28 which have the same ratios. These data indicate that for short, closed-system experiments, small variations in rock/water weight ratios have much less effect on the composition of the fluid than other parameters such as temperature.

The effect of time on the composition of fluids reacting with basalt cannot adequately be determined from this study alone since all of the runs are of similar duration. However, a comparison of experiments from this investigation (14 days duration) with two experiments (236–263 days duration) of Mottl *et al.* [30] provides some insight into this problem (Table 6). Somewhat surprising are the remarkably similar compositions even though the length of the experiments differ by a factor of 20. Note that, except for K, the fluids from the two 500° C runs are almost indistinguishable. The two 300° C runs have nearly identical Fe, Mn and Mg abundances whereas Na, K and Ca exchange has proceeded farther in the longer runs. These data demonstrate that substantial elemental exchange *and* approximate chemical equilibrium with respect to some elements are achieved rapidly, especially at high temperatures.

A cursory study of the effect of grain size on the fluid composition was conducted at 500° C. Two experiments were run that were identical except for the grain size of the starting material. The starting material is one run (#28) was ground to < 100 μm whereas the other run (#30) contained chips between

Table 7. Fluid compositions from two 500° C runs with starting material of vastly different grain sizes. NSW = natural seawater. Other symbols are defined in Table 1

I.D.	*T*	*P*	*t*	pH	Water/Rock	Grain size	Na	K	Ca	Mg	Fe	Mn
NSW	25	1	–	7.9	–	–	12200	420	430	1480	<0.1	<0.1
#28	500	800	14	3.2	1:1	<100 μ	10700	548	800	45	1100	131
#30	500	800	25	3.1	3:1	2–5 mm	12030	460	900	30	1500	63

2 mm and 5 mm. The two runs resulted in essentially identical fluid compositions (Table 7) and mineral assemblages (gray and green smectites, albite, tremolite-actinolite, hematite and Fe-sulfide phases). This similarity demonstrates that the chemical exchange reactions and precipitation processes can take place rapidly at high temperatures, even for relatively coarse-grained material.

To summarize, the experimental data indicate that seawater/basalt interactions at 200–500° C result in

1. the rapid removal of almost all of the available Mg in seawater probably by the formation of Mg-rich smectite,
2. the removal of a small amount of the Na in the seawater by the precipitation of analcite and albite,
3. the net addition of Ca to the seawater even though anhydrite precipitates at lower temperatures as long as sufficient $SO_4^=$ is available, and tremolite-actinolite and possibly cordierite precipitate at 400–500° C.
4. the slow addition of K to the seawater,
5. the addition of small amounts of heavy metals to the solution at low temperatures and exceedingly large amounts at high temperatures, accompanied by the formation of a green smectite, pyrrhotite, chalcopyrite and hematite, and
6. the lowering of the pH and f_{O_2} while increasing f_{S_2}.

In addition, temperature seems to be the most important parameter controlling the composition of the fluid and the type of alteration products formed. Variations in pressure, grain size, rock/water weight ratio and duration of the runs were less significant for most elements.

Submarine Metamorphism and Hydrothermal Activity: Application of Experimental Data

The experimental data add considerable support to the generalized model of submarine metamorphism and hydrothermal activity presented earlier. First, the experimentally altered seawater solutions and the hydrothermal fluids of seawater origin being discharged from the Reykjanes geothermal area in Iceland [28] show the same elemental exchange trends and, in most cases, similar compositions. Table 8, which compares elemental abundances in the experimental fluids with

Table 8. Comparison of the compositions of the experimental fluids with Reykjanes geothermal waters. NSW = natural seawater; all other symbols are defined in Table 1

I.D.	T	P	t	pH	Water/Rock	Na	K	Ca	Mg	Fe	Mn	Si
NSW	25	1	–	7.9	–	12200	420	430	1480	<0.1	<0.1	3
#20	200	500	14	6.3	5:1	11500	370	1400	2.9	0.5	0.1	268
#17	300	500	14	4.0	5:1	11150	425	1550	3.0	0.5	0.1	193
#19	400	700	14	4.8	5:1	11100	465	1700	10.8	3.5	4.0	340
#21	500	800	14	3.3	5:1	10950	435	900	15.7	960	88.0	625
[b] Reykjanes Drillhole #8 (277° C)						9610	1348	1530	16.0	0.5[a]	2.0[a]	374

[a] Mottl *et al.* [30].
[b] Bjornsson *et al.* [47].

the Reykjanes waters, demonstrates this point. In addition, pH and f_{O_2} decrease and f_{S_2} increases in the seawater solutions that have reacted with basalt, as predicted by the generalized model.

Bischoff and Dickson [26] observed that seawater which reacted with basalt at 200° C contained substantial Fe and Mn but had very little Al. The 200° C–500° C data presented here show the same elemental trends: large quantities of Fe and Mn can be removed from the rock without similar leaching of Al. This is inconsistent with Bostrom's [42] suggestion that significant amounts of Al would be mobilized along with Fe. Instead, the high metal/Al ratios in the fluids strongly imply that seawater/basalt interactions can easily provide Fe, Mn and other metals to the Al-deficient metalliferous sediments associated with mid-ocean ridges, as suggested by Corliss [10].

Also related to seawater's ability to leach Fe and Mn from the rock are the results from one 500° C run (#30, Table 7). This run, in spite of very coarse-grained starting material (2–5 mm), had 1500 ppm Fe and 63 ppm Mn in the altered fluid. These high metal concentrations were achieved by only 25 days contact with basalt of very limited surface area. This rapid increase in the metal concentrations implies that significant elemental exchange could take place between seawater and basalt during the relatively short initial cooling of a rock mass. That is, as a solid basaltic rock mass cools, thermal contraction fractures would develop and allow seawater to penetrate into the rock, leach Fe, Mn and other "incompatible elements" from the rock, and then re-enter the marine environment as transient, submarine hot springs [10].

The rapid nature of high-temperature exchange between seawater and basalt may be especially important if the residence times of the fluid in some submarine geothermal systems are substantially less than those of terrestrial systems, which are of the order of 10^4 years [1]. Spooner and Fyfe [24] pointed out that the gravitational constraint on fluid circulation imposed by the air/rock interface of terrestrial geothermal systems is removed in sub-sea-floor systems. In the deep submarine environment, the upper boundary condition is a water/water-rock interface which allows hydrothermal fluids to discharge freely into the overlying seawater, thus reducing the amount of recirculation and the residence time of the water in the system [24, 43].

Although an exact matching of minerals between the experiments and nature cannot be expected due to the large number of variables involved, the experimentally produced assemblages are impressively similar to those found associated with geothermal systems on Iceland, in metamorphic rocks from the oceanic crust, and in metamorphosed pillow basalts from ophiolite complexes. For example, minerals found in the experiments and in one or more of the three areas mentioned above include smectites, anhydrite, analcite, albite, tremolite-actinolite, prehnite, cordierite, hematite and Fe–Cu sulfides. Of particular importance is the formation of chalcopyrite and pyrrhotite in the seawater/basalt experiments. This unequivocally demonstrates the potential of seawater to leach, transport and precipitate the components of metallic sulfides as the fluid reacts with basaltic rocks at elevated temperatures. This means that, in essence, seawater can evolve into a hydrothermal *ore fluid* under reasonable geologic conditions. The data further substantiate the contention of others [23–24, 27] that such a process could be responsible for the stratiform Fe–Cu sulfide ore deposits associated with ophiolite complexes.

Chlorite, which is quite common in metabasalts from the oceanic crust, was not conclusively identified in any of the experimental runs. In the Reykjanes geothermal area of Iceland, chlorite first appears as a mixed-layer mineral with montmorillonite [28]; however, with increasing depth and temperature, montmorillonite gradually decreases in abundance as chlorite increases. It is possible that there is a small chlorite component mixed with the smectite in the experimental runs and that the formation of abundant chlorite is limited by kinetic factors. In any case, the absence of chlorite is puzzling.

Another problem recently discussed by Bischoff and Dickson [26] is the distribution of anhydrite. Anhydrite occurs in the 200–300° C seawater/basalt experiments [26, 29, 30] and in the Reykjanes area of Iceland where the hydrothermal fluid is of seawater origin [18]; however it has apparently not been dredged from the oceanic crust [26]. If seawater/basalt reactions constitute a significant process in the oceanic crust, anhydrite should be present, especially along fractures down which cold seawater travels toward the hot underlying rocks. It is possible that anhydrite 1) is not abundant in the oceanic crust, 2) has not been found due to inadequate sampling, or 3) has been destroyed by subsequent reactions. Spooner and Fyfe [24] suggested that $SO_4^{=}$ could be reduced in the production of magnetite and pyrite as follows:

$$11\,Fe_2SiO_4 + 2SO_4^{=} + 4H^+ \leftrightarrows 7Fe_3O_4 + FeS_2 + 11\,SiO_2 + 2H_2O.$$

Such a reaction could account for the absence of anhydrite in pyrite-magnetite bearing rocks from the oceanic crust. Bischoff and Dickson [26] suggested a similar reaction in which seawater Mg^{++} is substituted for H^+:

$$2H_2O + 11\,Fe_2SiO_4 + 2Mg^{++} + 2SO_4 \leftrightarrows Mg_2Si_3O_6(OH)_4 + FeS_2 + 8SiO_2 + 7Fe_3O_4.$$

Both reactions are consistent with the 400° C and 500° C experimental results presented here in that sulfides have been formed with little or no anhydrite by seawater/basalt reactions. The exact reactions that remove SO_4 from the fluid, however, are masked by the complexity of the system. In any case, it seems that lower temperature seawater/basalt reactions (200° C) would produce preservable amounts of anhydrite in the oceanic crust.

Implications and Speculations

The elemental exchanges accompanying seawater/basalt interactions may also affect the geochemistry of Na, Mg, Ca, Si and H in seawater and the oceanic crust [25]. The chemical exchange reactions are in the right direction to help maintain the steady-state composition of seawater by 1) removing excess Na and Mg from stream influx that is not removed by oceanic sedimentation [44], and 2) by providing the oceans with Ca and Si since the weathering of continental rocks does not provide enough of these elements to account for their abundances in sedimentary rocks [45]. In addition, the reduction of seawater $SO_4^=$ and subsequent formation of sulfides may be an effective way to remove sulfur from the oceans. The impact of seawater/basalt interactions on the geochemistry of these elements, however, depends entirely on how much of the oceanic crust is affected and to what degree. Reasonable estimates of these parameters are exceedingly difficult to achieve, if not impossible, and limit meaningful mass balance calculations. However, the potential magnitude of the process is indicated by seismic velocity studies of oceanic crustal layers. Fox [46] measured the sonic velocities of zeolite, greenschist and amphibolite facies metabasalts and metagabbros and found them to match the velocities from the lower 3–4 km of Layer II. Based on these data, he suggested that this portion of Layer II is metabasalt, mostly in the greenschist facies. This implies that hydrothermal circulation is indeed a large-scale process that may have involved as much as 60–75% of the oceanic crust. Such a process would surely be a major controlling factor of the geochemical mass balances of certain elements, notably Na, Mg, Ca, Si, Fe, Mn, and H.

Conclusions

The experimental results from this investigation support the following conclusions:

1. During short-termed seawater/basalt reactions between 200° C and 500° C, seawater tends to change from an oxygenated, slightly alkaline, Na, Mg, SO_4, Cl solution to a reducing, acidic(?), Na, Ca, Si, Cl solution.

2. Seawater/basalt reactions result in the formation of primarily gray smectite and anhydrite at 200–300° C; however, green smectite, albite, tremolite-actinolite, chalcopyrite, pyrrhotite, and hematite become the dominant alteration products at 400° C and 500° C.

3. As seawater reacts with basalt at elevated temperatures, large quantities of Fe, Mn and other metals are leached from the rock, concentrated in the seawater and transported by the solution. Some of the metals may be precipitated

as Fe–Cu sulfides in the upper levels of the oceanic crust, possibly forming economically significant ores. Fe and Mn would also be precipitated as the solution discharges into the oxygenated overlying seawater forming a major component of metalliferous sediments.

4. High-temperature seawater/basalt reactions could also exert substantial control on the steady-state composition of seawater by removing excess Na and Mg and adding Ca, Si and H^+ to the oceans. Such reactions would affect the geochemical budgets of these elements, especially if a large volume of the oceanic crust is involved.

Acknowledgements. I would like to thank my advisor and friend, Bob Scott, for suggesting and supporting this research project. Special gratitude is also extended to Tom Tieh, Dave Stearns, Dave Fahlquist and Bob Presley for their helpful discussions concerning the many aspects of this investigation. Jack Calhoun's aid in setting up the equipment and Mark DiStefano's help with the SEM work are especially appreciated.

In addition I would like to thank Mike Mottl and Jim Bischoff for their free exchange of ideas and data during the entire course of the study. Special appreciation is extended to Jim Bischoff, Enrico Bonatti and Bill Seyfried for critically reviewing the manuscript.

Oceanic basalts used in this study were provided by Peter Rona of the National Oceanic and Atmospheric administration.

This work was supported by National Science Foundation Grants GA29370 and DES 74-18567.

References

1. Elder, J.W.: Physical processes in geothermal areas. In: Terrestrial Heat Flow. Am. Geophys. Union Mon. **8**, 211–239 (1965)
2. Palmason, G.: On the heat flow in Iceland in relation to the Mid-Atlantic Ridge. In: Iceland and Mid-Ocean Ridges (S. Bjornsson, ed.), Soc. Sci. Islandica, Reykjavik. **38**, 111–127 (1967)
3. Talwani, M., Windisch, C.C., Langseth, M.G., Jr.: Reykjanes ridge crest: a detailed geophysical study. J. Geophys. Res. **76**, 473–517 (1971)
4. Bodvarsson, G., Lowell, R.P.: Ocean-floor heat flow and the circulation of interstitial waters. J. Geophys. Res. **77**, 4472–4475 (1972)
5. Lister, C.R.B.: On the thermal balance of a mid-ocean ridge. Geophys. J. **26**, 515–535 (1972)
6. Lister, C.R.B.: Water percolation in the oceanic crust. Eos Trans. Am. Geophys. Union **55**, 740–742 (1974)
7. Williams, D.L., Von Herzen, R.P., Sclater, J.G., Anderson, R.N.: The Galapagos spreading centre: lithospheric cooling and hydrothermal circulation. Geophys. J. **38**, 587–608 (1974)
8. Melson, W.G., Thompson, G., van Andel, T.H.: Volcanism and metamorphism in the Mid-Atlantic Ridge, 22°N latitude. J. Geophys. Res. **73**, 5925–5941 (1968)
9. Deffeyes, K.S.: The axial valley: a steady-state feature of the terrain. In: The megatectonics of continents and oceans (H. Johnson and B.L. Smith, eds), p. 194–222. New Brunswick, N.J.: Rutgers Univ. Press 1970
10. Corliss, J.B.: The origin of metal-bearing submarine hydrothermal solutions. J. Geophys. Res. **76**, 8128–8138 (1971)
11. Miyashiro, A., Shido, F., Ewing, M.: Metamorphism in the Mid-Atlantic Ridge near 24°N and 30°N. Phil. Trans. Roy. Soc. London Ser. A **268**, 589–603 (1971)
12. Muehlenbachs, K., Clayton, R.N.: Oxygen isotope geochemistry of submarine greenstones. Can. J. Earth Sci. **9**, 471–478 (1972)
13. Bonatti, E., Joensuu, O.: Deep sea iron deposits from the South Pacific. Science **154**, 643–645 (1966)
14. Bostrom, K., Peterson, M.N.A.: Precipitates from hydrothermal exhalations on the East Pacific Rise. Econ. Geol. **61**, 1258–1265 (1966)

15. Bostrom, K., Peterson, M.N.A.: The origin of aluminium-poor ferromanganoan sediments in areas of high heat flow on the East Pacific Rise. Marine Geol. **7**, 427–447 (1966)
16. Bostrom, K.: Submarine volcanism as a source for iron. Earth Planet. Sci. Letters. **9**, 348–354 (1970)
17. Bender, M., Broecker, W., Gornitz, V., Middel, U., Kay, R., Sun, S., Biscaye, P.: Geochemistry of three cores from the East Pacific Rise. Earth Planet. Sci. Letters. **12**, 425–433 (1971)
18. Dasch, E.J., Dymond, J.R., Heath, G.R.: Isotopic analysis of metalliferous sediment from the East Pacific Rise. Earth Planet. Sci. Letters. **13**, 175–180 (1971)
19. Dymond, J., Corliss, J.B., Heath, G.R., Field, C.W., Dasch, E.J., Veeh, H.H.: Origin of metalliferous sediments from the Pacific Ocean. Geol. Soc. Am. Bull. **84**, 3355–3372 (1973)
20. Piper, D.Z.: Origin of metalliferous sediments from the East Pacific Rise. Earth Planet. Sci. Letters. **19**, 75–82 (1973)
21. Bonatti, E., Zerbi, M., Kay, R., Rydell, H.: Metalliferous deposits from the Apennine ophiolites: Mesozoic equivalent of deposits from modern spreading centers. Bull. Geol. Soc. Am. In press (1975)
22. Corliss, J.B., Graf, J.L., Skinner, B.J., Hutchinson, R.W.: Rare earth data for Fe- and Mn-rich sediments associated with sulfide ore bodies of the Troodos massif, Cyprus (abstract). Geol. Soc. Am. Abstracts with Progam. **4**, 476 (1972)
23. Sillitoe, R.H.: Sulfide deposits at sites of sea-floor spreading. Trans. Inst. Mining Met. (Sect. B: Appl. earth sci.). **81**, B141–148 (1972)
24. Spooner, E.T.C., Fyfe, W.S.: Sub-sea floor metamorphism, heat and mass transfer. Contrib. Mineral. Petrol. **42**, 287–304 (1973)
25. Hart, R.A.: A model for chemical exchange in the basalt-seawater system of oceanic layer II. Can. J. Earth Sci. **10**, 799–815 (1973)
26. Bischoff, J.L., Dickson, F.W.: Seawater-basalt interaction at 200° C and 500 bars: Implications for origin of sea-floor heavy-metal deposits and regulation of seawater chemistry. Earth Planet. Sci. Letters. **25**, 385–397 (1975)
27. Bonatti, E.: Metallogenesis at oceanic spreading centers. In: Annual review of earth and planetary sciences (F.A. Donath, ed.), vol. 3, p. 401–431. Palo Alto, Calif.: Annual Reviews, Inc. 1975
28. Tomasson, J., Kristmannsdottir, H.: High temperature alteration minerals and thermal brines, Reykjanes, Iceland. Contrib. Mineral. Petrol. **36**, 123–134 (1972)
29. Hajash, A.: An experimental investigation of high-temperature seawater-basalt interactions. Geol. Soc. Am. Abstracts with Program. **6**, 771 (1974)
30. Mottl, M.J., Corr, R.F., Holland, H.D.: Chemical exchange between seawater and mid-ocean ridge basalt during hydrothermal alteration: an experimental study. Geol. Soc. Am. Abstracts with Program. **6**, 879 (1974)
31. Scott, M.R., Scott, R.B., Rona, P.A., Butler, L.W., Nalwalk, A.: Rapidly accumulating manganese deposit from the median valley of the Mid-Atlantic Ridge. Geophys. Res. Letters. **1**, 355–358 (1974)
32. Scott, R.B., Rona, P.A., McGregor, B.A., Scott, M.R.: The TAG hydrothermal field. Nature. **251**, 301–302 (1974)
33. Brown, G.: The X-ray identification and crystal structures of clay minerals, 544 pp. Norwich, Great Britain: Jarrold and Sons, Ltd. 1961
34. Dickson, F.W., Blount, C., Tunell, G.: Use of hydrothermal solution equipment to determine the solubility of anhydrite in water from 100° C to 275° C and from 1 bar to 1000 bars pressure. Am. J. Sci. **261**, 61–78 (1963)
35. Holland, H.D.: Gangue minerals. In: Hydrothermal ore deposits (H.L. Barnes, ed.), p. 382. Holt: Rinehart and Winston, Inc. N.Y. 1967
36. Hawkins, D.B., Roy, R.: Experimental hydrothermal studies on rock alteration and clay mineral formation. Geochim. Cosmochim. Acta. **27**, 1047–1054 (1963)
37. Marshall, W.L.: Aqueous solubility equilibria, 200 to 374° C (abstract). Am. Chem. Soc. Meeting Boston. Abstracts. 15R (1959)
38. Young, T.F.: Strong acids at high temperature (abstract). Am. Chem. Soc. Meeting Boston. Abstracts. 14R (1959)
39. Helgeson, H.C.: Complexing and hydrothermal ore deposition. In: International series of monographs on earth sciences (D.E. Ingerson, ed.), vol. 17, 128 pp. New York: Macmillan 1964

40. Helgeson, H.C.: Thermodynamics of hydrothermal systems at elevated temperatures and pressures. Am. J. Sci. **267** 729 804 (1969)
41. Krauskopf, K.B.: Introduction to geochemistry. Frank Press, consulting editor, p. 243-257. New York: McGraw-Hill Co. 1967
42. Bostrom, K.: The origin and fate of ferromanganoan active ridge sediments. Stockholm Contr. Geol. **24**, 149 243 (1973)
43. Elder, J.W.: Steady free convection in a porous medium heated from below. J. Fluid Mech. **27**, 29-48 (1967)
44. Broecker, W.S.: A kinetic model for the chemical composition of seawater. Quart. Res. **1**, 188-207 (1971)
45. Garrels, R.M., MacKenzie, F.T.: Evolution of sedimentary rocks (W.W. Norton, ed.), 397 pp. New York: Norton and Co. 1971
46. Fox, P.J.: The geology of some Atlantic fracture zones, Caribbean escarpments and the nature of the oceanic basement and crust. Unpubl. Ph.D. diss. Columbia Univ. N.Y. (1972)
47. Bjornsson, S., Aronoson, S., Tomasson, J.: Economic evaluation of Reykjanes thermal brine area, Iceland. Am. Assoc. Petrol. Geol. Bull. **56**, 2380 2391 (1972)

Accepted August 29, 1975

4

Original article prepared especially for this Benchmark volume

SUB-SEAFLOOR HYDROTHERMAL SYSTEMS ROCK-VS. SEAWATER-DOMINATED

Michael J. Mottl

Department of Chemistry
Woods Hole Oceanographic Institution

William E. Seyfried

Department of Geology and Geophysics
University of Minnesota

ABSTRACT

Experiments reacting seawater with basalt under hydrothermal conditions have shown that the water/rock ratio effective during alteration has a drastic effect on both the chemistry of the solution and the mineralogy of the solids produced. For the range of water/rock ratios likely to be encountered in sub-seafloor hydrothermal systems, two types of systems can occur: rock- and seawater-dominated. The distinction between the two holds over a temperature range of at least 150-350°C and results largely from the critical role of seawater Mg in generating and maintaining acidity. During reaction with basalt, H^+ is produced by the uptake of seawater Mg and OH^- in the form of a Mg $(OH)_2$-component in smectite or chlorite. As long as high concentrations of Mg and SiO_2 are maintained in solution, the rate of H^+ production exceeds that of H^+ consumption by silicate hydrolysis reactions, and the pH stays acid. As a result, concentrations of the heavy metals Fe, Mn, and Zn in solution are high. These conditions are maintained indefinitely in seawater-dominated systems, and the only minerals that form in equilibrium with the acid solution are smectite, mixed-layer clay, or chlorite, and quartz, hematite, and anhydrite. In rock-dominated systems, by contrast, seawater Mg is almost completely removed into solids and silicate hydrolysis reactions consume the earlier formed H^+, producing a neutral to slightly alkaline solution with low metal concentrations and pH controlled by equilibrium with a complex assemblage of secondary silicates. In the experiments the transition from rock- to seawater-dominated conditions occurred abruptly with increasing water/rock mass ratio at a value of 50 $\pm$ 5, the ratio at which the solids became "saturated" with Mg and depleted in the leachable cations Ca, Na, K, Mn, and Zn.

Whether rock- or seawater-dominated conditions prevail locally within a sub-seafloor hydrothermal system depends on the relative rates of fluid flow vs. reaction in that locality, which in turn depend on such factors as permeability, surface area, temperature, and type of rock being altered. Seawater-dominated systems are those with water/rock ratios $\geq$ 50 and possibly smaller, which produce acid, Mg- and metal-rich hot springs. Rock-dominated systems have lower water/rock ratios and produce hot spring solutions that prior to mixing with cold seawater, are heavily depleted in Mg and poor in heavy metals.

INTRODUCTION

Estimates of heat loss due to advection imply that seawater circulates through young oceanic crust at a rate of 10^{17}-10^{18} g/yr (Lister 1973; Williams and Von Herzen 1974; Wolery and Sleep 1976). This heated seawater reacts with the basaltic rocks of the oceanic crust. The estimated seawater flux may be compared with the rate of generation of new oceanic crust at mid-ocean ridges. Given a crust 5 km thick with density 3 g/cm^3, and the current rate of seafloor spreading of 3 km^2/yr (Chase 1972; Williams and Von Herzen 1974), this rate is 5 x 10^{16} g/yr, 2-20 times smaller than the seawater flux. The relative magnitudes of these two rates imply that the total mass of rock altered by convecting seawater is small compared with the mass of seawater with which it has reacted. This fact was first noted by Wolery and Sleep (1976), and it has important consequences for the chemistry of solutions produced by basalt-seawater interaction within the oceanic crust.

A large number of experiments have now been performed reacting seawater with mid-ocean ridge basalt of varying crystallinity over a range of conditions: 20-500°C, 1-1000 bars, and water/rock mass ratios of 1-125 (Bischoff and Dickson 1975; Hajash 1975; Seyfried and Bischoff 1977, 1979; Seyfried and Mottl 1977; Mottl and Holland 1978, 1979; Seyfried et al. 1978). In addition, the behavior of seawater on heating has been studied up to 350°C (Bischoff and Seyfried 1977, 1978). These experiments have shown that, over the range of conditions most likely to be encountered within the oceanic crust at mid-ocean ridges, the seawater/rock ratio effective during alteration is probably the most important parameter in determining the course of reaction and the resulting solution chemistry and mineralogy. In this paper we will summarize pertinent experimental data, both published and unpublished (Seyfried and Mottl, in preparation), to show that there are two types of basalt-seawater systems that can occur: rock-dominated and seawater-dominated; that the two types are characterized by distinctly different solution chemistry and alteration mineralogy; and that the transition between rock- and seawater-dominated conditions is discontinuous and relatively abrupt. Moreover, this transition occurs at a water/rock ratio that is intermediate within the range estimated for sub-seafloor hydrothermal systems, such that both types of systems could occur within the crust along the mid-ocean ridge system. This conclusion has important implications for the formation of submarine hydrothermal deposits.

As defined for the experiments, the seawater/rock ratio is the mass of seawater divided by the mass of fresh, powdered basalt initially contained within the reaction cell. How this operational definition relates to water/rock ratios in the natural setting will be discussed in a later section.

PROCEDURE

The experiments described in this paper were performed using the Dickson hydrothermal apparatus (Dickson et al. 1963), which allows solutions to be sampled during an experiment without disturbing P-T conditions. This apparatus consists of a gold sample cell that is connected to a Ti sampling valve via a gold filter and capillary tube. The gold cell is pressurized within a steel autoclave, which is heated in a furnace. This system virtually eliminates contamination during reaction or sampling. Its two chief advantages are that it permits monitoring of the reaction path by following the change in solution composition with time, and it eliminates back reaction between solution and solids during cooling. During an experiment the mass of solution in the sample cell decreases as samples are withdrawn for analysis. Because these samples are small (10 g) relative to the initial mass of solution (150-190 g), and because the greatest change in solution composition generally occurs early in an experiment, the resulting decrease in water/rock ratio is usually insignificant.

In order to speed reaction, the rocks were ground, and the 43-63 μm size fraction has been used in all experiments. The rocks have been described by Seyfried and Bischoff (1979). Whether all rock becomes altered during an experiment depends on its crystallinity and the temperature, water/rock ratio, and duration of the experiment.

RESULTS

Figure 1 summarizes the experimental data for three critical species. Plotted are the pH and the concentrations of Mg and Fe in solution with time during five experiments, all at 300°C and 500 bars.

The two experiments on the left (Seyfried 1977) were run at a water/rock mass ratio of 10. In these experiments seawater Mg was removed rapidly and efficiently into the alteration product smectite, such that less than 1 percent remained in solution after only four days. During the same period, pH initially dropped and then rose again to an intermediate value that is near neutrality under the conditions of the experiment (table 1). Simultaneously with the pH drop, the concentration of dissolved Fe increased dramatically. It then fell to only 2-3 ppm as the pH rose.

The final, steady-state solution, with a near-neutral pH and low concentrations of Mg and Fe, is nearly identical to that produced in experiments that used water/rock ratios 2-10 times smaller (i.e., 1-5) (Hajash 1975; Mottl and Holland 1978, 1979). It also is very similar to the solution produced in the basalt-seawater geothermal system at Reykjanes, Iceland, at 280°C (Bjornsson et al. 1972; Arnorsson et al. 1978). The water/rock ratio in this subaerial geothermal system is in the range 0.1-3 (Mottl et al. 1975; Mottl 1976; Hardie 1976). This solution chemistry apparently is determined by equilibrium with the alteration mineral assemblage, which for Reykjanes and the experiments variously includes smectite, zeolites, analcime, wairakite, prehnite, mixed-layer chlorite-smectite, chlorite, albite, epidote, quartz, tremolite-actinolite, sphene, anhydrite, and pyrite. It appears to be characteristic of solutions produced in basalt-seawater hydrothermal systems with low water/rock ratios, over a wide range of these ratios; we shall refer to such systems as rock-dominated.

Of the three experiments shown on the right side of figure 1, two were run at a water/rock ratio of 62 and the third at 125. A complete data set for one of these experiments was presented by Seyfried and Mottl (1977); the others will be presented in a later paper (Seyfried and Mottl, in preparation). Compared with reaction at lower water/rock ratios, these experiments show both some similarities and some significant differences. Mg was again removed rapidly from solution, but this time removal was considerably less than complete. For w/r = 62

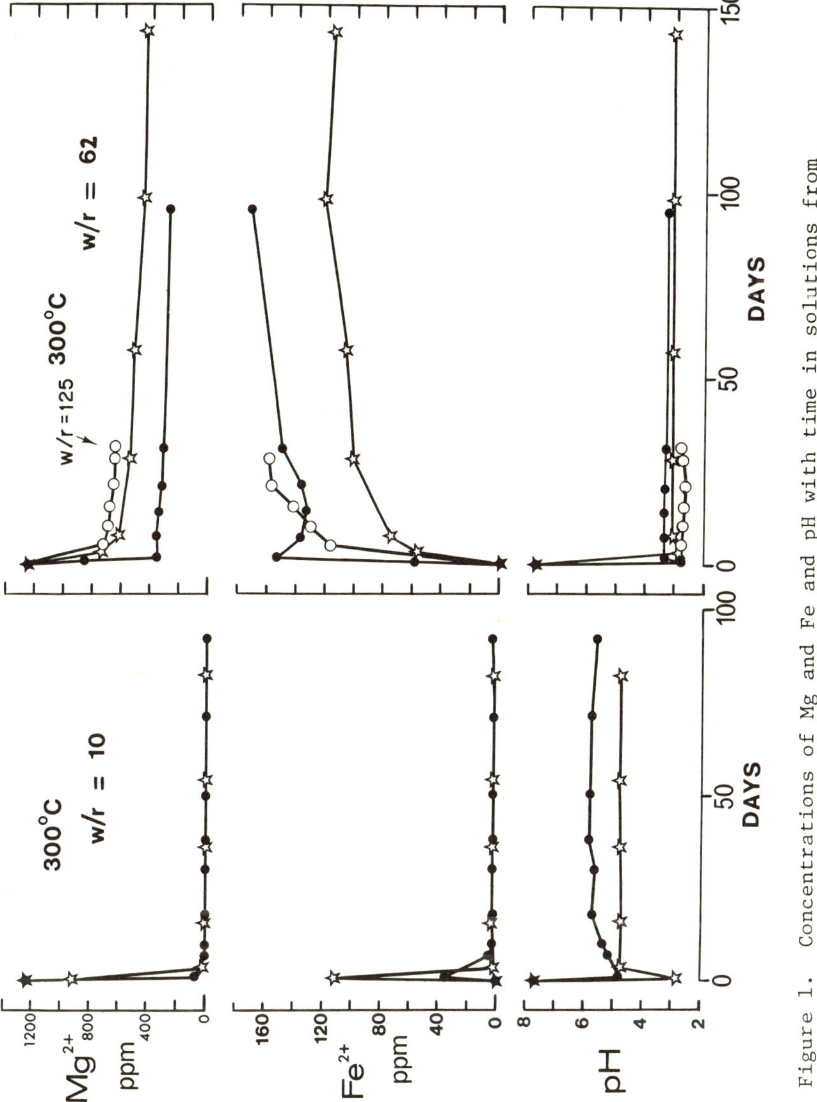

Figure 1. Concentrations of Mg and Fe and pH with time in solutions from five experiments reacting seawater with basalt at 300°C, 500 bars, and water/rock mass ratios of 10, 62, and 125. pH is that measured at 25°C. Circles = basalt glass, stars = holocrystalline basalt.

Table 1. Final pH of solutions produced by reaction of seawater with basalt, as measured at 25°C and calculated for the temperature of the experiments.

T°C	Basalt	w/r[b]	pH at 25°C	pH in situ	pH at neutrality[c]
150	glass	10	6.84	6.73	5.82
150	hx[a]	10	5.90	5.98	5.82
150	glass	62	5.36	5.46	5.82
200	glass	10	5.92	5.47	5.63
200	glass	62	4.41	4.67	5.63
300	glass	10	5.62	6.22	5.52
300	hx	10	4.82	5.32	5.52
300	glass	62	3.33	3.67	5.52
300	hx	62	3.16	3.51	5.52
300	glass	125	2.82	3.46	5.52

[a]holocrystalline basalt.

[b]seawater/rock mass ratio.

[c]from Helgeson (1969); for the temperature of the experiments.

the steady-state concentration of Mg remaining in solution was 300-450 ppm, and for w/r = 125, 650 ppm. The pH again dropped to a very acid value, while dissolved Fe rose, but the subsequent rise in pH that occurred at the lower water/rock ratio was either small or nonexistent in these experiments. The solutions stayed acid, and their Fe concentrations, as well as those of Mn and Zn, stayed high, with dissolved Fe exceeding 100 ppm. The alteration assemblage in these experiments consisted of mixed-layer chlorite-smectite, anhydrite, and minor quartz and hematite.

These experiments demonstrate that basalt-seawater systems with high water/rock ratios, which we shall refer to as seawater-dominated systems, produce solutions with similar chemistry over a wide range of ratios, and this chemistry is distinct from that in rock-dominated systems. As figure 1 shows, this distinction holds regardless of the crystallinity of the basalt. The chief difference between experiments using glass vs. holocrystalline basalt at 300°C is that the glass has altered completely to the respective alteration assemblage, whereas unreacted plagioclase and pyroxene remained in both experiments using holocrystalline basalt.

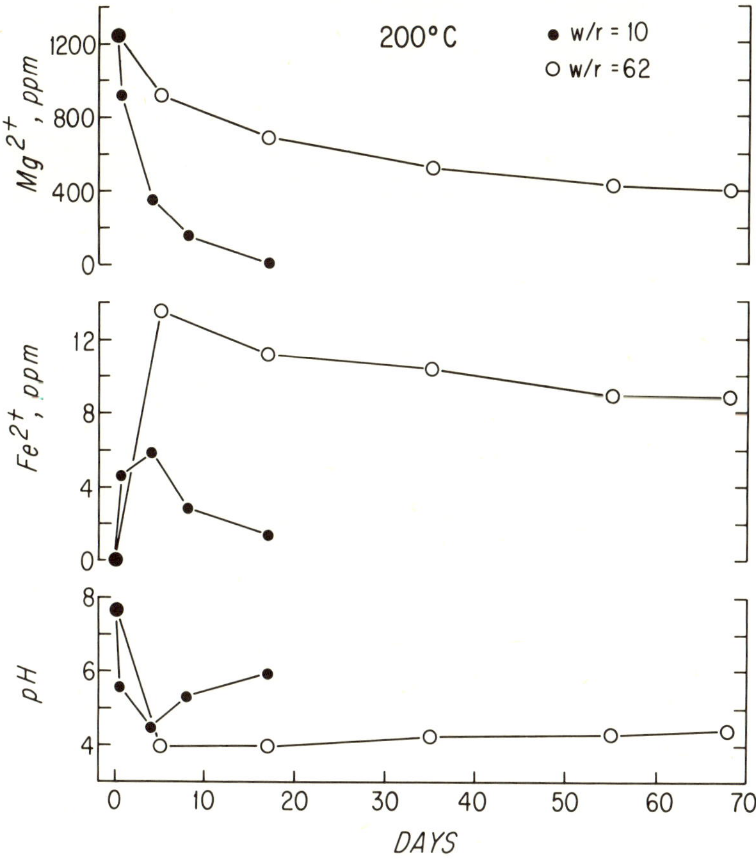

Figure 2. Concentrations of Mg and Fe and pH with time in solutions from two experiments reacting seawater with basalt glass at 200°C, 500 bars, and water/rock mass ratios of 10 and 62. pH is that measured at 25°C.

Figures 2 and 3 show that this distinction between rock- and seawater-dominated systems holds at 150° and 200°C as well. For a water/rock ratio of 10, seawater Mg was again almost completely removed, even during reaction with holocrystalline basalt at 150°C (Seyfried and Bischoff 1979). The chief difference is that removal of Mg from solution simply occurred more slowly at the lower temperatures. Also, some glass and olivine as well as plagioclase and pyroxene remained unaltered in these experiments. As at 300°C, reaction at 150° and 200° produced an initial drop in pH followed by a rise, which corresponded to an increase and subsequent decrease in the concentration of dissolved Fe. This same pattern was displayed by Bischoff and Dickson's (1975) experiment at 200°C, w/r = 10. At a water/rock ratio of 62, removal of seawater Mg was less than complete and the solutions stayed acid and relatively Fe-rich. Although the actual values of pH and dissolved Fe varied strongly with temperature, the inverse relationship between Fe and pH

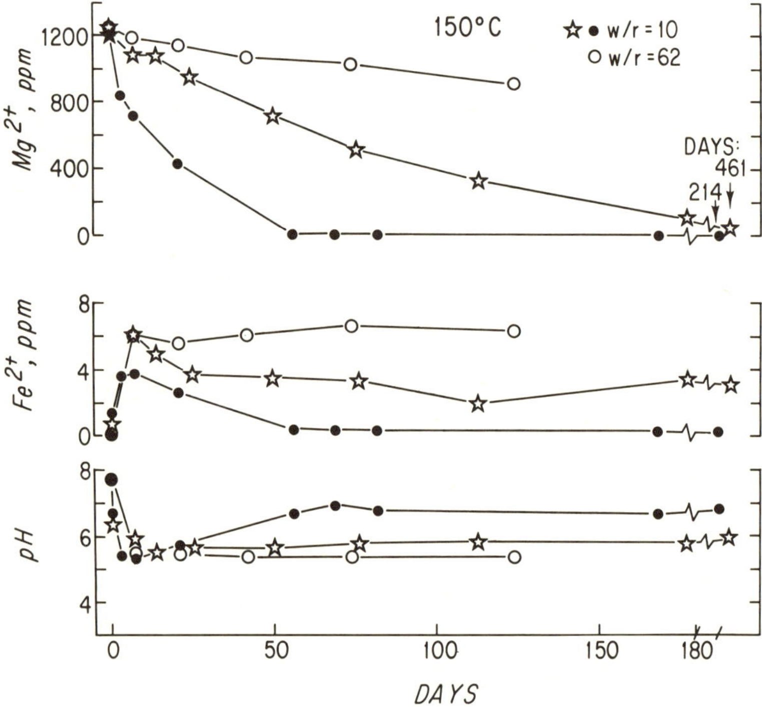

Figure 3. Concentrations of Mg and Fe and pH with time in solutions from three experiments reacting seawater with basalt at 150°C, 500 bars, and water/rock mass ratios of 10 and 62. pH is that measured at 25°C. Circles = basalt glass, stars = holocrystalline basalt.

is obvious, both with time in an individual experiment and for the final solutions among several experiments. The alteration mineral assemblage in all these experiments consisted of smectite and anhydrite.

The distinction made here between rock- and seawater-dominated systems holds at higher temperatures as well. Seawater reacted with basalt glass at 350°C, w/r = 10 (Seyfried et al. 1978) and produced a solution with nearly the same Mg and Fe concentrations as those found in the similar experiment at 300°C. Thus, over a range of at least 150-350°C, rock-dominated basalt-seawater systems produce neutral to slightly alkaline solutions with little ($\leq$ 3 ppm) Mg or Fe, while seawater-dominated systems produce acid, Mg- and Fe-rich solutions. Experiments by Hajash (1977) suggest that this distinction is meaningful at even higher temperatures: although the concentration of dissolved Fe becomes appreciable at 400°C even in low water/rock systems (∿100 ppm: Mottl and Holland 1979), it is higher still in high water/rock systems. At 500°C, seawater reacted with basalt at a low water/rock ratio (5) actually had a higher Fe concentration than that reacted at a high water/rock ratio (50), although the proportion of Fe initially in the rock that was leached into solution was much less (∿7% vs. ∿50%: Hajash 1977).

DISCUSSION

Reactions Affecting pH

Why are rock- and seawater-dominated systems chemically distinct? The answer lies in the mechanism that generates acid conditions.

Bischoff and Dickson (1975) first proposed that acidity is generated during basalt-seawater interaction by the incorporation of seawater Mg into smectite in the form of a $Mg(OH)_2$-component, accompanied by OH^- from the hydrolysis of water. Formation of brucite-type interlayers, an essential step in the transformation of smectite to mixed-layer clay to chlorite, also generates H^+ in solution if the Mg^{2+} and OH^- are derived from seawater. The clay produced in the seawater-dominated experiments at $300^{\circ}C$ is mixed-layered, as is that formed under rock-dominated conditions at $350^{\circ}C$.

Bischoff and Dickson attributed the subsequent rise in pH, which occurred in theirs and later experiments at a water/rock ratio of 10, to ordinary silicate hydrolysis reactions, which consume H^+ in exchange for cations such as Ca^{2+}. These reactions dissolve the primary igneous minerals or transform them into secondary hydrated silicates. Thus, basalt-seawater interaction consists of two types of reactions that compete to determine pH: reactions removing seawater Mg and generating H^+, and silicate hydrolysis reactions generating cations and consuming H^+. As long as the concentration of Mg in solution is high, the former set of reactions proceeds more rapidly, and the pH falls. As the Mg concentration decreases, however, so does the rate of H^+ production. Simultaneously, the lowered pH increases the rate of silicate hydrolysis, until at low concentrations of dissolved Mg this rate surpasses that of the Mg-removal reactions and the pH begins to rise, eventually reaching a steady-state value controlled by equilibrium with the alteration mineral assemblage. For rock-dominated systems this assemblage may consist of several Mg-, Ca-, and Na-aluminosilicates, and the corresponding pH is neutral to alkaline because silicate hydrolysis continues long after Mg has been removed to a low value.

Why, then, did the pH rise only slightly or not at all in the seawater-dominated experiments? The answer is that the concentration of dissolved Mg stayed high enough throughout the experiments for the rate of H^+ generation initially to surpass, and then nearly to equal, the rate of H^+ consumption as the primary minerals and glass were destroyed. Ultimately, for the experiments at $300^{\circ}C$ using basalt glass, the hydrolysis reactions went to completion and all glass was converted to chlorite-smectite and quartz. (Quartz was detected in only one of these experiments, but the aqueous SiO_2 concentration exceeded quartz saturation in both.) In the process the leachable cations Ca, Na, K, Mn, Cu, Zn, and Ba were largely removed into solution (Seyfried and Mottl 1977), leaving mainly Mg, Fe, Al, and Si in the clay. In these experiments, therefore, the final, acid pH was controlled in the presence of a high concentration of dissolved SiO_2 by equilibrium with chlorite-smectite, the only aluminosilicate that could form under the extreme conditions of high Mg and SiO_2 in solution and low pH. For the experiments using holocrystalline basalt, and for those using glass at $150\text{-}200^{\circ}C$, hydrolysis of the primary silicates was not complete. Nonetheless, the solutions appear to be very close to steady state, such that pH was kept low by equilibrium with smectite or chlorite-smectite and the requirement that silicate hydrolysis (H^+ consumption) proceed apace with Mg-removal (H^+ production).

It is apparent that seawater Mg plays a critical role during high-temperature basalt-seawater interaction because of its rapid removal into solids and its resulting capacity for generating acidity. The former trait derives from the well-known fact that hydrated Mg-aluminosilicates are extremely insoluble in SiO_2-rich hydrothermal solutions. The latter is due to the manner in which at least some Mg is incorporated into these silicates. In fact, the Mg in these silicates derives in part from seawater and in part from basalt. Smectite and chlorite therefore play a dual role during their formation from basalt in that they participate in both H^+-producing and H^+-consuming reactions. This is best illustrated by the experiments at 150-200°C. Smectite was the only silicate produced in these experiments, regardless of water/rock ratio. In the rock-dominated systems, pH first fell as the uptake of seawater Mg into smectite controlled pH, then rose as hydrolysis and the formation of smectite from basalt Mg predominated.

The capacity of seawater Mg to generate H^+ is not limited to interaction with basalt, however. Bischoff and Seyfried (1977, 1978) have shown that simple heating of seawater to 300°C and above generates a highly acid solution by homogeneous precipitation of a Mg-hydroxysulfate, analogous to the Mg-hydroxy-aluminosilicate (smectite or chlorite) produced when basalt is present. Thus seawater, by virtue of its high Mg content, has a "built-in" mechanism for becoming an acid hydrothermal solution, and a substantial mass of rock is required to "titrate" that acid.

It should be noted that one of the mechanisms proposed here to produce H^+, that of uptake of $Mg(OH)_2$-interlayers in smectite, has been shown to occur in seawater at 25°C as well (Deffeyes 1965; see discussion in Drever 1974). At this low temperature, however, pH is not buffered by this reaction until it reaches a much higher value (8.0-8.5).

Transition from Rock- to Seawater-Dominated Conditions

The foregoing discussion indicates that the critical difference between the rock- and seawater-dominated experiments is that the latter contained dissolved Mg in excess of what could be accommodated in the alteration products of the basalt. Presence of this Mg in turn produced H^+ in excess of what could be consumed by hydrolysis of the basalt silicates. For those seawater-dominated experiments that reacted to completion, the capacity of the solids to consume Mg or H^+ was exhausted. The solids became, in effect, "saturated" with Mg and H^+ and almost wholly depleted in leachable cations. In the rock-dominated experiments, by contrast, the solids took up nearly all seawater Mg and retained the capacity, chiefly in the form of Ca and Na remaining in the silicates, to take up still more.

If this explanation is correct, it follows that the transition from rock- to seawater-dominated conditions should occur at a seawater/rock ratio at which the solids become just "saturated" with Mg. This ratio can be estimated from the fact that 75-95 percent of the Mg uptake that occurred in the seawater-dominated experiments at 150-300°C was balanced by leaching of Ca and Na from basalt (Seyfried and Mottl 1977, and in preparation). Assuming that Mg is exchanged only for Ca and Na, each gram of mid-ocean ridge basalt has the capacity to remove all Mg from about 50 g of seawater. Thus, the transition should occur at a water/rock ratio near 50 in the temperature range 150-300°C.

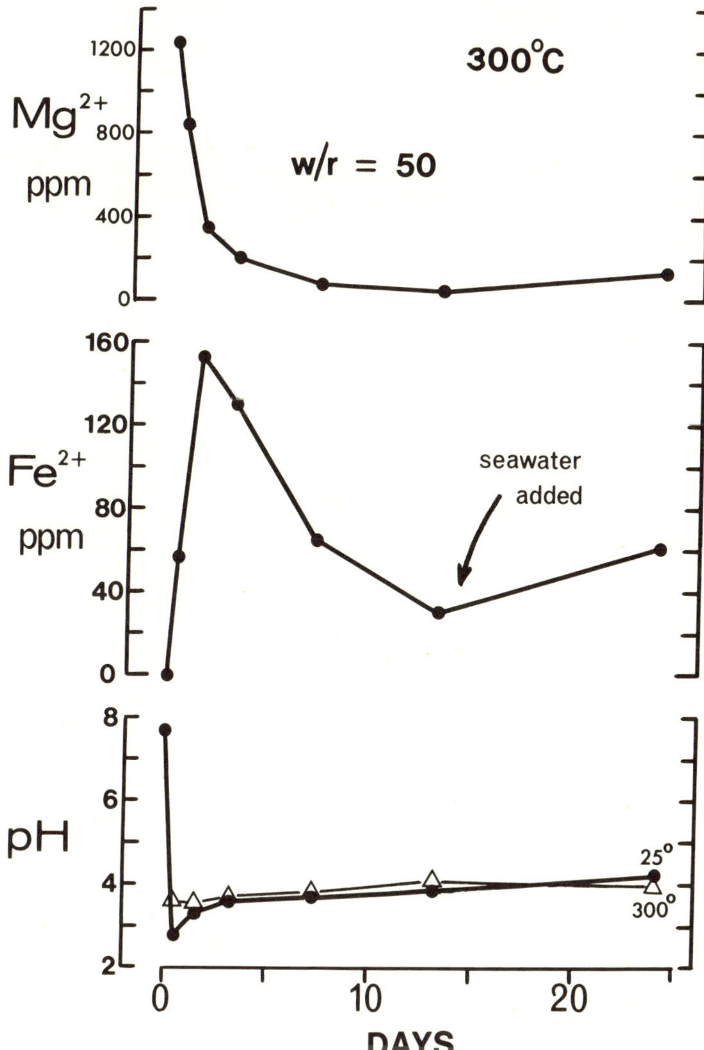

Figure 4. Concentrations of Mg and Fe and pH in solution with time in an experiment reacting seawater with basalt glass at 300°C, 500 bars, and an initial water/rock mass ratio of 50. Sampling lowered this ratio of 46 after thirteen days, at which time seawater was added so that the ratio increased to 51. pH is that measured at 25°C (circles) and calculated for 300°C (triangles).

To test this supposition we performed an experiment at 300°C and an initial water/rock ratio of 50 (figure 4). For the first thirteen days of this experiment, Mg, Fe, and pH in solution showed the behavior typical of a rock-dominated system: Mg decreased to 48 ppm while Fe increased to 150 ppm, then decreased to only 30 ppm as pH fell and then rose. Sampling during this time lowered the effective seawater/rock ratio to about 46. At this point we injected an amount of fresh seawater into the reaction cell sufficient to raise the water/rock ratio by only 5 units. Eleven days later, nearly half of the added Mg had been removed from solution into the solids, and, while the change in pH was small, the dissolved Fe concentration had doubled. The small amount of added seawater had been enough to reverse the previous trend toward lower Fe concentration and higher in situ pH and to shift the solution back toward the characteristic chemistry of a seawater-dominated system. The basalt glass used in the experiment was converted entirely to chlorite-smectite, anhydrite, and minor hematite, the typical assemblage produced in seawater-dominated systems.

This experiment demonstrates that the transition from rock- to seawater-dominated conditions occurs quite abruptly with increasing water/rock ratio. Moreover, it apparently is discontinuous: only a narrow intermediate range exists, in which the solution chemistry reflects rock-dominated conditions (low Mg, Fe, near-neutral pH) and the alteration mineral assemblage reflects seawater-dominated conditions. This is the range in which the altered basalt becomes just "saturated" with Mg. Any additional seawater Mg added to the system would remain largely in solution, producing an acid pH and high concentrations of dissolved Fe, Mn, and Zn. We estimate that this transition occurs at a seawater/rock mass ratio of 50 ± 5.

Additional Distinctions

Two additional distinctions between rock- and seawater-dominated systems are worth noting.

During high-temperature basalt-seawater interaction there are several elements whose concentration in solution is limited ultimately by their depletion from the rock. These are the so-called soluble elements (Ellis 1970), which are strongly partitioned into the solution rather than the solid phases. For the case of a closed system that has reacted to completion, the increase in the concentrations of these elements in solution will be inversely proportional to the water/rock ratio of the system. Ellis identified Cl, Br, B, Cs, F, Li, and As as soluble elements in most natural hydrothermal systems. Mottl and Holland (1978) showed that, for mid-ocean ridge basalt reacted at 300°C and above, K and probably Ba also behave as soluble elements, even at water/rock ratios as low as 1-3, while most other elements in rock-dominated systems are partitioned into solids and controlled by mineral equilibria.

In seawater-dominated systems, because of their acid solutions and relatively simple mineral assemblage, several other elements join K, Ba, B, Li, and the others in behaving as soluble elements. Thus, in the experiments at 300°C and water/rock ratios of 50-125, the steady-state concentrations of Mn and Zn in solution varied inversely with water/rock ratio. Cu, Ca, and Na also were almost completely removed from the silicate minerals, although much Ca ultimately resided in anhydrite, and Cu in the gold of the reaction cell walls. At 500°C and w/r = 50, Fe begins to approach "soluble" behavior, as shown by Hajash (1977).

A second major difference between rock- and seawater-dominated systems involves redox conditions and the behavior of sulfur. The main redox reaction that can occur during basalt-seawater interaction is that involving reduction of seawater sulfate and corresponding oxidation of Fe^{2+} from basalt. Redox conditions in the seawater-dominated experiments were controlled by a large excess of sulfate from seawater, and hematite was a common alteration product. In rock-dominated systems, Fe^{2+} and sulfide from basalt are relatively more important. Mid-ocean ridge basalts typically contain several hundred ppm S as sulfides, and this S is readily mobilized during reaction with seawater (Scott and Frank 1974; Mottl and Holland 1979). Once in solution it can have two effects: it can catalyze the reduction of seawater sulfate by Fe^{2+} from basalt and, if the water/rock ratio is low enough, it can saturate the solution with sulfides. Thus, sulfides such as pyrite and chalcopyrite were common alteration products in the rock-dominated experiments.

Application to Natural Hydrothermal Systems

In the experiments a given mass of seawater was reacted with a given mass of basalt at constant temperature in an essentially closed system. Natural hydrothermal systems are non-isothermal and more or less open. How do the results of these experiments apply to natural systems?

We have seen that the critical difference between rock- and seawater-dominated systems is that the latter maintain high concentrations of Mg and SiO_2 and a low pH in solution indefinitely, whereas the former do not. In the closed experimental systems with high water/rock ratios, these seawater-dominated conditions were maintained by the presence of dissolved Mg in excess of what the limited amount of rock could accommodate. In natural hydrothermal systems the amount of rock potentially available at any one time to a circulating fluid is much larger, and seawater-dominated conditions would have to be maintained by a continuous influx of seawater Mg at a rate that equaled or exceeded the rate of Mg removal from solution into alteration minerals. Thus, the relative rates of fluid flow vs. reaction within some volume of rock determine whether reaction occurs there under rock- or seawater-dominated conditions. Along with temperature, this rate determines the chemistry of the solutions produced, the nature of the alteration mineral assemblage, and the bulk composition of the altered rock.

The relative rates of fluid flow vs. reaction depend upon a number of interrelated physical parameters, such as permeability, size and configuration of flow channels, surface area and type of rock available for alteration, and temperature. Seawater-dominated conditions are favored by high permeability and flow rates, as in the case of flow through large open fractures, and/or slow reaction rates, as would occur if these fractures were lined with "Mg-saturated" alteration minerals, such as chlorite or smectite. Not only would such fractures have a relatively large ratio of volume/surface area, but the rates of reactions removing Mg would be limited by the rate of diffusion of components through the altered layer. Reaction rates also would be slower for alteration of less glassy, more crystalline rock (e.g., gabbro vs. basalt) and at lower temperatures. By contrast, rock-dominated conditions are likely to occur in settings of low permeability and high surface area, as may be the case for flow through numerous small, tortuous channels that pervade the rock and make it directly accessible for alteration by the flowing solution. Even in the latter case, however, seawater-dominated conditions would result if the rock became "Mg-saturated" (i.e., completely altered to smectite or chlorite and quartz).

It is evident that within a natural hydrothermal system, rock- and seawater-dominated conditions can exist at the same time in different places, and in the same place at different times, as a result of local variations in permeability and other factors. This latter phenomenon is demonstrated by the experiments at a low water/rock ratio: solutions produced early in the experiments reflected reaction under seawater-dominated conditions, which then gave way to rock-dominated conditions as more rock reacted. What does it mean, then, to speak of a natural system that is rock- or seawater-dominated?

The experimental systems have been defined as rock- or seawater-dominated on the basis of their seawater/rock ratios and the chemistry of the solutions produced at steady state. For a submarine hydrothermal system, the seawater/rock ratio may be defined as the total mass of water that has passed through the system (integrated through time) divided by the mass of rock within the system that has been altered. (The seawater/rock ratio could be defined similarly for a smaller volume of rock within a hydrothermal system, but such a definition would have to specify the Mg content of the incoming fluid, as this would not always be unreacted seawater for a smaller volume.) Using this definition, a submarine hydrothermal system may be considered to be seawater-dominated if it has a high enough water/rock ratio to produce solutions that still reflect seawater-dominated conditions as they exit from the system as seafloor hot springs. These hot springs would be acid and rich in Mg and metals. In contrast, a rock-dominated hydrothermal system would have a low water/rock ratio such that its hot spring solutions reflected reaction under rock-dominated conditions. Reaction could occur under seawater-dominated conditions in such a system, but just as these conditions were temporary in the experiments at a low water/rock ratio, so they would be localized in time and space in a natural rock-dominated system. The bulk of the circulating seawater would be slightly acid to slightly alkaline, relatively metal-poor, and almost completely depleted in Mg by the time it exited from the crust as a hot spring.

How high would the seawater/rock ratio have to be in order for a natural hydrothermal system to be seawater-dominated? The experiments have shown that, for reaction in a closed system, the transition from rock- to seawater-dominated conditions occurs at a water/rock ratio of 50 ± 5. If, in a natural hydrothermal system with this ratio, all solution that exited from the system as hot springs had lost nearly all its Mg, then the only secondary silicates present in the altered rock would be smectite or chlorite and quartz. This is because the altered rock necessarily would be "saturated" with Mg; such a system would be analogous to that in the experiment at w/r = 50 discussed earlier, in which the solution reflected rock-dominated conditions while the altered solids reflected seawater-dominated conditions. A system having precisely these characteristics would be unlikely to occur in nature, however, because natural systems are not so sharply bounded as was the experiment. As a consequence, other secondary silicates besides smectite or chlorite and quartz certainly would be present locally within the altered rock. It follows that for a natural system with a seawater/rock ratio of 50, some amount of solution containing appreciable Mg must have exited from the system. Thus, submarine hydrothermal systems with water/rock mass ratios $\geq$ 50 inevitably will produce some hot spring solutions that reflect reaction under seawater-dominated conditions, and such systems almost certainly will be seawater-dominated. Even in systems with much lower ratios, however, a sizable fraction of the exiting solution could reflect seawater-dominated conditions if at least a few percent of the altered rock reacted under seawater-dominated conditions. Thus, it is not possible to distinguish rock- from seawater-dominated natural systems precisely on the basis of their water/rock ratios, as was done for rock- vs. seawater-dominated conditions in the experiments. Natural systems with water/rock ratios $\geq$ 50 will be seawater-dominated, but so will some with smaller ratios.

Seawater/rock ratios in actual submarine hydrothermal systems have been estimated from several lines of evidence. Reasonable models based on heat flow data yield water/rock mass ratios in the range 1-100 for mid-ocean ridge hydrothermal systems (Wolery and Sleep 1976; Mottl 1976). The major and minor element composition of hot spring solutions from the Galapagos Spreading Center suggests values of 1-5 for the hydrothermal system there (J. M. Edmond, personal communication). On the finer scale of a hand sample, contamination of hydrothermally altered ocean floor rocks (Hart 1972; Bonatti et al. 1975) and ophiolite suites (Spooner et al. 1977) with isotopically heavy strontium from seawater indicates minimum seawater/rock mass ratios of 2-7. The Mg enrichment in submarine greenstones documented by Humphris and Thompson (1978) implies minimum ratios in the range 3-35 for these samples.

The bulk of the evidence thus suggests that most hydrothermal systems within the oceanic crust are rock-dominated, but that seawater-dominated systems probably exist as well. Because the two types of systems produce solutions with drastically different concentrations of Fe, Mn, Zn, and other heavy metals, this conclusion has far-reaching implications for the occurrence of submarine hydrothermal deposits. Experiments indicate that the high metal concentrations attained in solutions produced under seawater-dominated conditions are largely maintained when the solutions are cooled, even when cooling occurs slowly over a period of months (Mottl and Seyfried, in preparation). Thus, hot springs originating from seawater-dominated systems are likely to have much higher metal concentrations than those from rock-dominated systems. Of course, other processes may affect hot spring compositions, such as mixing with cold seawater at shallow levels within the crust. Nevertheless, the distinction between rock- and seawater-dominated systems may be important in explaining why hydrothermal deposits are accumulating so much more rapidly along some segments of the mid-ocean ridge system than along others. On the East Pacific Rise, for example, from about 0-30°S, heavy metals are accumulating at an extremely high rate. Calculations indicate that the relatively metal-poor solutions produced in rock-dominated systems are probably incapable of delivering Fe and Mn to the sea floor at the rates observed there (Mottl 1976). Thus, the equatorial East Pacific Rise is a likely locality for the presence of seawater-dominated hydrothermal systems. This hypothesis will be developed further in a later paper.

ACKNOWLEDGEMENTS

We wish to thank Drs. Frank W. Dickson and James L. Bischoff for their encouragement and active support. These experiments were carried out in their hydrothermal and analytical laboratories at Stanford University and at the U.S. Geological Survey, Menlo Park, with funding from National Science Foundation Grant ID 74-12880, as part of the Seabed Assessment program of the International Decade of Ocean Exploration. We also wish to thank Drs. Andrew Hajash and Jose Honnorez for critically reviewing the manuscript, which was written with support from NSF Grant OCE 77-26842 to the Woods Hole Oceanographic Institution.

REFERENCES

Arnorsson, S., K. Gronvold, and S. Sigurdsson. 1978. Acquifer chemistry of four high-temperature geothermal systems in Iceland. Geochim. et Cosmochim. Acta 42:523-536.

Bischoff, J. L. and F. W. Dickson. 1975. Seawater-basalt interaction at 200°C and 500 bars: implications for origin of seafloor heavy-metal deposits and regulation of seawater chemistry. Earth and Planetary Sci. Letters 25:385-397.

Bischoff, J. L. and W. E. Seyfried. 1977. Seawater as a geothermal fluid: chemical behavior from 25° to 350°C. Proc. Second Int. Symp. on Water-Rock Interaction, I.A.G.C., Strasbourg, France, pp. IV165-IV172.

Bischoff, J. L. and W. E. Seyfried. 1978. Hydrothermal chemistry of seawater from 25° to 350°C. Am. Jour. Sci. 278:838-860.

Bjornsson, S., S. Arnorsson, and J. Tomasson. 1972. Economic evaluation of Reykjanes thermal brine area, Iceland. Am. Assoc. Petroleum Geologists Bull. 56:2380-2391.

Bonatti, E., J. Honnorez, P. Kirst, and F. Radicati. 1975. Metagabbros from the Mid-Atlantic Ridge at 06°N: contact-hydrothermal-dynamic metamorphism beneath the axial valley. Jour. Geology 83:61-78.

Chase, C. G. 1972. The N plate problem of plate tectonics. Royal Astron. Soc. Geophys. Jour. 29:117-122.

Deffeyes, K. S. 1965. The Columbia River flood and the history of the oceans. Pacific NW Oceanographers Annual Meeting, Corvallis, Oregon.

Dickson, F. W., C. W. Blount, and G. Tunell. 1963. Use of hydrothermal solution equipment to determine the solubility of anhydrite in water from 100°C to 275°C and from 1 bar to 1000 bars. Am. Jour. Sci. 261:61-78.

Drever, J. I. 1974. The magnesium problem. In The Sea, E. D. Goldberg, ed., New York: John Wiley & Sons, pp. 337-357.

Ellis, A. J. 1970. Quantitative interpretation of chemical characteristics of hydrothermal systems. Geothermics Spec. Issue 2:516-528.

Hajash, A. 1975. Hydrothermal processes along mid-ocean ridges: an experimental investigation. Contr. Mineralogy and Petrology 53: 205-226.

Hajash, A. 1977. Experimental seawater/basalt interactions: effects of water/rock ratio and temperature gradient (abs.). Geol. Soc. America Abs. with Programs 9:1002.

Hardie, L. A. 1976. Origin of $CaCl_2$-rich geothermal brines by basalt seawater interaction beneath mid-ocean ridges: some simple mass balance calculations (abs.). Geol. Soc. America Abs. with Programs 8:903.

Hart, S. R. 1972. Strontium isotopic composition of the oceanic crust. Carnegie Inst. Washington Year Book 71:288-290. (Dept. of Terrestrial Magnetism).

Helgeson, H. C. 1969. Thermodynamics of hydrothermal systems at elevated temperatures and pressures. Am. Jour. Sci. 267:729-804.

Humphris, S. E., and G. Thompson. 1978. Hydrothermal alteration of oceanic basalts by seawater. Geochim. et Cosmochim. Acta 42:107-126.

Lister, C. R. B. 1973. Hydrothermal convection at seafloor spreading centers: source of power or geophysical nightmare? (abs.). Geol. Soc. America Abs. with Programs 5:74.

Mottl, M. J. 1976. Chemical Exchange Between Seawater and Basalt During Hydrothermal Alteration of the Oceanic Crust. Ph.D. diss., Harvard University, 188 pp.

Mottl, M. J., R. F. Corr, and H. D. Holland. 1975. Trace element content of the Reykjanes and Svartsengi thermal brines, Iceland (abs.). Geol. Soc. America Abs. with Programs 7:1206-1207.

Mottl, M. J. and H. D. Holland. 1978. Chemical exchange during hydrothermal alteration of basalt by seawater I. Experimental results for major and minor components of seawater. Geochim. et Cosmochim. Acta 42:1103-1115.

Mottl, M. J. and H. D. Holland. 1979. Chemical exchange during hydrothermal alteration of basalt by seawater II. Experimental results for Fe, Mn and sulfur species. Geochim. et Cosmochim. Acta 43. In press.

Scott, R. B., and D. J. Frank. 1974. Distribution of sulfur in the ocean crust (abs.). Geol. Soc. America Abs. with Programs 6:945.

Seyfried, W. E. 1977. Seawater-Basalt Interaction from 25°-$300^{\circ}C$ and 1-500 Bars: Implications for the Origin of Submarine Metal-Bearing Hydrothermal Solutions and Regulation of Ocean Chemistry. Ph.D. diss., University of Southern California, 242 p.

Seyfried, W. E., and J. L. Bischoff. 1977. Hydrothermal transport of heavy metals by seawater: the role of seawater/basalt ratio. Earth and Planetary Sci. Letters: 34:71-77.

Seyfried, W. E., and J. L. Bischoff. 1979. Seawater-basalt interaction-chemical exchange and mineral equilibria: An experimental study at 20°, 70°, and 150°C. Submitted to Geochim. et Cosmochim. Acta.

Seyfried, W. E. and M. J. Mottl. 1977. Origin of submarine metal-rich hydrothermal solutions: Experimental basalt-seawater interaction in a seawater-dominated system at 300°C, 500 bars. Proc. of the Second Int. Symp. on Water-Rock Interaction, I.A.G.C., Strasbourg, France, pp. IV173-IV180.

Seyfried, W. E., M. J. Mottl, and J. L. Bischoff. 1978. Seawater/basalt ratio effects on the chemistry and mineralogy of spilites from the ocean floor. Nature 275:211-213.

Spooner, E. T. C., H. J. Chapman, and J. D. Smewing. 1977. Strontium isotopic contamination and oxidation during ocean floor hydrothermal metamorphism of the ophiolitic rocks of the Troodos Massif, Cyprus. Geochim. et Cosmochim. Acta 41:873-890.

Williams, D. L. and R. P. Von Herzen. 1974. Heat loss from the earth: new estimate. Geology 2:327-328.

Wolery, T. J. and N. H. Sleep. 1976. Hydrothermal circulation and geochemical flux at mid-ocean ridges. Jour. Geol. 84:249-275.

Part III

FIELD

Editors' Comments on Papers 5 Through 31

5 LISTER
On the Thermal Balance of a Mid-Ocean Ridge

6 ANDERSON, LANGSETH, and SCLATER
The Mechanisms of Heat Transfer Through the Floor of the Indian Ocean

7 WILLIAMS and VON HERZEN
Heat Loss from the Earth: New Estimate

8 MUEHLENBACHS and CLAYTON
Oxygen Isotope Geochemistry of Submarine Greenstones

9 HUMPHRIS and THOMPSON
Abstract from *Hydrothermal Alteration of Oceanic Basalts by Seawater*

10 TÓMASSON and KRISTMANNSDÓTTIR
High Temperature Alteration Minerals and Thermal Brines, Reykjanes, Iceland

11 CORLISS
The Origin of Metal-Bearing Submarine Hydrothermal Solutions

12 SKORNYAKOVA
Dispersed Iron and Manganese in Pacific Ocean Sediments

13 BOSTRÖM and PETERSON
Precipitates from Hydrothermal Exhalations on the East Pacific Rise

14 BENDER et al.
Abstract from *Geochemistry of Three Cores from the East Pacific Rise*

15 VON DER BORCH and REX

Amorphous Iron Oxide Precipitates in Sediments Cored During Leg 5, Deep Sea Drilling Project

16 BISCHOFF

Red Sea Geothermal Brine Deposits: Their Mineralogy, Chemistry, and Genesis

17 BÄCKER and SCHOELL

New Deeps with Brines and Metalliferous Sediments in the Red Sea

18 SCOTT et al.

Rapidly Accumulating Manganese Deposit from the Median Valley of the Mid-Atlantic Ridge

19 WILLIAMS et al.

The Galapagos Spreading Centre: Lithospheric Cooling and Hydrothermal Circulation

20 WEISS et al.

Hydrothermal Plumes in the Galapagos Rift

21 LUPTON, WEISS, and CRAIG

Mantle Helium in Hydrothermal Plumes in the Galapagos Rift

22 KLINKHAMMER, BENDER, and WEISS

Hydrothermal Manganese in the Galapagos Rift

23 JENKINS, EDMOND, and CORLISS

Excess ^{3}He and ^{4}He in Galapagos Submarine Hydrothermal Waters

24 EDMOND et al.

Abstract from *Ridge Crest Hydrothermal Activity and the Balances of the Major and Minor Elements in the Ocean: The Galapagos Data*

25 EDMOND et al.

Abstract from *On the Formation of Metal-Rich Deposits at Ridge Crests*

Field investigations have concentrated on the various components of sub-seafloor hydrothermal systems at seafloor spreading centers, including heat flux, seawater-basalt alteration, solution chemistry, hydrothermal deposits, and tectonic framework. Measurements of heat flow constitute the basis for recognition of the magnitude of hydrothermal convection at seafloor spreading centers. Theoretical values of conductive heat flow calculated from models of heat produced by the generation of lithosphere at seafloor spreading centers (Le Pichon and Langseth 1969; Sclater and Francheteau 1970) exceeded the average measured values (Langseth and Von Herzen 1970). From the vantage point of Iceland, Palmason (1967) suggested that hydrothermal circulation may play a significant role in the distribution of oceanic heat flow, and Talwani et al. (1971) attributed a low measured in the overall high axial heat flow of the Reykjanes Ridge to water circulation. Lister (Paper 5) combined theoretical and field observations into a coherent formulation of the relative roles of conductive and convective heat transfer in the thermal regime of an oceanic ridge. The effectiveness of this approach in interpreting the heat flow distribution in the major ocean basins is demonstrated by Anderson et al. (Paper 6; 1979), and Hyndman et al. (1976). Williams and Von Herzen (Paper 7), Anderson and Hobart (1976), and Sleep and Wolery (1978) used the discrepancy between theoretical and average measured conductive heat flow to estimate the global magnitude of heat transfer by hydrothermal convection at seafloor spreading centers.

Convective transfer of heat produced by generation of lithosphere at seafloor spreading centers requires large-scale hydrothermal circulation. Hydrous metamorphosed oceanic crust (Miyashiro et al. 1971) and oceanic serpentine (Christensen 1972) evidence a voluminous source of water. Oxygen and certain other isotopes of the hydrated rocks demonstrate that the water is seawater (Paper 8; Muelenbachs and Clayton 1976; Spooner 1976).

Recognition of large-scale hydrothermal convection of seawater stimulated experimental (see Experiment section of this volume) and field research treating seawater-basalt interactions. One aspect of the field research concerned with the modulation of seawater chemistry by exchange of elements between seawater and basalt, a departure from the earlier influential views of Sillén (1967) that stressed the role of detrital sediments in this exchange, is presented by Drever in Benchmark volume 45 (1977), Hart (1973), and Edmond et al.(Paper 24). Another aspect of this field research is concerned with the hydrothermal alteration of basalts. Humphris and Thompson (Paper 9; 1978), and Elderfield et al. (1977) document the alteration chemistry of Mid-Atlantic Ridge basalts. Tómasson and Kristmannsdóttir (Paper 10) document the alteration mineralogy of a high-intensity hydrothermal system in Iceland (Björnsson et al. 1972; Olafsson and Riley 1978); alteration mineralogy has been documented at other sites as well (Aumento and Sullivan 1974; Browne 1978; Stone and Fan 1978; Hart and Staudigel 1978). Corliss (Paper 11) determined that seawater-basalt exchange could produce ore-bearing solutions adequate to account for the deposition of metallic mineral deposits at seafloor spreading centers.

Metalliferous deposits in sedimentary and solid form are the principal products of hydrothermal systems at seafloor spreading centers. Pure hydrothermal deposits precipitated from hot aqueous metal-bearing solutions and hydrogenous deposits precipitated from normal sea water appear to be opposite end members of a continuous series (Bonatti et al. 1972; Toth, 1980), but the position of certain members, like manganese nodules, within this series remains problematic (Arrhenius and Bonatti 1965; Lalou and Brichet 1976; Lalou et al., 1979; Skornyakova, 1979).

Reports of the *HMS Challenger* expedition (Murray and Renard 1891) identified a volcanogenic component in Pacific sediments that was later found to contain metals (Revelle 1944). Skornyakova (Paper 12) identified anomalously high contents of iron and manganese in sediment on the East Pacific Rise, which she attributed to hydrothermal activity related to volcanism. Boström and Peterson (Paper 13) also measured anomalously high metal content in East Pacific Rise sediments, which they attributed to "volcano-exhalative" processes involving mantle sources for the metals. Bender et al. (Paper 14) distinguished between "volcano-exhalative" and hydrogenous metal-bearing components of East Pacific Rise sediments (Dymond et al. 1973; Piper et al. 1975; Cronan 1976; Heath and Dymond 1977; Leinen and Stakes 1979). In their analysis of sediment cores recovered by the Deep Sea Drilling Project, von der Borch and Rex (Paper 15) identified a basal layer of metalliferous sediments overlying basaltic basement in the Pacific Ocean basin similar to the sediment forming at the crest of the East Pacific Rise, which demonstrated that metalliferous sediments have formed continuously during the seafloor spreading process (Horowitz and Cronan 1976).

The active high-intensity hydrothermal systems at seafloor spreading centers that are achieving classic status as prototype interdisciplinary study areas are Iceland, the Red Sea, the TAG Hydrothermal Field on the Mid-Atlantic Ridge, the Galapagos spreading center, and the East Pacific Rise at latitude 21°N. The Red Sea metalliferous sediments and hot brines documented by Bischoff (Paper 16) and systematically explored by Bäcker and Schoell (Paper 17) are products of the first active high-intensity submarine hydrothermal system discovered at a seafloor spreading center (Miller et al. 1966; Hunt et al. 1967; Degens and Ross 1969; Whitmarsh et al. 1974; Bäcker 1975; Schoell 1976; Stieltjes 1976; Bignell et al. 1976a, 1976b; Lupton et al. 1977; Shanks and Bischoff 1977, 1980). The thermal brines first appeared as an anomaly in the hydrographic data of the Swedish Deep Sea Expedition (Bruneau et al. 1953), which was missed at the time because the cruise geochemist, G. Arrhenius, is said to have taken leave for his marriage in Stockholm during passage of the research vessel through the Red Sea.

The question whether the hydrothermal systems of the Red Sea represented a special case related to the early opening stage of an ocean basin, or a general case representative of hydrothermal systems at seafloor spreading centers was raised at the time of discovery. The discovery of the TAG Hydrothermal Field on the Mid-Atlantic Ridge crest at latitude 26°N by the Trans-Atlantic Geotraverse (TAG) project of the National Oceanic and Atmospheric Administration (Rona and Scott 1974; Rona, 1980), the first active submarine hydrothermal system found at a seafloor spreading center in an open ocean basin, demonstrated that hydrothermal processes of metal concentration can act continuously from the Red Sea stage to the ocean stage in the opening of an ocean basin about a seafloor spreading center (Paper 31). Analysis of TAG deposits by Scott et al. (Paper 18) established geochemical methods for the recognition of solid hydrothermal precipitates (Scott et al. 1978). The first direct measurement of the magnitude and gradient of a hydrothermal water temperature anomaly at a seafloor spreading center in an open ocean basin was made at the TAG Hydrothermal Field (Rona et al. 1975) and led to the formulation of a model for hydrothermal heat flux calculations indicating a heat output equivalent to that at major geothermal fields on land (Lowell and Rona 1976; Rona 1978). Geochemical studies of seawater at the TAG Hydrothermal Field have resulted in the identification of a primordial helium isotope derived from mantle injection (Jenkins et al. 1980; Clarke et al. 1969; Jenkins et al. 1972; Jenkins and Clarke 1976), and of suspended particulate matter produced by hydrothermal activity (Betzer et al. 1974; Bolger 1976; Bolger et al. 1978; Betzer et al. 1980). Delineation of the tectonic framework has revealed structural and thermal controls that localize high-intensity hydrothermal convection (Paper 31; Rona et al. 1976; Temple et al. 1979).

Good weather and a sediment cover sufficiently thick to perform heat flow measurements initially attracted investigators to the Galapagos spreading center in the eastern equatorial Pacific. A systematic grid of heat flow measurements revealed a wave-like pattern interpreted in terms of hydrothermal convection cells and near bottom water temperature anomalies of possible hydrothermal origin (Paper 19). Hydrothermal deposits were identified near the spreading center (Moore and Vogt 1976). A group from the Scripps Institution of Oceanography succeeded, by using deep-towed instrumentation, in simultaneously sampling and measuring the temperature within a hydrothermal plume emanating from the Galapagos spreading center (Paper 20; Lonsdale 1977); their analyses of associated seawater samples reveal large anomalies of manganese (Paper 22) and an isotope of helium (Paper 21), which serve to elucidate hydrothermal processes (Clarke et al. 1969). Dives with the Deep Submergence Research Vehicle *Alvin* at the Galapagos spreading center enabled visual observation and sampling of hydrothermal sites (Corliss et al. 1978, 1979), mixing studies of the effluent (Papers 24 and 25), heat flow determinations (Williams et al. 1979), and the first experimental verification of the global magnitude of convective heat flux at seafloor spreading centers (Paper 23). Deeper crustal information was obtained by drilling at the hydrothermal deposits (Hekinian et al. 1978; Natland et al. 1979; Schrader et al. 1980). Dense populations of invertebrates were unexpectedly found surrounding the hydrothermal vents at the Galapagos spreading center and are inferred to be sustained by microbial primary production (Karl et al. 1980).

The crest of the East Pacific Rise at latitude 21°N near the mouth of the Gulf of California attracted attention when a French group diving there with the submersible *Cyana* sampled inactive deposits of massive metallic sulfides (pyrite, chalcopyrite, digenite, covellite, pyrrhotite, marcasite, melnicovite, wurtzite, sphalerite, goethite), a class of deposits of economic importance on land, which hitherto had not been found at a seafloor spreading center (*Cyamex* Scientific Team 1979; Hekinian et al. 1980; Picot and Fevrier 1980); nonmetallic sulfates are also present. Prior investigation with deep-towed instrumentation had produced suggestive evidence of hydrothermal temperature anomalies in the near-bottom water (Crane and Normark 1977; Crane 1979). Subsequent dives with DSRV *Alvin* along the spreading center at 21°N revealed hydrothermal solutions blackened by sulfide precipitates discharging from mineralized chimneys at temperatures up to 380 ±30°C (*Rise* Project Group 1980; Edmond 1980; MacDonald et al. 1980). The hydrothermal solutions incorporated significant quantities of the dissolved gases methane, hydrogen, and helium (Welhan and Craig 1979). Consideration of pressure-volume-temperature relations suggests limits on the temperature and effective water/rock mass ratio of the hydrothermal solution and emphasizes the role of adiabatic expansion in precipitating the sulfide minerals (Bischoff 1980).

Off-axial structures of seafloor spreading centers are also under investigation as possible loci of high-intensity hydrothermal systems. Off-axial structures where relict hydrothermal deposits have been found include transform faults and associated fracture zones that offset the axes of spreading centers (Paper 26; Rosanova and Baturin 1971; Hoffert et al. 1978), and an active strike-slip fault near a continental margin (Lonsdale, 1979). A number of active submarine sites where hydrothermal deposits are present at (Vidal et al. 1978; Lonsdale et al. 1980) and away from seafloor spreading centers (Zelenov 1964; Honnorez et al. 1973; Ferguson and Lambert 1972) exhibit variations of the same basic processes.

Synoptic studies consider small-scale (Paper 27) and large-scale distribution patterns of metalliferous sediments (Paper 28). Boström enunciates the problem of distinguishing between metalliferous components derived by injection from mantle sources and by hydrothermal concentration from crustal sources at seafloor spreading centers. Spooner and Fyfe (Paper 29), Spooner (1974), Spooner et al. (1974), and Spooner and Bray (1977) firmly consolidate various lines of evidence into a unified model for sub-seafloor metamorphism, heat and mass transfer at seafloor spreading centers. Integration of the results of theoretical, experimental, and field studies produces an overview of the magnitude of hydrothermal processes at spreading centers (Paper 30; Sleep and Wolery 1978). Rona (Paper 31) synthesizes knowledge of hydrothermal systems at seafloor spreading centers and translates this information into terms that can be applied to guide exploration for hydrothermal deposits in oceanic crust, a theme that is developed in other volumes of the Benchmark series (Walker 1976; Wright 1977; Moody, in preparation).

The nature of hydrothermal systems at seafloor spreading centers is considered to be strongly influenced by the rate of seafloor spreading through control of thermal regime and permeability. The Red Sea and the TAG Hydrothermal Field are situated on slow-spreading centers (spreading half-rate $\leqslant 2$ cm per year). The Galapagos spreading center and the East Pacific Rise at latitude 21°N are spreading at intermediate half-rates (~ 3.5 cm per year). The strategy guiding field investigations of hydrothermal systems at seafloor spreading centers is to extend exploration and sampling programs to those portions of spreading centers characterized by extreme spreading rates in order to compare and generalize results over the entire range of occurrence in terms of global heat and mass balances.

5

Reprinted from *Royal Astron. Soc. Geophys. Jour* **26**:515–535 (1972)

On the Thermal Balance of a Mid-Ocean Ridge

C. R. B. Lister

(Received 1971 June 18)

Summary

The northernmost expressions of the East Pacific Rise spreading centre, the Juan de Fuca Ridge and Explorer Trough, have unusual heat-flow distributions. Crestal values are only moderately high, but flank areas that are models of undisturbed sedimentation have mean heat flows near 7 HFU (290 mW m^{-2}). The values in these areas retain considerable scatter, but the detailed distribution is close to the opposite of that expected from conductive heat-flow refraction. The high mean values confirm that sea-floor spreading occurs by the upwelling of hot (magma temperature) material. The low crestal measurements and the non-refractive scatter in the flank areas strongly suggest that hydrothermal circulation is the dominant heat transfer process in fresh crestal material. Much of the heat loss from the new lithospheric material must occur through open circulation near the ridge crest and is never reflected in measurements of surface conduction.

1. Introduction

The investigation of the heat-flow anomaly at mid-ocean ridges has been an important aim of marine heat-flow measurement since the earliest studies were made (Bullard & Day 1961). A substantial body of data is now available, and has been summarized by several authors (Lee & Uyeda 1965; Langseth, Le Pichon & Ewing 1966; Le Pichon & Langseth 1969). Although the mean heat flow near ridge crests is usually higher than the global average, it is not dramatically so, and the scatter in the measurements is such as to compromise the validity of these means, both statistically and physically. The overall results are not in agreement with the generation of new oceanic crust at the ridges from material at a reasonable temperature (McKenzie 1967), and the recent reinforcement of the sea-floor spreading hypothesis by evidence from other fields (Vine & Matthews 1963; Peterson *et al.* 1970) requires that the disagreement between the heat flow and the other data be resolved.

It has been pointed out before (Lister 1970a; Sclater *et al.* 1970) that some environmental control is essential for any measurement whose scatter is sufficient to arouse suspicion of physical cause for the variation. In this study, standard surveying techniques were carefully applied to the area of interest to obtain enough topographic data at least for the *topological* situation of each heat-flow measurement to be estimated. The influence of thermal refraction cannot be separated from other causes of variation unless the sign and approximate magnitude of the expected correction are known *individually* for each station. This information has been obtained for almost all of the measurements quoted here, together with data that set an upper limit on the possible magnitude of recent surface disturbances. This last is important because of the suggestion by Sclater, Mudie & Harrison (1970) that the low heat-flow values often measured in valleys are caused by recent slumps of sediment from nearby hills or ridges. The assumption that the heat flow is governed solely by conduction in the crustal materials can therefore be tested directly without making alternative assumptions about the thermal regime at depth. Only when this test has been made can the overall thermal balance of a ridge crest be discussed.

The region of the North-east Pacific off the coast of North America is dominated tectonically by two long active transform faults and the most northerly expressions of the East Pacific Rise spreading axis. The San Andreas Fault, running from the Gulf of California to Cape Mendocino, is a right-lateral transform fault joining the equatorial East Pacific Rise to the ridge segments off the coasts of Oregon and Washington (see Fig. 1). The magnetic survey of Raff & Mason (1961) suggests that there are three segments of active spreading: the Gorda Ridge, the Juan de Fuca Ridge, and Explorer Trough (Vine 1966). They are joined by the Blanco Fracture Zone, and an ill-defined fault zone between the Juan de Fuca Ridge and Explorer Trough. The system ends in another long active transform fault, the Queen Charlotte Fault, and this extends toward the Alaskan end of the Aleutian Trench, completing the source-to-sink pattern (Atwater 1970).

Of the three spreading segments, the Juan de Fuca Ridge is the longest and the best defined physiographically. On the western side it has all the characteristics of a mid-ocean ridge: mountainous topography rising gradually to a crest at the centre of magnetic symmetry. The eastern side, however, appears dramatically different on the map, because the ridge has been inundated by the sediment fill of Cascadia Basin. Heavy continental sediment runoff through the Fraser and Columbia River systems has been ponded near the coast by the Juan de Fuca Ridge and by high topography on the Blanco Fracture Zone. Thus the Juan de Fuca Ridge is a unique example of a classic 'mid-ocean' ridge which has been inundated by terrigenous sediments almost to the crest on one side, but is sufficiently far from the continental margin to be structurally integral.

Freedom from structural interference is not a characteristic of Explorer Trough, however, for it is a small feature amid complex topography and structure just off the continental rise, north-west of Vancouver Island. The magnetic anomalies associated with it are poorly defined (Vine 1966), and interpretation is further complcated by the fact that it is a trough not a ridge. However, the lack of sediments in the topographic lows, relative to the surrounding ocean floor, confirms the magnetic and now thermal evidence that rifting of the crust has taken place recently (see Fig. 9). The feature is thus interesting as a spreading centre that has not, as yet, developed into a full-scale ridge, and will be useful for structural comparisons in the future.

This paper presents field results from two studies: one on the Juan de Fuca Ridge and one in the Explorer Trough region. The principal investigation is a transect across the Juan de Fuca Ridge at 47° N, and is shown on the location map (Fig. 1) as rectangle 'G'. Data for this area have been accumulated over several years from short cruises of RV *Thomas G. Thompson*, and include a substantial body of bathymetric surveying. The smaller study is within rectangle 'H', and consists of a heat-flow profile across the north-eastern extremity of Explorer Trough. This was completed on Phase *VII of Hudson* 70, a circumnavigation of the Americas by C.S.S. *Hudson*. No detailed surveying was carried out in this area, but the topographic features are large enough in scale to be delineated adequately by the bathymetric data of the *Pioneer* survey (cf. Fig. 1, Raff & Mason 1961).

2. Experimental techniques

Heat-flow measurements were made by examples of both generic types of marine apparatus. The corer-outrigger instrument has been described in Lister (1970a), and consists essentially of a 450 × 8 cm gravity corer equipped with a digital bridge

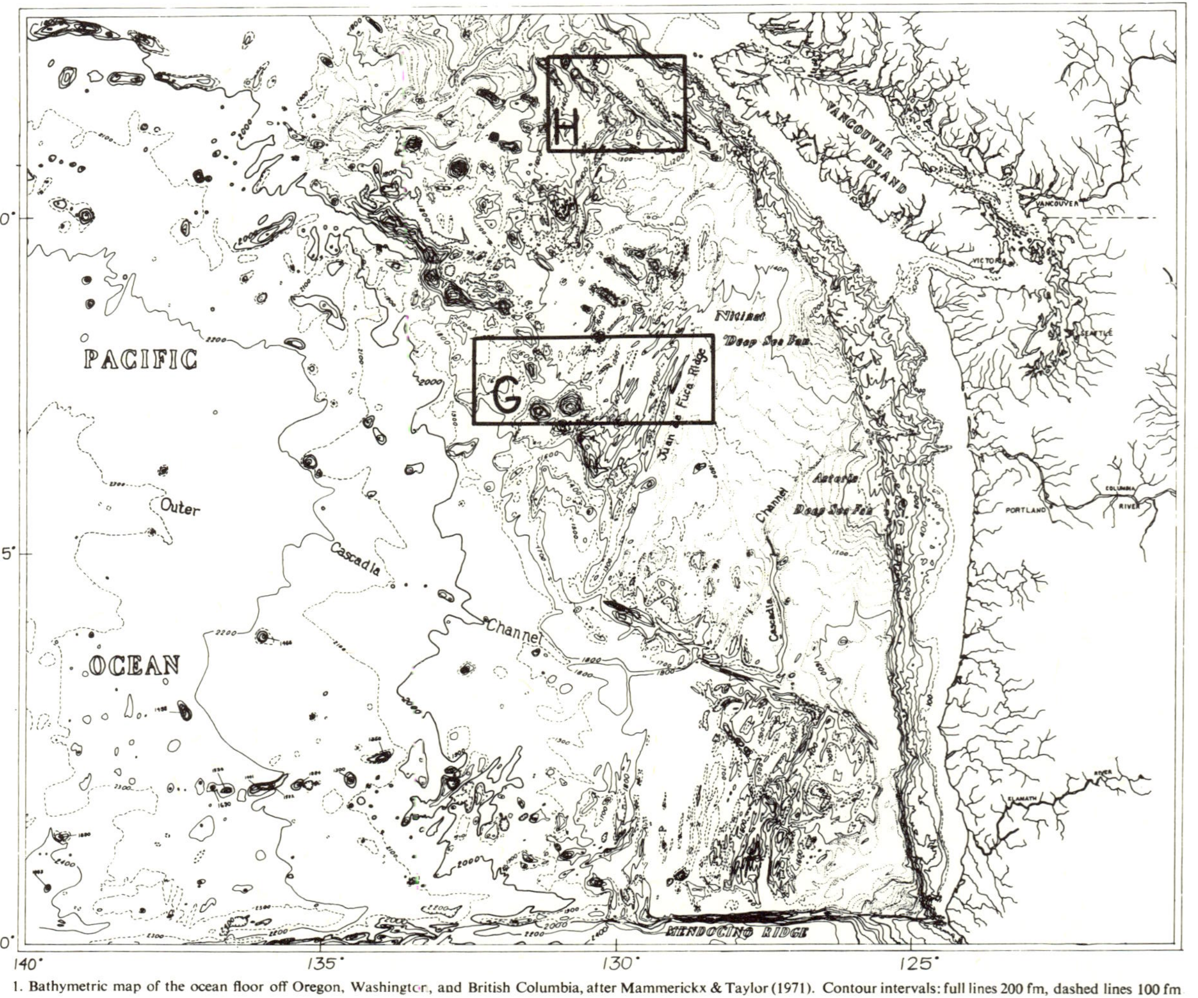

1. Bathymetric map of the ocean floor off Oregon, Washington, and British Columbia, after Mammerickx & Taylor (1971). Contour intervals: full lines 200 fm, dashed lines 100 fm

recorder and three thermistor outriggers. Sediment conductivities were estimated by oven-drying samples from the cores and using the water content versus conductivity nomograph of Ratcliffe (1960). The probe instrument uses the same digital recorder but a sensor arrangement modified to measure thermal conductivity *in situ* over the whole length of the probe. Geothermal gradient and conductivity are measured sequentially in 30 min on the ocean floor, and the probe dimensions have been made small to permit this (200 × 1 cm). The instrument and the interpretation techniques have been fully described in Lister (1970b). The corrections for initial-disturbance decay and the conductivity values were obtained by graphical methods from a large-scale analogue printout of the digital data. The technique has the advantage of permitting a selection of the data points to be computed, avoiding any disturbances experienced by the instrument and the occasional digitisation errors.

The cores obtained by the corer-outrigger apparatus were split and examined stratigraphically in addition to being used for the water content/ conductivity determinations. Total carbonate assays were made on selected parts of some to assist in distinguishing biogenic from terrigenic material, since the carbonate is exclusively of biogenic origin. The carbonate fraction of the cores was also used to obtain radiometric ages for samples from the uppermost 70 cm by measurement of the carbon-14 decay activity in derived methane (A. W. Fairhall*). Three normal cores were dated to establish the sedimentation rate and pattern, and the technique was then used to examine two disturbed cores that could have included slump or slump-exposed material.

The general distribution of sediment on the ocean floor in the areas was examined by acoustic reflection sounding. The systems used on the Juan da Fuca Ridge transect emitted short pulses of medium-frequency (150–300 Hz) sound that have a favourable penetration/resolution ratio in deep-ocean sediments. The early profiles were made with several models of a repulsion-type moving coil electromagnetic transducer energized with 11000 J, but the survey east of the ridge crest was made with an attraction-type electromagnetic transducer pulsed by 250 J. The same, somewhat noisy, 15 m line hydrophone was used throughout, and the signal was processed by automatic gain controlled amplifiers equipped with 12 db/octave variable filters. Full-wave variable density recording was used, but the attraction-transducer records were enhanced by a delay-line pulse shape correlator manually set to match the outgoing pulse. The ultimate resolution of both systems was about 10 ms (7·5 m) but the interpretability of the correlated records is better because of the symmetrical printed pulse free of multiples. Except in the flat-lying sediment fill of Cascadia Basion, resolution is limited by the roughness of the bottom rather than by the acoustic system.

The sediment distribution in the Explorer Trough region was estimated by reference to 40 cu in airgun profiles provided by R. L. Chase (University of British Columbia). In particular, the principal heat-flow profile was arranged along airgun profile UBC 70–16–14, that crosses the troughs about 15 km south-west of Paul Revere Ridge. Resolution on this profile was about 60 ms (45 m), as can be judged from Fig. 9.

A bathymetric background is necessary to identify the lateral extent of features observed on transect profiles. In the Explorer Trough region, it was obtained from a map contoured from previous survey data at 9 km line spacing (Mammerick & Taylor 1971). The features in this area appear to have diametral dimensions of about 15 km (Fig. 8), and are therefore adequately delineated. No significant discrepancies were observed between the contour map and echo-soundings on the cruise.

The topography on the western flank of the Juan de Fuca Ridge is, however, far more complex and of finer scale. No adequate map of the area was available at the time the investigation was started, and even the 9-km line spacing of the Pioneer Survey (Mammerick & Taylor 1971) cannot resolve the topographic structure. A bathymetric

*Professor, Chemistry & Geophysics, University of Washington

survey was therefore conducted as part of the investigation, to identify topographic trends near the transect line. Navigation was by Loran-A radiolocation which is good to ±2 km over most of the area, deteriorating steadily with increasing distance from the coast. Line crossings were used to improve the topological accuracy of the contour map by ensuring that track lines were plotted at the correct relative positions. A standard 12 kHz, 50° beam, echo-sounder gave the depth information, presented on the 1-s sweep of a precision dry-paper recorder. Two-way travel times were used in compiling the maps, to emphasize that sound velocity corrections were not made.

The concentration of stations near 47° N 132° W is in an ultra-detailed survey area, where navigation was assisted by taut-moored radar reflector buoys. Only the outlines of the detailed map have been transferred onto the figure used in this paper (Fig. 3). The survey was intended to establish the resolution limits and characteristic response of standard echo-sounding systems. The information has been valuable in interpreting the more widely-spaced lines employed in later stages of the study.

Heat-flow stations can be affected by small-scale local topography as well as by the large-scale features observable on surface echo-sounding records (Sclater *et al.* 1970). One method of obtaining information about the local irregularities to be expected in an area is photography from a camera drifting near the bottom. The method is severely limited by the loss of visual acuity in bottom water at ranges of more than 10 m, but can produce valuable information about local slope angles, sediment distribution, and bottom currents. An Edgerton stereo camera cycling every 16 s was drifted at 5–8 m above bottom in several places on the western flank of the Juan de Fuca Ridge. On some stations a narrow-band green interference filter was used to enhance the contrast of distant objects. On the Explorer Trough investigation a different technique was tried for the first time, and gave valuable information about the nature of the floor of the main trough. A slow frame-rate cine camera, synchronized to a pulsed Thallium Iodide arc light, was substituted for the stereo camera. The film gives almost continuous coverage of a considerable (c. 150 m) section of ocean floor.

3. Transect of the Juan de Fuca Ridge

A consolidated track chart of the investigations is presented in Fig. 2 to indicate the extent of bottom coverage. The background for the heat-flow results in Fig. 3 is bathymetry derived from these tracks and from the more general data available in earlier studies. The heat-flow measurements are listed by station in Table 1, except for one measurement from *Vema* 20 (Paul J. Grim, private communication), whose relationship to the topography on the map is not known with precision. The first fifteen entries in Table 1 have been listed in more detail in Lister (1970a), but the remainder are new. Gradient linearity and the quality of the conductivity measurements were taken into account in establishing the error bands for the heat-flow measurements.

The distribution of long gravity core samples of the sediment is shown in Fig. 2. There are wide variations in the appearance and layering of the sediments because both biogenic and terrigenic materials are deposited on the Juan de Fuca Ridge, sometimes in almost equal proportion. However, most cores exhibit a characteristic dark surface layer that probably represents postglacial deposition, since carbon-14 radiometric ages below it are greater than 15 000 years (A. W. Fairhall, private communication). Two important disturbed cores (measurements TT31–3 and TT31–9, Table 1) were dated radiometrically at more than one depth to make sure that recent slumping can be ruled out. One, discussed at length in Lister (1970a), is normal except for the loss of the top section on board ship. The other has older sediment in the top 20 cm than deeper within it, and dares confirm that a buried dark layer a 25 cm corresponds to normal surface material. It is probable that the corer caused

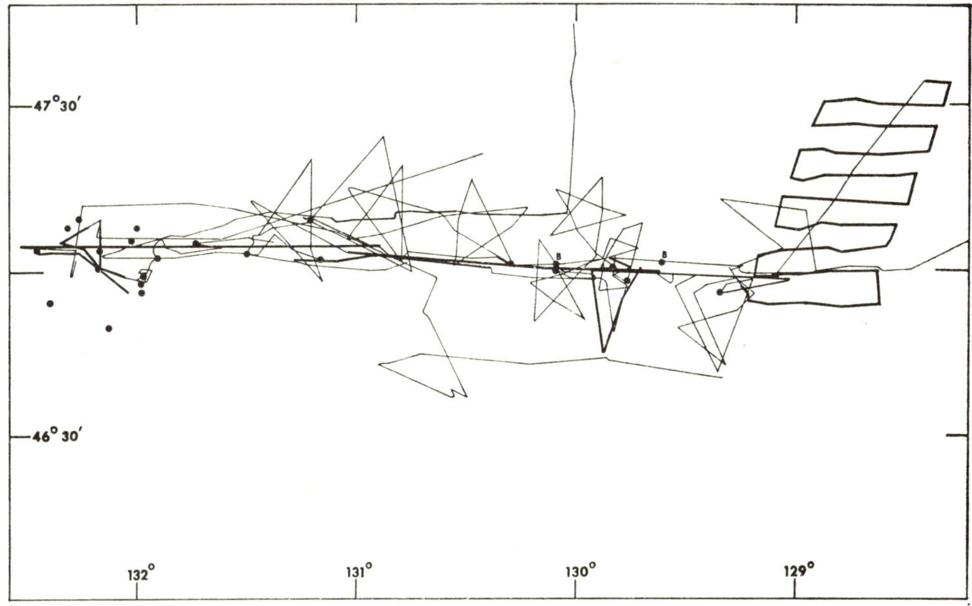

FIG. 2. Consolidated track chart of the Juan de Fuca Ridge investigations (not including the detailed survey area near 47° N 132° W). Core samples are shown by filled circles; superscript *B* means box core of surface material only.

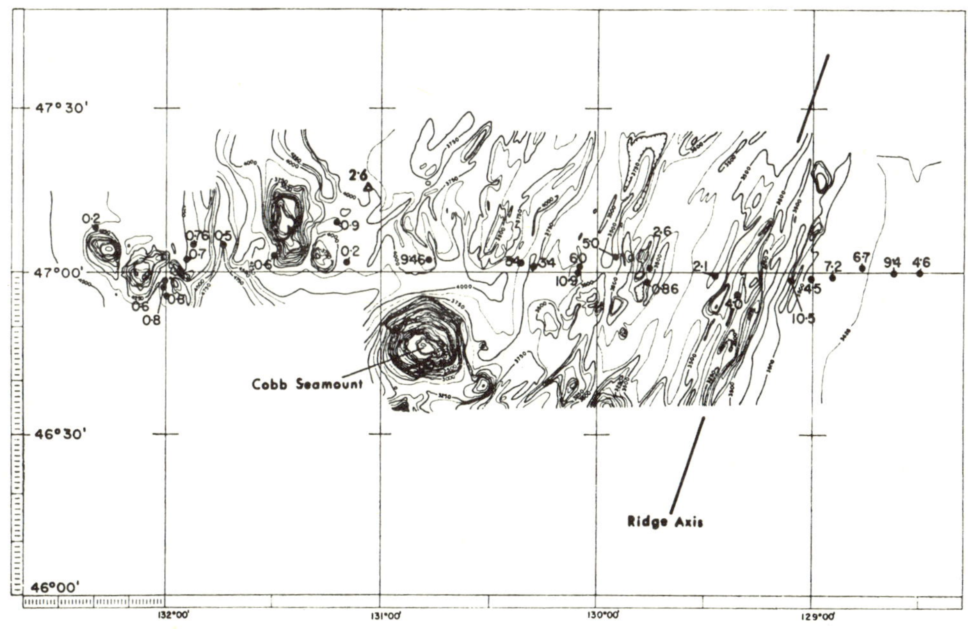

FIG. 3. Bathymetry and heat-flow data across the Juan de Fuca Ridge (axis shown). Map corresponds to area G in Fig. 1. Contour interval is in milliseconds of two-way travel time (×0·4 to convert to uncorrected fathoms). Heat-flow values are in microcalories per square centimetre per second (HFU).

Table 1

Heat-flow measurements: Juan de Fuca Transect

Station	Latitude (N)	Longitude (W)	Depth (sec, 2-way)	Conductivity ($\times 10^{-3}$ cal cg)	Heat flow ($\times 10^{-6}$ cal cm^{-2}—s^{-1})	Error band (±)	Heat flow (mW m^{-2})
TT3–1	46° 59·5′	131° 48·8′	4·24	2·0†	>0·6		25
TT3–2	46° 58·0′	131° 57·3′	4·29	1·9	0·8	0·06	34
TT3–3	46° 55·8′	131° 57·0′	4·32	1·89	0·9	0·1	38
TT3–8	47° 05·0′	131° 53·5′	4·31	1·93	0·7	0·06	29
TT17–1	46° 38·0′	134° 16·0′	5·15	2·13	0·2	0·06	8
TT17–5	47° 08·4′	132° 19·6′	4·34	2·08	0·2	0·06	8
TT31–1	47° 09·0′	131° 12·0′	4·09	1·86	0·9	0·04	38
TT31–2	47° 03·0′	131° 30·0′	3·96	1·96	0·6	0·03	25
TT31–3	46° 55·0′	129° 21·0′	3·40	2·04	4·0	0·10	168
TT31–4	46° 59·5′	129° 27·0′	3·61	2·14*	2·1	0·10	88
TT31–6	46° 58·0′	129° 46·0′	3·78	2·09	0·86	0·04	36
TT31–9	47° 00·0′	130° 06·0′	3·61	1·89	10·9	0·20	456
TT31–11	47° 02·0′	131° 10·0′	4·11	1·95	0·2	0·02	8
TT31–12	47° 10·0′	131° 44·0′	4·42	1·87	0·5	0·03	21
TT31–13	47° 05·0′	131° 52·5′	4·16	2·18*	0·76	0·04	32
TT40–5	47° 02·0′	130° 47·0′	3·75	1·93	9·46	0·20	396
TT40–6	47° 02·0′	130° 21·5′	3·71	2·17*	3·4	0·20	142
TT40–7	47° 01·0′	130° 17·5′	3·82	1·81	3·4	0·20	142
TT40–9	47° 01·0′	130° 05·0′	3·52	2·38*†	6·0	0·50	251
TT40–12	47° 01·0′	129° 45·0′	3·65	2·15*	2·6	0·30	109
TT40–15	47° 03·0′	129° 55·0′	3·57	2·15*	5·0	0·10	210
TT40–19	46° 58·5′	129° 06·0′	3·51	2·20	10·5	0·20	440
TT40–20	46° 59·0′	129° 00·5′	3·56	2·08*	4·5	0·30	189
TT40–22	46° 59·0′	128° 54·5′	3·56	2·21*	7·2	0·30	302
TT40–23	47° 01·0′	128° 46·0′	3·59	2·18*	6·7	0·30	281
TT40–24	47° 00·0′	128° 37·0′	3·65	2·06*	9·4	0·90	394
TT40–25	47° 00·0′	128° 30·0′	3·65	2·89*†	4·5	+0·60 −0·10	189

* *In situ* heated probe measurement (Lister 1970b)

† Assumed

‡ Partial penetration: lowest sensor

the disturbance by touching the sediment before penetrating fully, but even if a local slump has occurred here, the gradient is unlikely to have been affected. The therma time constant of 25 cm of sediment of normal diffusivity (0·0025 $cm^2 s^{-1}$) is only 0·01 yr, and it is only 1 yr for the whole 200 cm sediment column at this station.

The most striking feature of the heat-flow distribution west of the ridge crest is the zone of sustained high values, between 45 and 75 km (Fig. 3). An acoustic reflection profile across the western boundary of this zone is shown in Fig. 4, where the measured heat flow changes by a factor of 50 over a distance of 25 km. The high value is on a ridge, and the low value is in a valley, which is the reverse of what would be expected from simple heat-flow refraction. The core from the valley station does show evidence of turbidite deposition in the past, but it also has the dark-coloured surface layer indicative of slow deposition in the post-glacial environment. Total sediment thickness is 50 m in a valley 6 km wide: the steady-state refraction correction due to the sediment fill is negligible, and the time constant of the whole layer is 300 yr. There is no evidence of sediment disturbance in the last 13 000 yr, and the good constancy of the heat flow between the long and short probe intervals rules out any significant water temperature change.

The oceanic basement in the middle of the high heat-flow area is remarkably smooth on a small scale, so that the acoustic profiler was able to resolve 30–50 m of sediment draped over the abyssal hills. A section of profile near the two values of 3·4 HFU (Fig. 3) is reproduced in Fig. 5. Notable is the hard basement echo indicative of the absence of surface roughess greater than 10 m. In contrast to this, a section of the same acoustic profile, taken under identical recording conditions, but east of the high heat-flow zone, shows diffuse basement echoes except for short segments of valley floor (Fig. 6). The wider valley contains 25 m of sediment, but the narrower one, leading down to a ponded basin, contains 70 m. This latter valley is probably connected to the Vancouver Island turbidite source, but the 25 m cover is consistent with reasonable quasi-pelagic sedimentation rates and the 1 My magnetic spreading age of the ocean floor at this point (Vine 1966). The basement roughness on the hills is here greater than 25 m, so that bare rock may be expected to outcrop on the hills although they are no larger than those in the high heat-flow zone further west. The gross topographic relief of 0·2 km over a wavelength of 10 km can generate a refractive deviation of only 6 per cent, and the lowest measurement in the area (0·86) is in any case on the floor of a narrow valley (see Fig. 3).

East of the ridge crest, the heat-flow measurements are high, much more consistently than on the western side. The scatter is still considerable, but no longer

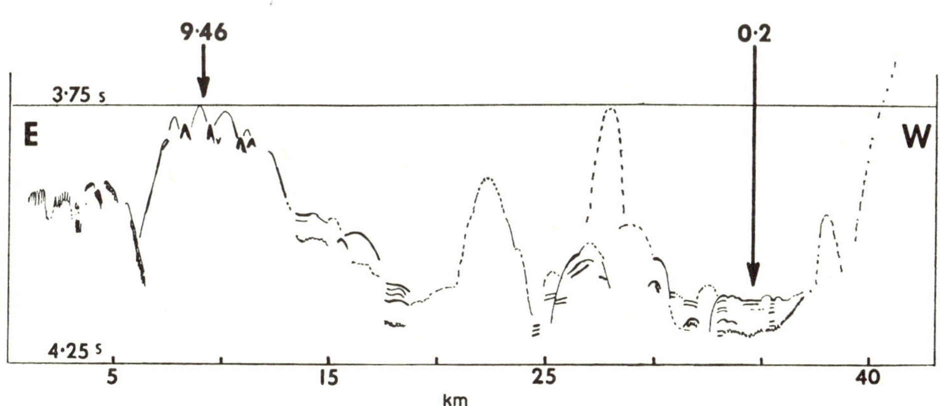

FIG. 4. Tracing of a seismic reflection profile between 130° 40′ W and 131° 15° W, at the western edge of the high heat-flow zone. Vertical exaggeration 30 : 1.

over an order of magnitude: the highest measurement is only 47 per cent above the 6 station mean of 7·15 HFU. Some, but not all, of the variation can be accounted for by the basement relief under the sediments, and the effect of the high turbidite sedimentation rates. The five innermost measurements fall on the acoustic reflection profile shown in Fig. 7. The value of 10·5, 12 km from the ridge crest, was taken at the extreme edge of a small sedimented valley, within 100 m of the outcropping rock slope that itself is 100 m high. The situation of the measurement is akin to a mixture of the cases treated by Lachenbruch (1968) and Von Herzen & Uyeda (1963). On the one hand the measurement is within one scale height of what may be a 45° slope, and on the other, it is in a sediment-filled trough that may not be too far from elliptical in cross-section. While the solutions referred to may not be added, a rough idea of the effect of the actual topography may be obtained by multiplying the corrections: the observed flow should be $\times 1{\cdot}17$ because of the slope and $\times 0{\cdot}9$ because of the trough, or only about 5 per cent too high. At the known spreading rate of the ridge (Vine 1966), the valley should be about 0·4 My old, and unless the 40 m of sediment were deposited recently, the sedimentation correction is only 3–5 per cent (Fig. 9 in Von Herzen & Uyeda 1963). Thus the expected net correction for this station is nearly zero, in so far as it can be estimated.

Another station that is notably out of line is the first on the abyssal plain proper, 4·5 HFU at 18 km from the crest. This is in a substantial sediment-filled trough, where both the refractive and sedimentation corrections are significant. A minimum sedimentation correction results if all 200 m of sediment has been deposited when the valley was formed: +7 per cent. A reasonable maximum corresponds to sedimentation 0·1 cm yr^{-1} for the last 200 000 years, or +15 per cent. The ellipsoidal trough solution, which should be a good approximation for a centrally located measurement, gives a further +10 per cent. Thus the measurement should be raised to between 5·2 and 6 HFU, not enough to be in line with a decreasing trend from the ridge crest, but at least changed in the right direction.

The last major out-of-line value is the 9·4 HFU at the extreme left of Fig. 7. There is some rise in the basement near the measurement, at least 100 m in 5 km, but it is less pronounced than the buried hill under the nearby 6·7 HFU value. In the case of a thin high-resistivity layer overlying topography on a medium of lower resistivity, the Jeffreys (1938) approximation must be modified to a surface thermal resistivity model. This shows that the effect of buried topography is *enhanced* by the ratio of the thermal resistivities minus one, provided the slopes are gentle enough for the basic Jeffreys approximation to be valid. Even with the $\times 1{\cdot}5$ enhancement, the correction to the measurement (on the sinusoidal ridge approximation) is only −5 per cent for the visible basement relief; it could be higher if basement deepened suddenly beyond where the profile terminates (cf. Fig. 2). There is some evidence that this may be the case, but basement returns on that section of Fig. 7 are not clear enough to be sure.

The prominent basement ridge under the measurement of 6·7 HFU (Fig. 7) should be taken into account. As it is flat topped, the inclined plane solutions of Lachenbruch (1968) may be used, and show that the −7 per cent correction is less than would be the case for a sinusoidal ridge of similar height. The enhancement factor is applied to the slope height in this case.

4. Explorer trough study

The pertinent measurements made from the *Hudson* are listed in Table 2. They are distributed over the ocean floor on, or south-west of, the prominent fault line associated with the upwarping of the Paul Revere Ridge (Fig. 8). This feature is capped by sediment layers that dip toward the north-east and outcrop on the south-west slope (R. L. Chase, private communication). It is therefore a piece of normal ocean floor

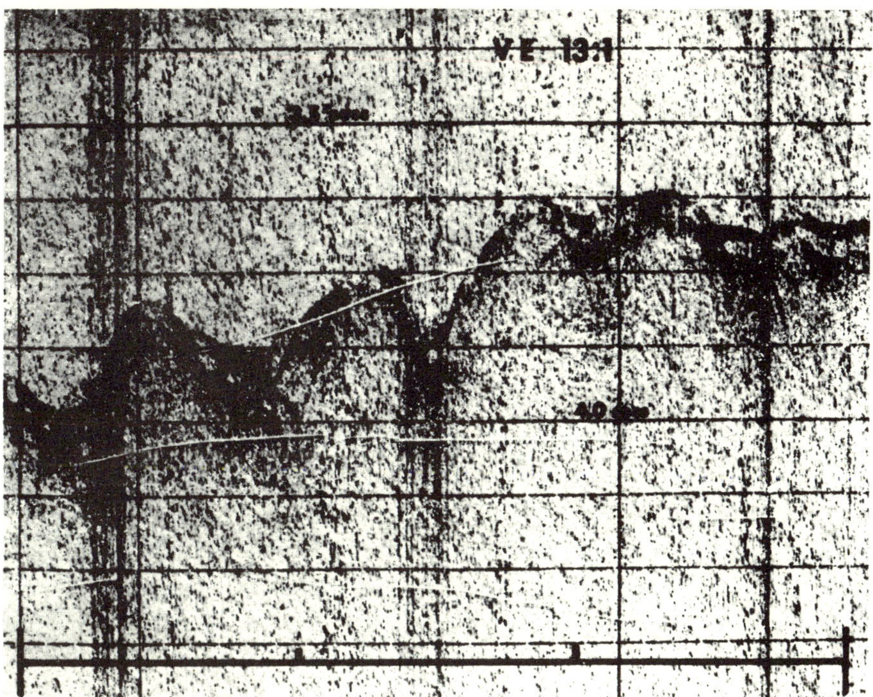

FIG. 5. Reflection profile of evenly sedimented basement within the high heat-flow zone, near 47° N and between 130° 23′ W and 130° 09′ W. Horizontal scale is 15 km long with 5-km divisions.

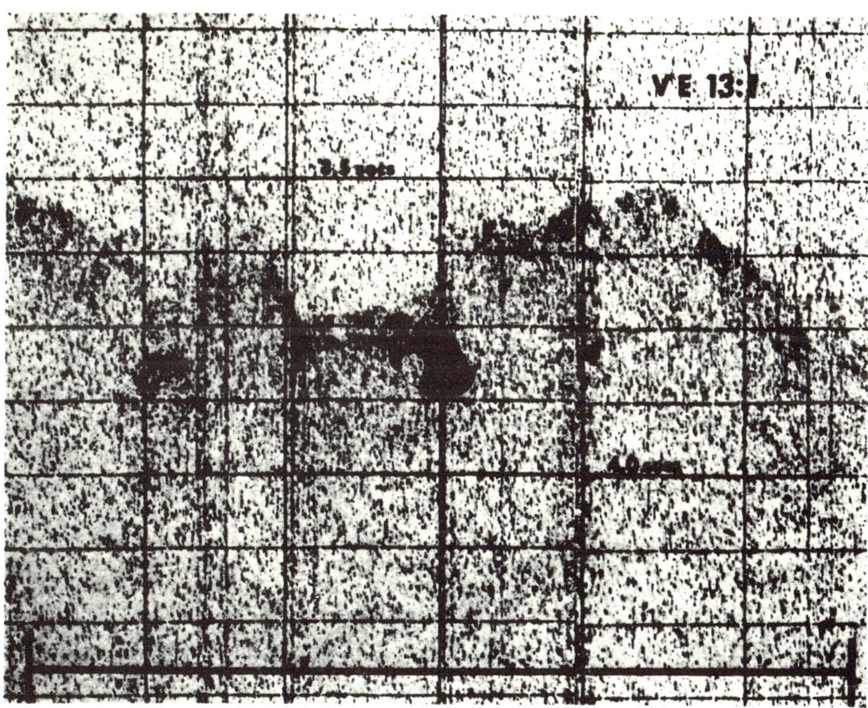

FIG. 6. Reflection profile near 47° N, from 139° 51′ W to 129° 37′ W, an area of moderate heat flow near the ridge crest. Horizontal scale is 15 km long with 5-km divisions. To the left of the 3·5-s identifier, a small peak reaching 3·35 s does not register on this record, due to strong scattering of the sound.

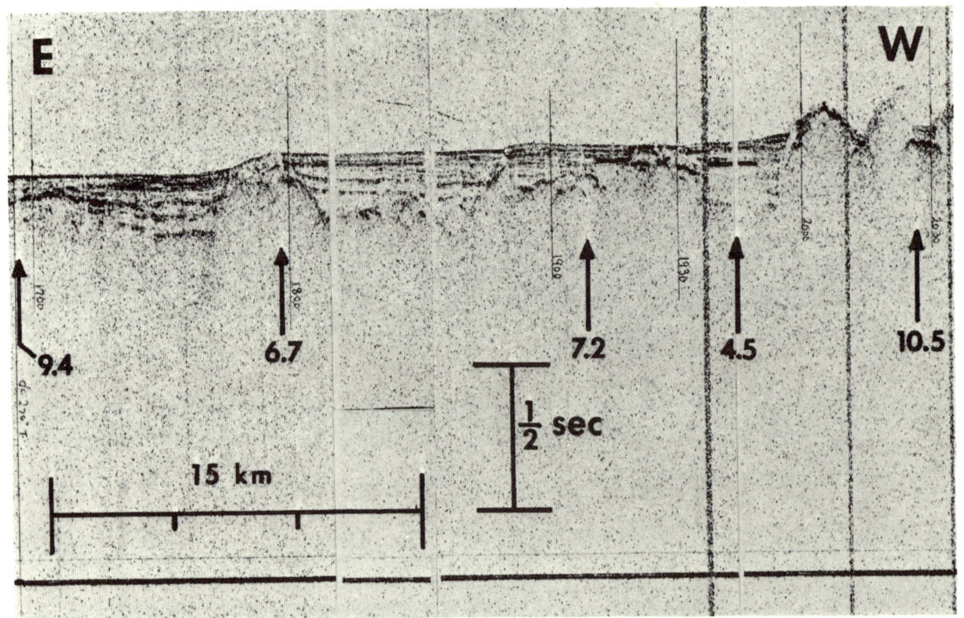

FIG. 7. Reflection profile east of the ridge crest near 47° N, with heatflow values (in HFU) superimposed. Vertical exaggeration 15 : 1.

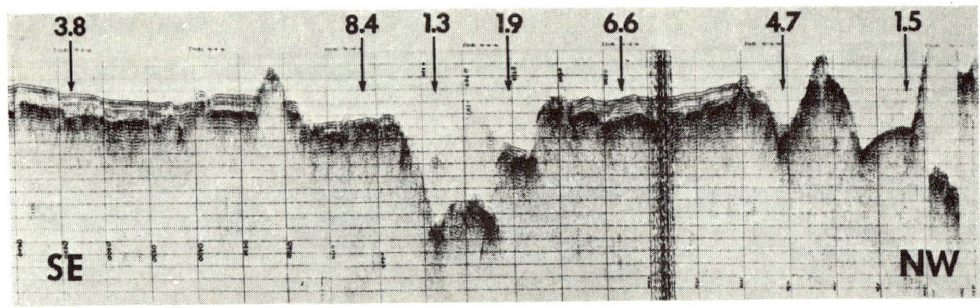

FIG. 9. The heat-flow measurements near the Explorer troughs projected onto UBC airgun profile 70–16–14. Profile by courtesy of Dr R. L. Chase, University of British Columbia. The vertical scale is from 2·5 to 5 s in 20 divisions, and the profile is 77 km long, so that the vertical exaggeration is 10 : 1.

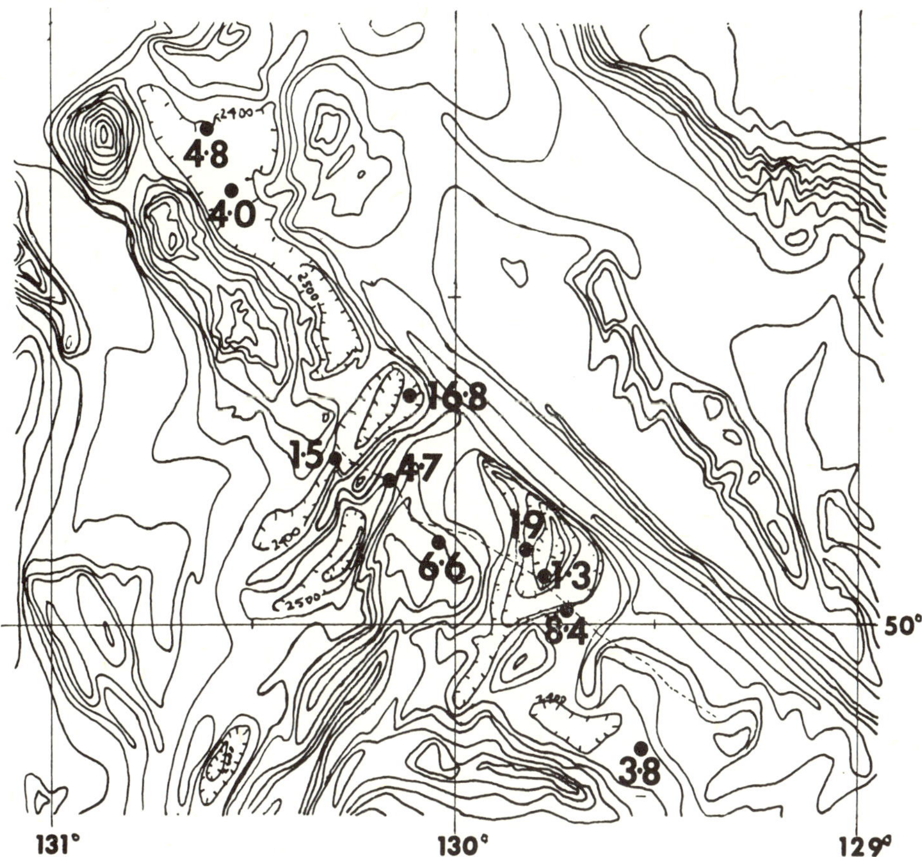

FIG. 8. Bathymetry and heat-flow data near Explorer Trough. The map is area H of Fig. 1 shown in more detail: contour interval 100 m. Paul Revere Ridge is the strong NW–SE lineation across the middle right of the area. UBC profile 70–16–14 is the faint dashed line running across the troughs almost parallel to the P.R. ridge.

that has been upwarped, rather than a linear volcanic pile, and represents a clear boundary to the spreading centre. The first seven measurements form a profile across the two troughs that abut the upwarped scarp, and were sited as close as possible to airgun reflection profile UBC–70–16–14 (R. L. Chase, private communication).

The measurements have been projected onto the profile in Fig. 9, and it is striking that the floors of the troughs are almost bare of sediment although the surrounding topography is well covered: in the case of south-easterly trough, to the very brim. The measurement in the north-westerly trough obtained only partial penetration with the 2-m probe, and a cine camera station was run nearby to determine the structure of the smooth, acoustically-hard floor. It is indeed quite flat, and consists of sediments interrupted every few metres by outcropping, though sediment-covered, boulders. A lava flow covered by 1 m of sediment would fit the appearance precisely. The relative absence of sediments strongly suggests that the floors of *both* troughs were formed recently, and the distribution of heat-flow measurements is therefore extraordinary. All the measurements on the sedimented basement near the troughs are high, but those in the troughs themselves are not, even though the bench in the middle of Fig. 9 is sediment-covered. The trend is exactly opposite to what would be expected from refraction through the topography, and so refraction corrections have not been made.

Table 2

Heat-flow measurement: Explorer Trough area

Station	Latitude (N)	Longitude (W)	Depth (sec, 2-way)	Conductivity ($\times 10^{-3}$ cal cg)	Heat flow ($\times 10^{-6}$ cal cm^{-2}—s^{-1})	Error band (±)	Heat flow (mW m^{-2})
H70–4	49° 58·60′	129° 32·60′	2·78	2·19	3·8	0·04	159
H70–5	50° 01·57′	129° 43·04′	3·13	2·08	8·4	0·30	352
H70–6	50° 04·36′	129° 46·36′	4·36	1·94	1·3	0·10	54
H70–7	50° 07·70′	129° 49·80′	3·70	1·92	1·9	0·20	80
H70–8	50° 07·50′	130° 01·80′	2·79	2·19	6·6	0·20	276
H70–9	50° 13·20′	130° 09·20′	3·22	2·34	4·7	0·10	197
H70–10	50° 15·26′	130° 17·84′	3·25	1·71‡	1·5	0·30	63
H70–11	50° 21·00′	130° 06·60′	3·53	2·43	16·8	0·40	704
H70–12	50° 45·90′	130° 36·60′	3·28	2·27	4·8	0·10	201
H70–13	50° 39·50′	130° 33·00′	3·27	2·06	4·0	0·06	168

All these measurements were made with the *in situ* conductivity probe.

‡ Incomplete penetration: lowest sensor value.

The remaining three measurements are essentially on the Paul Revere fault line (Fig. 8). One, at the end of the north-westerly trough, is in a small shallow sediment pond at the base of the ridge slope. It is unusually high, even for this area. The other two are in a larger and deeper sediment pond to the north-west. This straddles a possible final spreading rift leading to the Queen Charlotte Fault proper, which essentially follows the continental margin off the Queen Charlotte Islands (Fig. 1). In view of the exceedingly high value farther down the Paul Revere Fault, the two measurements of 4 and 4·8 HFU do not confirm that this trough is the site of further spreading, but neither do they rule it out.

5. Discussion

The heat-flow measurements of this study differ from previous results (cf: Le Pichon & Langseth 1969) primarily in the presence of numerous zones of high values. Near Explorer Trough, both the plateaus have high heat flow (Fig. 9) and so does the fault line associated with the Paul Revere Ridge (Fig. 8). The more extensive transect of the Juan de Fuca Ridge establishes two major high zones: the sedimented flank east of the crest, mean 7·1 HFU, and between 50 and 130 km west of the crest, mean 6·4 HFU (Fig. 3). It is notable that all the areas where the heat flow is high have an even sediment cover thicker than local basement roughess (Figs 5, 6, 7 & 9), though this *by itself* is not sufficient to ensure a high measured heat flow (Fig. 4).

If an even sediment cover is necessary to measure consistently high heat flow, one must test the possibility that local sediment ponds separated by outcropping rocks can result in substantially depressed heat flow in those ponds. Von Herzen & Uyeda (1963) have shown that an isolated circular pond depresses the heat flow most, and if the conductivity contrast were as high as 3 and the basin were hemispherical, a measurement in it would be 0·4 of the true value. If the rock topography between basins is elevated, the depression is reduced, and the interference between neighbouring basins also reduces the effect. The 0·4 factor therefore represents a maximum for any reasonable configuration of lightly sedimented ocean floor, and cannot be expected to apply on an average basis to randomly located measurements. If the extreme correction factor were applied to all the inner western flank values (mean 2·4) it would bring them up to the same level as the others—yet all the heat-flow measurements can hardly have been made in deep local sediment pockets. The heat-flow contrast shown in Fig. 4 is orders of magnitude larger than the 2·5 : 1 expected, and moreover the *high* value is on the ridge where outcropping rocks may be common, not in the valley. The explanation cannot be applied at all to the measurements in the Explorer Troughs themselves: the local ponds would have effects smaller than the gross topographic effects of opposite sign. Therefore, while local refraction is likely to have some depressant effect on heat-flow measurements in rough topography, it cannot explain the gross contrasts observed in this study.

The next question that arises is whether even the highest values are high enough to be consistent with a conductively cooled sea-floor spreading model fed by material at observed lava temperatures (1400 °C). The simplest spreading model has been analytically solved by McKenzie (1967), but an even simpler model is adequate for use here: cooling of a half-space from uniform initial temperature, neglecting lateral conduction entirely. The solution is given in Carslaw & Jaeger (1959, p. 59) and the surface gradient is $T_0/\sqrt{(\pi\kappa t)}$, where T_0 is the initial temperature, referred to surface zero, κ is the diffusivity of the rock material, and t is time. The time axis can be translated into spreading distance, and although the solution is clearly inapplicable close to the ridge crest, where lateral heat flow is important, the approximation is good once the spreading distance substantially exceeds the penetration depth of conductive cooling $2\sqrt{(\kappa t)}$, at which depth $T = 0{\cdot}84T_0$. The Juan de Fuca spreading velocity is 3 cm yr^{-1} (Vine 1966) and if a diffusivity of 0·007 cm^2 s^{-1} may be assumed

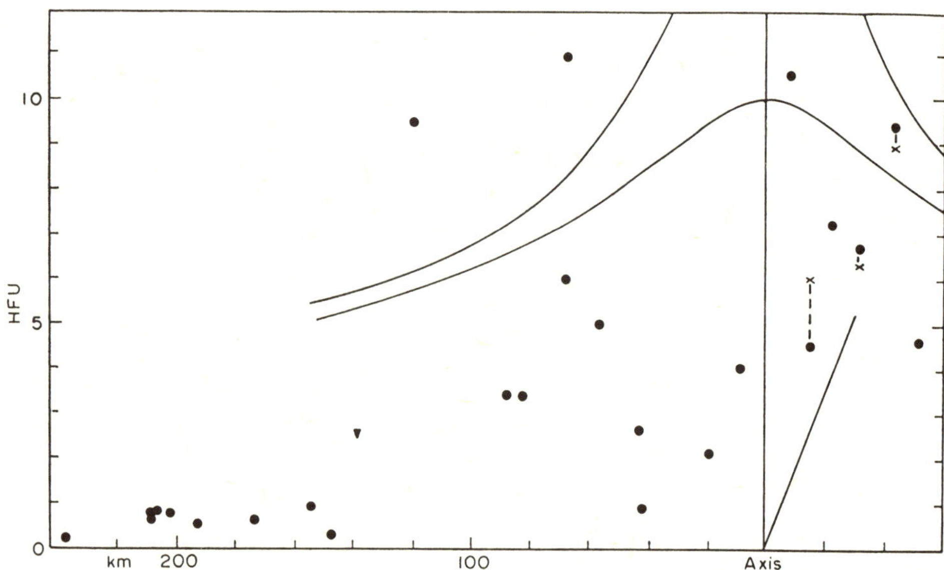

FIG. 10. Heat flow in HFU, versus distance from the crest of the Juan de Fuca Ridge. Filled circles: raw measurements from this study; inverted triangle: measurement from previous study projected onto profile; crosses: values corrected for sedimentation and refraction. The angled line from the crest origin shows the angle of the ridge crest relative to the profile. The solid curves are half-space conductive solutions for constant initial temperature (upper) and an initial linear gradient to magma temperatures. Spreading times have been corrected for the angle of the profile to the spreading direction.

for the rocks, the spreading distance is three times the penetration depth at a distance of 30 km from the ridge crest. The reason this is considered adequate is apparent from Fig. 10, where the upper curve is the heat flow predicted by the cooling model: it comes on scale at 30 km. All the heat-flow values of the ridge crest study are also plotted in Fig. 10 by their distance from the ridge crest along the 47th parallel. Only the highest values west of the ridge crest are above the conductive cooling curve; the mean in the high heat-flow zone, at 6·4, is not too different from the mean of the cooling curve in that region, 7·5. However, none of the measured values less than 50 km from the crest reach the curve, and the discrepancy increases as the ridge crest is approached.

The lower curve in Fig. 10 has been obtained from another simplistic model. In this case it is assumed that a linear gradient corresponding to a surface heat flow of 10 HFU is somehow established by cooling of the rocks as they are emplaced. With the figures used ($K = 0{\cdot}007\ \mathrm{cal\,cm^{-1}\,s^{-1}\,{}^\circ C^{-1}}$, $T_0 = 1400\,{}^\circ\mathrm{C}$), the penetration of this cooling is 10 km. The model was chosen because it gives a monotonically decreasing surface heat flow and a simple form for it: $q_0 \operatorname{erf} d/2\sqrt{(\kappa t)}$, where q_0 is the initial heat flow and d is the penetration depth. The solution was obtained from Carslaw & Jaeger (1959, p. 59), and it again ignores lateral conduction, but in this case it is a good approximation throughout. It is apparent from Fig. 10 that this model cannot be distinguished from purely conductive cooling in the western flank area of high heat flow, but it comes much closer to agreeing with the values on the eastern flank. The negligible effect of drastically different crestal conditions on the flank heat flow is important: not only do flank measurements give no insight into dyke injection or other ridge spreading mechanisms, but the explanation for depressed flank heat-flow values must be sought in locally operating mechanisms of non-conductive heat loss. A comparison between the theoretical and measured heat flow at 3 My on this ridge flank, 7 HFU, and the normalized means for the East Pacific Rise, 2·6 HFU, and the Mid-Atlantic Ridge, 1·9 HFU (Le Pichon & Langseth 1969) shows that there is something to be explained.

At some distance from the ridge crest in any convecting model the upper boundary reaches a state that corresponds closely to a lithospheric layer cooling slowly by conduction. The total heat output between the crest and such a point on the flank must be equal to the heat lost by the lithospheric material in cooling from its initial state. If the temperature distribution at this flank point is similar to that of a half-space cooling from an initial uniform temperature, the total heat output can be obtained simply by integrating the surface heat flux of the one-dimensional model through time. This is only a fair approximation for the more imaginative convective models, but, since the real situation is complicated by processes of partial melting, magma percolation and differentiation, a more elaborate calculation is hardly justified. The interestingly simple result is that the mean heat flow between the ridge axis and a point on the flank is just twice the heat flow at that point computed from the one-dimensional cooling model. For example, in Fig. 10 the mean heat for flow the 100 km (full-width) central zone is nearly 16 HFU, and, so long as volcanic activity is confined to this zone, and cooling is conductive, the result is not affected by the detailed spreading model. In fact, the real heat loss should be substantially greater than this, since a proportion of the uppermost 10 km of material comes up molten, and the latent heat of freezing must be added to the integrated specific heat. The addition could be considerable: the latent heat of melting for dry Forsterite can be estimated from the Clausius–Clapeyron equation and data in Clark (1966); it is 800 cal cm^{-3}.

The thermal balance of the ridge crest zone is thus incompatible with the conductive cooling of material upwelling at a reasonable magmatizing temperature, and the observed level of the heat-flow measurements. The presence of high values in rough agreement with the conductively expected heat flow on the western flank suggests that the discrepancy is caused by failure of the assumption of conductive cooling at the crest rather than a more general failure in the sea-floor spreading hypothesis. It is then not necessary to postulate unusual local heat sources for an area of oceanic

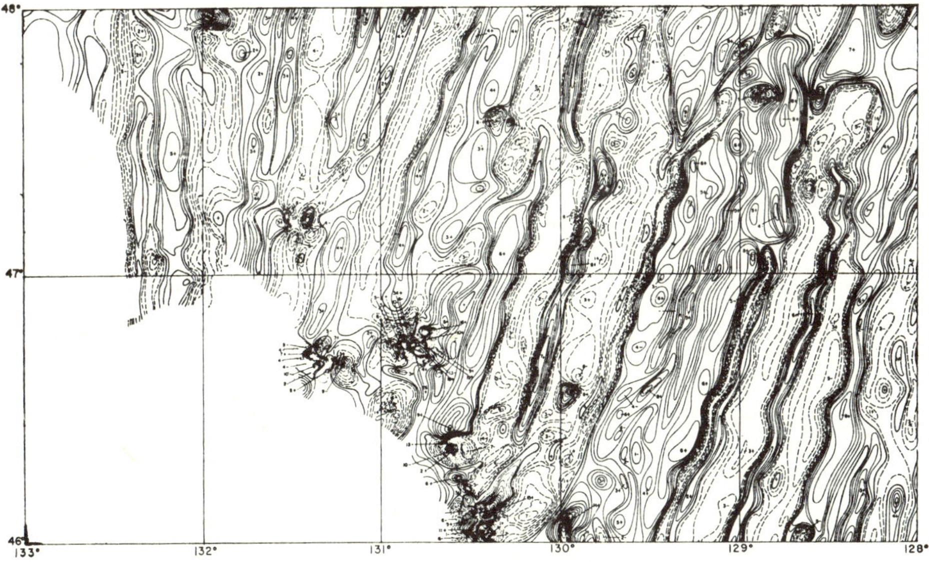

Fig. 11. A portion of the Raff & Mason (1961) magnetic survey, with an addition near 47° N 132° W, reproduced to cover the same area as Fig. 3. Positive anomalies are contoured as full lines, negative as dashed lines; contoui interval 100γ.

crust that is a classic example of undisturbed sedimentation (Fig. 5). It may be argued that the abrupt decrease in heat flow beyond 131° W (Fig. 3) is not in agreement with this idea, but a consideration of the magnetic evidence makes it even more plausible. Fig. 11 is a reproduction of a portion of the Raff & Mason (1961) magnetic anomaly map corresponding to the area of this investigation (Fig. 3). The abrupt decrease in heat flow to more usual flank values occurs west of 131° W. This is the point at which the long anomaly trends begin to diverge from parallelism with the ridge crest: at 131° 30′ W the broad negative lineation is almost NS in strike, and the three positive lineations immediately east of it at the northern edge of the map have been compressed into a broad positive band at 47° N. The special conditions of ocean floor-spreading that produced the smooth basement of the high heat-flow zone cannot be expected to extend into the disturbed region probably indicative of the birth of the Juan de Fuca spreading epoch. Note that the well surveyed and sampled area near 47° N 132° W lies on true NS lineations of an earlier spreading era, so that the well established low heat flow could be due to a greater age of the lithosphere there.

6. Hydro-thermal convection

Once it is established that a substantial part of the surface heat loss at the ridge crest is non-conductive, no great imagination is needed to select the most likely mechanism: hydrothermal circulation in the oceanic crust. Iceland is a part of the Mid-Atlantic Ridge, and it is laced with hot springs and geysers. Other tectonically active regions of the world contain large-scale geothermal areas: New Zealand, Italy, California, to name a few. The mechanisms involved in these areas have been discussed at length by Elder (1965) and others; the important question is whether a hypothesis of hydrothermal circulation can help to explain other aspects of the ridge heat-flow observations beside the gross heat loss. The purpose of the following discussion is to show that it *does* explain both the valley effect in rough topography and the relatively uniform heat flow observed where basement topography has been buried by sediments.

If hydrothermal circulation is to occur, the convective driving forces must exceed the viscous drag of water percolation through the rock. Although both the thermal expansion and the fluidity of water will be decreased somewhat by the high ambient pore pressure beneath the ocean, the principal unknown is the permeability to be expected of the oceanic crustal rocks. Borehole tests in the New Zealand geothermal area (Elder 1965) suggest bulk permeabilities of 0·01 Darcy ($0{\cdot}987\times10^{-8}\,\mathrm{cm}^2$), but it is possible that the rarity of major geothermal areas is due more to lack of permeability than to lack of heat sources in other regions. If the oceanic basaltic layer is assumed to have a permeability k of 10^{-4} Darcy, an effective circulation depth h of 7 km, a diffusivity κ_m of $7\times10^{-3}\,\mathrm{cm}^2\,\mathrm{s}^{-1}$, and a temperature difference ΔT of 1400 °C, while the percolating water has a thermal expansion coefficient α of 8×10^{-4}/°C and kinematic viscosity ν of $1{\cdot}5\times10^{-3}\,\mathrm{cm}^2\,\mathrm{s}^{-1}$, corresponding to water at 200° C (Elder 1965); then the Rayleigh stability number $\mathscr{A} = k\alpha g\Delta T h/\kappa_m\nu$ is approximately 80. It has been shown (Lapwood 1948) that convection can occur for Rayleigh numbers greater than 40 if the pattern is forced by uneven heating from below. Very conservative values have been assumed for the permeability and for the expansion and fluidity of water, since the latter increase with temperature in the range (<300 °C) where measurements have been made (Elder 1965). Therefore convective overturn of the pore water in the crustal layer near a ridge crest is highly likely, since the requirement of uneven heating is almost certainly met.

On the flanks of a ridge, both the thermal drive and the forcing effect of uneven heating must decrease as the lithosphere becomes older, and eventually convection will cease. The situation is complicated by the unknown effect of weathering and hydrous alteration of the rock on its permeability, and the natural variability in the large-scale structure of the crust. The case where convection ceases as soon as cooling has penetrated to 10 km has been treated above, and the lower heat-flow curve of Fig. 10 describes the subsequent conductive cooling. The heat flow predicted by this curve is greater than the mean of most ridge flank values until the lithosphere is about 30 My old (Le Pichon & Langseth 1969), but although the *means* are thereafter in rough agreement, there is still more scatter in the measurements than can be expected from conductive refraction. Only in the older ocean basins does the heat flow reach the uniformity purely conductive flow should have (cf. Reitzel 1963). This is, perhaps, to be expected: low permeability regions will cease to convect while comparatively young, but where the permeability is high, say 0·01 Darcy as in geothermal areas, the flow can continue until the thermal drive drops below 140 °C over 10 km, a gradient corresponding to less than normal basin heat flow in the absence of convection.

At the ridge crest the hydrothermal circulation must be open, since an essentially infinite reservoir of fluid abuts the free surface of the crest. The free boundary should slightly reduce the Rayleigh number required for the onset of flow, but its most

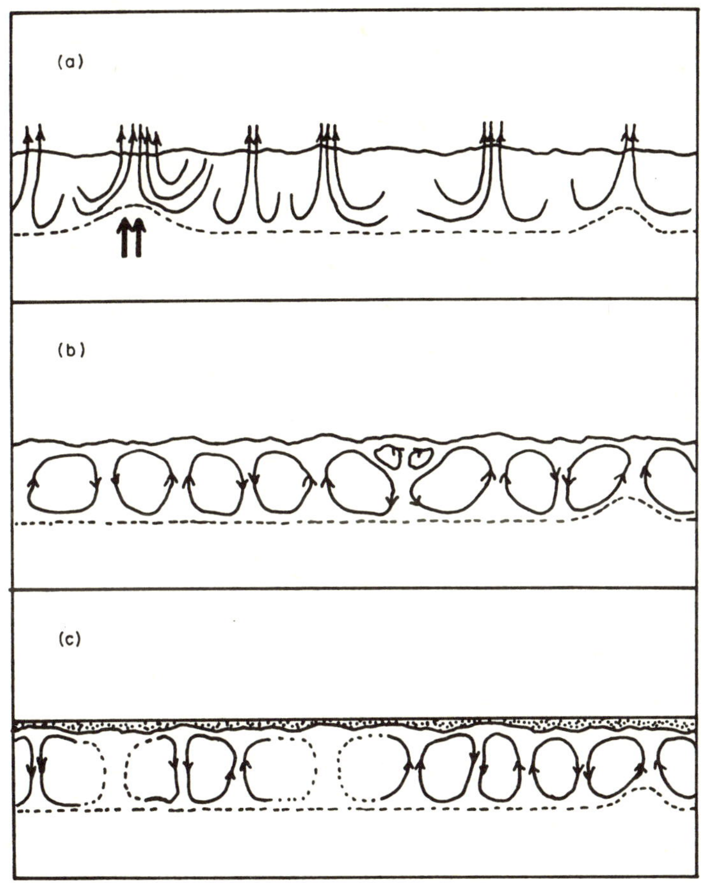

FIG. 12. (a) Pattern of convection to be expected near a ridge crest, where the permeable layer is open to the ocean. (b) Convective pattern forced by topography in an area of permeable crust bounded by a thin impermeable blanket of sediment. (c) The case of permeable crust with a rough boundary, buried by flat-lying sediments. The varying thermal resistance of the sediment blanket overpowers the direct effect of the topography.

important effect is on the thermal gradient at the boundary. Elder (1965) presents the results of one experiment on convection in a permeable layer heated along an infinite strip of width $2l$ below it. The heat output of the system is sharply concentrated over a band of width $0{\cdot}8l$, outside of which there is almost no heat output at all. The importance of this can be gauged when the effect of topography on the convective pattern is considered. Ridges and seamounts in the permeable material will act as chimneys for the hot flow, while valleys, representing the deepest penetration of cold bulk seawater, are natural sites for the ingoing return flow. The type of circulation to be expected is shown in Fig. 12(a) for two-dimensional topography, and in a perfect, or uniform percolation, system, zero heat flow should observed in the down-draught zones.

The plausibility of a circulation system for removing the initial heat of the fresh crust can be tested by a basic thermal output calculation. Suppose that at a ridge spreading at 3 cm yr^{-1} (one side), the top 10 km of lithosphere is cooled an average of 700° C, releasing 700 cal cm^{-3}. The continuous thermal output of 100 m of crest is $1{\cdot}4 \times 10^6$ cal s^{-1} (or 6 MW), and if the circulation mechanism results in a hot spring at about 50 °C, it must run at 28 l. s^{-1}, equivalent to a 20-cm diameter pipe discharging at 1 m s^{-1}. This would be a large spring, but by no means exceptional, and it could surface anywhere over the central 1km of the ridge crest. If the heat were dispersed into a cross-ridge flow of 1 cm s^{-1}, and 100 m thick, the temperature rise of the ocean water would be about 0·014 °C. Since the thermal gradient in the near-bottom water observed on lowering the heat probe is commonly 0·01 °C/30 m, it would not be easy to detect the heating in the topographically disturbed flow near a ridge crest. Only a careful areal study of absolute temperature at a fixed depth, on the lines of Chung *et al.* (1969), would be able to test for the overall thermal output of the ridge crest.

The second test of the hydrothermal circulation hypothesis is the distribution of measured heat flow over topography bounded by an impermeable layer, such as the high heat-flow region on the western flank of the Juan de Fuca Ridge. There are two reasons why this area can be considered bounded by an impermeable layer of sediment: the mean heat flow is close to what it should be for conductive cooling (Fig. 10), and the pelagic sediment, besides being of low inherent permeability (Appendix), can can readily plug the discrete vents of the hard rock structure at the interface. If the heat flow in the region were purely conductive, the even sediment cover over gently undulating basement (Fig. 5) would require the measurements to have a low scatter, and to be higher in the valleys and on the flanks of the ridges or hills than on the crests. Although the sample in Fig. 3 is a small one, the tendency is clear: the highest values are on top of hills or ridges, and lower values are observed on the flanks or in valleys. The high value of 10·9 HFU at 130°06′W, discussed at length in Lister (1970a), is a striking case.

It is situated on the brow of a hill, a place where conductive refraction would generate a minimum of surface heat flux, yet the measurement is nearly twice as high as one made nearby in a small valley or hollow. The values measured near Explorer Trough (Fig. 9) are an even more telling example of heat-flow correlation with topography opposite to that expected from conductive refraction. As the area is complicated by an apparent split in the spreading centre, and thus has an ill-defined recent history, it would be premature to attempt a detailed interpretation. The effect of an undulating impermeable boundary on hydrothemal convection in a permeable layer is illustrated diagrammatically in Fig 12(b). Valleys or surface depressions tend to depress the isotherms beneath them and therefore are places of preferential separation for the cold columns. By elimination, hills and ridges will become the sites for the hot rising columns, and should thus show the highest conductive heat flow through the impermeable sediment blanket. The forcing effect of the topography can be visualised by considering the stability of a cooled boundary layer on an inclined plane: the denser boundary layer will tend to slide down the plane until separation occurs. The ability of the surface topography to control the cell geometry will depend on the ratio between the inverse wavenumber of the most pronounced topography and the cell size, and the relative effect of variations in the lower boundary of the permeable layer. If the permeability decreases gradually with depth, only large variations in the height of the lower boundary will affect the convection pattern, but such large variations could occur. Variations in the permeability of the crustal material are also likely to distort the cells, so high heat flow should by no means be 100 per cent correlated with the tops of hills or ridges. The evidence certainly suggests a positive correlation between high heat flow and topographic elevations, as opposed to the negative correlation that would be expected from conductive refraction.

The third test of the convection hypothesis is the prediction of the heat-flow variation on a flat sediment plain overlying topography on the basement. The contrast in thermal conductivity between moderately compacted sediment and basaltic rock is about 1 : 2·5, so that, in the absence of convection, the isotherms in the basement will be distorted by $-1{\cdot}5$ times the topography, provided that the Jeffreys (1938) approximation is valid. The coldest areas of the upper boundary of the permeable material are now the highest topographically, and the sense of any convective flow should be reversed from its original configuration before the sediments were laid down (Fig. 12(c)). The geometrical effect aiding the motion and separation of a cold boundary layer is no longer operative, and the forcing effect of the buried topography is probably somewhat weaker than that of similar topography coated with an impermeable skin (see second test, above).

Now, the hottest water is in contact with the thickest sediment blanket, and the coolest water occurs where the sediment blanket is thinnest. Whether the highest heat flow will be observed over basement lows or highs depends on whether the temperature difference between the hot and cold columns of the cell is greater or less than the basal temperature difference between the sediment colums for a uniform heat flow equal to the regional average. In general, when the sediment thickness is much less than 0·4 times the convection penetration into the crust, and the Nusseldt number is not too high, the highest heat flows should be observed over troughs in the basement. If convection is vigorous, and the temperature difference across the cell is a small fraction of that required for equal conductive flow, it is possible for the heat flow to be quite uniform even over rough buried basement. Conversely, if the basement relief is too low to force the cell pattern, and the Nusseldt number is low, there will be 'hot spots' over the rising columns, uncorrelated with the topography.

The four Cascadia Plain stations in Fig. 7, and the one further east (Fig. 10), bear out the prediction in a general way. The mean heat flow for the region is in fair agreement with the lower cooling curve in Fig. 10: this corresponds to strong initial convection cooling the crest to considerable depth, followed by a resumption of conductive flow. Surface sealing and a reduction in the vigour of convection would produce similar results to true bulk conduction, and there are enough variable parameters to produce a range of models with reasonable mean heat flows. The chief discrepancy from prediction is the low value at the edge of the plain. It is within 3 km of outcropping basement, and it is not too implausible to suggest that the low value is due to forcing of an eddy under the plain by vigorous open cells near the ridge crest. The odd high (9·4 HFU) value over no particular basement disturbance corresponds well with the idea of an upwelling hot spot over a cell not correlated with topography.

7. Conclusions

The unusual heat-flow distribution across the Juan de Fuca Ridge at 47° N, and, to a lesser extent, the heat-flow distribution near Explorer Trough, can be interpreted to agree with a simple model of sea-floor spreading. The crucial requirement is that the rock material at the ridge crest be immediately cooled to a depth of 5–10 km by open hydrothermal circulation. If the permeable crustal layer is subsequently sealed off by sedimentation, then the measured heat flow increases to values compatible with conductive cooling models. On the western flank of the Juan de Fuca Ridge, an area of unusually smooth basement appears to have been adequately sealed off by hemipelagic sedimentation; on the eastern side the encroaching turbidities have completely buried the basement topography. In both these areas, the mean heat flow is near 7 HFU, and is consistent with either conductive flow after cooling by hydrothermal circulation, or continued circulation beneath an impermeable boundary. The *distribution* of the measured values over the topography is more consistent in all areas with continued hydrothermal circulation, forced in part by the topography, than with conductive refraction. The continuation of hydrothermal flow in inadequately sealed areas is also necessary to explain the low values commonly found on ridge flanks (cf. Le Pichon & Langseth 1969).

The hypothesis can be further tested by examining the relationship between measured heat flow and topography in areas where topological information is complete for large-scale features, and the distribution of sediments is known. Eventually, the ridge crest hot springs should be found (and possibly tapped for geothermal energy), but it has been shown that the hot water will be hard to detect in an open ocean with significant bottom current flow. The principal value of the data presented in this paper is to confirm the long-held suspicion that the thermal balance of a ridge crest is dominated by circulatory cooling rather than by conductive processes.

Acknowledgments

This paper summarizes the work of several years, supported by Grants GA–577, GA–1640 and GA–27947 of the National Science Foundation, and Contract Nonr–477(37), Project NR 083012 of the Office of Naval Research, both of the United States of America. The work off C.S.S. *Hudson* was possible through the generosity of the Department of Energy, Mines, and Resources of Canada, at the Bedford Institute of Oceanography, and through a travel grant made by the Department of Oceanography, University of Washington. I wish to thank Dr R. L. Chase, of the University of British Columbia, for making available one of his seismic profiles, Dr A. W. Fairhall, of the University of Washington, for running carbon–14 dates on critical core samples, and Mr E. E. Davis for his invaluable assistance, both on the *Hudson* and subsequently with the data reduction.

Geophysics Group and Department of Oceanography,
University of Washington, WB-10,
Seattle,
Washington 98105

References

Atwater, T., 1970. Implications of plate tectonics for the Cenozoic tectonic evolution of western North America, *Bull. geol. Soc. Am.*, **81,** 3513.

Bullard, E. C. & Day, A., 1961. The flow of heat through the floor of the Atlantic Ocean, *Geophys. J. R. astr. Soc.*, **4,** 282.

Carslaw, H. S. & Jaeger, J. C., 1959. *Conduction of Heat in Solids*, Oxford University Press.

Chung, Y., Bell, M. L., Sclater, J. G. & Corry, C., 1969. *Temperature data from the Pacific abyssal water*, Scripps Institute of Oceanography, University of Calfornia, La Jolla, Data Report, **69,** 17.

Clark, S. P. (Editor), 1966. *Handbook of Physical Constants*, Geological Society of America, New York.
Elder, J. W., 1965. Physical processes in geothermal areas, *Am. geophys. Un. Mon.*, **8**, 211.
Hamilton, E. L., 1964. Consolidation characteristics and related properties of sediments from experimental Mohole (Guadalupe site), *J. geophys. Res.*, **69**, 4257.
Jeffreys, H., 1938. The disturbance of the temperature gradient in the Earth's crust by inequalities of height, *Mon. Not.. R. astr. Soc., Geophys. Suppl.*, **4**, 309.
Lachenbruch, A. H., 1968. Rapid estimation of the topographic disturbance to superficial thermal gradients, *Rev. Geophys.*, **6**, 365.
Langseth, M. G., Le Pichon, X. & Ewing, M., 1966. Crustal structure of the mid-ocean ridges, 5, Heat flow through the Atlantic Ocean floor and convection currents, *J. geophys. Res.*, **71**, 5321.
Lapwood, E. R., 1948. Convection of a fluid in a porous medium, *Proc. Camb. phil. Soc.*, **44**, 508.
Lee, W. H. K. & Uyeda, S., 1965. Review of heat-flow data, in *Terrestrial Heat Flow*, Geophysical Monograph, **8**, 87.
Le Pichon, X. & Langseth, M. G., 1969. Heat flow from the mid-ocean ridges and sea-floor spreading, *Tectonophysics*, **8**, 319.
Lister, C. R. B., 1970a. Heat flow west of the Juan de Fuca Ridge, *J. geophys. Res.*, **75**, 2648.
Lister, C. R. B., 1970b. Measurement of *in situ* sediment conductivity by means of a Bullard-type probe, *Geophys. J. R. astr. Soc.*, **19**, 521.
Mammerickx, J. & Taylor, I. L., 1971. *Bathymetry of the Pioneer survey area, north of 45° Latitude, Special chart No.* 1, Scripps Institue, of Oceanography, University of California, La Jolla.
McKenzie, D. P., 1967. Some remarks on heat flow and gravity anomalies, *J. geophys. Res.*, **72**, 6261.
Peterson, M. N. A. *et al.*, 1970. Initial reports of the deep-sea drilling project, U.S. Government Printing Office, Washington, D.C., **2.**
Raff, A. D. & Mason, R. G., 1961. Magnetic survey off the west coast of North America, 40°N to 52°N latitude, *Bull. geol. Soc. Am.*, **72**, 1267.
Ratcliffe, E. H , 1960. The thermal conductivities of ocean sediments, *J. geophys. Res.*, **65**, 1535.
Reitzel, J. S., 1963. A region of uniform heat flow in the North Atlantic, *J. geophys. Res.*, **68**, 5191.
Sclater, J. G., Mudie, J. D. and Harrison, C. G. A., 1970. Detailed geophysical studies on the Hawaiian Arch near 24°25'N, 157°40°W: a closely spaced suite of heat-flow stations, *J. geophys. Res.*, **75**, 333.
Vine, F. J. & Matthews, D. H., 1963. Magnetic anomalies over oceanic ridges, *Nature*, **199**, 947.
Vine, F. J., 1966. Spreading of the ocean floor: new evidence, *Science*, **154**, 1405.
Von Herzen, R. P. & Uyeda, S., 1963. Heat flow through the eastern Pacific Ocean floor, *J. geophys. Res.*, **68**, 4219.

Appendix I

The permeability k of a representative sample of deep-sea sediment has been estimated from the transient loading curve of Mohole Phase I, sample EM8–10, published by Hamilton (1964). Axial loading was applied to a circular disk of sediment by a pair of porous plugs, and the pore fluid was squeezed out into these plugs. During the ' primary consolidation ' phase (9 min), the rate of shrinkage of the sample should be controlled by the permeability of the sediment, the permeability of the interface between sediment and plug, and the permeability of the plug itself. It will be assumed that the permeability of the plug is much higher than that of the sediment, that the interface resistance is negligible compared to the bulk sediment resistance after the first few seconds of loading, and also that the dimensional changes are small enough to be ignored in setting up the flow equations. Since the total primary consolidation in the case considered is 4 per cent, a simple constant-dimension treatment is adequate.

The actual equilibrium response of the sediment fabric to loading is non-linear (Fig. 13(a), points), and varies from sample to sample. The effect of this can be estimated by considering the transient response for two cases: the fully linear compaction OP, and the sudden compaction model OYP where the sediment fabric shows no strength until the final level of compaction is reached. The first case corresponds to a diffusion equation solution, while the second results in a sharp state transition boundary that moves away from the free surface at a steadily decreasing rate.

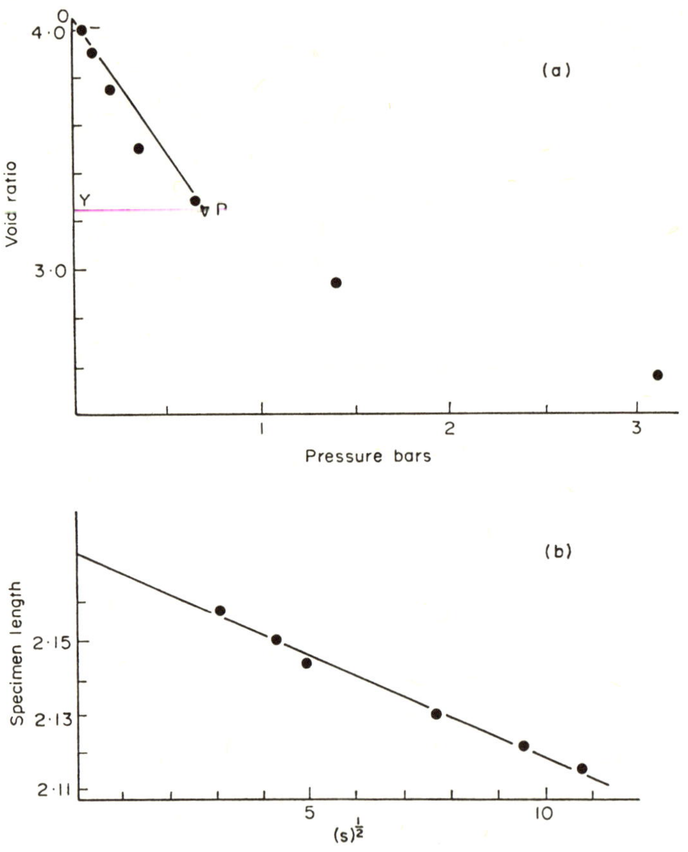

Fig. 13. (a) Steady-state compaction of Mohole sample EM–10 versus increasing pressure, after Hamilton (1964). (b) Transient primary compaction of the same sample with an applied pressure of 0·7 bars, plotted against square root of time. Replotted from Hamilton (1964, Fig. 9).

The linear model

If a compression state Θ is defined by $\Theta = (V_0 - V)/(V_0 - V_p) = (V_0 - V)/\Delta V$, where V_0 = initial volume, V_p = final volume, V is the actual volume, and ΔV is the final volume change, then the pore pressure at any point $P = P_0(1-\Theta)$, where P_0 is the loading pressure. Since it is the volume change that provides the flow, a one-dimensional diffusion equation $d\Theta/dt = [kP_0/\rho v\Delta V]\,d^2\Theta/dx^2$ results for the axial loading case, where x is axial distance, ρ is density and v is kinematic viscosity of the pore fluid. If the sample is long enough, the initial compaction will follow the solution for the semi-infinite solid $1-\Theta = \text{erf}\,[x/2(\kappa' t)^{\frac{1}{2}}]$, where $\kappa' = kP_0/\rho v\Delta V$, *at both ends*, so that $dy/dt = [2(d\Theta/dx)_{x=0}]\,P_0 k/\rho v$ and the displacement

$$y = -(4k\,P_0 t^{\frac{1}{2}})/(\rho v\sqrt{(\pi\kappa')}) = -(4/\sqrt{\pi})\,[kP_0\Delta V/\rho v]^{\frac{1}{2}}\,t^{\frac{1}{2}}$$

The moving transition model

In this model the compacted boundary layer has a linear pressure gradient across it P_0/x, and the velocity of the sharp transition is given by $dx/dt = (k/\rho v)\,(P_0/x)\,(1/\Delta V)$ since the flow is provided by collapse of more material. The single boundary solution for this is $x = \sqrt{((2kP_0 t)/(\rho v\Delta V))}$ and therefore $y = 2\Delta V x = 2[\sqrt{((2kP_0\Delta V)/\rho v)}]\,t^{\frac{1}{2}}$.

It is interesting that the two models both give a $y \propto t^{\frac{1}{2}}$ curve for the transient displacement in the compaction machine, and the value of k that would be calculated for the extreme non-linear model is $2/\pi$ of that for the linear model. The actual compaction data have been picked off Fig. 9 in Hamilton (1964), and are plotted in Fig. 13(b). The points lie close to a $y/t^{\frac{1}{2}}$ compaction line of 0·0054 cm s$^{-\frac{1}{2}}$, and using the values $P_0 = 0{\cdot}7$ bars, $\Delta V = 0{\cdot}042$, $\rho v = 0{\cdot}01$ poise, the values of k that result are

$$k_{\text{linear}} = 1{\cdot}95\times10^{-12}\ \text{cm}^2$$

$$k_{\text{transition}} = 1{\cdot}24\times10^{-12}\ \text{cm}^2.$$

The points in Fig. 13(a) represent *equilibrium* rather than *primary* compaction, but there is no reason to suspect that the primary compaction response would be more non-linear than the equilibrium response. Since the k calculated for the extreme OYP path differs little from k_{linear}, it is safe to take the latter as a good approximation to the steady-state permeability of the sample, some 2×10^{-12} cm^2 or 2×10^{-4} Darcy. Note that this value is for a sediment sample consisting of 60 per cent water by weight: closer packed material may have up to 10 times less permeability.

6

Reprinted from *Jour. Geophys. Research* **82**:3391-3409 (1977)

The Mechanisms of Heat Transfer Through the Floor of the Indian Ocean

ROGER N. ANDERSON AND MARCUS G. LANGSETH

Lamont-Doherty Geological Observatory, Columbia University, Palisades, New York 10964

JOHN G. SCLATER

Department of Earth and Planetary Sciences, Massachusetts Institute of Technology, Cambridge, Massachusetts 02139

We present 206 new heat flow measurements in the Indian Ocean. These and approximately 300 previously published heat flow values are individually evaluated for sedimentary environment and instrumental performance. The relationship between average heat flow and age is found to be little affected by selection of the most reliable experiments, although the scatter about the mean is significantly lowered. The variation of mean heat flow with age is found to be very similar to that in the eastern Pacific and Atlantic oceans: There is a crestal low heat flow zone with large variability, a transition zone within which the heat flow increases from values considerably below to values in agreement with predictions from thermal models of the oceanic lithosphere, and a region where heat flow values are in accord with theoretical predictions. However, the transition zone occurs over different crustal ages from ocean to ocean: 40–60 m.y. in the Indian Ocean, 4–6 m.y. in the Galápagos spreading center, 10–15 m.y. on the East Pacific Rise, and 50–70 m.y. on the Mid-Atlantic Ridge. The transition zone generally corresponds to a sea floor age where (1) sedimentary thickness increases to ≥300 m, (2) sea floor roughness is significantly smoothed by sediment blanketing, and (3) the carbonate content of surface sediments decreases to ≤40%. The transition zone occurs where water circulation in the oceanic crust stops affecting the surface heat flow strongly. There are two possible explanations for the transition. First, a change in composition from carbonate to siliceous sediments results in a decrease in bulk permeability. This combined with general thickening of the sedimentary blanket with aging results in the deposition of an impermeable layer which prevents the convective exchange of heat from the oceanic crust to the ocean. Second, hydrothermal flow within the oceanic crust is plugged by filling of circulation cracks in the oceanic crust. The fact that in several basins of the Indian Ocean the heat flow transition corresponds with the carbonate-siliceous boundary is support for the former mechanism. However, the fact that locations of increases in velocity of seismic layer 2A generally correspond to the transition regions in the Atlantic and Pacific oceans provides support for the latter mechanism. Heat flow measurements in the world's oceans allow us to calculate the variations of bulk permeability and basal temperature in the oceanic crust as a function of age and to evaluate the geochemical implications of the variation in these parameters between oceans. The combination of conductive heat flow and elevation versus age observations in old lithosphere demonstrates the deviation from $t^{1/2}$ cooling in the Indian Ocean and indicates that the Mozambique and western Somali basins are considerably older than preliminary deep-sea drilling results suggest.

INTRODUCTION

Modeling the oceanic lithosphere as a plate that is conductively cooling by loss of heat through its surface successfully explains many geophysical features of the ocean floor [*McKenzie*, 1967; *Sclater and Francheteau*, 1970]. Specifically, the variation in elevation of the sea floor away from midocean ridges [*Sclater et al.*, 1971] closely follows predictions from models which consider simple density increases caused by thermal contraction within an isostatically compensated slab. This observation suggests that thermal models of the oceanic lithosphere closely predict the total heat content of the plates. When in fact we compare measurements of surface heat flow, we observe a considerably lower heat flux than is predicted by the cooling slab models near midocean ridges. Heat flow observations near the ridge are complicated by large scatter. Frequently, the standard deviation is of the same order as the mean of observed values. This scatter is partially caused by topography, refraction of heat flow at sloping interfaces, and sedimentary processes [e.g., *Von Herzen and Uyeda*, 1963]. However, these surficial disturbances in the conductive heat transfer at the sea floor have proved too small to account for the magnitude of the scatter in the observations [e.g., *Langseth and Von Herzen*, 1970]. Exclusion of data based upon a subjective evaluation of the success of each heat flow measurement can significantly lower the scatter in any given region. However, a large component of the scatter near midocean ridges is real; that is, the conductive heat flow does indeed vary by up to an order of magnitude for closely spaced measurements. The only process capable of explaining such variation is thermally driven seawater convection in the oceanic crust and sediment. *Talwani et al.* [1971], *Lister* [1970, 1972, 1974], *Hyndman and Rankin* [1972], *Anderson* [1972], *Sclater and Klitgord* [1973], *Sclater et al.* [1974], and *Anderson and Hobart* [1976] have presented data which suggest that the scatter in heat flow measurements on a midocean ridge is due to widespread circulation of seawater into the oceanic crust. *Williams et al.* [1974] conducted a closely spaced survey on the Galápagos spreading center and mapped a large-amplitude systematic variation in conductive heat transfer physically and geologically explainable only by convection cells in the oceanic crust. In addition, *Hyndman et al.* [1976] measured low heat flow in deep-sea drill holes on the Mid-Atlantic Ridge and concluded that cellular convection was the dominant mechanism of heat transfer. Since most heat flow stations have been made at spacings of ~10 times the thickness of the oceanic crust, they do not resolve the spatial distribution of heat flow due to convection; instead, it is expressed as large scatter about the mean.

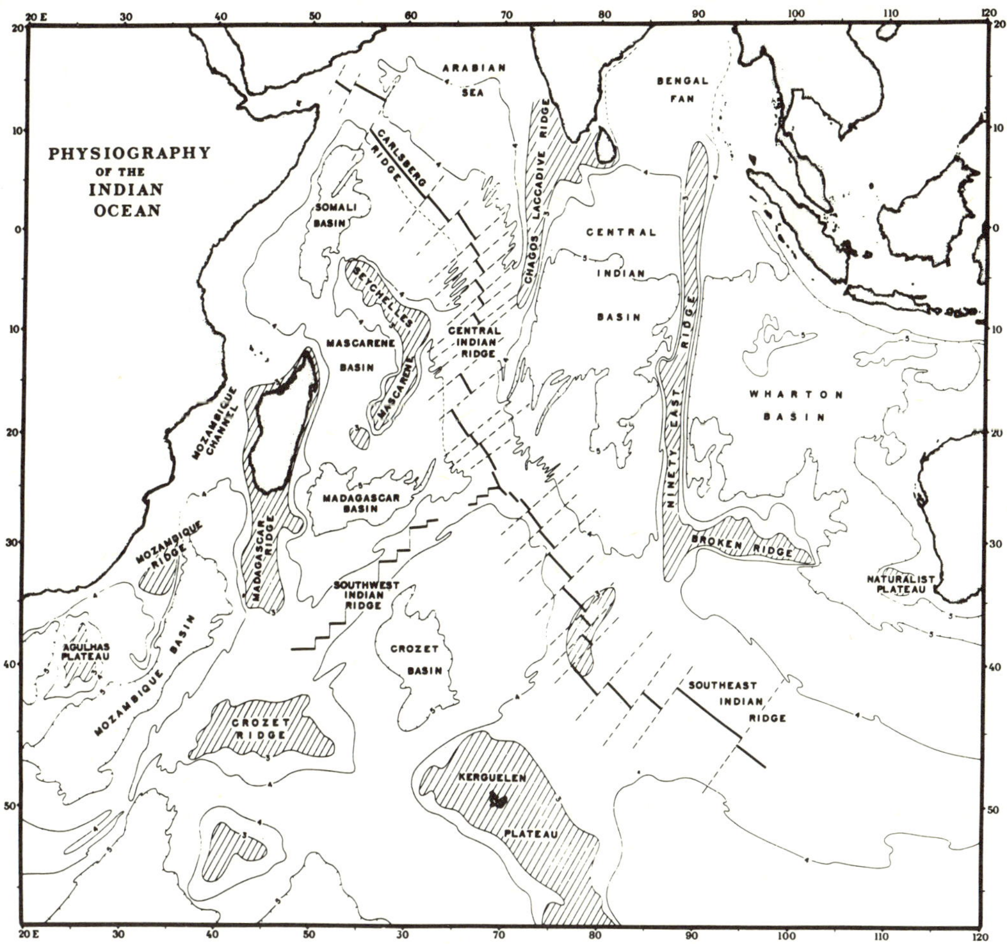

Fig. 1. Physiography of the Indian Ocean. The contours, in kilometers, are from a bathymetric chart published by the *National Academy of Sciences of the USSR* [1975]. Hachured regions are depths less than or equal to 3 km.

In addition to the scatter in heat flow attributable to convection, *Lister* [1972], *Williams et al.* [1974], *Sclater et al.* [1974], and *Anderson and Hobart* [1976] noted that since the mean of measured heat flow in oceanic ridge areas was significantly below that predicted by thermal models, convection must occur by the exchange of heat by seawater directly from the oceanic crust to the ocean. In this paper we analyze the more than 500 values in the Indian Ocean, principally in terms of their variation with age, and consider the implication of this variation on heat transfer mechanisms. Over the past year, studies by *Anderson and Hobart* [1976] and *Herman et al.* [1977] have revealed that on a large scale, the departure of observed mean heat flow from the mean heat flux versus age relation predicted by cooling lithosphere models is not random but follows a systematic pattern that is repeated in all major ridge systems. These studies suggest that the degree of seawater circulation within the oceanic crust changes as the lithosphere evolves and that the departure of the observed mean heat flow from the theoretically expected heat flux sets constraints on both the pattern of the seawater circulation and the resultant physical and chemical alteration of the oceanic crust.

HEAT FLOW AND THERMAL CONDUCTIVITY MEASUREMENTS IN THE INDIAN OCEAN

Figure 1 shows the main physiographic and tectonic features in the Indian Ocean, and Figure 2 shows the values at all reliable heat flow stations. Prior to this study, approximately 300 heat flow values in the Indian Ocean were published by *Von Herzen* [1963], *Von Herzen and Langseth* [1966], *Von Herzen and Vacquier* [1966], *Sclater* [1966], *Langseth and Taylor* [1967], *Vacquier and Taylor* [1966], *Popova et al.* [1972], *Hyndman et al.* [1974], and *Marshall and Erickson* [1975]. Equipment used for these measurements consisted of three types: outrigger probes strapped to the Ewing piston corer [*Gerard et al.*, 1962], the Bullard short probe consisting of probes within a 3-m lance [*Corry et al.*, 1968], and the Joides

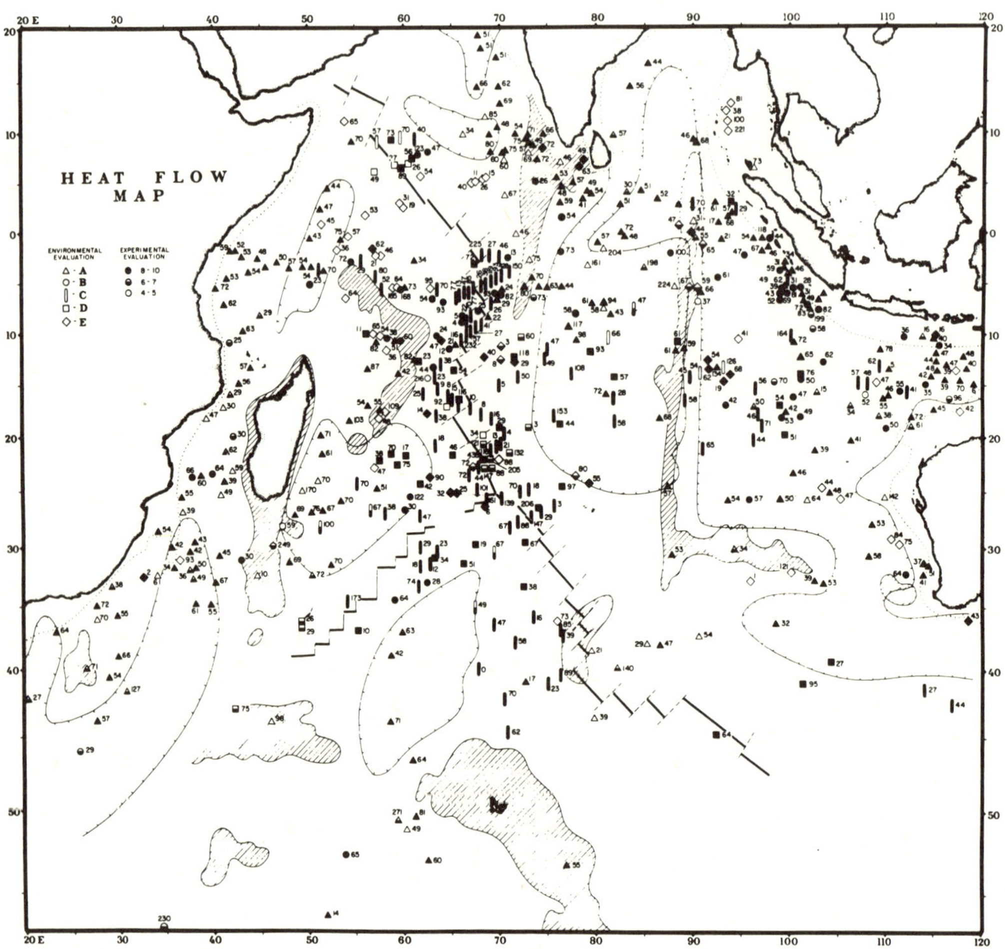

Fig. 2. Heat flow values (in milliwatts per square meter) in the Indian Ocean from the microfiche table [*Langseth and Taylor*, 1967; *Von Herzen and Langseth*, 1966; *Von Herzen and Vacquier*, 1966; *Vacquier and Taylor*, 1966; *Bookman et al.*, 1972; *Sclater*, 1966]. Siliceous sedimentary regions of Indian Ocean (<30% $CaCO_3$) from *National Academy of Sciences of the USSR* [1975] enclosed by hatched contour lines. Environmental evaluation as in Figure 3. Experimental evaluation as in text and microfiche table.

heat flow apparatus used in Deep-Sea Drilling Project (DSDP) holes. Thermal conductivity measurements were either made on deck by using a hypodermic needle [*Von Herzen and Maxwell*, 1959] or approximated from nearby stations.

The 206 new heat flow measurements in the Indian Ocean are listed in the microfiche table.[1] Of these measurements, 28 were made with the Bullard probe (Scripps Institution of Oceanography stations), and the rest were made with the Ewing thermograd (Lamont-Doherty Geological Observatory stations).

In order to evaluate the reliability of both new and old heat flow values, three independent studies of all stations were made.

Experimental performance. The experimental results of each station have been evaluated according to the subjective scheme of *Langseth and Taylor* [1967] (Figure 2, microfiche table). Each station is ranked on a scale of 0–10 (poor to excellent, respectively) according to the following criteria: verticality and stability of the probe in the sediment; number and depth of temperature measurements in the mud; whether the conductivity value was measured or assumed; and instrumental performance during the measuring interval. Measurements with the evaluations from 0 to 3 are considered too unreliable to report; those from 4 to 6 are stations with only one sediment temperature measurement or those with transient cooling disturbances and are possibly unreliable; those

[1] Supplementary table is available with entire article on microfiche. Order from American Geophysical Union, Suite 1000, 1909 K Street, N. W., Washington, D. C. 20006. Document J77-003; $1.00. Payment must accompany order.

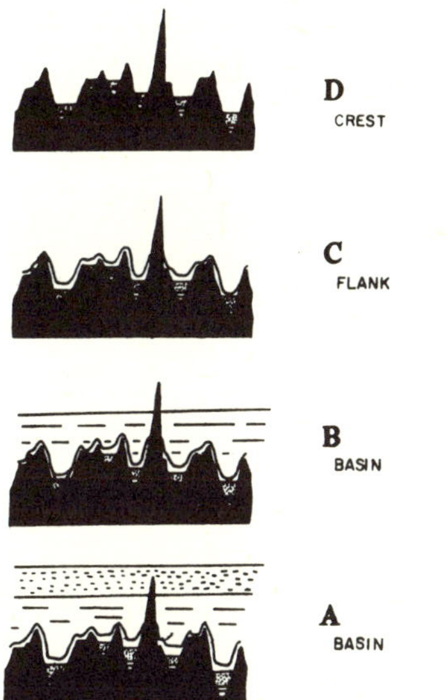

Fig. 3. Schematic diagram of the sedimentary environmental evaluation from *Sclater et al.* [1976*b*].

from 7 to 10 have two or more accurate sediment temperature measurements and are considered reliable measurements of heat flow through the sea floor (microfiche table).

Sedimentary environment. We have evaluated the local environment of each station by examining nearby seismic reflection and echo sounder profiles. The results are summarized in Figure 2 and the microfiche table. It has been shown [e.g., *Lister,* 1972] that the degree of scatter of heat flow values in any given region is inversely correlated with the degree of sediment cover over basement. Although no seismic reflection profiling accompanied most Scripps heat flow measurements, the sediment structure in the vicinity of those stations was estimated from the degree of smoothing of sea floor relief indicated on 12-kHz echo sounder records near each station. Each environment has been classified in the manner proposed by *Sclater et al.* [1974, 1976*a*] (see Figure 3).

The types of environment are as follows: A indicates a generally thick and uniform sediment cover draping all basement relief with no outcrops within 10 km of the station; B, thick sediment but with isolated basement outcrops within 10 km of the station; C, rough topography with a thin sediment cover usually less than 100 m over most basement relief; D, a station in rough terrain with isolated sediment ponds between outcropping basement highs; and E, no subbottom geophysical information available near the station.

Sclater et al. [1976*a*] argue that measurements classified A (above) consistently produce heat flow values representative of the total heat loss from the lithosphere with little scatter about the mean, whereas stations classified B–D show much greater variability because seawater circulation in the oceanic crust is more likely to occur in these locales.

Distribution of thermal conductivity. In addition to evaluating the experimental error associated with thermal conductivity K measurements at each station we have examined the regional trends in thermal conductivity to further understand the distribution and variation. The regional thermal conductivity trends are well enough established to allow contouring (Figure 4). Relatively high K measurements are found along the mid-Indian Ocean ridges except over the flanks of the narrow Southwest Indian Ridge, where low values are observed. The deep basins have generally low K as illustrated in the histograms in Figure 5. Near continental margins the coarser terrigenous component in the sediment results in higher thermal conductivity. This general decrease in K with sea floor age probably reflects the change of surface sediments from coarse-grained carbonates to finer-grained clays in the deep basins (as indicated in Figure 17, to be discussed later in the paper).

Contouring proved to be an excellent means of editing K values. A few measurements were inconsistent with the regional distribution (the circled dots in Figure 4). Heat flow values associated with these measurements were found to be consistently different from either nearby stations or regional heat flow means. In deep basins the heat flow values were recomputed for stations with anomalous K values by using the average K values in the region, and the resulting heat flow values were then compared with the original 'local K measurement' values (microfiche table).

The variation of K within each of the A–D sedimentary environment classifications was examined, and it was found that A environments have a nearly normal distribution with a large standard deviation. The C environments have a strong mode at ~0.85 W m^{-1} K^{-1} and a smaller mode of lower values at ~0.70 W m^{-1} K^{-1}. The D environment stations show a strong bimodal distribution. The values forming the higher mode for both C and D environments are found over the younger more elevated portions of the mid-Indian Ocean ridges. There is no correlation between thermal conductivity and heat flow values; however, high scatter in heat flow is associated with higher K values in the areas of C and D environments.

Distribution of heat flow. Each station has been evaluated as described above and shown in Figure 2. Physiographically, the Indian Ocean can be divided into a midocean ridge region and nine deep basins. Ridge segments and basins are sometimes separated by aseismic ridges (Figure 1). The heat flow coverage is generally good on the ridges and in all the deep-sea basins except for the Mascarene Basin. Coverage south of 40°S is sparsely scattered over a large region; however, coverage is sufficient north of this latitude to analyze the variation of heat flow in the Indian Ocean versus age.

HEAT FLOW VERSUS AGE

In order to construct a sea floor isochron map of the Indian Ocean we have compiled basement ages determined by Joides drilling [*Initial Reports of the Deep-Sea Drilling Project,* 1974–1975] and published magnetic anomaly identifications from *McKenzie and Sclater* [1971], *Sclater and Fisher* [1974], *Schlich* [1974], *Sclater et al.* [1976*b*], *Markl* [1974], and *Larson* [1977]. This compilation is summarized in Figure 6. There are many segments of the Indian Ocean for which we presently have no magnetic anomaly identification. In regions of no magnetic lineations, age gradients were determined from a study of the elevation decrease away from the mid-Indian Ocean ridges (see Figure 7). For example, in the ridge crestal province of the Central Indian Ridge the high density of fracture zones makes identification of magnetic anomalies difficult, and no magnetic anomalies have been identified beyond anomaly 5 (10 m.y.). Yet the elevation away from the crest

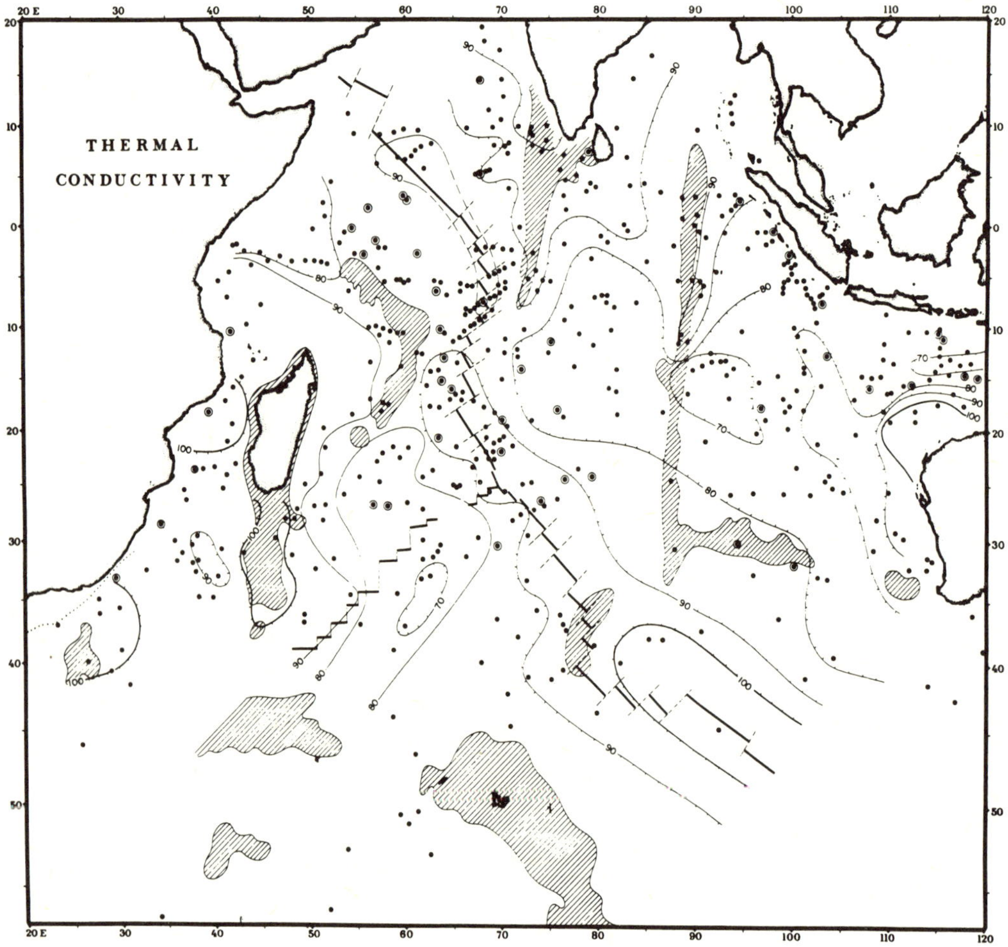

Fig. 4. Distribution of the thermal conductivity of sea floor sediments in the Indian Ocean shown by contours. Units are 10^{-2} W/m K. Circled dots indicate stations with values well outside the indicated contour interval.

shows uniform monotonic decay from the ridge crest to the Chagos-Laccadive and Seychelles ridges as shown in profiles B, C, and D (Figure 7), with the exception of an anomalously shallow region of sea floor that appears just to the east of the Mascarene Plateau (profiles C and D, Figure 7). However, Figure 2 shows that there are few heat flow stations between 15° and 20°S in this area. On the basis of the age-depth curves of *Sclater et al.* [1971] we identified ages on the topographic profiles of Figure 7. Note the excellent correspondence between deep basin ages predicted by the topography on the 8°N, 15°S, 20°S, and 30°S profiles and identified magnetic anomalies in these areas from Figure 6. Although the tectonic history of the Central Indian Ridge has been quite complex, this study is consistent with the interpretation presented by *McKenzie and Sclater* [1971]. Prior to 40 m.y. B.P. the midocean ridge generating the Arabian Sea and Somali Basin crust was separated from the Southeast Indian Ridge by a N-S transform fault. A relatively continuous sequence of spreading from 90 m.y. to present took place on the segment of the Southeast Indian Ridge between 65° and 125°E. At~40 m.y. B.P. a major plate reorganization began the current phase of opening of the Carlsberg, Central Indian, and Southeast Indian ridges and possibly the Southwest Indian Ridge. The

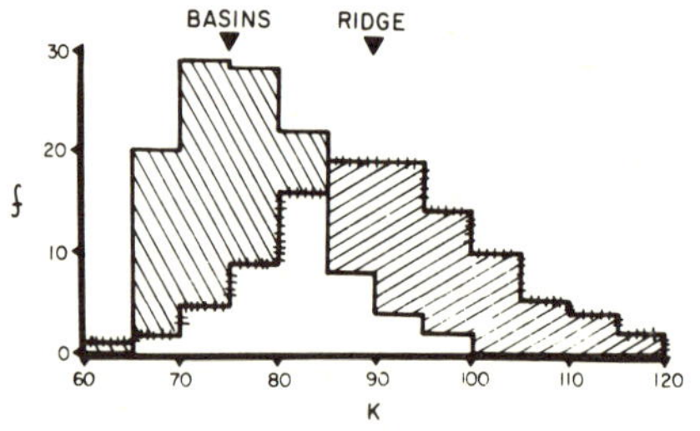

Fig. 5. Histogram of thermal conductivity values in the Indian Ocean grouped in basin or ridge province. White is overlap of two plots. Units are 10^{-2} W/m K.

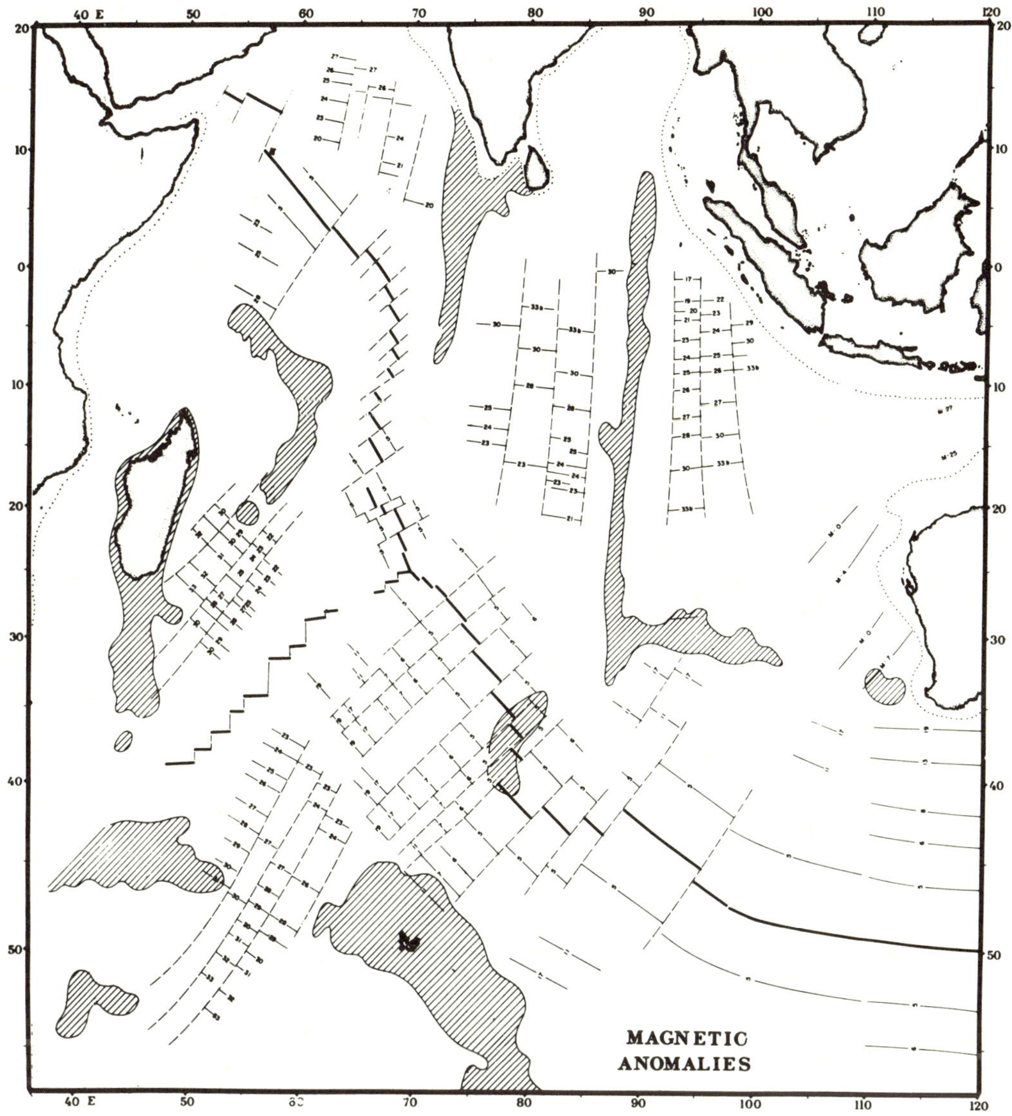

Fig. 6. Identifiable magnetic anomalies in Indian Ocean. For sources, see text. Hachures represent shoal regions of Figure 1.

elevation discontinuities, reflecting large age differences between abutting regions caused by this shift in spreading direction, are easily seen in Figure 7, profiles A, C, and F.

In basins such as the western Somali, Mozambique, and eastern Wharton, deep-sea drilling data can be used to identify ages. Below, we will show that age estimates based upon heat flow and depths in the Mozambique and eastern Somali basins may be older than those suggested by drill results. We then assume rigid plates to further extrapolate ages where other geophysical information is still lacking (Figure 8). Figure 8 shows the age distribution in the Indian Ocean based on the above sources. The details of many areas such as the southwest branch, basins south of Africa, and the eastern Wharton remain uncertain.

The age map of Figure 8 is the basis for the heat flow versus age plot (Figure 9*a*). Shown also on this plot is the range in heat flow versus age predictions from lithospheric thermal models. The *Parker and Oldenburg* [1973] model, which includes effects of latent heat, predicts higher heat flow at the ridge but lower heat flow in the old basins than the *Sclater and Francheteau* [1970] models. It is apparent that the ridge crest is a region of high scatter, but most of the values are equal to or less than predicted heat flow. There is a segment of sea floor in the Indian Ocean (from 40 to 60 m.y. old) where the observed heat flow actually increases with increasing age. For ages greater than 60 m.y. the thermal character of the Indian Ocean changes dramatically. In sea floor older than 60 m.y. B.P. the heat flow values vary about the predicted heat flow curves. To further investigate the relation between these two different provinces of heat flow (and the intervening transition zone) and convection in the oceanic crust, we have plotted only A environmental stations with experimental evaluations of 7–10 (Figure 9b). Since all but six of these stations are in sea floor older than 50 m.y., the values are differentiated by basin. In the crestal region we plotted all stations with experimental evaluations of 6–10 ('6' stations were included because these were frequently distinguished from '7' stations by nonlinear gradients, which should not be excluded in an area of suspected seawater circulation). Stations with C environments are distinguished from D environments within this crestal region (Figure 9b).

Heat flow distribution in sea floor older than 60 m.y. The scatter is significantly lowered by filtering out unreliable and unrepresentative stations in the old basins, but the mean is not changed significantly. This suggests that the scatter is normally distributed about the mean and results from surficial conductive geometries or random experimental errors rather than from convective interchange between crust and seawater. Since only the conductive heat flux is measured, lower heat flow than that predicted should be observed if convective interchange took place. Convection confined to the crust below the sediment is still possible. The point is further illustrated by the fact that both the A stations and the A–D (or all) stations in each of the Indian Ocean basins are

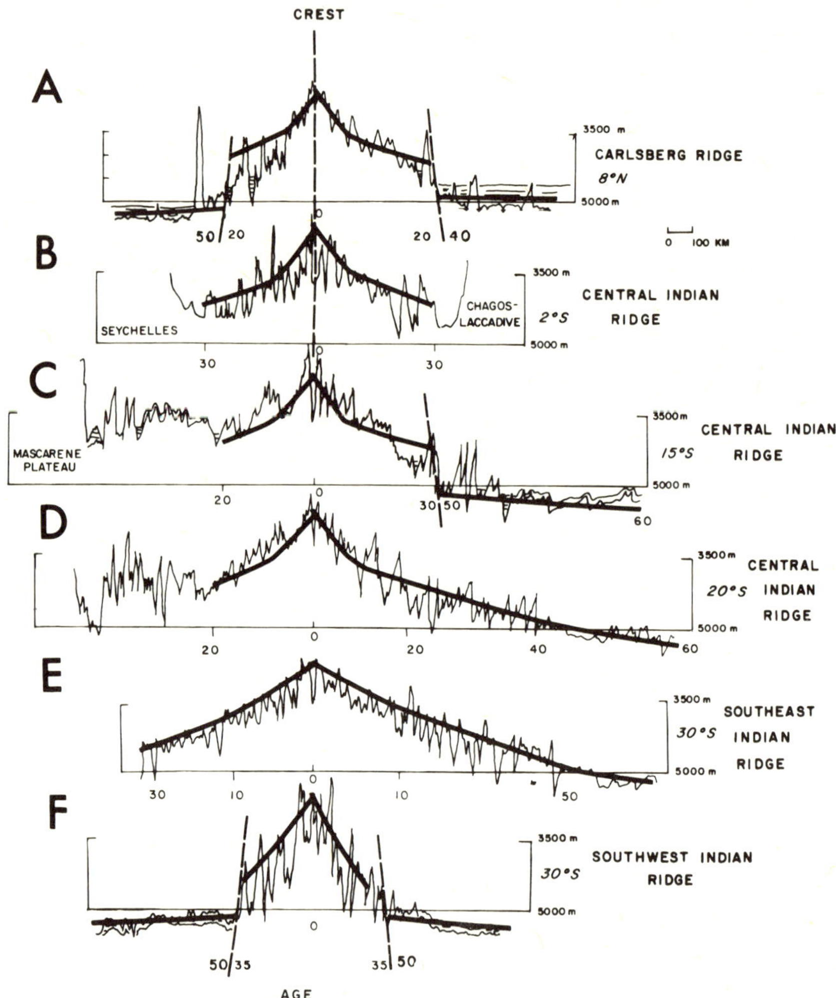

Fig. 7. Bathymetric profiles in corrected meters of Lamont-Doherty Geological Observatory tracks (Figure 16) projected perpendicular to ridge crest at 8°N, 2°S, 15°S, 20°S, and 30°S (point at which ridge crest crosses that latitude). Sediment thickness from seismic profiler records is shown layered. Solid lines are predicted depths for ages (in millions of years) indicated at base of each profile.

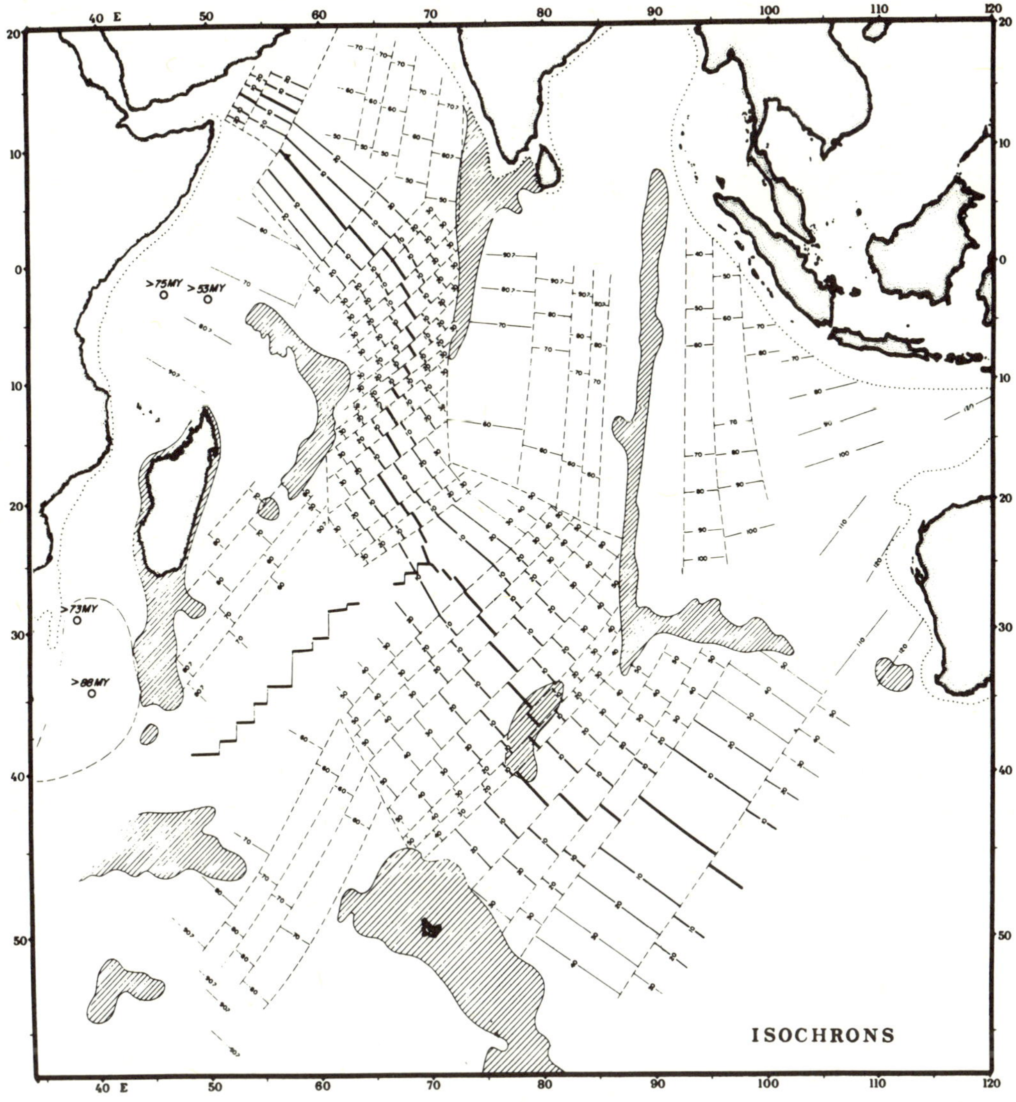

Fig. 8. Isochron map of the Indian Ocean, from Figures 6 and 7 and the assumption of rigid plates.

normally distributed about the mean, with the possible exception of only the Somali Basin (Figure 10). Thus in the old segments of the Indian Ocean, selection of A environments over C or even D stations does not improve the fit of the mean heat flow to the theoretical, and the average heat flow of all the basins in the Indian Ocean appears to be well constrained by the data. This is particularly true for the Argo Abyssal Plain and the Arabian, Somali, and Mozambique basins.

Heat flow in sea floor younger than 40 m.y. On the ridge crest the opposite appears the case. There, selection of C stations gives a mean that is different from that of either the D environments or the theoretical heat flow (Figure 9*b*). In the ridge crest the mean heat flow for all data is significantly lower than that predicted by the theoretical lithospheric plate models. This, combined with the fact that the mean comes into agreement with theoretical models in old basins, can only indicate that a major component of heat transfer is being lost in the 0- to 40-m.y. region of the Indian Ocean, which is not being measured by our current techniques. One of the interesting observations coming from Figure 9*b* is the apparent difference in heat flow between stations with C versus D environments. The C environmental measurements appear to yield consistently low heat flow values. This is somewhat puzzling because C environments are those with thin but continuous sedimentary cover draping all the basement, whereas D measurements were made in sedimentary ponds surrounded by rock outcrops. Most measurements were made in the thin sediment of the C environment with a short 2-m Bullard-type heat flow probe, whereas most D measurements were accompanying Ewing-type piston core stations. The short-probe measurements were made in much more variable terrains, since regions of thinner sedimentary cover could be sampled. Near ridge crests the deeper penetrating piston corer was used only in relatively large sedimentary ponds.

Thus in terms of geothermal observations the Indian Ocean sea floor can be divided into two large regions: (1) the convective region, the ridge crest and flanks where convection dominates; and (2) the conductive region, the basins where conduction is the dominant heat transfer mechanism. A transition zone separates the two, in which the mean observed heat flow actually increases with the age of the sea floor. The remainder of this paper will be concerned with three major problems: the constraints that heat flow in the conductive region places upon the evolution and age of Indian Ocean lithosphere, the mechanisms that cause the convective heat transfer to stop or be masked, and the mode of heat transfer in the convective region suggested by the heat flow observations.

ELEVATION AND HEAT FLOW VERSUS AGE IN CONDUCTIVE REGIONS OF THE INDIAN OCEAN

Current thermal models of oceanic lithosphere generate an adequate fit to the empirical elevation versus age relation for midocean ridges from the crest to ~80 m.y. [*Sclater et al.*, 1971]. The sea floor subsides in direct proportion to $t^{1/2}$ over this age span [*Parker and Oldenberg*, 1973; *Davis and Lister*, 1974], which results from the fact that during this period of its evolution the lithosphere cools as a homogenous half space. *Parsons and Sclater* [1977] have recently demonstrated that for ages greater than 80 m.y. in the Atlantic and Pacific oceans the elevation versus age curve departs from a simple linear relationship with $t^{1/2}$ and begins to flatten relative to the half-space model. There has been much discussion as to whether this flattening is significant or not because it has an important

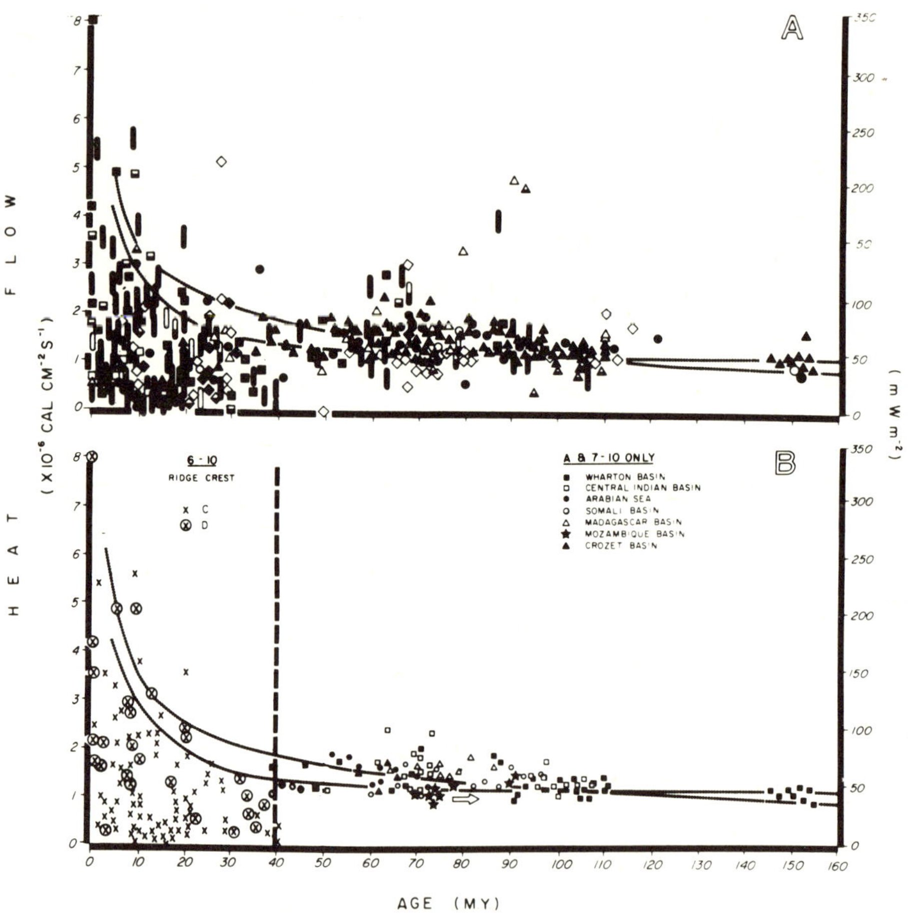

Fig. 9. Heat flow versus age in the Indian Ocean. (*a*) All the data. (*b*) Filtered data. Notice that the filtering removes much scatter but does not change the mean of values older than 50 m.y. B.P. The arrow refers to the Mozambique Basin values that appear to be from older sea floor than that indicated by DSDP hole ages in the basin. Solid curves are from *Sclater and Francheteau* [1970] and *Parker and Oldenburg* [1973].

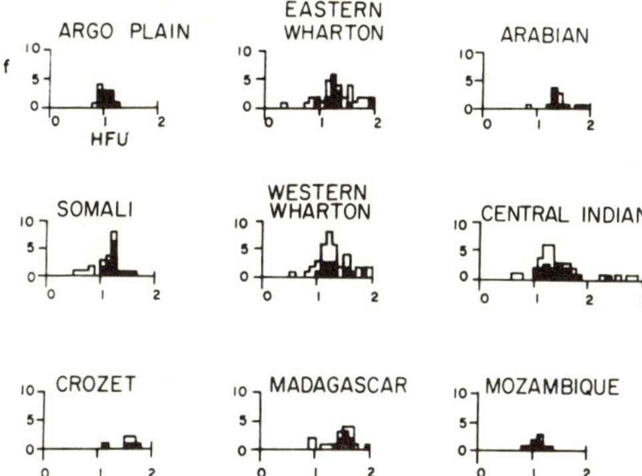

Fig. 10. Histograms of heat flow values from A sediment environment stations (solid squares) and all measurements (open and solid squares) in various basins in the Indian Ocean.

bearing on thermal processes occurring at the base of the lithosphere [e.g., *Forsyth,* 1977]. Flattening of the elevation versus $t^{1/2}$ curve would require the addition of heat from below the lithosphere which would begin to have an appreciable effect on the surface elevation at about 80 m.y. (50–100 km). Whether this heat comes from efficient upward transport of heat in the asthenosphere by small-scale convection [*Richter,* 1973], by radioactive sources buried deep in the mantle [*Forsyth,* 1977], or by stress heating at the base of the lithosphere [*Schubert et al.,* 1976] is not known.

Elevation versus age. Figure 11 shows an elevation versus age plot for the Indian Ocean compiled from bathymetry along tracks of Lamont-Doherty Geological Observatory ships (Figure 16) and the ages shown in Figure 8 (Table 1). The basement depths are adjusted for sediment loading by using the sediment isopach map (Figure 15). Superimposed on Figure 11 are elevation versus age data compiled by *Parsons and Sclater* [1977] for the North Pacific. A linear fit to data out to 80 m.y. is shown by the dashed line. It is clear that the elevation in all oceanic areas departs from the simple curve predicted by the half-space model beyond 80 m.y. Thermal models that assume an additional input of heat from below fit the data more closely (the curved line of Figure 11). It is also interesting to note that most of the elevation points in the Indian Ocean lie consistently shallower than those in Pacific Ocean floor of similar age.

The elevations of basement in the Mozambique Basin (open squares) lie well below the other best fitted points in the Indian Ocean. The ages used to make the plot were basalts cored in DSDP holes 248 and 250 in the basin (Figure 11); however, the elevation suggests that the age in this basin is considerably greater, about 130 m.y. This age is consistent with the Mozambique Basin being formed as the Falkland Plateau sheared away from the Mozambique Ridge and Agulhas Plateau. The Somali Basin basement elevations fit the curve in Figure 11 well with ages predicted in Figure 8 by assuming that it was formed contemporaneously with the northern Arabian Sea.

Heat flow versus age. If the scatter in the conductive portion of the Indian Ocean can be sufficiently reduced, perhaps heat flow can add to the understanding of the thermal evolution of old lithosphere. Specifically, does surface heat flow over the lithospheric plate continue to decay as predicted by a simple half-space model? (See Figure 11, bottom.) Although on the logarithmic scale the scatter is exaggerated, it is still too large to discriminate between models. By correcting the thermal conductivity of two locally anomalous measurements in the Argo Abyssal Plain of the eastern Wharton Basin (microfiche table) the scatter in heat flow is reduced to ~10% of the mean. The region is believed to be one of the oldest in the Indian Ocean. The Argo Abyssal Plain has a mean heat flow of 44.8 ± 4.2 mW m^{-2}; the age of the Argo Abyssal Plain is estimated to be 150 m.y.

Since the heat flow is determined from near-surface gradients, it takes much longer for heat flow at the base of the lithosphere or heat production deep in the mantle to affect the heat flow than it does for it to affect the elevation. The average values in each basin fit a $t^{-1/2}$ relation well. However, the cooling half-space model that best fits the elevation data (Figure 11, top) predicts a heat flow at 150 m.y. of 37.7 mW m^{-2} The Argo Abyssal Plain data lie significantly above this prediction.

Finally, notice that the reliable heat flow data from the western Somali and Mozambique basins would fit better with the trend of other basin averages if their age were much older (Figure 11*b*).

HEAT FLOW IN THE CONVECTIVE REGIONS OF THE MIDOCEAN RIDGE

The general pattern of the heat flow versus age distribution in the convective regions of the Indian Ocean is repeated on all other ridges so far examined. The analyses of heat flow versus age on ridge flanks in the eastern Pacific [*Anderson and Hobart,* 1976] and in the eastern Atlantic [*Herman et al.,* 1977] indicate that the mean heat flow and standard deviation are consistent and age dependent, suggesting that the effect of water convection on surface flux is age dependent. However, the age dependence varies from ridge to ridge.

Figure 12 compares the variation of mean heat flow with age for four midocean ridges with predicted heat flow from a thermal model of the lithosphere. The results from these areas, which were previously published analyses, are discussed below more fully.

1. The distribution of heat flow versus age on the Galápagos spreading center (GSC) illustrates the low zone on the crest and high heat flow on the flanks of that midocean ridge, but the transition occurs over 4- to 6-m.y.-old sea floor. The pattern of the curve is similar to that in the Indian Ocean, but all boundaries are compressed in time.

2. The detailed distribution of heat flow has been studied on the Costa Rica Rift on the GSC [*Anderson and Hobart,* 1976], where the low heat flow on the crest was found to qualitatively correlate with thin sediment. However, over the whole ridge, both heat flow and sedimentary thickness appear to vary uniformly with age.

3. The distribution of heat flow versus age on the East Pacific Rise from 20°N to 20°S again shows features similar to those on the GSC and in the Indian Ocean except that the transition is encountered at 10–15 m.y. as opposed to 4–6 m.y. on the GSC [*Anderson and Hobart,* 1976] and 40–60 m.y. in the Indian Ocean.

4. *Anderson and Hobart* [1976] examined the relation between heat flow and sedimentary thickness in the eastern Pacific and found that the ratio between sedimentary thickness and topographic relief is a more useful parameter to correlate with the transition from low heat flow to that which is in better agreement with theoretical prediction on the flanks of these two ridges.

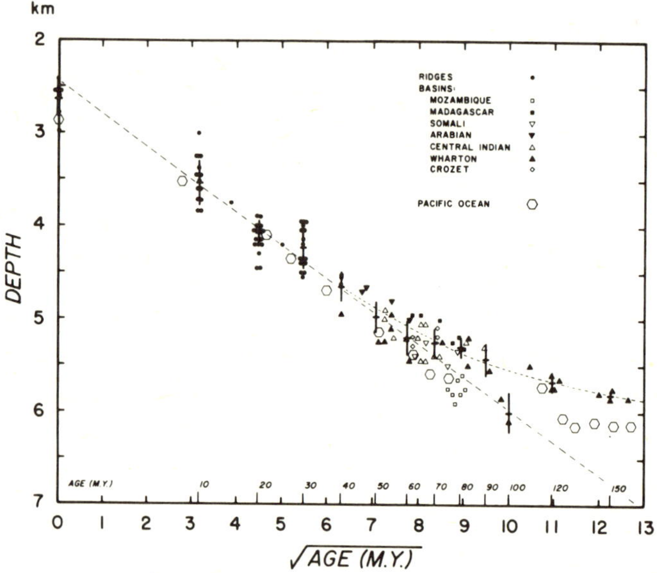

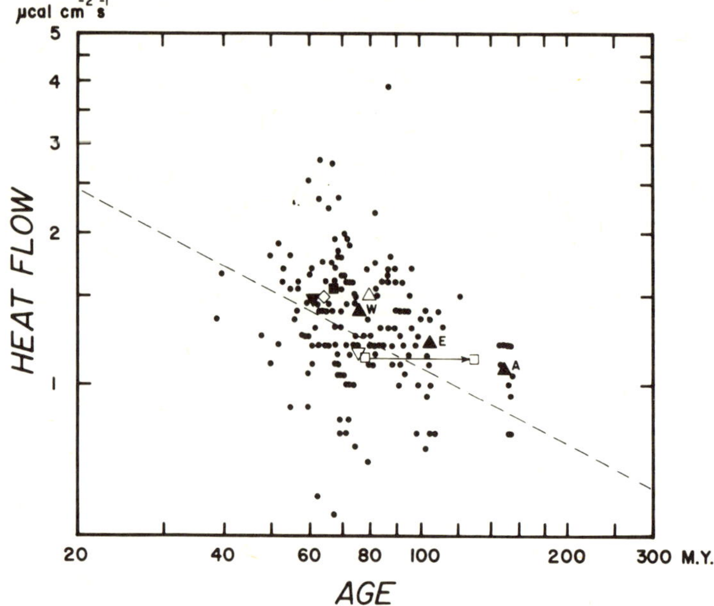

Fig. 11. (Top) Basement depth versus the square root of age in the Indian Ocean (see Table 2). The dashed straight line corresponds to a half-space cooling model D = 2420 m + 354 m · (age)$^{1/2}$. The curved dashed line is a slab model with the same thermal parameters and a thickness of 100 km. Pacific Ocean points are from *Parsons and Sclater* [1977]. (Bottom) Log (heat flow) versus log (age) for Indian Ocean floor older than 40 m.y. B.P. The larger symbols are basin averages. The inverted triangle and squares connected by an arrow show where the Mozambique and western Somali basin averages would lie if their ages were 130 m.y. The W and E next to solid triangles differentiate western and eastern parts of the Wharton Basin. The dashed line corresponds to the same half-space model that best fits the depth data, Q = 10.9 (HFU)/age (m.y.), where 1 HFU = 1 μcal cm^{-2} s^{-1} or 42.87 mW m^{-2}.

TABLE 1. Elevation Versus Age for the Indian Ocean

Age, m.y.	Number of Values	Mean Depth*	Standard Deviation, m
0	8	2630	187
10	16	3540	239
20	16	4080	142
30	17	4230	228
40	6	4650	161
50	10	4970	169
60	13	5200	188
70	8	5250	156
80	6	5300	110
90	2	5430	177
100	2	6000	212
120	5	5670	104
150	4	5810	48

*Corrected meters; depths adjusted for sediment loading effects.

5. Heat flow versus age relations over the eastern Atlantic [*Herman et al.*, 1977] show the same general distribution of values; however, the transition from low to predicted heat flow occurs at 70 m.y.

Combining the mean heat flow from oceanic regions described above confirms that a consistent pattern of the variation of mean heat flow versus age exists in convective regions of the ocean. That is, at any given age it appears that the same amount of heat is being lost on all ridges; but as the effects of convective exchange of heat are masked or stopped at different ages on each ridge, the mean conductive heat flow rebounds to that predicted by theory.

HEAT TRANSFER IN YOUNG OCEANIC CRUST

Our regional analysis augments the increasing evidence that convection of seawater through the pores and fractures of the oceanic crust is the dominant mode of heat transfer in young ocean regions. The variation and intensity of heat flow through the sediment are profoundly affected by this convection. Conversely, the heat flow observations can provide constraints on temperature and flow regimes taking place in the crust. Assume that the theoretical surface heat flow calculated from thermal models of the lithosphere, Q_{theor}, is the true heat flow coming up from the mantle below the crust. The difference $Q_{\text{theor}} - Q_{\text{obs}}$ gives a crude measure of the heat flow lost by convection.

The Rayleigh number is the critical parameter determining the onset and rate of convection. In terms of the vertical temperature difference ΔT across the convection layer it is

$$R = k\alpha g \Delta T L/\kappa_m \nu_w \quad (1)$$

(ΔT is positive when the temperature at the lower boundary is higher), where k is bulk permeability, α is the thermal expansion coefficient of the fluid, g is acceleration due to gravity, ΔT is temperature difference between top and bottom of convecting region, L is the depth of convecting region, κ_m is the effective thermal diffusivity of the medium (equal to thermal conductivity of the porous medium (K_m) divided by ρc_p of the fluid), and ν_w is the viscosity of seawater.

Convection is possible when the Rayleigh number reaches a critical value R_c. It has been shown empirically that the Nusselt number ($\mathfrak{N}_u$), the ratio of convective plus conductive heat flux to that heat flow which would occur in the absence of convection, is approximately equal to the ratio between the Rayleigh number and the critical Rayleigh number [*Silverston*, 1958; *Elder*, 1965, 1967]. The ratio between observed and theoretical heat flow corresponds to a measurement of the Nusselt number only if it is assumed that the theoretical heat flux is a good approximation to the heat flow that would occur in the ridge environment if the convection were to be artificially stopped. Our measurements actually yield a temperature gradient difference, over the layer in question, between that expected from conduction only and that observed from conduction plus convection. Then

$$\mathfrak{N}_u \simeq \Delta T_{\text{theor}}/\Delta T_{\text{obs}} \simeq R/R_c \quad (2)$$

where $\Delta T_{\text{theor}} = Q_{\text{theor}}\Delta Z/K_m$ and $\Delta T_{\text{obs}} = Q_{\text{obs}}\Delta Z/K_m$.

The temperature at the base of the convection layer is constrained roughly by the heat flow observations (see Figure 13), if we take $K_m \sim 2.36$ W m^{-1} K^{-1}, approximated from the measurements of *Robertson and Peck* [1974] and *Hyndman et al.* [1974] of 1.93 W m^{-1} K^{-1} for crushed and water-saturated basalt and a value for dry gabbro of 3.00 W m^{-1} K^{-1} *Ribando et al.* [1976] have argued, on the basis of the wave-

Fig. 12. Composite of heat flow versus age for all midocean ridges. The bottom curves are the observed means of the ridges. The Galápagos spreading center has low heat flow in the 0- to 5-m.y. region. Then, the observed curve returns to the theoretical or top curve. On the East Pacific Rise this increase in mean heat flow occurs at 15 m.y., in the Indian Ocean at 50 m.y., and in the Atlantic Ocean at 70 m.y.

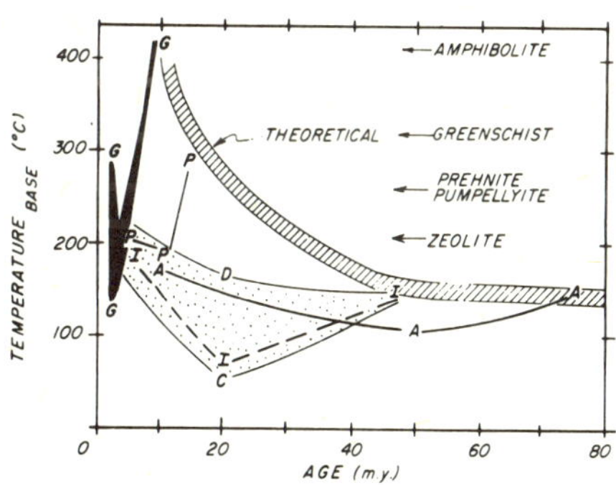

Fig. 13. Predicted temperatures at the base in an assumed constant 5-km-thick convecting layer in the oceanic crust predicted by observed heat flow taken from Figure 12. Here G represents the Galápagos spreading center; P, the East Pacific Rise; A, the Mid-Atlantic Ridge; and I, Indian Ocean ridges. The C and D are sediment environment distinctions as in text and Table 2. Stippled area is variation in Indian Ocean.

length of heat flow variation observed by *Williams et al.* [1974] on the Galápagos spreading center, that the depth of circulation is probably of the order of 5 km [*Ribando et al.*, 1976; *Lister*, 1974].

Combining (1) and (2), we can determine the permeability by

$$k = Q_{\text{theor}} R_c \nu_w / \rho c_p \alpha g (\Delta T)^2 \quad (3)$$

assuming that we can estimate the other parameters in (3) with sufficient accuracy. *Kassoy and Zebib* [1975] have shown that the temperature dependence of the viscosity of seawater greatly lowers R_c. If we take $\alpha_{0°C}$ from *Helgeson and Kirkham* [1974] to be 3.1×10^{-4}/°C and $\nu_{w0°C} = 1.3 \times 10^{-2}$ g/cm s from *Dorsey* [1968] (R_c has been determined for various ΔT by *Kassoy and Zebib* [1975] for temperature-dependent viscosity flow models), we can calculate k as a function of age (see Figure 14). Observed values of age, $\Delta T_{\text{theor}}/\Delta T_{\text{obs}}$, and the resulting values of other parameters in the ridges where measurements are available are shown in Table 2. These values of permeability are probably accurate to an order of magnitude and are on the high side for two reasons: first, 0°C properties, which promote stability, have been used instead of the more common $\Delta T/2$ properties, and second, thermal expansion of water and the saline nature of the hydrothermal fluid were not considered.

If the permeability decreases with age in the ocean crust, then R approaches R_c, and the bulk permeability of the oceanic crust drops to a value of $\sim 10^{-11}$ cm². This is a one-layer model for flow in an oceanic crust with permeability which does not vary with depth. If, however, sediments with a different permeability than that calculated above are deposited on the sea floor, the model considered must become two layered. We will discuss a two-layered model later in this paper.

Geochemical variations with age. In the transition regions of the Pacific ridges, ΔT increases with age, and thermal rebound produces significant reheating of the oceanic crust at depth (Figure 13), whereas the base of the Atlantic and Indian ocean crusts is essentially isothermal to at least 75 m.y. from the crest. It is possible that the lower layers of the oceanic crust in the Pacific are retrograde-metamorphosed in the heat flow transition region from zeolite to greenschist and possibly amphibolite facies metamorphic suites. Temperature ranges of common metamorphic facies are also shown in Figure 13. Resultant dehydration would release bound H_2O, and a transient excess head could develop in the oceanic crust. Upward flow of this water would result, possibly producing extremely low heat flow [*Sclater et al.*, 1974]. Since dehydration reactions are endothermic, a significant portion of the heat flow from below would be required to drive the water from the lower-graded metamorphic facies and thus further reduce the surface heat flow. This mechanism may play a role in forming very low heat flow zones with relatively low scatter such as the ones

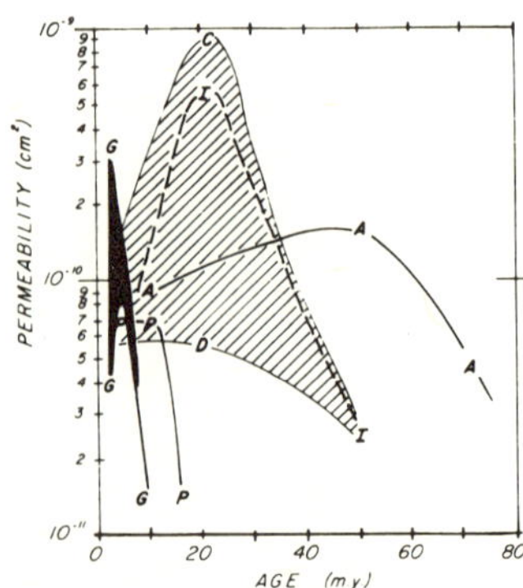

Fig. 14. Predicted permeabilities for the oceanic crust based upon observed ΔT base from Table 2 and the relation $Q_{\text{obs}}/Q_{\text{theor}} \simeq R_c/R$, where R is the Rayleigh number. See text. Symbols are the same as those in Figure 13. Hachured area is variation in Indian Ocean.

TABLE 2. Physical Parameters for Convective Flow in the Oceanic Crust

Ocean	Age, m.y.	$\mathfrak{N}_u$	R_c*	k Cold Boundary, 10^{-11} cm²	ΔT, °C	v, 10^{-8} cm/s
Indian	5	3	11	8.4	182	2.6
	20	3	26	59	60	2.3
	50	1	11	2.4	182	...
D environment	5	2.6	10	5.7	210	2.2
	20	1.3	13	5.7	140	3.5
C environment	5	4	13	18	136	3.6
	20	4	29	120	45	3.4
Pacific						
Galápagos	2	2.5	5	0.8	291	1.8
	4	3	11	8.4	182	2.5
	6	1	5	0.2	455	...
Costa Rica Rift	2	6	15	34	121	5.8
	4	4	13	20	136	3.5
	6	2	9	3.7	227	1.2
	8	1	5	0.2	455	...
East Pacific Rise	5	2.5	11	7	182	1.8
	10	2	12	7	159	1.2
	15	1	6	1	273	...
Atlantic	10	2.4	12	8.7	152	1.7
	50	1.8	18	16	90	0.9
	70	1	13	4.4	136	...

$\mathfrak{N}_u$ is observational approximation to Nusselt number based upon the ratio of Q_{obs} to Q_{theor}.
*From *Kassoy and Zebib* [1975].

found near the Galápagos spreading center [*Sclater et al.*, 1974].

Flow parameters in the oceanic crust. In the Indian Ocean the net fluid flow rate can be calculated (Table 2), since the mean convective heat flow is $Q_{\text{theor}} - Q_{\text{obs}}$ and

$$v = Q_{\text{conv}}/\rho c_p \Delta T$$

For convection in the oceanic crust we calculate flow rates of the order of $1\text{-}6 \times 10^{-8}$ cm/s using ΔT of Table 2 (in close agreement with the calculation of *Ribando et al.* [1976] based upon the *Williams et al.* [1974] survey data on the crest of the Galápagos spreading center). This flow rate is 3–4 times that calculated by *Anderson and Hobart* [1976] from the crack models of *Lowell* [1975] and *Bodvarsson and Lowell* [1972].

MECHANISMS THAT STOP OR MASK CONVECTION ON MIDOCEAN RIDGES

There have been at least two different but interrelated mechanisms presented to date to explain sealing of convective heat transfer on midocean ridges; both relate convective heat exchange between the oceanic crust and the water column to 'effective permeability' within the oceanic crust. The first suggests that when sediment of sufficient thickness to completely blanket basement is deposited, it forms an impermeable cap above a convecting and permeable oceanic crust. Then the mean heat flow will represent the true flux from depth [*Lister*, 1972; *Sclater et al.*, 1974; *Anderson and Hobart*, 1976]. The second suggests that hydrothermal precipitation within the oceanic crust seals pores and stops the convection by lowering the permeability of the oceanic crust to a level below that needed to sustain convection [*Sclater et al.*, 1974].

In order to examine whether either or both are applicable in the Indian Ocean we have constructed a sediment isopach map (see Figure 15) based on a similar map prepared by J. Ewing et al. [*National Academy of Sciences of the USSR*, 1975], recent Lamont-Doherty Geological Observatory seismic profiler records, and work by *Houtz et al.* [1977]. Note that the Central Indian and southern Wharton basins have a very thin sediment cover.

We have constructed a similar map of the areal distribution

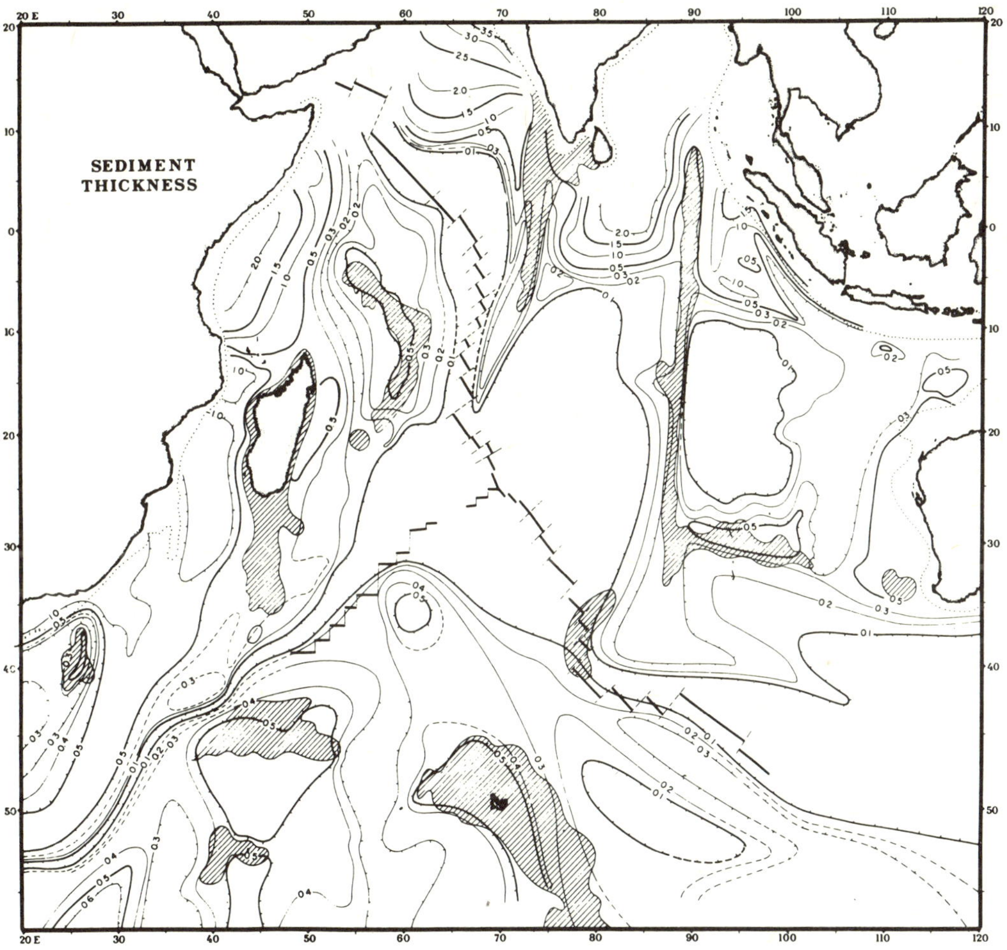

Fig. 15. Sedimentary isopachs (in kilometers) in the Indian Ocean from *National Academy of Sciences of the USSR* [1975], *Kolla et al.* [1976], and *Houtz et al.* [1976].

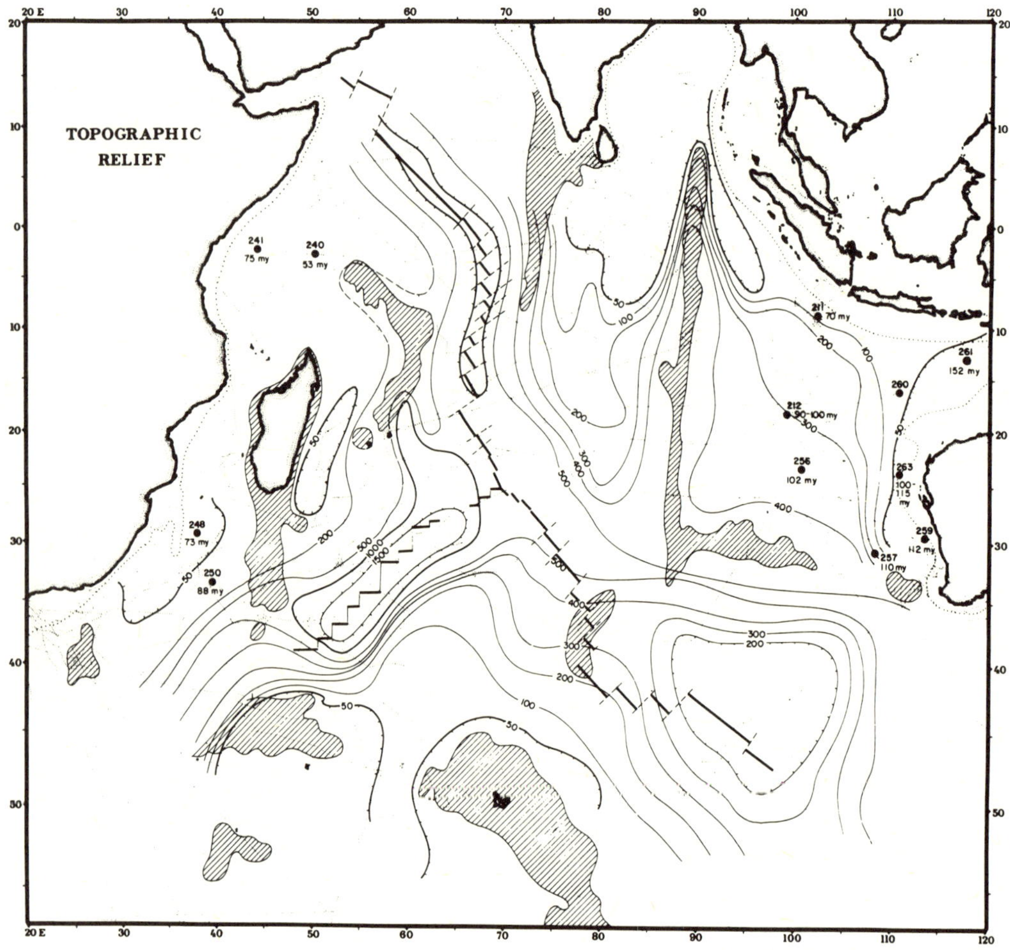

Fig. 16. Topographic relief in the Indian Ocean measured by taking the average peak-to-trough elevation change at 1° intervals along the tracks shown in light dots. These averages were then contoured in 100-m intervals. Dotted lines show Lamont-Doherty Geological Observatory ship tracks in the Indian Ocean.

of topographic roughness in the Indian Ocean (Figure 16) by measuring peak-to-trough bathymetric relief along the tracks shown in the figure. With the aid of these maps we then examined the variation with age of (1) the ratio of the conductive to the theoretical or total heat flow in various basins of the Indian Ocean, (2) the increase of sediment thickness, (3) the ratio of sediment thickness to topographic relief, and (4) the surface-sediment carbonate content in these basins. When Q_{cond}/Q_{theor} approaches 1, the effects of convection are masked or stopped. The age of this transition in the various basins should correspond to the age at which the sediment thickness becomes ~300 m [*Sclater et al.*, 1974] or the ratio of sediment thickness to topographic relief increases to >1 [*Anderson and Hobart*, 1976] if either of these parameters adequately represents the sealing mechanism. These relations fit in some but not all of the seven Indian Ocean basins (Figure 17). For example, in the southern Wharton Basin (where sediment thickness actually decreases with increasing age), the Central Indian Basin (with its thin sediment cover), and the Madagascar Basin (which is very rough), convection has already been sealed by the time the sediment thickness to topography ratio becomes large. The basic problem with intercomparisons of these sedimentary parameters as they vary with age is that the parameters examined are all age dependent and therefore cannot be separated from each other with existing data. Thus age itself, in addition to a decrease in heat flow, becomes a parameter. Qualitatively, the effect that obviously is very important is related to the increase in sediment thickness and its smoothing effect, but the fact that these parameters do not vary in a quantitative manner similar to the heat flow transition from basin to basin suggests that other parameters may play a role in stopping or masking convection on the flanks of midocean ridges.

Hydraulic resistance of the sedimentary cover. Numerous measurements of heat flow made in the Crozet Basin, where there is a fairly complete but thin sediment cover (i.e., C type

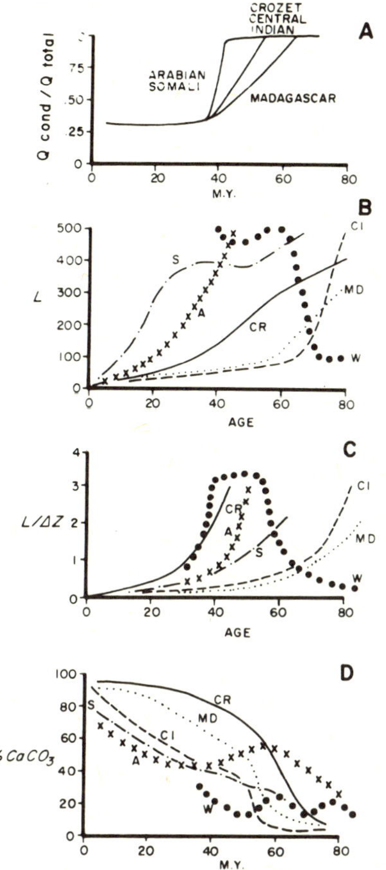

Fig. 17. (*a*) Ratio of conductive heat flow to predicted heat flow as a function of age in the various basins of the Indian Ocean. (*b*) Sediment thickness in meters versus age in the Indian Ocean basins. For sources, see text. (*c*) Sediment thickness to topographic relief ratio as a function of age. (*d*) Percentage of $CaCO_3$ versus age from *Kolla et al.* [1976] plotted for various basins of Indian Ocean. In Figures 17*b*, 17*c*, and 17*d* the basins are designated as follows: the dot-dashed line represents Somali (S); crosses, Arabian (A); heavy dotted line, Wharton (W); solid line, Crozet (CR); dashed line, Central Indian (CI); and light dotted line, Madagascar (MD).

environment) over crust with an age of about 20–40 m.y., yield values that are consistently one half to one fourth of the theoretical predictions (Figure 9*b*). These observations suggest that a complete sediment cover is not a sufficient condition to cap the circulation; otherwise, we should see variability in the values with a mean equal to the theoretical value. This suggests that large-scale convective flow may be continuous throughout the sediment layer. In addition, the fact that in some regions (for example, the Central Indian and Somali basins, Figure 17) the transition zone corresponds to the zone where the surface sediment changes from carbonaceous to siliceous suggests that sediment composition, as it affects permeability, may be a factor in sealing the effects of convection.

Bryant et al. [1974] have reported laboratory measurements of the permeability of deep-sea sediments. They found variability of from 10^{-11} to 10^{-16} cm^2 ($10^{-13\pm1}$ cm^2 for deep-sea clays to $10^{-12\pm1}$ cm^2 for carbonate-rich sediments).

If we assume that the layer of sediment over an oceanic crust in which convection takes place acts as a simple hydraulic resistance to the vertical flow of seawater, we can calculate the criteria for formation of an impermeable lid (suggested by D. L. Turcotte, personal communication, 1976). If the sediment layer acts as a simple resistance in series with the flow, then when $K_{sed}/L_{sed} \ll K_{crust}/L_{crust}$, the sediment layer will be impermeable. When we assume that L_{crust} = 5 km, K_{crust} = 10^{-10} cm^2, and K_{sed} = 10^{-12} cm^2 (typical for carbonates), the thickness required to form an impermeable blanket is ~1000 m; whereas if K_{sed} = 10^{-13} cm^2 (more typical of clays), then the thickness must be only of the order of 100 m to seal convection. For this model the ratio of K to L in the oceanic crust and sedimentary lid are the critical parameters that affect convection.

The *Bryant et al.* [1974] measurements show that the permeability of sediment might vary by several orders of magnitude from locale to locale. However, the mean of measured permeabilities shows a clear decrease from carbonate to clay. The scatter in heat flow values in the convecting region may to some extent reflect this local variation in sediment permeabilities, but the correspondence between the ratios of (1) convective to conductive heat flow in a given region and (2) the mean sediment permeability to sediment thickness may indicate regional sealing of convection. Figure 17*d* shows the general locale of the carbonaceous-siliceous transition in the various basins of the Indian Ocean from *Kolla et al.* [1976] (see also Figure 2). As expected, the percentage of $CaCO_3$ alone does not directly correlate with the heat flow transition, but note that in the basins with thick sequences of low carbonate sediments (Wharton, Arabian, and Somali), convection appears to be stopped or masked earlier than in basins with thin sediments of higher carbonate content. This crude hydraulic resistance concept has been explored further by using a two-dimensional two-layer model of convection in porous media by *Skilbeck and Anderson* [1977].

Plugged plumbing within the oceanic crust. There is general observational support for a correlation between the age of sealing of convection and the variation in crustal velocities from seismic refraction studies of the oceanic crust. It is not easy to dismiss the 'plugged plumbing' model as yet another mechanism that can stop convection in crust. *Houtz and Ewing* [1976] observed an increase in the velocity of seismic layer 2A with increasing age at sonobuoy stations in the Atlantic and Pacific oceans. They interpreted this as an effective thinning of layer 2A, since the velocity of 2A increased until it was the same as that of layer 2B. Further, the increase in seismic layer 2A velocity in the Atlantic occurs at 70 m.y., an age corresponding to that at which the heat flow transition zone occurs (Figure 18). Velocity studies of basalts suggest that metabasalts have higher velocities than unaltered basalts [*Fox et al.*, 1973]. *Fox and DeLong* [1976] noted the correspondence between this increase in velocity of layer 2A and the increase in heat flow away from the East Pacific Rise and attributed it to metamorphic redeposition which plugged circulation cracks in the oceanic crust and lowered its permeability. We further note that *Shor et al.* [1970] show an increase in layer 3 velocity with age in the Pacific that might also correspond to the increase in heat flow on the East Pacific Rise in 10- to 15-m.y.-old crust. We are thus at this time unable to say what factor or factors control sealing

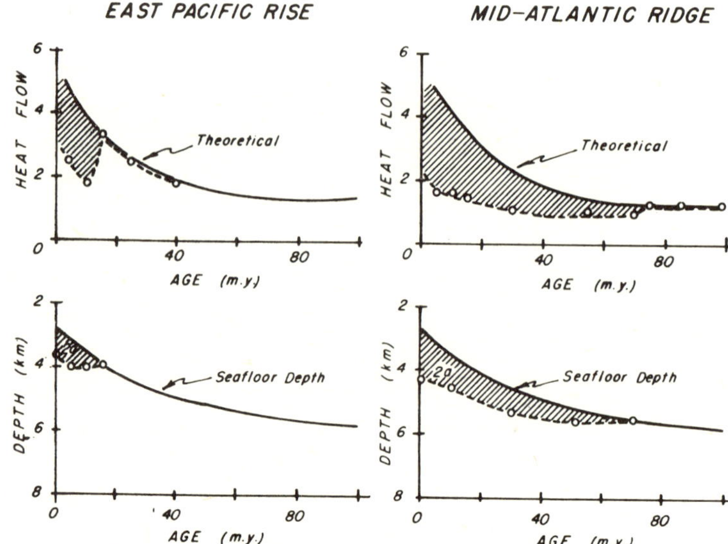

Fig. 18. Heat flow versus age and thickness of crustal layer 2A versus age in the Pacific and Atlantic oceans from *Anderson and Hobart* [1976], *Herman et al.* [1977], and *Houtz and Ewing* [1976]. The shaded portions of the heat flow figures are the quantity of heat transferred convectively. Note that the convective heat transfer and the thickness of layer 2A go to zero at approximately equal ages.

of convection in the ocean floor. Very possibly it depends on a number of age-dependent parameters acting together.

Convective heat loss on midocean ridges. *Williams and Von Herzen* [1974] estimated the total heat loss from the earth to be 10.2×10^{12} cal/s. On the basis of the heat loss on the crest of the Galápagos spreading center at 86°W described by *Williams et al.* [1974] they estimated that the total heat loss from convection on midocean ridges is at least 20% of the earth's total heat loss or 2×10^{12} cal/s.

On the basis of our Indian Ocean heat flow analysis (Figure 9) and similar studies in the Pacific [*Anderson and Hobart*, 1976] and the Atlantic [*Herman et al.*, 1977] (Figure 12) we can calculate the convective heat loss on the midocean ridges more directly than by extrapolating the GSC rate as *Williams and Von Herzen* [1974] did.

For ridges on which we have heat flow analyses we find that the total heat loss due to convection per centimeter of midocean ridge length is ~170 cal/s and that the total convective heat loss from these ridges is 6×10^{11} cal/s. These values were obtained by integrating under the curves of Figures 9 and 12 (Table 3). However, we have to date analyzed only a small part of the 48,000-km midocean ridge system. We have also made no attempt to place error limits on our estimates. Assuming that the average heat loss rate given above holds for these additional ridges, we calculate a total convective heat loss from midocean ridges of 8×10^{11} cal/s. This is only 8% of the total heat loss of the earth, but it is still a significant component. Our calculated heat loss of 170 cal/s compares with the predicted convective heat loss of the *Sclater and Klitgord* [1973] lithospheric plate model, which included hydrothermal circulation at midocean ridges of ~200 cal/s.

CONCLUSIONS

We began with the basic premise that since the elevation versus age relation predicted by thermal models of oceanic lithosphere is in close agreement with empirical depth-age relations, the thermal evolution of oceanic plates is adequately modeled; thus the heat flow with age of the oceanic lithosphere

TABLE 3. Heat Loss by Convection on Midocean Ridges

Ridge	Length, km	Spreading Rate, cm/yr	Generation Rate, km²/yr	Q_{conv}, cal/cm²	Heat Loss (Total), cal/s	Heat Loss (per cm of Ridge Length), cal/s
East Pacific Rise	8,800	6	0.53	7×10^{8}	1.2×10^{11}	136
Galápagos spreading center	2,200	3	0.10	3.8×10^{8}	1.2×10^{10}	54
Mid-Atlantic Ridge	13,100	2	0.26	2.5×10^{9}	2.1×10^{11}	160
Central and Southeast Indian Ridge	11,300	3	0.34	2.5×10^{9}	2.6×10^{11}	230
Total above	35,400	...	...	...	6×10^{11}	170
Total world	48,100	...	...	...	8.2×10^{11}	...

is predicted. If true, our analyses of heat flow measurements in the Indian Ocean show that a significant quantity of that heat flow is not measured by conduction experiments on midocean ridges. The missing heat flow is removed convectively by seawater circulation in the oceanic crust. A bulk permeability of 10^{-10} cm^2 is required for this convection to occur. The effects of this convective heat exchange stop 40–60 m.y. away from the ridge crest. A number of mechanisms may play a role in this, for example, the deposition of an impermeable lid over the oceanic crust or the deposition of metamorphic minerals in cracks within the oceanic crust. Correspondence between (1) the carbonate-clay transition and general increase in thickness of the sediment blanket with aging and (2) the increase in conductive heat flow in the various basins of the Indian Ocean supports the former mechanism; whereas the correspondence between the increase in seismic layer 2A velocity and the heat flow transition on the East Pacific Rise and in the Atlantic Ocean supports the latter. There is a great need for closely spaced heat flow, coring, and sonobuoy profiles across the transition zones from dominantly convective to conductive heat flow on the midocean ridges. If it can be established that a region is sealed to convective interchange of heat between the lithosphere and water column, the scatter of heat flow can be reduced to such an extent that crude age constraints can be placed upon sea floor where magnetic anomalies have yet to be identified. In this regard the Mozambique and western Somali basins appear to be much older than DSDP sites in the basins would indicate.

Acknowledgments. Though the specific analysis reported here was funded by NSF grant DES 74-24112, the data were gathered through a decade of support by the National Science Foundation and the Office of Naval Research. To the many chief scientists, captains, crews, and technicians responsible for this wealth of data we are particularly grateful. A. Lachenbruch and D. Turcotte were particularly helpful during early stages of this project. We thank D. Forsyth, J. Lawrence, W. Pitman, and J. Skilbeck for reviewing the manuscript for us and offering many useful improvements. Contribution 2472 of Lamont-Doherty Geological Observatory.

REFERENCES

Anderson, R. N., Petrological significance of low heat flow on the flanks of slow-spreading midocean ridges, *Geol. Soc. Amer. Bull., 83*, 2947–2956, 1972.

Anderson, R. N., and M. A. Hobart, The relation between heat flow, sediment thickness, and age in the eastern Pacific, *J. Geophys. Res., 81*, 2968–2989, 1976.

Bodvarsson, G., and R. P. Lowell, Ocean floor heat flow and the circulation of interstitial waters, *J. Geophys. Res., 77*, 4472–4475, 1972.

Bookman, C. A., I. Malone, and M. G. Langseth, Jr., Sea floor geothermal measurements from *Conrad* cruise 13, *Tech. Reps. 5-CU-72, 1-CU-1-72*, 279 pp., Columbia Univ., New York, 1972.

Bryant, W. R., A. P. Deflache, and P. K. Trabant, Consolidation of marine clays and carbonates, in *Deep Sea Sediments: Physical and Mechanical Properties*, edited by A. L. Interlützen, Plenum, New York, 1974.

Corry, C., C. Dubois, and V. Vacquier, Instrument for measuring terrestrial heat flow through the ocean floor, *J. Mar. Res., 26*, 165, 1968.

Davis, E. E., and C. R. B. Lister, Fundamentals of ridge crest topography, *Earth Planet. Sci. Lett., 21*, 405–413, 1974.

Dorsey, N. E., *Properties of Ordinary Water Substance*, Hafner, New York, 1968.

Elder, J. W., Physical processes in geothermal areas, in *Terrestrial Heat Flow, Geophys. Monogr. Ser.*, vol. 8, edited by W. H. K. Lee, pp. 211–239, AGU, Washington, D. C., 1965.

Elder, J. W., Steady free convection in a porous medium heated from below, *J. Fluid Mech., 27*, 27–48, 1967.

Forsyth, D. W., The evolution of the upper mantle beneath midocean ridges, submitted to *J. Geophys. Res.*, 1977.

Fox, P. J., and S. E. DeLong, Healing of oceanic layer 2A: Geophysical and geological consequences (abstract), *Eos Trans. AGU, 57*, 329, 1976.

Fox, P. J., E. Schreiber, and J. J. Peterson, The geology of the oceanic crust: Compressional wave velocities of oceanic rocks, *J. Geophys. Res., 78*, 5155–5172, 1973.

Gerard, R., M. Langseth, and M. Ewing, Thermal gradient measurements in the water and bottom sediment of the western Atlantic, *J. Geophys. Res., 67*, 785–803, 1962.

Helgeson, H. C., and D. H. Kirkham, Theoretical prediction of the thermodynamic behavior of aqueous electrolytes at high pressures and temperatures, 1, Summary of the thermodynamic/electrostatic properties of the solvent, *Amer. J. Sci., 274*, 1089–1198, 1974.

Herman, B. M., M. G. Langseth, and M. A. Hobart, Heat flow in the oceanic crust bounding west Africa, *Tectonophysics*, in press, 1977.

Houtz, R., and J. Ewing, Upper crustal structure as a function of plate age, *J. Geophys. Res., 81*, 2490–2498, 1976.

Houtz, R. E., D. E. Hayes, and R. G. Markl, Kerguelen Plateau: Bathymetric sediment distribution and crustal structure, *Geol. Soc. Amer. Mem.*, in press, 1977.

Hyndman, R. D., and D. S. Rankin, The Mid-Atlantic Ridge near 45°N, 18, Heat flow measurements, *Can. J. Earth Sci., 9*, 664–670, 1972.

Hyndman, R. D., A. J. Erickson, and R. P. Von Herzen, Geothermal measurements on DSDP leg 26, in *Initial Reports of the Deep-Sea Drilling Project*, vol. 26, pp. 451–462, U.S. Government Printing Office, Washington, D. C., 1974.

Hyndman, R. D., R. P. Von Herzen, A. J. Erickson, and J. Jolivet, Heat flow measurements in deep crustal holes on the Mid-Atlantic Ridge, *J. Geophys. Res., 81*, 4053–4060, 1976.

Initial Reports of the Deep-Sea Drilling Project, vols. 22–26, U.S. Government Printing Office, Washington, D. C., 1974–1975.

Kassoy, D. R., and A. Zebib, Variable viscosity effects on the onset of convection in porous media, *Phys. Fluids, 18*, 1649–1651, 1975.

Kolla, V., A. W. H. Bé, and P. Biscaye, Calcium carbonate distribution in the surface sediments of the Indian Ocean, *J. Geophys. Res., 81*, 2605–2616, 1976.

Langseth, M. G., and P. T. Taylor, Heat flow in the Indian Ocean, *J. Geophys. Res., 72*, 6249–6260, 1967.

Langseth, M. G., and R. P. Von Herzen, Heat flow through the floor of the world oceans, in *The Sea*, vol. 4, edited by A. E. Maxwell, pp. 299–352, Interscience, New York, 1970.

Larson, R., Early Cretaceous breakup of Gondwanaland off western Australia, *Geology, 5*, 57–60, 1977.

Lister, C. R. B., Heat flow west of the Juan de Fuca Ridge, *J. Geophys. Res., 75*, 2648–2654, 1970.

Lister, C. R. B., On the thermal balance of a midocean ridge, *Geophys. J. Roy. Astron. Soc., 26*, 515–535, 1972.

Lister, C. R. B., On the thermal balance of a midocean ridge, *Geophys. J. Roy. Astron. Soc., 39*, 465–509, 1974.

Lowell, R. P., Circulation in fractures, hot springs, and convective heat transport on midocean ridge crests, *Geophys. J. Roy. Astron. Soc., 40*, 351–365, 1975.

Markl, R. G., Evidence for the breakup of eastern Gondwanaland by the Early Cretaceous, *Nature, 251*, 196–200, 1974.

Marshall, B. V., and A. J. Erickson, Heat flow and thermal conductivity measurements, leg 25, in *Initial Reports of the Deep-Sea Drilling Project*, vol. 25, pp. 349–355, U.S. Government Printing Office, Washington, D. C., 1975.

McKenzie, D. P., Some remarks on heat flow and gravity anomalies, *J. Geophys. Res., 72*, 6261–6273, 1967.

McKenzie, D. P., and J. G. Sclater, The evolution of the Indian Ocean since the Late Cretaceous, *Geophys. J. Roy. Astron. Soc., 25*, 437–528, 1971.

National Academy of Sciences of the USSR, *Geological-Geophysical Atlas of the Indian Ocean*, Moscow, 1975.

Parker, R. L., and D. W. Oldenburg, A thermal model of oceanic ridges, *Nature, 242*, 137–139, 1973.

Parsons, B., and J. G. Sclater, An analysis of the variation of ocean floor heat flow and bathymetry with age, *J. Geophys. Res., 82*, 803–827, 1977.

Popova, A. R., Ya. B. Smirnov, E. I. Suvilov, G. B. Udintsev, and B. V. Shekhvatov, Geothermal studies on the largest tectonic structure of the bottom of the Indian Ocean, in *Investigations on the Problem Rift Zones of the World Oceans*, vol. 2, National Academy of Sciences of the USSR, Moscow, 1972.

Ribando, R. J., K. E. Torrance, and D. L. Turcotte, Numerical models for hydrothermal circulation in the oceanic crust, *J. Geophys. Res.*, *81*, 3007-3012, 1976.
Richter, F. M., Convection and large-scale circulation of the mantle, *J. Geophys. Res.*, *78*, 8735-8745, 1973.
Robertson, E. C., and D. L. Peck, Thermal conductivity of vesicular basalt from Hawaii, *J. Geophys. Res.*, *79*, 4875-4888, 1974.
Schlich, R., Sea floor spreading history and deep-sea drilling results in the Madagascar and Mascarene basins, western Indian Ocean, leg 26, in *Initial Reports of the Deep-Sea Drilling Project*, vol. 26, pp. 663-678, U.S. Government Printing Office, Washington, D. C., 1974.
Schlich, R., E. S. W. Simpson, and T. L. Vallier, Regional aspects of DSDP in western Indian Ocean, leg 25, in *Initial Reports of the Deep-Sea Drilling Project*, vol. 25, pp. 743-759, U.S. Government Printing Office, Washington, D. C., 1975.
Schubert, G., C. Froidevaux, and D. A. Yuen, Oceanic lithosphere and asthenosphere: Thermal and mechanical structure, *J. Geophys. Res.*, *81*, 3525-3540, 1976.
Sclater, J. G., Heat flow in the northwest Indian Ocean and Red Sea, *Phil. Trans. Roy. Soc. London, Ser. A*, *259*, 271-278, 1966.
Sclater, J. G., and R. L. Fisher, Evolution of the east central Indian Ocean with emphasis on the tectonic setting of the Ninety East Ridge, *Geol. Soc. Amer. Bull.*, *85*, 613-702, 1974.
Sclater, J. G., and J. Francheteau, The implications of terrestrial heat flow observations on current tectonic and geochemical models of the crust and upper mantle of the earth, *Geophys. J. Roy. Astron. Soc.*, *20*, 509-542, 1970.
Sclater, J. G., and K. D. Klitgord, A detailed heat flow, topographic, and magnetic survey across the Galápagos spreading center at 86°W, *J. Geophys. Res.*, *78*, 6951-6975, 1973.
Sclater, J. G., R. N. Anderson, and M. L. Bell, Elevation of ridges and evolution of the central eastern Pacific, *J. Geophys. Res.*, *76*, 7888-7915, 1971.
Sclater, J. G., R. P. Von Herzen, D. L. Williams, R. N. Anderson, and K. D. Klitgord, The heat flow low on the flank of the Galápagos spreading center, *Geophys. J. Roy. Astron. Soc.*, *38*, 609-626, 1974.
Sclater, J. G., J. Crowe, and R. N. Anderson, On the reliability of oceanic heat averages, *J. Geophys. Res.*, *81*, 2997-3006, 1976*a*.
Sclater, J. G., B. P. Luyendyk, and L. Meinke, Magnetic lineations in the southern part of the Central Indian Basin, *Geol. Soc. Amer. Bull.*, *87*, 371-378, 1976*b*.
Shor, G. G., H. W. Menard, and R. W. Raitt, Structure of the Pacific Basin, in *The Sea*, vol. 4, edited by A. E. Maxwell, pp. 3-28, Interscience, New York, 1970.
Silverston, P. L., Wärmedurchgang in waagerechten Flüssigkeitsschichten, 1, *Forsch. Ingenieurw.*, *24*, 29-32, 51-69, 1958.
Skilbeck, J. N., and R. N. Anderson, Heat flow and a two-layer model for convection in porous media: The oceanic crust and overlying sediments, submitted to *J. Geophys. Res.*, 1977.
Talwani, M., C. C. Windisch, and M. G. Langseth, Reykjanes Ridge crest: A detailed geophysical study, *J. Geophys. Res.*, *76*, 473-517, 1971.
Vacquier, V., and P. T. Taylor, Geothermal and magnetic survey off the coast of Sumatra, *Bull. Earthquake Res. Inst. Tokyo Univ.*, *44*, 531-540, 1966.
Von Herzen, R. P., Geothermal heat flow in the gulfs of California and Aden, *Science*, *140*, 1207-1208, 1963.
Von Herzen, R. P., and M. G. Langseth, Present status of oceanic heat-flow measurements, *Phys. Chem. Earth*, *6*, 365-407, 1966.
Von Herzen, R. P., and A. E. Maxwell, The measurement of thermal conductivity of deep-sea sediments by a needle probe method, *J. Geophys. Res.*, *64*, 1557-1563, 1959.
Von Herzen, R. P., and S. Uyeda, Heat flow through the eastern Pacific Ocean floor, *J. Geophys. Res.*, *68*, 4219-4250, 1963.
Von Herzen, R. P., and V. Vacquier, Heat flow and magnetic profiles on the Mid-Indian Ocean Ridge, *Phil. Trans. Roy. Soc. London, Ser. A.*, *259*, 262-270, 1966.
Williams, D. L., and R. P. Von Herzen, Heat loss from the earth: New estimate, *Geology*, *2*, 327-328, 1974.
Williams, D. L., R. P. Von Herzen, J. G. Sclater, and R. N. Anderson, the Galápagos spreading center: Lithospheric cooling and hydrothermal circulation, *Geophys. J. Roy. Astron. Soc.*, *38*, 587-608, 1974.

(Received October 28, 1976;
revised February 23, 1977;
accepted February 24, 1977.)

7

Reprinted from *Geology* 2:327-328 (1974)

HEAT LOSS FROM THE EARTH: NEW ESTIMATE

David L. Williams

Woods Hole Oceanographic Institution

Richard P. Von Herzen

Woods Hole Oceanographic Institution

ABSTRACT

The average heat loss from the Earth's interior is calculated from heat-flow values and tectonic models of sea-floor spreading. The value of 10.2×10^{12} cal/s (±15 percent) is about 32 percent larger than previous estimates, mainly due to the previously ignored contribution from the cooling lithosphere. The principal uncertainty derives from unknown parameters of the oceanic lithosphere.

INTRODUCTION

Most hypotheses that deal with the origin and development of the Earth and its surface features utilize or imply stresses that derive from the release of thermal energy. The magnitude and distribution of heat loss at the Earth's surface place limits on the energy available for tectonic processes, because most of this energy eventually appears as heat within the Earth.

Calculations of the total rate of heat loss from the Earth (Q_e) have been based on the world-wide average of conductive heat-flow measurements. A recent estimate is $1.47 \pm 0.08 \times 10^{-6}$ cal/cm^2/s (HFU) (Lee, 1970), or 7.7×10^{12} cal/s over the whole Earth (Lee and Uyeda, 1965). World-wide data shows that heat loss is not uniformly distributed over the Earth. Even before the tectonic hypotheses of sea-floor spreading and moving lithospheric plates were widely accepted, investigators had observed that the conductive heat flux near mid-ocean ridges, although variable, is significantly higher than the mean heat flow measured in ocean basins (Bullard and others, 1956). From physical models based on these hypotheses, various theoretical conductive cooling models (Sclater and Francheteau, 1970; Parker and Oldenburg, 1973; D. L. Williams and K. Poehls, in prep.) were developed to predict heat flux versus distance from ridges. The models indicate that the heat loss that results from sea-floor spreading could be a substantial portion of the Earth's total heat flux (McKenzie and Sclater, 1969).

A careful analysis of heat-flow measurements does not entirely substantiate these models of cooling lithospheric plates. When reasonable physical parameters are used, these models predict significantly higher heat flux through young crust (<50 m.y.) than is observed.

We and other investigators have recently suggested that hydrothermal circulation is a major, if not dominant, mode of heat transfer through the surface of young lithosphere (see references in Williams and others, 1974). It appears that the discrepancy between measured and theoretical heat flux may be due mainly to this physical process. Existing techniques measure only that portion of the heat flow that is conducted through oceanic sediments. The lack of sediments near mid-ocean ridge, where heat flux is highest, accentuates the magnitude of this unmeasured component. Hydrothermal heat transfer tends to reduce and distort measured values. As a result, we must use more indirect, theoretical methods to estimate the total heat loss through young oceanic crust.

CONTINENTAL COMPONENT

We consider the total heat loss of the Earth as separable into three components continental, oceanic background (steady-state), and oceanic due to lithospheric formation (transient). Including the continental shelves and slope, the continental component is derived from an area of 1.48×10^{18} cm^2 and an average heat flux of 1.46×10^{-6} HFU (Lee, 1970), resulting in a heat loss of 2.16×10^{12} cal/s ±10 percent. The heat loss from volcanoes, hot springs, and geysers, although sometimes spectacularly evident on a local scale, is estimated to be only 5×10^{10} cal/s (Elder, 1965) and is insignificant when compared with the total (Q_e).

OCEANIC BACKGROUND

The second component, the oceanic background, is the heat flux through the surface of the ocean floor after spreading lithospheric plates have cooled. Here we estimate it be examining heat-flow values from the oldest identified parts of the sea floor. Mesozoic-age magnetic anomalies have been postulated in the northwest Atlantic and west Pacific Oceans (Larson and Pitman, 1972). The average heat flux for these regions is 1.12 ± 0.04 (standard error) HFU for the Atlantic, and 1.17 ± 0.03 HFU for the Pacific. The Wharton Basin of the Indian Ocean, which from deep sea drilling appears to have an age greater than 90 m.y. (von der Borch and others, 1972; Veevers and Heirtzler, 1973), has an average heat flow of 1.21 ± 0.06 HFU (McKenzie and Sclater, 1969). From these data, we obtain an over-all average for this background component of 1.15 ± 0.05 HFU. However, the theoretical cooling time of a 100-km-thick lithospheric plate would require us to correct this background to a value of 1.10 ± 0.05 HFU; whereas for a 75-km-thick lithospheric plate, this correction is <1 percent; here we adopt a corrected value of 1.12 ± 0.06 HFU. The area of the oceans (Menard and Smith, 1966), including marginal seas and continental rises, is 3.62×10^{18} cm^2, which results in a total rate of heat loss from the background component of 4.06×10^{12} cal/s ±10 percent. Most of this heat originates below the lithosphere. The higher average heat flux from volcanoes (exclusive of actively spreading ocean ridges) and marginal seas has not been considered, because the effect on the total (Q_e) is insignificant (<1 percent).

SEA-FLOOR SPREADING

The final component is the heat (Q_{ℓ}), which is continually being released as a result of sea-floor spreading. Because this quantity is estimated from a theoretical expression, its accuracy is mainly dependent on the physical model and the associated parameters assumed for lithospheric formation. The seismic, topographic, and thermal data (reviewed in D. L. Williams and K. Poehls, in prep.) indicate that most oceanic lithosphere attains near-thermal equilibrium conditions prior to subduction. After 50 m.y., a lithosphere lost more than 95 percent of its transient heat and over 99 percent after 100 m.y. Although some subduction zones (for example, the Peru and Middle America trenches) are subducting relatively young lithosphere at a relatively rapid rate, the effect on the calculation of total (Q_e) heat flux is small (<1 percent).

We have utilized a simplified expression, $Q_{\ell} = P\ell dvT_A f$, in which the principal uncertainties are the equilibrated lithospheric thickness (ℓ) and basal temperature (T_A). In this expression, we ignore effects (such as adiabatic compression, thermal expansion, and heat sources within the lithosphere) that would reduce Q_ℓ, as well as effects of latent heat of crystallization and hydration/dehydration of crustal rocks that would increase Q_ℓ. These are not insignificant effects but their over-all influence on Q_ℓ is much smaller than uncertainties in ℓ and T_A; thus we omit further discussion of them here.

We have assumed values for the basal temperature (T_A) ranging from 1000° to 1400°C (Forsyth and Press, 1971; Verhoogen, 1973). The lower value is based on the temperature required to melt the rock, and the upper temperature is constrained by seismic velocities. The range of values chosen for equilibrated lithospheric thickness (ℓ) is 76 to 100 km and derives primarily from seismic data (Forsyth and Press, 1971). We have measured the total length (d) of all oceanic spreading centers as 53,700 km ±5 percent, and have estimated the average half-spreading rate (v) as 2.74 cm/yr ±5 percent. The specific volumetric heat capacity (P), taken as 0.90 ± 0.05 cal/cm^3 °C, is deduced from Schatz and Simmons' (1972) measurements of lattice thermal conductivity (K_L) and diffusivity (K_L) and diffusivity (k_L) where $P = K_L/k_L$. Their results indicate that this value varies little for the minerals and temperatures encountered in the lithosphere.

The factor f accounts for effects that lithospheric creation has on the oceanic background component of heat loss. If the heat flux into the bottom of the lithosphere is constant, then $f = 1.0$. This model would hold if sublithospheric heat derives primarily from internal heat sources (for example, radiogenic heat) and if all this heat is efficiently transported to the base of the lithosphere. In contrast, if the presence of a high-temperature young lithospheric plate suppresses some of the asthenospheric heat flux (for example, McKenzie's model, McKenzie and Sclater, 1969), f approaches 2/3 (Sleep, 1969). We have taken $f = 0.84 \pm 0.16$ as a compromise between these models, where the estimated error covers the range of uncertainty of the extreme models.

DISCUSSION

Table 1 shows the calculations for three different sets of parameters. We assume the median value between the two extremes represented by solutions 1 and 3; the total of the three components of the Earth's heat loss is then 10.2×10^{12} cal/s, which represents a 32 percent increase over Lee and Uyeda's (1965) estimate. This results in a mean heat loss per unit area of over 2.2 HFU for the oceans and about 2.0 HFU for the Earth. Lithospheric cooling represents about 40 percent of this total. More than 50 percent of the lithospheric cooling

TABLE 1. SUMMARY OF SEA-FLOOR SPREADING HEAT LOSS CALCULATIONS

Solution no.	ℓ (km)	T_A (°C)	K	Q_ℓ
1	76	1050	7.5	2.9
2	85	1300	6.8	3.9
3	100	1400	7.5	5.0

Note: K is in 10^{-3} cal/cm/sec °C. Q_ℓ is in 10^{12} cal/sec.

occurs before the sea floor is 2 m.y. old. Measurements of conductive heat flow on young crust (Williams and others, 1974) indicate that only a small part of this heat is released by thermal conduction through rocks and sediment at the sea floor. Sea-water quenching of extrusive lavas (seismic layer 2) can also only account for a small portion of the heat loss. If hydrothermal circulation in the oceanic crust out to 2 m.y. is primarily responsible for removing the remaining heat, this process accounts for approximately 20 percent of the total heat loss of the Earth. There is some evidence for hydrothermal circulation in much older crust (Williams and Poehls, in prep.), which would make this a minimum estimate.

It is interesting that, because most of the heat is released through the components of lithospheric formation and oceanic background, the heat flow through the Earth's surface is significantly biased toward the Southern (ocean) Hemisphere. Due to the probable complexity of the kinematics and dynamics of material in the Earth's interior (McKenzie, 1968), this does not necessarily indicate long-term asymmetry in the cooling of the Earth. However, models of the Earth are no longer constrained by the previously accepted equality of oceanic and continental heat flux.

For comparison, we estimate that the total rate at which man uses energy (coal, oil, and natural gas) is approximately 2×10^{12} cal/s. The energy dissipated by tides in the Earth, water, and atmosphere is approximately 0.7×10^{12} cal/s (Hendershott, 1973). Because only a small part of tidal dissipation occurs in the solid earth, it contributes a negligible part of the total heat loss. Finally, the heat intercepted by the Earth from solar radiation is about 4×10^{16} cal/s, about 4,000 times the rate of heat loss from the Earth's interior. From first-order error analyses, we estimate the accuracy of this new value for the Earth's rate of heat loss, 10.2×10^{12} cal/s, at ±15 percent. The principal uncertainty derives from the unknown parameters of the oceanic lithosphere (Table 1). A substantial fraction (40 percent) of the heat loss derives from the transient cooling of new oceanic lithosphere, and it seems probable that this mechanism is the "safety valve" for excessive heat generated in the Earth's interior.

REFERENCES CITED

Bullard, E. C., Maxwell, A. E., and Revelle, R., 1956, Heat flow through the deep sea floor: Advances in Geophysics, v. 3, p. 153-181.

Elder, J. W., 1965, Terrestrial heat flow: Am. Geophys. Union Mon. 8, p. 211-239.

Forsyth, D. W., and Press, F., 1971, Geophysical tests of petrological models of spreading lithosphere: Jour. Geophys. Research, v. 76, p. 7963-7979.

Hendershott, M. C., 1973, Ocean tides: EOS (Am. Geophys. Union Trans.), v. 54, p. 76-86.

Larson, R. L., and Chase, C. G., 1972, Late Mesozoic evolution of the western Pacific Ocean: Geol. Soc. America Bull., v. 83, p. 3627-3644.

Larson, R. L., and Pitman, W. C., 1972, World-wide correlation of Mesozoic magnetic anomalies, and its implications: Geol. Soc. America Bull., v. 83, p. 3645-3662.

Lee, W.H.K., 1970, On the global variations of terrestrial heat-flow, *in* Geophysical studies on the evolution of the Earth's deep interior: Physics Earth and Planetary Interiors, v. 2, p. 332-341.

Lee, W.H.K., and Uyeda, S., 1965, Terrestrial heat flow: Am. Geophys. Union, Geophys. Mon. 8, p. 87-190.

McKenzie, D. P., 1968, The influence of the boundary conditions and rotation on convection in the Earth's mantle: Royal Astron. Soc. Geophys. Jour., v. 15, p. 457-500.

McKenzie, D. P., and Sclater, J. G., 1969, Heat flow in the eastern Pacific and sea floor spreading: Bull. Volcanol., v. 33, p. 101-118.

Menard, H. W., and Smith, S. M., 1966, Hypsometry of ocean basin provinces: Jour. Geophys. Research, v. 71, p. 4305-4325.

Parker, R. L., and Oldenburg, D. W., 1973, Thermal model of ocean ridges: Nature, v. 242, p. 137-139.

Schatz, J. F., and Simmons, G., 1972, Thermal conductivity of Earth materials at high temperatures: Jour. Geophys. Research, v. 77, p. 6966-6983.

Sclater, J. G., and Francheteau, J., 1970, The implication of terrestrial heat flow observations on current tectonic and geochemical models of the crust and upper mantle of the Earth: Royal Astron. Soc. Geophys. Jour., v. 20, p. 509-542.

Sleep, N. H., 1969, Sensitivity of heat flow and gravity to the mechanism of sea-floor spreading: Jour. Geophys. Research, v. 74, p. 542-549.

Veevers, J. J., and Heirtzler, J. R., 1973, Deep Sea Drilling Project, Leg 27: Geotimes, v. 18, no. 4.

Verhoogen, J., 1973, Possible temperatures in the oceanic upper mantle and the formation of magma: Geol. Soc. America Bull., v. 84, p. 515-521.

von der Borch, C. C., Sclater, J. G., and others, 1972, Deep Sea Drilling Project, Leg 22: Geotimes, v. 17, no. 6.

Williams, D. L., Von Herzen, R. P., Sclater, J. G., and Anderson, R. N., 1974, Royal Astron. Soc. Geophys. Jour. (in press).

ACKNOWLEDGMENTS

Reviewed by Roger N. Anderson and David Blackwell.

Supported by Woods Hole Oceanographic Institution and by National Science Foundation Grant GA-16078.

Contribution No. 3282, Woods Hole Oceanographic Institution.

MANUSCRIPT RECEIVED MARCH 25, 1974

MANUSCRIPT ACCEPTED APRIL 8, 1974

8

Reprinted from *Canadian Jour. Earth Sci.* 9:471-478 (1972)

Oxygen Isotope Geochemistry of Submarine Greenstones

K. MUEHLENBACHS[1]
Department of Chemistry, University of Chicago, Chicago, Illinois 60637

AND

R. N. CLAYTON
Departments of Chemistry and Geophysical Sciences and the Enrico Fermi Institute, University of Chicago, Chicago, Illinois 60637

Manuscript received November 19, 1971
Revision accepted for publication January 28, 1972

Most submarine greenstones recovered from the North Mid-Atlantic Ridge range in $\delta^{18}O$ from 2.8 to 6.8‰ (SMOW). An extremely light rock. ($\delta^{18}O = -7.1‰$), was also analyzed, but it might be an ice rafted erratic. The heavier ones, up to 8‰, are probably weathered. On the average, the $\delta^{18}O$ of 14 greenstones is slightly lower than that of fresh unmetamorphosed submarine basalt. The ^{18}O fractionations between coexisting metamorphic minerals of the greenstones are consistent with the independently inferred range of metamorphic temperatures of 200–300 °C. The $\delta^{18}O$ of the possible metamorphic water can be estimated from this temperature range and the $\delta^{18}O$ of the metamorphic minerals to be −2 to +2‰. This argues strongly that little or no 'juvenile' water (water outgassed from the mantle) participated directly in the metamorphism. The $\delta^{18}O$ data are consistent with the participation of sea water in the metamorphism. Similar conclusions can be drawn from the ^{18}O analyses of a submarine diabase and two gabbros. Disseminated carbonates in the greenstones have $\delta^{13}C$ values typically found in carbonatites or partially decarbonated marine limestones.

La plupart des roches vertes de la crête Midio-Atlantique Nord ont des valeurs $\delta^{18}O$ variant de 2.8 à 6.8‰ (SMOW). Une roche extrêmement légère, probablement un "ice rafted boulder", a une valeur $\delta^{18}O = -7.1‰$. Les roches les plus lourdes, dont les valeurs atteignent 8‰ sont probablement altérées. Les valeurs $\delta^{18}O$ moyennes des 14 roches vertes analysées sont légèrement inférieures à celles des basaltes océaniques non altérés ni métamorphisés.

L'intervalle de température du métamorphisme (200–300 °C) déterminé par la répartition de l'^{18}O entre les minéraux métamorphiques coexistants concorde avec celui déterminé par d'autres méthodes. Le même intervalle permet d'avoir une idée de la valeur $\delta^{18}O$ de l'eau de métamorphisme et d'estimer celle des minéraux métamorphiques (−2 à +2‰). D'après ces valeurs, l'eau de mer, plutôt que l'eau juvénile (eau directement issue du manteau) aurait participé au métamorphisme. Les analyses de l'^{18}O de diabases sous-marines et de deux gabbros mènent aux mêmes conclusions. Les carbonates disséminés dans les roches vertes ont des valeurs $\delta^{13}C$ identiques à celles des carbonatites et des calcaires marins partiellement décarbonatés.

Introduction

The vast majority of rocks recovered from mid-ocean ridges are tholeiitic basalts (Engel *et al.* 1965). However, a wide variety of metamorphosed basic rocks have been dredged from the lower reaches of seamounts, fracture zones, and fault blocks. There is little doubt that these rocks are metamorphosed equivalents of the common basalts. Dredged diabases and gabbros are invariably metamorphosed at least to a slight degree (Miyashiro *et al.* 1970; Barrett and Aumento 1970). Metavolcanics are often thought to make up the oceanic crust (Barrett and Aumento 1970; Cann 1970).

---

[1]Present address: Carnegie Institution of Washington, Geophysical Laboratory, Washington, D.C. 20008.

The lowest grade of metamorphism observed in submarine basalts is the zeolite facies (Miyashiro *et al.* 1971; Aumento *et al.* 1971). The oxygen isotope geochemistry of 2 such metabasalts was discussed by Muehlenbachs and Clayton (1972). The most common metabasalts recovered are greenstones: submarine basalts metamorphosed to the greenschist facies (Melson and Van Andel 1966; Hekinian 1968; Cann 1969; Aumento *et al.* 1971; Miyashiro *et al.* 1971; Bonatti *et al.* 1970; Chernysheva 1971). Igneous texture is preserved in these rocks, although some are brecciated or mylonitized. The primary minerals and glass have been reconstituted to quartz, albite, chlorite, actinolite, and epidote. Some relic plagioclase and pyroxene remain. In contrast with continental occurrences, only rare carbonates are found (Cann 1969).

Compared to fresh submarine basalts, many of the greenstones have undergone marked changes in Na_2O, CaO, H_2O, and boron contents (Melson and Van Andel 1966; Cann 1969; Miyashiro *et al.* 1971; Thompson and Melson 1970; Aumento *et al.* 1971), indicating metamorphism in the presence of abundant water. Magnetic properties of the greenstones as well as the location of their dredge sites imply that they were metamorphosed at low pressure by burial of only a few km below the sea floor (Barrett and Aumento 1970; Miyashiro *et al.* 1971; Aumento *et al.* 1971). From heat flow measurements on ridges as well as from estimates of mineral stabilities it has been inferred that the temperature of the greenstone metamorphism ranged from 200–300 °C (Melson and Van Andel 1966; Cann 1969; Aumento *et al.* 1971).

Garlick (1969) has reviewed previous oxygen isotope studies of metamorphic rocks. No studies on the metamorphism of submarine basalts have been published. Only a few $\delta^{18}O$ analyses of continental greenstones exist. An albite-quartz-chlorite-epidote greenstone from the Lake St. Joseph area, S.W. Ontario, Canada (C63–35; Clifford 1969) was found to have $\delta^{18}O = 10.2‰$. Four albite and chlorite-containing spilitic rocks from the Pinde Ophiolite, Greece, ranged from 9.2–11.7‰ (Javoy 1970).

This study is concerned with the oxygen isotope geochemistry of submarine greenstones from the Mid-Atlantic Ridge. We have determined the isotopic composition of oxygen in a number of submarine greenstones and gabbros as well as in some of their constituent minerals. From these data we attempt to determine temperature of metamorphism and the identity of the fluid involved. The isotopic composition of carbonates from greenstones has also been determined.

Experimental Procedure

All of the submarine rocks analyzed in this study were obtained by dredging mid-ocean ridges. The location and geophysical descriptions of the dredged areas and some chemical and petrographic descriptions of individual samples can be found in the literature reviewed with the isotopic data.

Many of the samples were obtained in powdered form from the collector and were analyzed without further preparation except drying for one hour at 100 °C. Powders of rocks from the Mid-Atlantic Ridge near 45° N, designated by the prefix AG, were prepared in this laboratory by crushing and grinding rock chips using a steel mortar and pestle. Mineral separates were made from crushed samples (100–200 mesh) by density and chemical separation and by hand picking. The purity of the mineral separates was checked by optical means and was above 95%. The anorthite content of the plagioclase was estimated from the index of refraction. The impurities were for the most part inclusions in the grains.

Oxygen was extracted from the rocks and minerals by the BrF_5 method (Clayton and Mayeda 1963). Oxygen was recovered quantitatively from all samples including the water-rich greenstones. Carbon dioxide was liberated from the carbonates by treating whole rock powders with 100% phosphoric acid at 25.3 °C for 24 hr.

Oxygen extracted from the rocks was converted to CO_2 and analyzed in a 15 cm, 60° sector, double-collecting, mass spectrometer (McKinney *et al.* 1950). Corrections were made for valve mixing, $\delta^{13}C$ variations, mass-44 tail, and background. The isotopic compositions are reported in the usual δ notation, where for oxygen (and similarly for carbon) δ_x is defined as:

$$\delta_x = \left[\frac{(^{18}O/^{16}O)_x}{(^{18}O/^{16}O)_{standard}} - 1 \right] \times 1000$$

The oxygen isotope analyses are reported relative to the SMOW standard (Craig 1961) while carbon isotope analyses are reported relative to the PDB standard (Craig 1957). All of the silicate analyses were performed at least in duplicate except, because of lack of sample, for AG-168-3Ab, Pc, Qtz; AG-168-1 Qtz; AG-198-6 Qtz; and AG-198-2 Px. The pooled variance of silicate analyses performed during the course of this work is .08‰.

Results and Discussion

Greenstones

The oxygen isotope analyses of submarine greenstones and gabbros are presented in Table 1. One is struck by the wide range of $\delta^{18}O$ values, −7.1 to +8.0‰, of the greenstones compared to the very narrow range of +5.5 to +5.9‰ of fresh unmetamorphosed submarine basalts (Taylor 1968; Muehlenbachs and Clayton 1972). The higher values might reflect submarine weathering and zeolitization processes that are known to increase markedly the $\delta^{18}O$ of submarine basalts (Garlick and Dymond 1970; Muehlenbachs and Clayton 1972). Except for the isotopically heavy layer of the metamorphosed pillow basalt, A150-20-Am52B, to be discussed later, the heavier rocks do show signs of such low temperature alteration unrelated to the greenschist facies metamorphism. The heaviest rock is CH 44-2-7, a rather oxidized and zeolitized metamorphosed tuff (Fe_2O_3/FeO = 1.5). The next heaviest are TW-1-2-9 and CH 44-3-6, a mylonitized basalt and a greenstone, respectively.

Several of the greenstones as well as the gabbros are significantly lighter: −7.1 to +5.2‰. These light rocks could not have been produced by metamorphism of a "normal" submarine basalt in a closed system. Their oxygen atoms must have exchanged with an external reservoir which decreased the $^{18}O/^{16}O$ ratio. This corresponds with the observations by others (Melson *et al.* 1968; Cann 1969; Miyashiro *et al.* 1971) of changes in chemical composition and indications of hydrothermal alteration. All the greenstones exhibit water enrichment, but in contrast with the water enrichment noted in submarine weathering and palagonitization (Garlick and Dymond 1970; Muehlenbachs and Clayton 1972), no correlation between water content and $\delta^{18}O$ is found.

Three concentric layers from a supposedly unweathered metamorphosed pillow were analyzed, A150-20-Am 52. It is composed mainly of plagioclase and chlorite, the latter of which increases in abundance towards the rim (Miyashiro 1969, pers. comm.). The metamorphic derivative of the original glassy rim, A (10 mm thick, Fe_2O_3/FeO = 0.15) has a $\delta^{18}O$ near that of the majority of greenstones, 5.54‰. However, the interior, C (Fe_2O_3/FeO = 0.21) has a $\delta^{18}O$ of 6.06‰, slightly heavier than the presumed parent rock, while the intermediate layer, B (10–20 mm thick, Fe_2O_3/FeO = 0.45) is much heavier, 8.02‰. Such inhomogeneities within one rock are difficult to explain, but have also been observed in unmetamorphosed but weathered submarine basalts (Garlick and Dymond 1970; Muehlenbachs and Clayton 1972). Thus A150-20-Am 52 might have been slightly weathered, establishing an isotopic inhomogeneity in the pillow, and then metamorphosed, with only the outer layer having exchanged oxygen with its surroundings.

Coexisting Minerals

Specific information about the conditions of metamorphism can be obtained from the oxygen isotope analyses of coexisting minerals. In the most favorable case, where equilibrium has been established, and where isotopic fractionations between minerals have been calibrated to give temperatures, one can calculate the temperature of metamorphism as well as the isotopic composition of any fluids involved (Clayton and Epstein 1961). Quartz, albite, chlorite, epidote, calcic plagioclase, and pyroxene were isolated from submarine greenstones and analyzed and the results are also reported in Table 1.

The isotopic fractionations between quartz and albite in the submarine greenstones are of the expected magnitude and direction. The rather large fractionation of 2.87‰ observed in AG-168-3 is nearly the same as that found by O'Neil and Taylor (1967) in a quartz-adularia vein from Hot Springs, Arkansas. They calculated a temperature of 265 °C for this fractionation. In the same greenstone a fractionation of 7.62‰ between quartz and epidote is observed. A quartz-epidote pair from the I.I.D. No. 2 well in the Salton Sea geothermal field (California) gave a fractionation of 5.4‰ at a measured temperature of about 300 °C. Thus if both fractionations reflect

TABLE 1. Isotopic composition of greenstones and metagabbros from the mid-Atlantic ridge

Sample Number	Description	$\delta^{18}O$ Rock	$\delta^{18}O$ Qtz[5]	$\delta^{18}O$ Ab	$\delta^{18}O$ Pc	$\delta^{18}O$ Other	$\delta^{18}O$ Cc	$\delta^{13}C$ Cc	Yield[1]
Greenstones									
A-150-20 Am-52	1								
Rim A		5.54							
Intermediate Layer B		8.02							
Center C		6.06							
AG-22-68-112-2	2,3	4.53							
AG-22-68-120-3	2,3	2.81	7.15			1.12 (Ch)			
AG-22-68-165-12	2,3	6.52							
AG-22-68-168-1	2,3	5.61	9.39	7.18	6.52 (An_{30})	5.04 (Px)	12.67	−4.48	0.68
AG-22-68-168-3	2,3	5.79	10.23	7.36	7.08 (An_{35})	2.61 (Ep)			
AG-22-68-187-1	2,3	3.21			4.26 (An_{45})		7.38	−5.46	0.97
AG-22-68-198-6	3	5.66	9.02		5.54 (An_{80})		22.50	−3.93	0.07
Th-6-3	2,3	−7.13					−0.74	−5.58	2.18
AG-22-134-22	3	4.60					8.82	−6.58	3.02
CH-44-2-5	4,5	5.62							
CH-44-3-2	4,5	5.33							
CH-44-3-3	4,5	5.66							
CH-44-3-6	4,5	6.80							
CH-44-3-7	4,5	5.28							
CH-44-2-7[2]	5	7.61							
TW-1-2-9[3]	5	6.47							
C 63-35[4]	6	10.20					23.21	4.01	10.8
Diabase and metagabbros									
CH-44-3-10	5	5.34			5.58 (An_{70})	5.26 (Px)			
AG-22-68-198-1	2	4.92			4.57 (An_{80})	5.11 (Px)			
AG-22-68-198-2	2	5.00			4.97 (An_{70})	4.89 (Px)			

(1) Miyashiro *et al.* 1971.
(2) Unpubl. rep., Atlantic Oceanographic Lab., Bedford Inst., N.S.
(3) Aumento *et al.* 1971.
(4) Melson and Van Andel 1966.
(5) Melson *et al.* 1968.
(6) Schwarcz 1969, pers. comm.
[1]Yield measured as micromoles CO_2 per mg. of rock.
[2]Metamorphosed tuff.
[3]Mylonitized basalt.
[4]Greenstone, Lake St. Joseph Area, S. W. Ontario (Clifford 1969).
[5]Abbreviations: Qtz = Quartz, Ab = Albite, Pc = Plagioclase, Cc = Carbonate, Px = Pyroxene, Ch = Chlorite, Ep = Epidote.

isotopic equilibrium, then the metamorphic minerals in AG-168-3 must have formed substantially below 300 °C. The 6.03‰ quartz-chlorite fractionation in AG-120-3 is about 1.5‰ smaller than that found in regionally metamorphosed biotite grade pelitic schists (Garlick 1969). This would indicate a higher temperature of metamorphism for the greenstone. However, quartz-chlorite isotope fractionations are not good geothermometers, because chlorite can be either a primary or secondary mineral and one does not know if its isotopic chemistry might not depend on its wide compositional range. The isotopic temperatures are within the range proposed independently by other studies of submarine metamorphism (Melson and Van Andel 1966; Cann 1969; Aumento *et al.* 1971). They are, however, below the temperature of the greenschist facies as deduced from the experimental work of Liou (1971) and Thompson (1971). Our estimates are probably compatible with their studies since the combination of low pressure and saline metamorphic fluid lowers the temperatures of the pertinent phase transitions.

It should be noted that the plagioclase phenocrysts and xenocrysts in unmetamorphosed submarine basalts are quite calcic (An_{70-90}) and uniform in $\delta^{18}O$, 5.5–6.1‰ (Taylor 1968; Muehlenbachs and Clayton 1972). Thus the more sodic relic plagioclase in some greenstones (An_{30-50}) must have undergone changes in both anion and cation composition. The observed isotopic differences between coexisting albite and relic plagioclase are a little less than would be expected on the basis of differences in their chemical composition, as shown by O'Neil and Taylor (1967).

If isotopic equilibrium is established among coexisting minerals then the order in which they concentrate ^{18}O should be the same as observed elsewhere (reviewed by Garlick 1969). From laboratory experiments as well as natural occurrences it is known that at equilibrium quartz concentrates ^{18}O relative to coexisting calcite. Such a relationship is not observed in the submarine greenstones. Clearly the carbonates are much too heavy to be in equilibrium with other minerals of the same rock, except possibly for AG-187-1 and AG-134-22.

There is a tendency for the more abundant carbonates to be lighter, i.e., nearer to the expected equilibrium value. This might mean that the metamorphic carbonates are contaminated or partially exchanged with low temperature diagenetic carbonate. A carbonate vein containing foraminiferal tests from AG-168-3 yielded a typical marine carbonates $\delta^{13}C$ value of +2.4‰ and a $\delta^{18}O$ of +35.7‰, values indicative of ocean floor conditions. The lightest, presumably least altered $\delta^{18}O$ and $\delta^{13}C$ values are similar to those observed in carbonatites. This would appear to be significant in view of possible carbonatitic volcanism along the ocean ridges (Boström *et al.* 1968). Schidlowski *et al.* (1970) argued that such $\delta^{13}C$ values found in carbonates of continental spilites indicate a magmatic origin for the carbonates. However, carbonates from regionally metamorphosed greenschists from Otago, New Zealand, commonly have $\delta^{13}C$ values typical of carbonatites (Devereux 1968). Deines and Gold (1969), Hoefs and Epstein (1969) and Shieh and Taylor (1969) have also shown that decarbonation of sedimentary carbonates can give rise to "igneous" isotopic ratios.

Coarse-grained Rocks

As was mentioned earlier, the $\delta^{18}O$ values of two submarine gabbros and one diabase are lower than those typical of mafic rocks. This is not surprising since these rocks are known to be slightly metamorphosed (Melson *et al.* 1968; Barrett and Aumento 1970). Table 1 also lists the oxygen isotopic composition of plagioclase and pyroxene from these rocks. No carbonates were found. None of these rocks display the plagioclase-pyroxene fractionations typical of continental gabbros (~1.2‰). AG-198-1 is striking in that its pyroxene is heavier by 0.4‰ than the coexisting plagioclase. This reversal of direction of the fractionation is a sure sign of disequilibrium. A similar occurrence has been described from Skye (Taylor 1968) where the disequilibrium was ascribed to the interaction of the hot igneous rock with isotopically light meteoric water. Plagioclase exchanges oxygen atoms with hot water more readily than does pyroxene, thus causing the disequilibrium. The observation that calcic plagioclase of some submarine metamorphic rocks has undergone extensive exchange of oxygen atoms without alteration of its chemical composition strongly supports the recent hypothesis of Miyashiro *et al.* (1971) that such calcic plagioclase need not be "relic", but may be a metamorphic mineral of a distinct metamorphic facies characterized by the mineral assemblage actinolite-chlorite-calcic plagioclase.

The Metamorphic Fluid

The greenstones and metagabbros have acquired from 2 to 4% water and also show signs of hydrothermal alteration (Melson *et al.* 1968; Cann 1969; Aumento *et al.* 1971; Miyashiro *et al.* 1971). Thus it may be inferred that an aqueous phase played a role in their metamorphism. The fluid might have participated in the metamorphism only as a phase of small volume which facilitated the reconstitution of the basaltic glass and minerals. At first glance, this appears to have happened in the greenstones whose isotopic composition is near that of fresh basalts. However, the coincidence between the $\delta^{18}O$ of fresh basalts and that of some of the greenstones does not necessarily imply closed system metamorphism, because whole-rock isotopic compositions similar to igneous values can result due to the modal composition of the greenstones. Quartz and albite are relatively heavy, while chlorite and epidote are light. If, by analogy with hornblende (Taylor 1968), actinolite has an isotopic composition similar to pyroxene, and if the minerals had been equilibrated with water of sea water composition during metamorphism, then one would expect whole-rock $\delta^{18}O$ values for the greenstones to be similar to those of unaltered basalts.

Insight into the nature of the metamorphic fluid can be gained from the data in Table 1 by estimating the isotopic composition of water in equilibrium with quartz, albite, calcite, and calcic plagioclase at various temperatures within the range considered reasonable from the isotopic, petrographic, and geophysical evidence. The isotopic composition of water in equilibrium with the average of the quartz samples is −2.9‰ at 200 °C, 0.0‰ at 250 °C, and +2.6‰ at 300 °C (Becker 1971). Likewise, at the same temperatures, water in equilibrium with the albite would be −2.3‰, +0.1‰, and +2.6‰ (O'Neil and Taylor 1967). It can be estimated (O'Neil *et al.* 1969) that water in equilibrium with the relatively abundant carbonates of AG-187-1 and 134-22 would have an isotopic composition of −1.3‰ at 200 °C, +0.9‰ at 250 °C, and +3.0‰ at 300 °C. Similar estimates based on the plagioclase analyses of the coarse grained rocks would lead to the same conclusion.

The above observations indicate that some of the greenstones and metagabbros have undergone extensive alteration of their oxygen isotopic composition and that any water involved in the metamorphism would have had a $\delta^{18}O$ of −2 to about +2‰.

The ocean, with $\delta^{18}O$ of 0.0‰, is a possible source of water which may be involved in metamorphism in the oceanic crust. By comparing heat flow measurements on Iceland and mid-ocean ridges, Palmason (1967) proposed that widespread, deep-seated convection of sea water occurs in the oceanic crust under ridges. The average $\delta^{18}O$ of the submarine greenstones in Table 1 is about 0.5‰ less than that of fresh submarine basalt. By using mass balance considerations, and accepting the observed depletion of $\delta^{18}O$ in the greenstones as real and significant, the $\delta^{18}O$ of the inferred metamorphic water implies that the volume ratio of sea water to rock during metamorphism must have been at least 1:8 and probably greater. Taylor (1970) estimated that some shallow igneous intrusions had exchanged oxygen with a volume of meteoric water equal to the volume of rock.

It has been suggested that 'juvenile' water (water outgassed from the mantle for the first time) might be debouched along the oceanic ridges (Boström and Peterson 1969). One expects such water to have a $\delta^{18}O$ of 7 to 9‰ as a result of exchange with the silicate reservoir at temperatures >1000 °C. Water of this isotopic composition could not have participated to any great extent in the metamorphism under study. If it had, the $\delta^{18}O$ of the quartz, albite, and plagioclase would be much heavier than is observed. If 'juvenile' water were involved, its imprint would be most visible in the gabbros (Cann 1970), but some of the strongest evidence for isotopically light metamorphic water comes from the coarsest grained gabbro, AG-198-1.

Conclusions

A notable feature of the oxygen isotopic composition of submarine greenstones is their variability. The extremes of the range of values observed, −7.1 to +7.6‰, need not reflect metamorphic processes on the mid-ocean ridges, because the ultra-light rock might be an ice-rafted erratic while the heavier values are probably due to submarine zeolitization and weathering. Disregarding these, one still has a range for 14 greenstones of 2.8 to 6.8‰, with an average of 5.2 ± 1.1‰. This must imply that at least in some cases metamorphism in the oceanic crust was not closed with respect to oxygen, and that on the average it tended to lower the $\delta^{18}O$ of the rocks. Three coarse-grained basic rocks also showed ^{18}O depletion.

Isotopic compositions of disseminated carbonates in submarine greenstones are variable and the carbonates are not in equilibrium with their hosts. They show evidence of low temperature exchange with sea water or addition of low temperature marine carbonates. In some cases they are similar to those found in carbonatites or decarbonated marine limestones.

Oxygen isotope fractionations between coexisting metamorphic minerals in the greenstones are consistent with the independently inferred range of metamorphic temperatures of 200°–300 °C (Melson and Van Andel 1966; Cann 1969; Aumento *et al.* 1971). It can be estimated that water in equilibrium with the metamorphic minerals in this temperature range has a $\delta^{18}O$ near 0‰. This argues strongly that little or no 'juvenile' water, or other water equilibrated at very high temperatures with rocks of the mantle, participated directly in metamorphism of greenstones.

The oxygen isotope data are consistent with the postulate of participation of sea water in the metamorphism of submarine greenstones. If the metabasalts and metagabbros analyzed in this study are representative of large parts of the oceanic crust, then the crust was equilibrated with sea water at the time of its formation. If this is indeed the case, it is important to note that metamorphism in a presumably open system causes only slight alteration of the average $\delta^{18}O$ of the greenstones. It follows that the interaction of sea water with basic rocks at 200–300 °C in the oceanic crust, acting over geologic time, may be the dominant control on the oxygen isotopic composition of sea water. Deffeyes (1970) and Mackenzie and Garrels (1971) have suggested that the formation of greenstones affects the Mg, Ca, K, and Fe budgets of the ocean and sediments. Diagenesis of sediments and the weathering of igneous rock on the continents and the ocean floors will of course affect the $\delta^{18}O$ of the oceans as well. The relative importance of these processes will be discussed in another paper.

Acknowledgments

We wish to thank Dr. B. D. Loncarevic of the Atlantic Oceanographic Laboratory, Dartmouth, Nova Scotia for the opportunity for one of us to participate on cruise 22-68 of the CSS Hudson. We also wish to thank Drs. F. Aumento of Dalhousie University, A. Miyashiro of the Lamont-Doherty Geological Observatory, and G. Thompson of the Woods Hole Oceanographic Institution for donation of samples and unpublished data. Mrs. T. K. Mayeda helped with the experimental procedures.

Financial support was provided by the National Science Foundation through grants GA-1390, GA-22711, and a graduate traineeship.

AUMENTO, F., LONCAREVIC, B. D., and ROSS, D. I. 1971. Hudson geotraverse, geology of the Mid-Atlantic Ridge at 45° N. Phil. Trans. Roy. Soc. London, A, **268**, pp. 623–650.

BARRETT, D. L. and AUMENTO, F. 1970. The Mid-Atlantic Ridge near 45° N. XI. Seismic velocity, density, and layering of the crust. Can. J. Earth Sci., **7**, pp. 1117–1124.

BECKER, R. 1971. A study of oxygen and carbon isotopic ratios in rocks from banded iron formations. Ph.D. Thesis, University of Chicago, 238 p.

BONATTI, E., HONNOREZ, J., and FERRARA, G. 1970. Equatorial Mid-Atlantic Ridge: Petrologic and Sr isotopic evidence for an Alpine-type rock assemblage. Earth Planet. Sci. Letters, **9**, pp. 247–256.

BOSTRÖM, K. and PETERSON, M. N. A. 1969. The origin of aluminum-poor ferromanganoan sediments in areas of high heat flow on the East Pacific Rise. Mar. Geol., 7, pp. 427–447.

BOSTRÖM, K., TALLBACKA, L., and JOENSUU, O. 1968. Evidence for carbonatitoid volcanism of active oceanic ridges. (Abstr.). Trans. Am. Geophys. Union, **49**, p. 364.

CANN, J. R. 1969. Spilites from the Carlsberg Ridge, Indian Ocean. J. Petrol., **10**, pp. 1–19.

——— 1970. New model for the structure of the ocean crust. Nat., **226**, pp. 928–930.

CHERNYSHEVA, V. I. 1971. Greenstone-altered rocks of rift zones in median ridges of Indian Ocean. Internat. Geol. Rev., **13**, pp. 903–913.

CLAYTON, R. N. and EPSTEIN, S. 1961. The use of oxygen isotopes in high-temperature geological thermometry. J. Geol., **69**, pp. 447–452.

CLAYTON, R. N. and MAYEDA, T. K. 1963. The use of bromine pentafluoride in the extraction of oxygen from oxides and silicates for isotopic analysis. Geochim. Cosmochim. Acta, **27**, pp. 43–52.

CLIFFORD, P. M. 1969. Geology of the Lake St. Joseph Area. Ont. Dep. Mines, Geol. Rep. 70, 55 p.

CRAIG, H. 1957. Isotopic standards for carbon and oxygen and correction factors for mass-spectrometric analysis of carbon dioxide. Geochim. Cosmochim. Acta, **12**, pp. 133–149.

——— 1961. Standard for reporting concentrations of deuterium and oxygen-18 in natural waters. Sci., **133**, pp. 1833–1834.

——— 1963. The isotopic geochemistry of water and carbon in geothermal areas. Proc. Spoleto Conf. on Nuclear Geol. (F. Tongiorgi (*Ed.*)), pp. 17–53.

DEFFEYES, K. S. 1970. The axial valley: A steady state feature of the terrain. *In*: The Megatectonics of Continents and Oceans (H. Johnson and B. L. Smith (*Eds.*)), Rutgers, pp. 194–222.

DEINES, P. and GOLD, D. P. 1969. The change in oxygen and carbon isotopic composition during contact metamorphism of Trenton limestones by the Mount Royal pluton. Geochim. Cosmochim. Acta, **33**, pp. 421–424.

DEVEREUX, I. 1968. Oxygen isotope ratios of minerals from regionally metamorphosed schists of Otago, New Zealand. N.Z. J. Sci., **11**, pp. 526–548.

ENGEL, A. E. J., ENGEL, C. G., and HAVENS, R. G. 1965. Chemical characteristics of oceanic basalts and the upper mantle. Geol. Soc. Am., Bull. 76, pp. 719–734.

GARLICK, G. D. 1969. The stable isotopes of oxygen. *In*: Handbook of Geochemistry, II (K. Wedepohl (*Ed.*)), Springer-Verlag, New York, 8-B1-B27.

GARLICK, G. D. and DYMOND, J. R. 1970. Oxygen isotope exchange between volcanic materials and ocean water. Geol. Soc. Am., Bull. 81, pp. 2137–2142.

HEKINIAN, R. 1968. Rocks from the Mid-Oceanic Ridge in the Indian Ocean. Deep Sea Res., **15**, pp. 195–213

HOEFS, J. and EPSTEIN S. 1969. $^{18}O/^{16}O$ ratios of

minerals from migmatites, rapakivi granites and orbicular rocks. Lithos, **2**, pp. 1–8.

JAVOY, M. 1970. Utilisation des isotopes de l'oxygène en magmatologie. Ph.D. Thesis, A la Faculté des Sciences de Paris, 196 p.

LIOU, J. G. 1971. Stilbite-laumontite equilibrium. Contrib. Mineral. and Petrol., **31**, pp. 171–177.

MACKENZIE, F. T. and GARRELS, R. M. 1971. Chemical history of the oceans. (Abstr.). Am. Assoc. Petrol. Geol., Bull. 55, p. 351.

MCKINNEY, C. R., MCCREA, J. M., EPSTEIN, S., ALLEN, H. A., and UREY, H. C. 1950. Improvements in mass spectrometers for measurement of small differences in isotope abundance ratios. Rev. Sci. Instr., **21**, pp. 724–730.

MELSON, W. G. and VAN ANDEL, TJ. H. 1966. Metamorphism in the Mid-Atlantic Ridge, 22° N latitude. Mar. Geol., **4**, pp. 165–186.

MELSON, W. G., THOMPSON, G., and VAN ANDEL, TJ. H. 1968. Volcanism and metamorphism in the Mid-Atlantic Ridge 22° N latitude. J. Geophys. Res., **73**, pp. 5825–5941.

MIYASHIRO, A., SHIDO, F., and EWING, M. 1970. Crystallization and differentiation in abyssal tholeiites and gabbros from Mid-Oceanic Ridges. Earth Planet. Sci. Letters, **7**, pp. 361–365.

——— 1971. Metamorphism in the Mid-Atlantic Ridge near 24° and 30° N. Phil. Trans. Roy. Soc. London, A, **268**, pp. 589–603.

MUEHLENBACHS, K. and CLAYTON, R. N. 1972. Oxygen isotope studies of fresh and weathered submarine basalts. Can. J. Earth Sci., **8**, pp. 1591–1594.

O'NEIL, J. R. and TAYLOR, H. P., JR. 1967. The oxygen isotope and cation exchange chemistry of feldspars. Am. Mineral., **52**, pp. 1414–1437.

O'NEIL, J. R., CLAYTON, R. N., and MAYEDA, T. K. 1969. Oxygen isotope fractionation in divalent metal carbonates. J. Chem. Phys., **51**, pp. 5547-5558.

PALMASON, G. 1967. On heat flow in Iceland in relation to the Mid-Atlantic Ridge. *In*: Iceland and Mid-Ocean Ridges (S. Björnsson (*Ed.*)). Vis. Isl. Reykjavik, pp. 111–117.

SCHIDLOWSKI, M., STAHL, W., and AMSTUTZ, G. C. 1970. Oxygen and carbon isotope abundances in carbonates of spilitic rocks from Glarus, Switzerland. Naturwiss., **57**, pp. 542–543.

SHIEH, Y. N. and TAYLOR, JR., H. P. 1969. Oxygen and carbon isotope studies of contact metamorphism of carbonate rock. J. Petrol., **10**, pp. 307–331.

TAYLOR, JR., H. P. 1968. The oxygen isotope geochemistry of igneous rocks. Contrib. Mineral. Petrol., **19**, pp. 1–71.

——— 1970. Oxygen isotope evidence for large-scale interaction between meteoric ground waters and tertiary diorite intrusions. Western Cascade Range, Oregon. (Abstr.). Trans. Am. Geophys. Union, 51, p. 453.

THOMPSON, A. B. 1971. Analcite-albite equilibria at low temperatures. Am. J. Sci., **271**, pp. 79–92.

THOMPSON, G. and MELSON, W. G. 1970. Boron contents of serpentinites and metabasalts in the oceanic crust: Implications for the boron cycle in the oceans. Earth Planet. Sci. Letters, **8**, pp. **61–65**.

9

Reprinted from *Geochim. et Cosmochim. Acta* **42**:107 (1978)

Hydrothermal alteration of oceanic basalts by seawater

SUSAN E. HUMPHRIS*
Woods Hole Oceanographic Institution, Massachusetts Institute of Technology

and

GEOFFREY THOMPSON
Woods Hole Oceanographic Institution/Massachusetts Institute of Technology

(*Received* 4 *April* 1977; *accepted in revised form* 6 *September* 1977)

Abstract—Hydrothermally altered pillow basalts dredged from the Mid-Atlantic Ridge, and belonging to the greenschist facies, have been studied in order to determine the mineralogical and corresponding chemical changes, that result from basalt–seawater interaction at elevated temperatures.

The mineralogical transformations are predominantly to albite–actinolite–chlorite–epidote assemblages. Quartz and pyrite are common accessory minerals. On the basis of their mineralogy, the samples may be divided into chlorite-rich and epidote-rich assemblages. The chlorite-rich assemblages, which are the predominant variety, show the greatest chemical changes, while the epidote-rich samples show very little change in composition compared with their basaltic precursors.

Mass balances across individual pillows in which the central portions are relatively unaltered allow the directions and ranges of elemental fluxes to be calculated. In general, SiO_2 and CaO are leached from the basalt, while MgO and H_2O are taken up. No consistent trends are observed for Na_2O and K_2O although they do show some variations in the core-and-rim analyses.

Consideration of the elemental fluxes in terms of steady-state geochemical mass balances for oceanic inputs and outputs indicates that hydrothermal alteration provides a sink for Mg, which may be extremely important in solving the problem of apparent excess Mg input to the oceans. The amount of Ca that is leached from the rock may be of significance in the geochemical budget of that element. The amount of SiO_2 in the circulating fluid is controlled by the solubility of quartz or amorphous silica, depending on temperature, and considerable redistribution of silica takes place within the basaltic pile. The changes in redox conditions during hydrothermal alteration do not affect the present day oxidation state of the atmosphere and hydrosphere.

[*Editors' Note:* In the original, material follows this abstract.]

10

Reprinted from *Contrib. Mineralogy and Petrology* **36**:123–134 (1972)

High Temperature Alteration Minerals and Thermal Brines, Reykjanes, Iceland

Jens Tómasson and Hrefna Kristmannsdóttir

National Energy Authority, Dept. of Natural Heat, Laugavegur 116, Reykjavík, Iceland

Received May 10, 1972

Abstract. The geothermal area on Reykjanes, Iceland has been investigated mineralogically. The temperature within the studied area is very variable from 30–300° C. Mineral zones corresponding to the temperature conditions in the area are found. Accidental changes in the geothermal system are also reflected in the mineralogy by formation of anhydrite. Changes in temperature conditions in the field are indicated by epidote occurrence at 40° C and retrograd formation of montmorillonite.

Introduction

The active volcanic belt in Iceland is the subaerial continuation of the crest of the Mid-Atlantic Ridge. The Reykjanes high temperature area lies within the active volcanic zone on the tip of the Reykjanes peninsula (Fig. 1). The Reykjanes thermal area is one of 17 active high temperature areas in Iceland all which are located within the active volcanic zone. The high temperature fields are characterized by fairly restricted area of hot ground with fumaroles, mud pools and sometimes solfataras. Subsurface temperatures exceed 200° C at depths of less than 1 km (in the area drilled to 1972). The groundwater in the area is of seawater origin. When the saline groundwater is heated it reacts with the surrounding rocks such that SO_4^{2-} and Mg^{2+} disappear nearly completely but Ca^{2+}, K^+ and SiO_2 increase (see Table 1).

The different types of basaltic rocks have a very varying susceptibility to alteration depending on the degree of crystallization and permeability. The glass,

Table 1. Concentration in ppm

Locality	Depth of drill-hole	Temp. °C	pH/°C	SiO_2	Na^+	K^+	Ca^{++}	Mg^{++}	Total carb as CO_2	SO_4^{--}	Cl
1. Reykjanes drillhole 2	300	190	7.2/23	374	1380	1607	1915	8	754	60	21610
2. Njardvíkurheidi drillhole 1	500	30	—	54	8900	294	2140	442	—	2280	19800
Sea Water	—	—	—	3	10520	416	386	1282	—	2640	13800

This data are from Björnsson *et al.* (1970 and 1971).

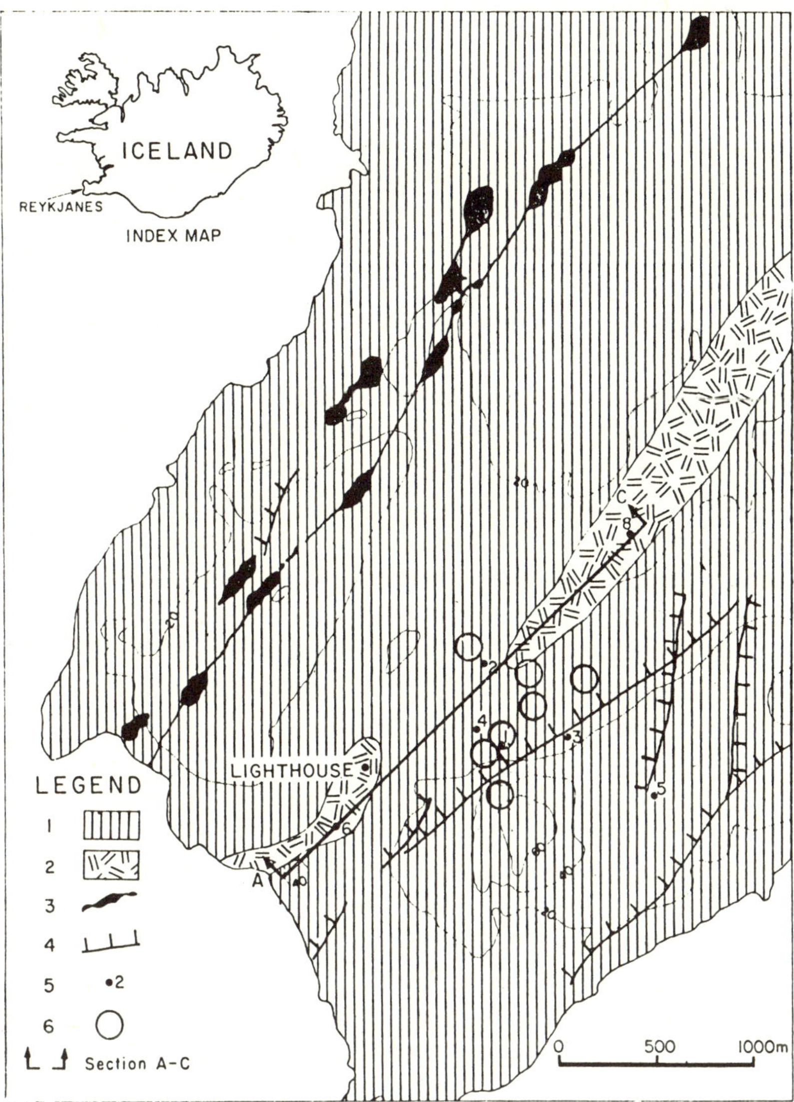

Fig. 1. A geological map of the area studied. *1* Basalt; *2* Palagonite breccia and tuffs; *3* Volcanic fissures; *4* Faults; *5* Drillholes (numbered); *6* Hot springs and fumaroles. The location of section A—C in Fig. 2 is shown on the map

the partly crystalline matrix and the dark minerals are mostly replaced by montmorillonite and chlorite. The glass is totally altered, but even where alteration is most intense about half of the original pyroxene is still left in the basalts, but the olivine is nearly completely altered. The plagioclase in the basalt is fairly resistant.

According to Kristjánsson (1972) it appears that most of the original magnetite in the deep continuous basalt formation (below 1 500 m) have survived the alteration.

Geology and Hydrology of the Area

The Reykjanes thermal area is mostly covered by postglacial lava flows, but Pleistocene hyaloclastic ridges protrude through the lavas in some places. A section through the area is shown in Fig. 2. Hyaloclastic tuffs and breccias and tuffaceous sediments dominate in the uppermost 1000 m. Beneath 1000 m about 50 percent of the rock is basalt and the rest tuffaceous rocks, mainly sediments.

The ground water in Reykjanes approximately 30 km inland is seawater and there are no significant differences between the two, except when the ground water is heated within the thermal area. In the Reykjanes thermal area high heat flow heats the saline ground water and causes it to rise nearly to the surface. The water table within the thermal system is at similar depth as that of the ground water table surrounding the system. The pressure within the thermal system is therefore lower than outside. A pressure difference of 10 atmospheres has been measured at 1700 m. Cold water therefore flows into the area from the sides replacing the water lost at the surface. Accompanying this circulation substantial precipitation of secondary minerals takes place on the boundaries of the thermal system forming an impervious cap. This leads to separation of the hydrothermal system from the surrounding cold ground water, This separation is most advanced close to the surface and is considered to decrease progressively downwards. An example of this is a 100 m deep hole 200 meters from the margin of the thermal area showing 30% of the tide. The water is seawater of temp.

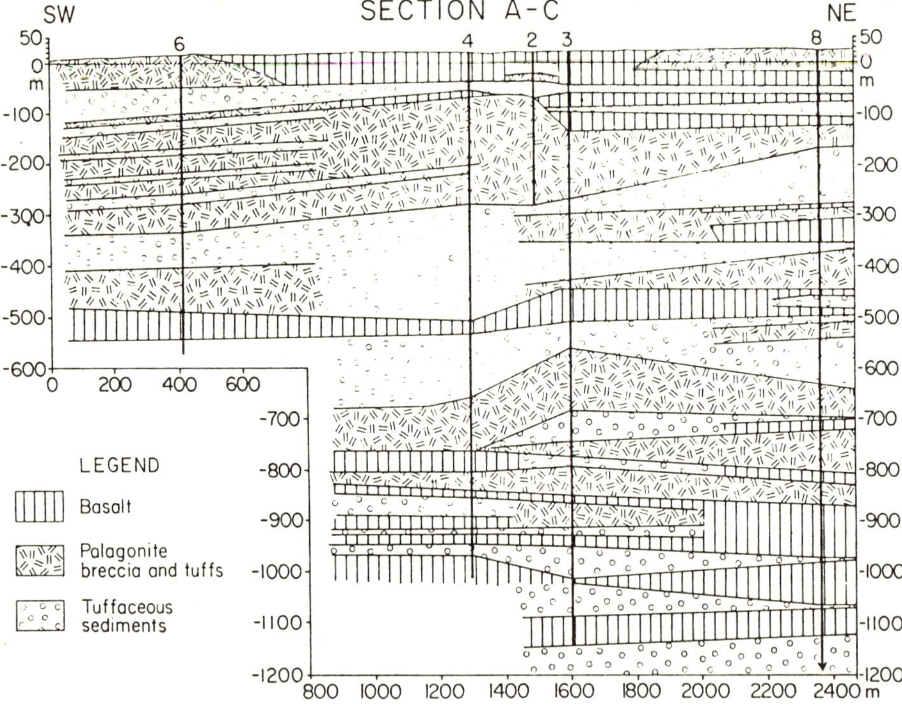

Fig. 2. A simplified cross section through the area, with the drillholes projected into the profile

12° C which is the same as the seawater temp. surrounding the peninsula but just inside the thermal area the temp. is 150° C at the same depth.

Formation and Distribution of Alteration Minerals

As seen from Fig. 2 the dominating rocks are breccias, tuffs and tuffaceous sediments. The primary or unaltered rocks are treated as composed of four main components; glass, crystalline matrix, dark minerals (olivine and pyroxene) and plagioclase. In Fig. 3a–c the distribution of the main alteration minerals is shown.

Anhydrite

Anhydrite ($CaSO_4$) has precipitated from the salty ground water and there is hardly any SO_4^{--} left in the thermal brine. The seawater is undersaturated in $CaSO_4$. As the solubility of $CaSO_4$ decreases with increasing temperature the saturation point will be reached during heating of the seawater within the geothermal system. It is difficult to calculate the temperature at which saturation occurs because of increasing Ca by percolation of seawater in the ground. Thus anhydrite should not be present in the upstreaming zone but should precipitate during the deep inflow of ground water into the system. Anhydrite is however present in high local concentrations in the upflow zone as seen in holes 2, 3 and 4. This anomalous presence of anhydrite is explained by occasional invasions of cold salty ground water into the hydrothermal system, which happened when the cap rock was fractured by tectonic movements. Earthquakes are frequent in the area and the last major earthquake occured in 1967 when substantial changes in the fumarole activity were observed. In earthquakes like these gaps form in the cap rock and the cold sea water invades the hydrothermal system because of the pressure difference. There is no anhydrite in the cold belts in the upper part of drillholes no. 3 and 4, where the temperature is less than 60° C and most often only about 40° C, indicating that the saturation point is not reached. If there has been any anhydrite in this zone (in hole 6 and the cold zones of 3 and 4) it has been dissolved by the cold seawater. Beneath the cold belt where the temperature gradient is very steep the greatest concentrations of anhydrite are found. The local concentrations in holes 2, 3 and 4 are approximately 2–3% anhydrite. The concentration of 0.2% SO_4^{--} in seawater corresponds to approximately 0.3% anhydrite. The maximum porosity of the rocks is 30%. If we assume that this volume is filled during the seawater invasion then the heating up of this seawater would give the maximum of 0.1% anhydrite for each event. This means that the 2–3% anhydrite was formed by 20–30 seawater invasions.

In drillhole 8 there are no local concentrations of anhydrite and the sporadic occurences of anhydrite are formed by the more gradual percolation and heating of the salty ground water.

Zeolites

Zeolites are found in the uppermost part of the drillholes but the zeolite zone extends to different depths in the different holes. The depth of occurence appears to be temperature dependent and zeolites disappear at about 230° C. In hole

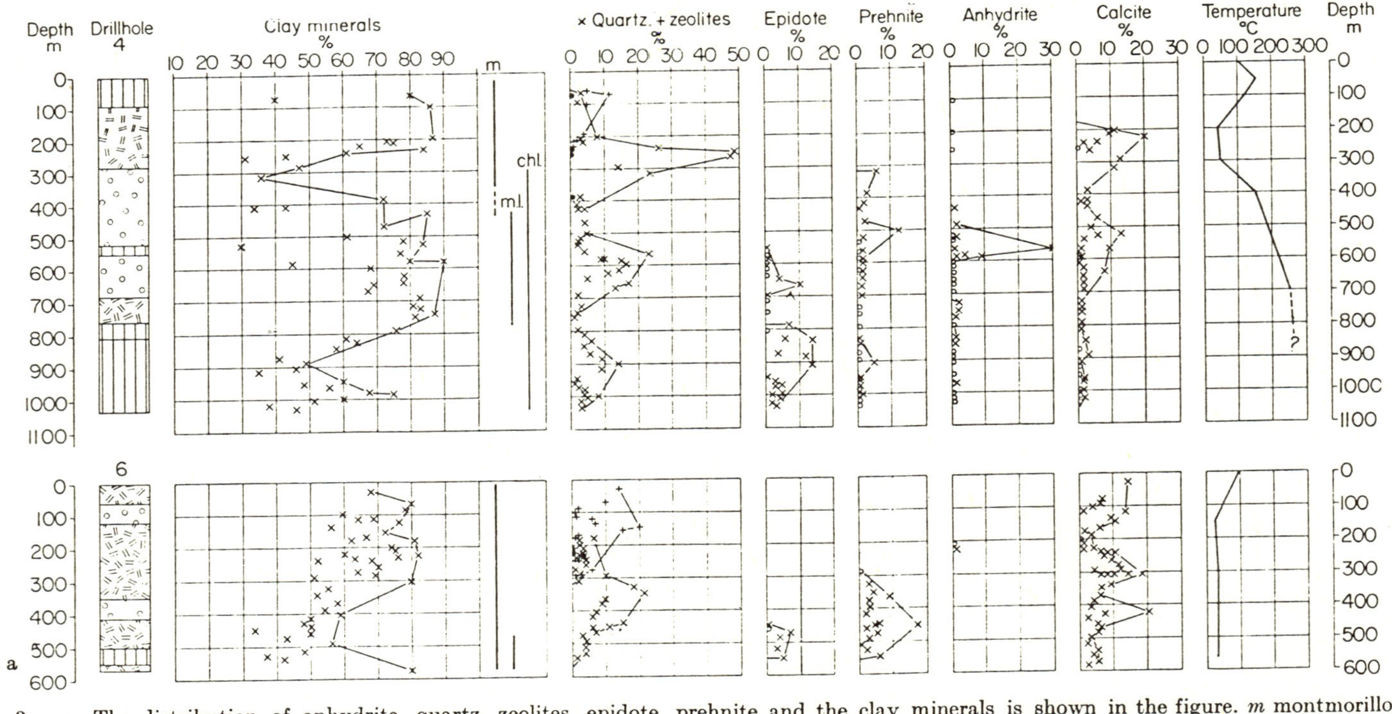

Fig. 3a—c. The distribution of anhydrite, quartz, zeolites, epidote, prehnite and the clay minerals is shown in the figure. *m* montmorillonite; *m.l.* mixed layer minerals; *chl* chlorite. The minerals were identified with common optical methods combined with X-ray diffraction analysis. In the case of clay minerals this was supplemented by D.T.A. analysis and chemical treatments. The amount is volume percent, obtained by point counting in thin sections made from the cuttings. The amount varies substantially even at similar depths but the line joins the points representing the maximum amount encountered. Also shown in the figures are predrilling thermal gradients mainly obtained by bottom measurements obtained during drilling intervals and a simplified geological section

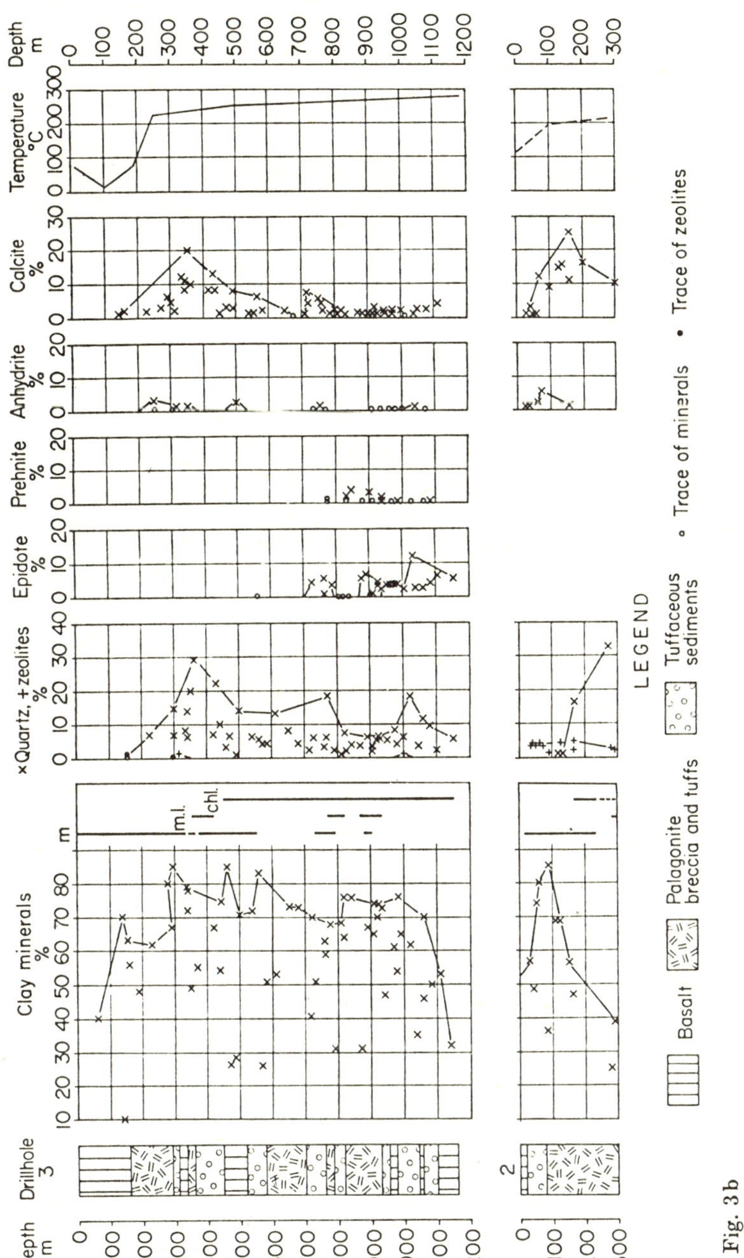

Fig. 3b

3 only minor amounts of zeolites are found. This is apparently caused by the fact that the temperature below the cold top layer increases so rapidly that no zeolites can form.

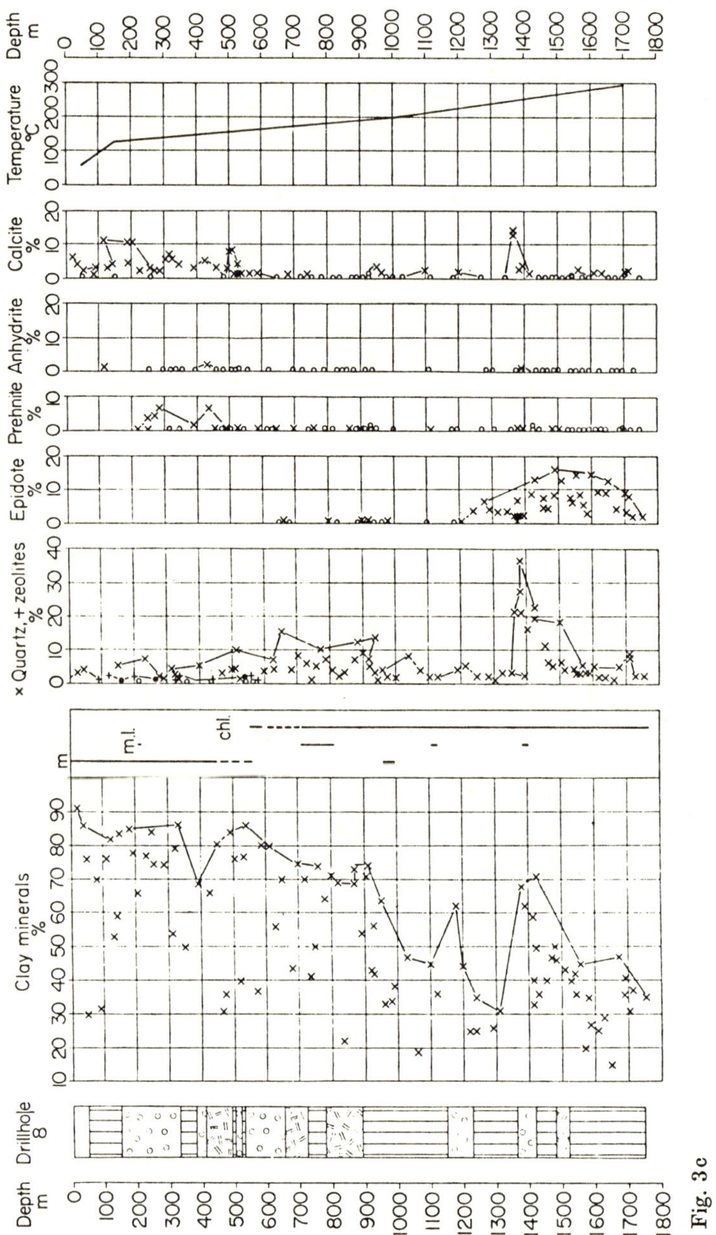

Fig. 3c

The main zeolites are mordenite, stilbite, mesolite and analcime. Analcime is found below the main zeolite belt in hole 8. Minor amounts of wairakite were observed.

Quartz and Opal

Opal occurs in the uppermost meters and is replaced by quartz which extends to the deepest levels penetrated by drillholes (1750 m). Opal is occasionally found with quartz indicating local disequilibrium, presumably due to supersaturation resulting from rapid cooling. Temperatures at the opal quartz boundary are about 100° C. The amount of quartz depends on the rock type and degree of alteration. The maximum amount occurs in hyaloclastites and tuffaceous sediments.

Calcite

Calcite is found throughout the altered rock but its amount varies much. It is most abundant in the tuffs. It is also obvious that the maximum amount occurs in the uppermost approx. 500–700 m. The most likely explanation of the greater abundance in the upper parts is the change in pH caused by boiling and the $CaCO_3$ zone extends to greatest depth in the hottest holes.

Prehnite

Prehnite occurs first in the zone between the zeolite zone and the epidote zone and is in varying content to the bottom of the holes.

Epidote

The formation of epidote is temperature dependent. Possibly the permeability influences its growth. Epidote first appears in tuffaceous sediments and breccias but disappears in the less permeable layers. At 260–270° C the epidote occurrence is continuous. Above the main epidote zone epidote forms from the clay fraction only and possibly calcite. The minimum temp. for epidote formation is approx. 200° C. In the main epidote zone the epidote forms mainly from plagioclase and less from the sheet silicates.

Other Minerals

Beginning albitization of plagioclase is found in several samples and it appears to be nearly unrelated to depth. Newly formed potash feldspar is found sporadically in all the holes. Pyrite occurs almost everywhere in the rocks in small concentrations. Hematite is most common in the uppermost tuffaceous layers.

Clay Minerals

The main types of sheet-silicates are montmorillonite minerals, chlorite and mixed-layer minerals of chlorite and montmorillonite. Small amounts of illite appear to be incorporated in some of the mixed-layer mineral structures. In Table 2 X-ray data are shown for all the types of sheet-silicates found in the samples.

Up to temperatures of about 200° C montmorillonite is the dominating sheet silicate. The chlorite appears in the temperature range 230–280° C. In the transition zone between the zones of montmorillonite and chlorite at temperature of 200–230° C the most common sheet silicates are random-mixed-layer minerals of chlorite and montmorillonite.

Table 2. X-ray diffraction data for the sheet silicates

Mineral	d (001) Å (35% rel moisture)	d (001) Å glycolated	d (001) Å heated at 600° C for 2 h	d (060) Å
Montmorillonite				
type 1 and 2	14.1–15.3	16.9–17.1	9.6–9.9	1.53
Montmorillonite				
type 3a	12.3–14.3 broad with top badly defined	16.7–17.0	9.7	1.51
type 3b	12.3–15.0 broad with top badly defined	16.5–17.0	9.6	1.53
Chlorite				
type 1	14.0–14.6	unchanged	13.9–14.0	1.53
type 2	14.0–14.6	15.0–15.5	13.9–14.0	1.53
type 3	14.0–14.6	15.0–15.5	none	1.53
type 4	14.0–14.6	16.5–17.0	13.9–14.0	1.53
type 5	14.0–14.6	16.5–17.0	none	—
Mixed layer minerals				
type 1	14.0–14.6	15.3–15.8	12–14 broad	not recorded
type 2	14.0–14.6	15.3–15.7	12.3–12.9 sharp	not recorded

The montmorillonites form zoned microcrystalline aggregates together with calcite and zeolites. In a top zone of 50–100 m and also at deeper levels in some breccias and pillow lavas there occurs a brown montmorillonite with low birefringence and variable refraction indices (^{n}mean = 1.510–1.570) (type 1 in Table 2). Most common beneath is a strongly green coloured montmorillonite with higher (mean) refraction indices and variable birefringence (type 2).

X-ray methods reveal no difference between those two types, which are identified as saponite with Ca as dominating interlayer cation. The optical characteristics suggest that the green saponite is a more Fe rich variety. Rock analysis show slightly higher MgO in upper levels possibly as a consequence of invasion of cold sea and subsequent loss of Mg to the rocks.

The ion exchange capacity of the montmorillonite is ca. 70 meq/100 g. In the cold shallow drillhole no. 6 and partly in drillhole no. 2 most of the montmorillonite (type 3 in Table 2) is interpreted as being in a state of irregular hydration, with more Na in interlayer positions. In the uppermost 100 m of drillhole 6 the montmorillonite is of normal dioctahedric type, replaced at greater depths by dominating trioctahedric montmorillonite.

The chlorite minerals are classified into five groups according to the X-ray diffraction behaviour (Table 2). The strongly expanding chlorite (type 4) is the most common. Normal non-swelling chlorite is only found at depths of 1200 to

1600 m in drillhole 8. The swelling chlorites (types 3 and 5) which totally disintegrate after heat treatment are almost restricted to the transition zone. Optically it can be difficult to see any difference between the chlorite minerals and the montmorillonites. All the chlorite minerals are green to brownish, often pleochroic. The thermally instable chlorites have lower refractive indices and higher birefringence than the others. They often resemble the mixed-layer minerals in optical character. The expanding chlorite of types 2 and 4 and the normal non-swelling chlorite have all similar optical features. Their refractive indices are about ^{n}mean = 1.600–1.610.

The mixed layer minerals are randomly mixed-layer structures of chlorite and montmorillonite. Illite occurs as minor component in the structure of some of them. Chlorite-illite mixed-layer minerals replacing pyroxene were also found in one restricted layer of basalt. The minerals have been roughly divided in two types (Table 2). Most common are minerals of type 1. The optical properties of the minerals are variable. They are brown or yellow to brownish. The birefringence is lower than that of montmorillonite but higher than that of chlorite, while the refractive indices are higher than for montmorillonite and usually lower than that of the chlorite.

On a mineral stability diagram for calcium and sodium minerals (Browne and Ellis, 1970) the Reykjanes water falls in a position close to the Ca-Montmorillonite-Wairakite join. The composition of the water shows very slight variation within the geothermal field itself. Outside the geothermal field where the temperature is much lower the Ca/Na activities are changed and in hole 6 Na becomes the dominating interlayer ion. Swelling chlorite beeing the main sheet silicate in the high temperature epidote/chlorite zone is an interesting feature. Swelling chlorites have not been recorded from the few other investigated geothermal fields in Iceland. Sporadic occurrence of swelling chlorites is known in other geothermal areas (Steiner, 1967). In clays from several places such minerals are found (Brown, 1961; Rich, 1968). Some artificial chlorites showing similar features as the natural swelling chlorites have been prepared from montmorillonite (Calliere and Hénin, 1949). Formation of chlorite-like structures from montmorillonite by precipation of $Mg(OH)_2$ occur in strongly alkaline milieu, while precipation of $Al(OH)_3$ probably could occur in less alkaline milieu (Gupta and Malik, 1969). In the Reykjanes geothermal field the formation of swelling chlorite has occurred in nearly neutral milieu. The minerals are trioctahedric although the nature or composition of the "brucite" interlayers are not known. The glass, fine-crystalline matrix and partly the dark silicate minerals have altered to trioctahedric sheet-silicate minerals. Fixation of discontinuous and imperfect "brucite" layers between the silicate layers has resulted in chlorite-like structures of swelling behaviour. The transformation from trioctahedric montmorillonite to a chlorite-like structure appear to have been possible without major structural changes. The swelling chlorites are most likely a metastable phase in transformation into a stable sheet silicate. Why they are so common in this system is not clear as there is little experimental knowledge about this mineral-group. This might be another consequence of the frequent accidental changes in the geothermal system.

Conclusion

Three vertical zones have been defined displaying progressive alteration of the rocks.

1) Montmorillonite-zeolite-calcite zone;
2) mixed-layers-prehnite zone;
3) chlorite-epidote zone.

The zones are not always clearly defined and the type minerals for low temperature zones can occur in high temperature zones. This is probably a consequence of the complex thermal history of the area with invasions of cold sea and reheating. No succession of the zeolite species is found after depth and temperature. As the chemical composition shows only slight variation, the variation in temperature, rock permeability and porosity must be responsible for the zones of different alteration minerals. The temperature is considered to be the main factor. The occurrence of epidote and chlorite at considerably lower temperature than in the main zones is most likely the consequence of higher permeability in those zones than in the rocks below. The greater concentrations of the anhydrite is not directly connected to the three zones, as it forms by a non-continuous process, connected with invasion of cold seawater.

The original chemical composition of rocks from investigated geothermal fields in New Zealand (Ellis, 1967; Brown and Ellis, 1970; Steiner, 1967) and U.S.A. (Muffler and White, 1969) is quite different from Icelandic geothermal fields. This is reflected by the alteration minerals, but many of the characteristic minerals are the same and appear in the same temperature range. Epidote usually occurs at lower temperatures in the Reykjanes area.

The only high-temperature geothermal areas in Iceland thoroughly investigated sofar are Reykjanes and Hveragerdi. Results from the geothermal field in Hveragerdi (Sigvaldason, 1962; White and Sigvaldason, 1962) show the same general pattern as in Reykjanes, but the chlorite-epidote zone reaches higher up even though the maximum temperature is lower. A mixed-layer minerals + prehnite zone appear to be lacking. The composition of thermal water in Hveragerdi is common hydrothermal water with much less dissolved solids. The resemblance of alteration in those two geothermal areas shows that the temperature is more important than chemical composition of the geothermal water in the alteration process.

References

Björnsson, S., Arnórsson, S., Tómasson, J.: Exploration of the Reykjanes Brine Area. Report to the U.N. Symposium on Geothermal Energy. Pisa 1970, 1–25.

Björnsson, S., Tómasson, J., Arnórsson, S., Jónsson, J., Ólafsdóttir, B., Sigurmundsson, S. G.: Reykjanes, Orkustofnun, An internal report (1971).

Brown, G.: The X-ray identification and crystal structures of clay minerals. Min. Soc. (Clay min. group) London 1961, 540 p.

Browne, P.R.L., Ellis, A.J.: The Ohaki-Broadlands hydrothermal area, New-Zealand: Mineralogy and related geochemistry. Am. J. Sci. **269**, 97–131 (1970).

Callière, S., Hénin, S.: Experimental formation of chlorites from montmorillonite. Mineral. Mag. **28**, 612–620 (1947–1949).

Gupta, G.C., Malik, W.U.: Chloritization of montmorillonite by its coprecipitation with magnesium hydroxide. Clays Clay Minerals **17**, 331–338 (1969).

Kristjánsson, L.: Magnetism of basaltic material recovered by deep drilling in Iceland. Earth and Planetary Sci. Letters (in press).

Muffler, L.J.M., White, D.E.: Active metamorphism of upper zenozoic sediment in the Salton Sea geothermal field and the Salton trough, Southeastern California. Bull. Geol. Soc. Am. **80**, 157–182 (1969).

Rich, C.I.: Hydroxy interlayers in expansible layer silicates. Clays Clay Minerals **16**, 15–30 (1968).

Sigvaldason, G.E.: Epidote and related minerals in two deep geothermal drill holes, Reykjavík and Hveragerdi, Iceland. U.S. Geol. Surv. Profess. Papers **450-E**, 77–79 (1962).

Steiner, A.: Clay minerals in hydrothermally altered rocks at Wairakei, New Zealand. Clays Clay Minerals **16**, 193–213 (1967).

White, E.D., Sigvaldason, G.E.: Epidote in hot-spring systems, and depth of formation of propylitic epidote in epithermal ore deposits. U.S. Geol. Surv. Profess. Papers **450-E**, 80–84 (1962).

Dr. J. Tómasson
H. Kristmannsdóttir
National Energy Authority
Department of Natural Heat
Laugavegur 116
Reykjavik, Iceland

Errata

Table 1.

1. Reykjanes drillhole 2 Na^+ 11380 ppm
   Seawater Cl^- 19800 ppm

Drillhole 2. Gypsum has been found above 100 m depth (reported as anhydrite in the paper). Potash feldspar identified optically, but has not been confirmed by X-ray.

The mineral referred to as montmorillonite in the paper is now designated smectite owing to a change in clay mineral classification.

11

Reprinted from *Jour. Geophys. Research* 76:8128-8138 (1971)

The Origin of Metal-Bearing Submarine Hydrothermal Solutions

JOHN B. CORLISS[1]

Department of Oceanography, Oregon State University, Corvallis, Oregon 97330

Instrumental activation analyses for 16 major and trace elements in a suite of mid-Atlantic ridge basalts reveals that the slowly cooled interior portions of these submarine extrusions are depleted, relative to the quenched flow margins, in several elements that are enriched in pelagic sediments and manganese nodules (Mn, Fe, Co, the rare earth elements, and others). Many of these elements are excluded from the solid phases that crystallize from the melt, and thus are concentrated in residual liquids. Additional elements are mobilized during the deuteric alteration of early-formed olivine and the formation of immiscible sulfide liquids. It is suggested that these components of melt occupy accessible sites (e.g., intergranular boundaries) in the hot solid rock mass, and are mobilized by dissolution as chloride complexes in sea water introduced along contraction cracks that form during cooling and solidification. These solutions may be the metal-bearing 'hydrothermal exhalations' or 'volcanic emanations' that accompany submarine volcanism, which are often cited as a source of metals into the pelagic environment. Reasonable estimates of the amount of material involved suggest that a significant fraction of the mass of these elements that reside in pelagic sediments could have been supplied by this process. Recently described amorphous iron-manganese-silica material from the east Pacific rise may form by direct precipitation from these hydrothermal solutions following their introduction into sea water.

Submarine volcanic activity has often been cited as a source of metals in the pelagic environment. Murray [*Murray and Renard*, 1891] first suggested a relationship between modern pelagic iron and manganese deposits and volcanic eruptions. *Park* [1948] cited the pillow basalts, pelagic sediments, and manganese deposits on the Olympic Peninsula as evidence of such a correlation. *Hewett* [1963] documented the connection between continental Mn accumulations and volcanic activity, and suggested a similar relationship for pelagic Mn nodules and encrustations.

Several processes have been proposed for the transfer of elements from submarine basalt magmas into the deep marine environment [e.g., *Bostrom*, 1967]. One such mechanism is the long-term, low-temperature weathering (halmyrolysis) of volcanic rock exposed at the sea floor. Another is the mobilization of elements during the violent interaction of molten lavas with sea water to form palagonite tuffs [*Bonatti and Nayudu*, 1965]. A third mechanism, often cited but never explicitly defined, is the action of 'hydrothermal exhalations' or 'volcanic emanations' that accompany submarine volcanic activity, which transport elements from the magma into sea water.

Bostrom and Peterson [1966, 1969] and *Bostrom* [1970] describe the enrichment of Fe, Mn, Cu, Cr, Ni, and Pb as metal oxide precipitates in pelagic sediments on the crest of the east Pacific rise. The zone of enrichment coincides with the zone of high heat flow over the crest of the rise. *Bender et al.* [1970] have determined accumulation rates in sediment cores from the east Pacific rise by the excess Th-230 method. They also measured the abundance of U, Th, Mn, Fe, and Cu in the cores. They find that the Mn accumulation rate is some 30 times greater at the crest compared to the flanks, and attribute the high rate to volcanic activity on the crest.

Bostrom and Peterson [1966] attribute these precipitates to 'ascending solutions, of deep seated origin, related to magmatic processes.' Several studies of ferromanganese nodules have cited these hypothetical solutions as sources for at least part of the metals found in these deposits [e.g., *Bonatti and Joensuu*, 1966; *Skornyakova et al.*, 1962; *Cronan*, 1967]. Layers of

[1] Present address: Department of Geology and Geophysics, Yale University, New Haven, Connecticut 06520.

metal-rich amorphous iron oxides in piston cores from near the east Pacific rise [*Sayles and Bischoff*, 1971] and in Deep Sea Drilling Project cores [*von der Borch and Rex*, 1970] and *von der Borch et al.* [1970] have also been attributed to the action of these hydrothermal solutions.

A recent review of the genesis of base metal ore deposits by *White* [1968] integrates experimental work and critical field observations that are particularly relevant to the present discussion if submarine 'volcanic emanations' are to be proposed as geologically significant sources of trace metals for marine environments. White has defined four critical aspects of the generation of base metal ore deposits, based on observations of two active metal-bearing hydrothermal systems (the Red Sea and Salton Sea areas) and on studies of fluid inclusions in ore and gangue minerals in three mining districts. To form such an ore deposit [*White*, 1968] the following must be true: (a) there must be a source of the ore constituents; (b) the ore constituents must dissolve in a hydrous phase (a Na-Ca-Cl brine in the cases he considered); (c) these solutions must migrate in directions controlled by local pressure gradients; and (d) the ore constituents must be selectively precipitated from these solutions in response to physical and chemical changes as the fluid moves into a new environment.

In this discussion we are concerned with the source of the metals and their dissolution and transport as 'hydrothermal exhalations.' A detailed study of the major and trace element distribution and petrography of a suite of basalts dredged from 22°N on the mid-Atlantic ridge samples has revealed evidence for both the mode of origin and the composition of such mineralizing solutions. The precipitation of these metals to form ore deposits requires an additional set of conditions; these are beyond the scope of the present paper.

Pillow Basalts and Holocrystalline Basalts from the 22°N Suite

Seven dredge hauls containing abundant basalt fragments were obtained in the course of a detailed survey of the mid-Atlantic ridge at 22°N on cruise 1965-1 of the R.V. *Thos. Washington*. Results of this survey are reported in *Melson et al.* [1966, 1968] and *van Andel and Bowin* [1968]. Fifty-one samples of 31 rocks from this suite were analyzed by instrumental activation analysis by methods described by *Gordon et al.* [1968]. Details of the analyses reported here as well as additional work on these samples, which relates them to other suites of mid-Atlantic ridge basalt, are presented in *Corliss* [1970].

The most abundant rock types in the 22°N suite, as described by *Melson et al.* [1968], are pillow basalts and holocrystalline basalts, contrasted in Table 1. Thin sections of the pillow fragments reveal rare to common microphenocrysts of plagioclase or olivine, or both, set in a matrix ranging from variolitic glass at the outer pillow margins to a hyalopilitic ground mass largely crystallized to plagioclase and augite in the pillow centers, 20 to 30 cm from the glassy rims. In contrast, the massive, holocrystalline fragments are uniform in texture from margin to center; intersertal to subophitic with no glass. Fresh plagioclase laths and granular augite form the bulk of the rock, with minor amounts of olivine and saponite. The saponite was identified by *Melson et al.* [1968], who suggest it results from deuteric alteration of olivine. The petrography and chemistry of these rocks are discussed in more detail in *Corliss* [1970].

Melson et al. [1968] interpreted the holocrystalline rocks to be samples of the slowly cooled interiors of submarine basalt flows whose margins were quenched to form the pillows. The important distinctions in the present context are the presence of abundant glass and the lack of deuteric alteration effects in the pillows, contrasted with the absence of interstitial glass coupled with extensive deuteric alteration of the olivine in the holocrystalline rocks.

In addition to the textural differences, the high precision of the activation analysis data reveals significant chemical distinctions between the two rock types (Figure 1). The Na, Al, and Ca abundances from the unweathered cores of the holocrystalline fragments are identical to those in the pillows. In the holocrystalline rocks Cr is slightly higher, while Sc is slightly lower. On the other hand, Mn, Fe, Co, and REE and Hf (not shown), are significantly lower in the holocrystalline rocks. A similar relationship for Cu, Pb, and Ga is indicated by the results

TABLE 1. Contrast of Basalt Fragments Derived from Pillow Lavas with Those Derived from Massive Interiors of Flows in the 22°N Area (modified from *Melson et al.*, [1968])

Characteristics of:	Pillow	Massive
Glassy rinds	Abundant	Absent
Jointing	Radial	Columnar
Groundmass	Glassy	Holocrystalline
Texture	Hyalopilitic, intergranular	Subophitic, intersertal
Vesicles	Common	Rare
Deuteric alteration	Absent	Extensive olivine alteration

of emission spectrographic analysis reported by Thompson [*Melson et al.*, 1968]. I attribute these differences to mobilization and loss of some components of the initial magma from the interiors of the flows. In justifying this interpretation we must first explore other possible explanations for this difference.

It is possible that the holocrystalline rocks lose these elements during low temperature weathering after exposure by faulting. This would, of course, reduce their importance as sources of trace elements in the deep marine environment; only these few flows, whose interiors have been disrupted and exposed by faulting or slumping, would contribute these elements to sea water. Several lines of evidence suggest that this does not happen.

First, the dredged holocrystalline basalt fragments have altered zones or bands parallel to their broken surfaces. This alteration, visible in thin sections, is principally an oxidation of the Fe in the green saponite ('bowlingite') changing its color to reddish brown ('iddingsite'). In addition, analysis of these bands in one sample reveals that the Fe and K abundance increases sharply in these narrow bands, compared to the interiors of these holocrystalline fragments [*Corliss*, 1970]. The orientation of these bands parallel to the irregular surfaces of the broken fragments indicates that they formed after the rocks were broken and exposed to sea water, and the increase in Fe content outward suggests that this element, as well as K, was added to the rock from sea water rather than removed from it. The presence of significant amounts of saponite and related clay mineral phases, which have large ion-exchange capacities, presumably accounts for this addition of Fe from sea water to the holocrystalline rocks.

A similar observation has also been made by *R. A. Hart* [1970], who reports on a study of published dredged basalt analyses. He suggests that Fe and K, and other elements, are removed from sea water and added to these rocks when they are exposed at low temperatures over long periods of time.

Second, *Garlick and Dymond* [1970] have examined H_2O and ^{18}O values in a holocrystalline fragment with similar alteration banding. The outside and intermediate zones showed effects of hydration and exchange of oxygen with sea water while the center did not.

Third, *S. R. Hart's* [1969] data, as interpreted by *Corliss* [1970], indicate that Cs abundances in the centers of holocrystalline fragments are often very low, identical to those in the erupted magma, as indicated by the abundance in quenched glass. This suggests that the centers of these holocrystalline rocks are often unaffected by sea-water alteration at low temperature.

These three observations suggest that interiors of holocrystalline fragments with deuteric minerals which do not show visible effects of low-temperature alteration by sea water (such as oxidation of iron) have not been significantly affected by this process. Thus this type of alteration cannot be the cause of the depletions relative to pillow basalts shown in Figure 1.

Another possible cause for lower abundance of some metals in holocrystalline rocks relative to associated pillows would be enrichment of these elements in the glassy pillows by reaction with sea water. Two observations indicate this is not the case. First, nearly all of the pillow samples show no visible effects of alteration either in hand specimen or in thin section. A few pillow fragments do have traces of iron oxides on the exposed glassy surfaces, but the Fe content of interior samples of these rocks is identical to that of the completely unaltered pillow fragments. Second, comparison of samples from the pillow margins and interiors reveal higher and more variable K abundances in the interiors, where the glassy matrix is partially crystallized and sea water has presumably penetrated. Fe, on the other hand, is homogeneously distributed in the pillows [*Corliss*, 1970].

Another explanation for the chemical differences between pillows and holocrystalline rock is possible. It could be argued that the two compositions represent two distinct magma types present at 22°N, which have essentially identical Na, Al, Ca, and Sc contents, but differ in Fe, Mn, Co, and others (Figure 1). This explanation cannot be rigorously eliminated, but it requires an explanation of the observation that one of these magma types, which was found in six dredge hauls, only occurred as pillow basalt fragments, and the other, present in three dredge hauls, was found only as deuterically altered holocrystalline rocks, never as pillow fragments. Moreover, in two of the dredge hauls the two rock types occurred together. The suggestion that they are distinct magma types requires acceptance of a model significantly more complex than that proposed here.

As an alternative hypothesis, the favored one, to explain the chemical distinction between these two rock types, we may consider that some components of the magma are mobilized and lost during the slow crystallization and subsequent cooling of the flow interiors.

Fractionation of Submarine Basalt Flows during Cooling

The composition of the magma when it reaches the sea floor is the end product of a series of geochemical processes, the most recent being partial melting in an upper mantle source region followed by transport to the surface accompanied by some degree of fractional crystallization. If we were to sample this lava as it arrives at the sea floor, and cool separate identical portions of it in closed containers (closed to gain or loss of matter) at various cooling rates, the texture and modal composition of each rock so formed would be different, but their bulk chemical compositions would necessarily still be the same. Very rapid quenching would produce homogeneous glass containing only those phenocrysts or microphenocrysts present in the liquid when sampled. In the more slowly cooled samples, the crystallizing solid phases would proceed to alter the composition of the residual liquid, markedly increasing the concentration of those elements rejected by the crystal lattices, the 'incompatible elements.' The residual liquids might differ considerably from the liquid from which the initial crystals formed, and could react with the solid phases that are now unstable in their presence. On further cooling in our hypothetical 'closed' system the residual components would form solid phases in the interstices of the previously formed crystalline mesh, and the composition of the total system would remain unchanged. I propose that the natural system is 'open' and that residual components of the erupted magma can be lost from the slowly cooling portions of the rock mass.

The distribution of radiogenic noble gases in pillow basalts provides strong evidence that when cooling is slow enough to allow crystallization, the rock is open to loss of mobile components. Radiogenic argon and helium, produced by the decay of ^{40}K, and U and Th

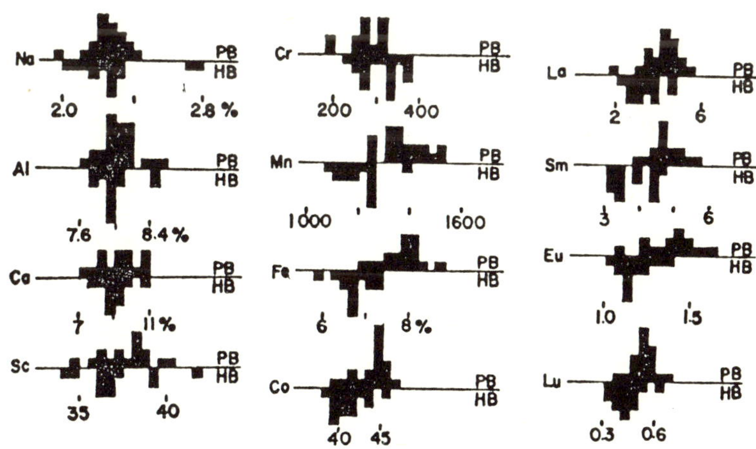

Fig. 1. Histograms contrasting the compositions of the pillow basalts from the flow margins (PB) with the holocrystalline basalts from the flow interiors (HB). Abundances are ppm except where noted.

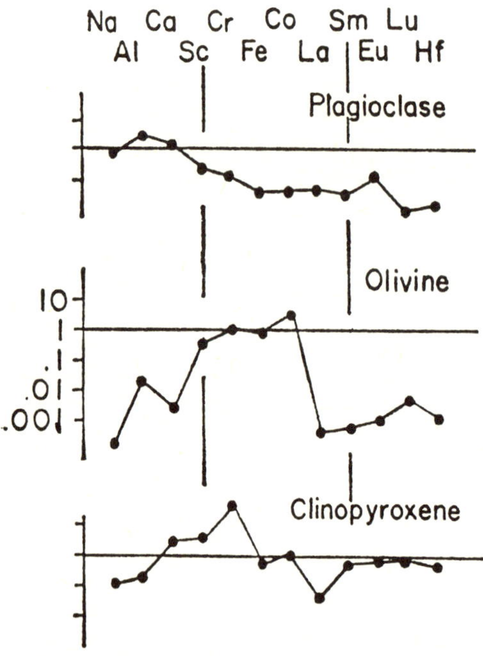

Fig. 2. Solid-liquid partition coefficients for 12 elements are shown in 3 major phases, which crystallize in the flow interiors. A partition coefficient is the ratio of the abundance of an element in the solid phase divided by its abundance in the melt.

isotopes in the upper mantle source of the magmas, are trapped by quenching in the glassy pillow margins, but have been lost from the pillow interiors where extensive crystallization has taken place [*Dalrymple and Moore*, 1968; *Funkhouser et al.*, 1968; *Dymond*, 1970].

Evidence that slow crystallization will lead to the formation of residual liquids with high concentrations of several 'incompatible elements' can be seen in the partition coefficients presented in Figure 2. (The data plotted here are discussed in detail in *Corliss* [1970]). The three major phases that crystallize from oceanic tholeiite magmas at sea floor pressures are plagioclase, olivine, and clinopyroxene. (subcalcic augite). In the present context, the important features of the partition coefficient data are that Cr and Sc tend to be located in the clinopyroxene lattice and Co in the olivine; that the REE and Hf are fractionated into the liquid; and that Fe enters into olivine while Na, Al, and Ca do not. Data on Mn from *DeVore* [1955] and *Onuma et al.* [1968], not tabulated here, indicate that Mn is fractionated into both clinopyroxene and olivine. In addition, the separation of plagioclase, clinopyroxene, and olivine from a tholeiitic liquid tends to enrich the residual liquid in silica, as *Yoder and Tilley* [1962] have pointed out.

The deuteric alteration of olivine might be expected to mobilize additional elements. Data tabulated by *Baker and Haggerty* [1967] indicate that the green 'bowlingite', produced during deuteric alteration of olivine under nonoxidizing conditions, is depleted in Fe relative to the original olivine. It is reasonable to assume that other transition elements in the olivine lattice, which substitute for Fe^{2+} (Co and Mn), are also mobilized during this process.

An important set of observations relevant to these residual liquids, formed during the solidification of a tholeiitic magma, is reported by *Peck et al.* [1966] and *Skinner and Peck* [1969]. They document a detailed program of drilling and sampling during the cooling and crystallization of a tholeiitic basalt flow in Alae Lava Lake, Hawaii. The following observations are considered relevant:

1. The flow was 14 meters thick. The last interstitial melt solidified (at 980°C) 13 months after eruption. The entire flow cooled to 100°C in four years. (A submarine flow would cool faster; its boundaries are initially at 2°C and its upper surface is maintained at this temperature.)

2. As the melt solidified downward a zone of crystalline mesh and interstitial fluid formed between the solid rock above and the liquid below. The interstitial liquid in this zone was isolated from the bulk liquid below and could not mix with it by convection. These interstitial fluids were sampled several times by allowing them to flow into open drill holes. Analysis showed them to be depleted relative to the initial liquid, in Mg, Ca, Cr, and possibly Ni; and enriched in Fe, Ti, Na, K, P, F, Ba, Ga, Li, Y, and Yb, and in most samples, SiO_2 (Figure 3). *Peck et al.* [1966] note that this trend is distinct from the olivine fractionation trend of the Hawaiian tholeiites.

3. In some samples of the interstitial fluids, an immiscible sulfide-rich phase was found in the siliceous residual melt. The sulfur is apparently fractionated into the siliceous residual

liquids until it reaches a concentration adequate to form an immiscible phase. (The S content of the siliceous liquid from which the sulfides separated was 380 ± 20 ppm compared to ~ 100 ppm in the liquid below the zone of crystallization). Several elements are strongly fractionated into the sulfide phases. They contain 59% Fe, 5% Cu, 1% Ni, and presumably concentrate other trace metals such as Co, Pb, etc.

4. Deuteric alteration of olivine was observed at about 700°, corresponding to an increase in the ferric to ferrous iron ratio.

The formation of sulfide segregations has also been observed in dredged submarine pillow basalts from the mid-Atlantic ridge and Juan de Fuca ridge by *Moore and Calk* [1971]. They occur as 'spherules,' which form on vesicles walls during the quenching of the pillows, and larger 'globules,' which perhaps form during volatile fractionation prior to eruption. They contain abundant Fe, Cu, and Ni, which are strongly fractionated from the melt into the sulfide phase.

Another interesting set of observations relevant to this discussion of the origin of submarine hydrothermal metal-bearing solutions has been presented and interpreted by *Mackin and Ingerson* [1960]. They propose a model for the origin of magnetite mineralization in the Iron Springs district in southern Utah. In this area, erosion has exposed Tertiary laccoliths of granodiorite porphyry surrounded by replacement ore bodies of magnetite and hematite in Jurassic limestone. They suggest that the source of the iron was biotite and hornblende, crystallized in the melt prior to intrusion. In the rapidly cooled margins of the intrusive body, these minerals are unaltered. In the interior, the biotite and hornblende phenocrysts 'were largely or completely destroyed by deuteric alteration, and the iron contained in them was released into the interstitial fluid of a slowly consolidating crystal mush.' They cite tension cracks produced in the semisolid crystalline mush by subsequent intrusion and uplift, as the pathways along which the fluids migrated into the adjacent limestone.

Mackin and Ingerson [1960] suggest that this 'deuteric release' mechanism is important in all ore deposits related to hypabyssal intrusives in which early formed mafic phenocrysts move to regions of lower pressure where they are unstable and are deuterically altered. This suggests a parallel to the olivine in oceanic tholeiite lavas. Olivine is commonly present as microphenocrysts in the erupted liquid (they appear in quenched pillows); these are unstable, and when cooled slowly they are altered.

The observations on the Alae Lava Lake and

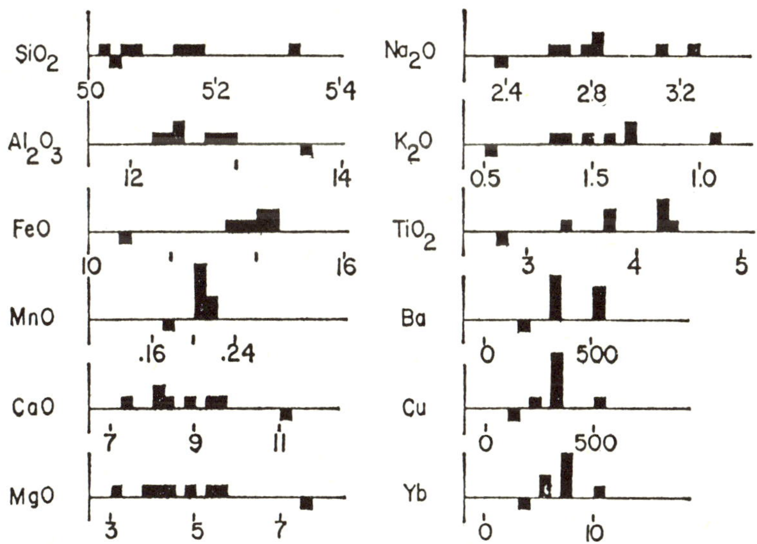

Fig. 3. Comparison of the composition of the initial liquid and residual liquids in the Alae Lava Lake tholeiite. The initial liquid composition is plotted below the lines, the compositions of the residual liquids above. (Data from *Peck et al.* [1966]).

on sulfide segregations in submarine pillows, and the model proposed for the mineralization of Iron Springs, coupled with the major and trace elements data on the 22°N pillows and holocrystalline basalts, suggest that during the cooling of the basalt flows, several elements, previously postulated as those transported by submarine 'hydrothermal solutions' or 'volcanic emanations,' are concentrated in residual phases during slow cooling and are subsequently lost from the flow interiors. The mechanisms that might account for this migration and loss must now be explored.

Dissolution of Metals and Their Transport in Hydrothermal Solutions

The parallel joint surfaces of many of the dredged holocrystalline basalt fragments indicate that the flows in which they are formed are typical columnar basalts. The formation of contraction cracks has been observed in some detail in cooling lava lakes in Hawaii by *Peck and Minakami* [1968].

They conclude that the maximum temperature at which cracks initiate within the crust of the lava lake is probably about 900°C, or ~ 80°C below the solidus. Citing *Lachenbruch* [1962, p. 37], they point out that these tension cracks, once initiated, can propagate freely into a zone of no stress at higher temperatures. Through several lines of argument they argue that 'the maximum temperature at depths to which cracks will propagate near the center of the lava lakes must be close to 1000°C and perhaps as high as 1040°C,' into the zone where the residual melt has not solidified. In submarine flows, sea water is necessarily admitted into such fractures, and as the fracture systems develop, convective flow must occur.

Helgeson [1964], in a detailed discussion of hydrothermal ore deposition, supports the conclusion that 'most hydrothermal solutions are alkali chloride-rich electrolyte solutions containing predominant Na and Cl, lesser amounts of K and Ca, small amounts of SO_4, CO_3, HCO_3, Li, Rb, Cs, and other minor constituents . . .' (p. 80). This is, of course, a reasonable description of sea water. *White* [1968] suggests that intimate contact of a Na-Ca-Cl brine with solid phases that contain base metals, which are crystallizing or reconstituting at temperatures from 100° to 900°C, is 'the most favored circumstance for concentration of these metals in the liquid phase,' probably as chloride complexes. *Krauskopf* [1967] presents experimental data which suggest that Fe and Mn would be readily dissolved by sea water which has been acidified by solution of the gases accompanying volcanic extrusions.

This model for the formation of metal-rich hydrothermal solutions requires that those components of the melt which are fractionated into residual solutions as discussed in the previous section, are available for dissolution in sea water when it is admitted by fracturing into their vicinity. This may occur initially at temperatures close to the solidus (~980°C) along the contraction fractures. In the bulk of the rock the interchange of matter between the sea-water phase and the residual magmatic phases must take place by migration along intergranular boundaries. As sea water dissolves the residual magmatic components, it also gains heat, and the temperature gradients so established drive a convective flow in the vertical fractures which brings fresh sea water to contact with the hot rocks and allows these solutions to enter the deep marine environment as submarine hot springs. These fluids are the hydrothermal solutions, or volcanic emanations, to which some submarine deposits are attributed.

This phenomenon has been directly observed on the submarine Banu Wuhu volcano in Indonesia [*Zelenov*, 1964]. The site of the 1919 eruption, as examined in 1963, was a submarine bank of dacite-andesite that ranged in depth from sea level to about 30 meters, from which jets of hot water streamed. The hot waters were of sea water composition with added Fe, Mn, and Si. 'Suspended iron and manganese hydroxides can be seen precipitating right under the water, . . . the rising jet starts becoming yellow and turbid about 1 meter above the bottom.'

These solutions precipitated 100 to 140 mg/liter of Fe and Mn hydroxides on cooling. Brown iron-manganese-silica sediment that is rich in trace metals covers the sea floor around these vents and is also carried away from the area in the current. It is reasonable that similar submarine hot springs occur in areas of volcanic activity along mid-ocean ridges.

Geochemical Balances

If this model for the origin of the metal-bearing submarine hydrothermal solutions is fect, the proposed mechanism can contribute gnificant amounts of Fe, Mn, the REE, and presumably silica and other elements not measured, into the deep sea sedimentary environment. The magnitude of the effect can be estimated by comparing the loss from a given thickness of basalt underlying the ocean floor to the mass of these elements present in the overlying sea water, manganese nodules, and amorphous iron oxide beds. Table 2 presents the results of such calculations. The data show that basalt flows a few hundred meters thick which have generated hydrothermal mineralizing solutions according to the model presented here, may yield (1) amounts of Fe, Mn, Co, and the REE which are significantly greater than the amounts in the steady-state sea-water reservoir; (2) amounts of Mn, Co, and Cr which are of the same order of magnitude as the amounts in Mn nodules; (3) enough Ce to supply the anomalously high Ce content in manganese nodules although some excess of Sm (and the other REE) is indicated; and (4) a considerable excess of Fe over that present in Mn nodules to be accounted for. Evidence for the existence of this iron in pelagic sediments is found in data both on Pacific pelagic clays and recently discovered 'amorphous iron-oxide' beds.

Chester and Hughes [1969] have described results of selective leaching of samples of a pelagic clay core from the North Pacific. They interpret their results to indicate that there is 'excess iron over that required to form nodules' in the hydrogenous fraction of the core. *Goldberg and Arrhenius* [1958] also observed this 'excess' iron and suggested that it originates as 'free' colloidal iron, which is incorporated into the sediment and converted to goethite with time.

In the core studied by *Chester and Hughes* [1969], this 'excess' iron has an average concentration of ~900 ppm in five samples of the core, ranging from 600 to 1300 ppm. This concentration is too low to account for all of the excess iron produced by the model proposed here—enough excess iron for about 125 meters of sediment would be supplied by each meter of basalt—but it does suggest the existence of such excess iron.

More dramatic occurrences of non-nodular, presumably hydrogenous iron, have been discovered by the Deep Sea Drilling Project (DSDP), and in piston cores taken near the east Pacific rise described by *Sayles and Bischoff* [1971]. In DSDP cores from the western North Atlantic, abundant hematite is reported at sites 8 and 9A. The hematite colors the sediment brick red, and makes up 11.7% of the sediment in one sample. Siderite, a high-iron carbonate, is a significant component in several cores taken on leg III in the South Atlantic [*Rex*, 1970]. *Von der Borch and Rex* [1970] and *von der Borch et al.* [1970] report the occurrence, in several core holes in the Pacific, of thick sections of 'amorphous iron oxide precipitates.' At three sites on leg V, (37, 38, and 39) a 'basal amorphous iron oxide facies' immediately overlies basalt. In two holes (37 and 39) this facies consists of 5 to 6 meters of 'dusky yellowish-brown amorphous goethite mud.' In the third (38), it is made up of 9 meters of mixed amorphous iron oxide and calcified nannoplankton ooze overlain by 6 meters of '100% amorphous iron oxide sediment.' Overlying this basal facies at all three sites are 5- to 25-meter thick 'mixed amorphous iron oxide-detrital facies.' *Von der Borch et al.* [1970] report the occurrence of amorphous iron-oxide

TABLE 2. Geochemical Budget of Elements Contributed to the Pelagic Environment by Submarine Volcanism

	Lost from Flow Interiors	Sea Water[1,2]	100-meter Equivalent[3]	Mn Crust[4]	100-meter Equivalent[5]	Amorphous Fe-Oxide Beds[6]	100-meter Equivalent
Fe	10,000 ppm	0 01 ppm	100,000 km	125,000 ppm	800 cm	160,000 ppm	625 cm
Mn	120 ppm	0 001 ppm	12,000 km	160,000 ppm	7 5 cm	50,000 ppm	24 cm
Co	4 ppm	0 0001 ppm	4,000 km	3,100 ppm	7 8 cm	160 ppm	250 cm
Cr	0	0 00005 ppm		10 ppm		10 ppm	
Sm	1 ppm	50×10^{-8} ppm	200,000 km	50 ppm	200 cm		
Ce	3 ppm	200×10^{-8} ppm	150,000 km	1,500 ppm	20 cm		

1. *Goldberg* [1965, pp. 164–165] (Cr, Sm, Ce).
2. *Brewer and Spencer* [1970] Pacific Deep Water (Fe, Mn, Co).
3. Depth of sea water containing amount of element lost by 100 meters of holocrystalline basalt.
4. [*Cronan*, 1967, p. 108].
5. Thickness of Mn crust containing same amount of element lost by 100 meters of holocrystalline basalt.
6. *Sayles and Bischoff* [1971].

layers interbedded with non-iron-rich calcareous nannoplankton ooze in the basal portions of holes 78 and 79 taken on DSDP leg X. These sediments are Eocene and Oligocene in age.

More recently, F. Sayles and J. L. Bischoff (unpublished manuscript, 1971) report on the occurrence of amorphous iron-manganese-silica beds cored east of the crest of the east Pacific rise at 9-10° South. This material is characterized by high concentrations of Ba, Cu, Pb, Zn, and Ni, and low $CaCO_3$. Three cores contained from ~1 to nearly 9 meters of this material, which they suggest 'is best explained by deposition of colloidal precipitates produced through the introduction of hot mineralized solutions into the ocean water.' They cite acoustic profiler (3.5 kHz) data to suggest that as much as 35 meters of this material may be present at one site.

Summary and Conclusions

It is observed that fine-grained holocrystalline basalts dredged along with glassy pillow fragments from a surveyed area on the MAR, are depleted relative to the pillows in Fe, Mn, Co, the REE, Cu, and Pb. The holocrystalline basalt fragments are assumed to be samples of the slowly cooling component of submarine extrusions. Pillow basalts are the rapidly quenched component of these flows, and their composition is close to that of the erupted liquid. These observations, along with information on the cooling behavior of tholeiitic lava masses in Hawaii and on hydrothermal mineralization on the continents, suggest a model for the origin of submarine hydrothermal solutions.

When basaltic rocks are erupted onto the sea floor, the margins of the extrusions are quenched to form pillows, which closely approximate the composition of the erupted liquid, including volatile components. The flow interiors are insulated by this shell of pillows, and cool slowly. Elements that do not readily enter the crystallizing phases are fractionated into residual silica-rich fluids, ultimately residing in residual solid phases in accessible sites along intergranular boundaries. Somewhat below solidus temperatures the early formed olivine is deuterically altered making additional elements accessible. Contraction fractures originate in the solidified shell at temperatures slightly below the solidus and propagate into the interior as it cools, allowing volatile components to escape and admitting sea water into the hot crystallizing rock mass. This sea water is heated, dissolves escaping volatile components, and mobilizes metals in residual phases as chloride complexes. Convective flow in fracture systems should allow these hydrothermal solutions to emerge at the sea floor as submarine hot springs. Reasonable estimates of the amount of several elements mobilized in this way lead to the conclusion that a significant fraction of the mass of these elements that reside in pelagic deposits could have been supplied by this process.

Although it is presumed that the holocrystalline rocks studied here are from flow interiors near the sea-water interface, it is reasonable that the submarine ground water (sea water) could penetrate to considerable depths. Oxygen isotope datacited by *Muehlenbachs and Clayton* [1971] suggest that sea water is involved in the reactions producing greenstones from basalts beneath the mid-Atlantic ridge. *Taylor* [1970] cites oxygen isotope data on diorite and granodiorite intrusions in the western Cascades in Oregon to suggest that 'much so-called deuteric hydrothermal activity is probably caused by heated ground waters rather than by H_2O released during magmatic crystallization.' He suggests that a volume of water about equal to the volume of rock has been flushed through these intrusive stocks by convective circulation of ground water during much of the cooling and crystallization history.

The amorphous iron oxide deposits near the east Pacific rise and in DSDP cores and the 'excess' hydrogeneous iron in pelagic clays presumably form when these iron-rich solutions emerge from submarine hot spring vents and mix with sea water, which cools and oxidizes then precipitates a ferric-hydroxide floc which may settle in topographic lows near the vents or be widely dispersed, depending on bottom current conditions. *Dasch et al.* [1971], who reported data on the amorphous iron-manganese-silica sediments described by Sayles and Bischoff (1971), find that the isotopic composition of the abundant lead in this material is typical of mid-ocean ridge basalt lead. The strontium isotopic composition is predominantly that of sea water, with a small volcanogenic component. *Bender et al.* [1971] have found that the REE abundance patterns and the isotopic composition of Sr and U in an east Pacific rise core indicate that these elements are predominantly derived from sea water while the Pb isotopic composition indicates that the Pb is of local volcanic origin. These observations combined with the well-known tendency of many elements to co-precipitate with ferric hydroxide suggest that the mechanism proposed here can concentrate elements not only by mobilization of the residual components of submarine volcanic extrusions, but also by the extraction of trace constituents from sea water.

The model proposed here implies that we can determine the contribution of submarine volcanism to sea water for many elements by comparing the quenched pillow basalts with slowly cooled holocrystalline rocks from the same extrusions. In addition to inherited radiogenic helium and argon and the elements studied in this work, it is likely that other elements present in the magma and trapped on quenching are released to sea water from the slowly cooling portions of the flow. These might include other trace metals, silica, halogens, sulfur, rare gases, and nitrogen. Further studies of these rocks should provide data on the rate of supply of these elements to the hydrosphere from the upper mantle.

Acknowledgments. Sincere thanks are due Tj. H. van Andel, G. G. Goles, G. R. Heath, and J. R. Dymond for encouragement and critical discussions in the course of this work.

This work was done as part of a Ph.D. dissertation at Scripps Institution of Oceanography, University of California, San Diego. Principal support was provided by an American Chemical Society, Petroleum Research Fund grant to Tj. H. van Andel. Additional financial assistance was provided by the National Science Foundation and the National Aeronautics and Space Agency.

References

Baker, I., and S. E. Haggerty, The alteration of olivine in basaltic and associated lavas, *Contrib. Mineral. Petrol., 16,* 258–273, 1967.

Bender, M., W. Broecker, V. Gornitz, and U. Middel, Accumulation rate of manganese and related elements in sediments from the East Pacific Rise (abstract), *Eos, Trans. Amer. Geophys. Union, 51,* 327, 1970.

Bender, M., W. Broecker, V. Gornitz, and U. Middel, R. Kaye, S. S. Sun, and P. Biscaye, Geochemistry of three cores from the east Pacific rise, *Earth. Planet. Sci. Lett.,* in press, 1971.

Bonatti, E., and O. Joensuu, Deep sea iron deposit from the South Pacific, *Science, 154,* 643–645, 1966.

Bonatti, E., and Y. R. Nayudu, The origin of manganese nodules on the ocean floor, *Amer. J. Sci., 263,* 17–39, 1965.

Bostrom, K., The problem of excess manganese in pelagic sediments, in *Researches in Geochemistry,* edited by P. Abelson, p. 421–452, John Wiley, New York, 1967.

Bostrom, K., Submarine volcanism as a source for iron. *Earth Planet. Sci. Lett., 9,* 348–354, 1970.

Bostrom, K., and M. N. A. Peterson, Precipitates from hydrothermal exhalations of the East Pacific Rise, *Econ. Geol., 61,* 1258–1265, 1966.

Brewer, P. G., and D. W. Spencer, Trace element intercalibration study, *Tech. Rep. 70-62,* Woods Hole Oceanographic Institution, 1970.

Chester, R., and M. J. Hughes, The trace element geochemistry of a North Pacific pelagic clay core, *Deep Sea Res., 16,* 639–654, 1969.

Corliss, J. B., Mid-ocean ridge basalts: 1. The origin of submarine hydrothermal solutions, 2. Regional diversity along the Mid-Atlantic Ridge, Ph.D. dissertation, University of California, San Diego, 1970.

Cronan, D. S., The geochemistry of some manganese nodules and associated pelagic deposits, Ph.D. thesis, University of London, London, England, pp. 108–125, 1967.

Dalrymple, G. D., and J. C. Moore, Argon 40: Excess in submarine pillow basalts from Kilaula Volcano, Hawaii, *Science, 161,* 1132–1135, 1968.

Dasch, E. J., J. R. Dymond, and G. R. Heath, Isotopic analysis of metalliferous sediment from the East Pacific Rise (abstract), *Geol. Soc. Amer. Abstr., Programs,* in press, 1971.

DeVore, G., Crystal growth and the distribution of elements, *J. Geol. 63,* 471, 1955.

Dymond, J., Excess argon in submarine basalt pillows, *Geol. Soc. Amer. Bull., 81,* 1229–1232, 1970.

Funkhouser, J. G., E. D. Fisher, and E. Bonatti, Excess argon in deep-sea rocks, *Earth Planet. Sci. Lett., 5,* 95–100, 1968.

Garlick, G. D., and J. Dymond, Oxygen isotope exchange between volcanic materials and ocean water, *Geol. Soc. Amer. Bull., 81,* 2137–2142, 1970.

Goldberg, E. D., and G. Arrhenius, Chemistry of Pacific pelagic sediments, *Geochim. Cosmochim. Acta, 13,* 153–212, 1959.

Gordon, G. E., K. Randle, G. G. Goles, J. B. Corliss, M. H. Beeson, and S. S. Oxley, Instrumental activation analysis of standard rocks with high resolution x-ray detectors, *Geochim. Cosmochim. Acta, 32,* 364–396, 1968.

Hart, R. A., Chemical exchange between sea water and deep-ocean basalts, *Earth Planet. Sci. Lett., 9,* 269–279, 1970.

Hart, S. R., K, Rb, Cs contents with K/Rb, K/Cs ratios of fresh and altered submarine basalts, *Earth Planet. Sci. Lett., 6,* 295–303, 1969.

Helgeson, H. C., *Complexing and Hydro-*

Thermal Ore Deposition, Macmillan, New York, 1964.

Hewett, D. F., M. Fleischer, and N. Conklin, Deposits of the manganese oxides-supplement, *Econ. Geol.*, *58*, 1–50, 1963.

Krauskopf, K. B., Separation of manganese from iron in sedimentary processes, *Geochim. Cosmochim. Acta*, *12*, 61–84, 1967.

Lachenbruch, A. H., Mechanics of thermal contraction cracks and ice-wedge polygons in permafrost, *Geol. Soc. Amer. Spec. Pap. 70*, 1962.

Mackin, J., and E. Ingerson, A hypothesis for the origin of ore-forming fluid, Geol. Surv. Res., 1960, *U.S. Geol. Surv. Prof. Pap. 400-B*, B1-B2, 1960.

Melson, W. G., and Tj. H. van Andel, Metamorphism on the Mid-Atlantic Ridge, 22° N latitude, *Mar. Geol.*, *4*, 165–186, 1966.

Melson, W. G., G. Thompson, and Tj. H. van Andel, Volcanism and metamorphism in the Mid-Atlantic Ridge, 22° N latitude, *J. Geophys. Res.*, *73*, 5925–5941, 1968.

Moore, J. G., and L. Calk, Sulfide spherules in vesicles of dredged pillow basalt, *Amer. Mineral.*, *56*, 476–488, 1971.

Muehlenbachs, K., and R. N. Clayton, The oxygen isotope geochemistry of submarine greenstones (abstract), *Eos, Trans. Amer. Geophys. Union*, *52*, 365, 1971.

Murray, J., and A. F. Renard, Deepsea deposits, report of the scientific results of H. M. S. Challenger, 1873–1876, 372–378, 1891.

Onuma, N., H. Higuchi, H. Wakita, and H. Nagasawa, Trace element partition between two pyroxenes and the host lava, *Earth Planet. Sci. Lett.*, *5*, 47–51, 1968.

Park, C. F., The spilite and manganese problems of the Olympic Peninsula, Washington, *Amer. J. Sci.*, *244*, 305–323, 1948.

Peck, D. L., T. L. Wright, and J. G. Moore, Crystalliaztion of tholeiitic basalt in Alae Lava Lake, Hawaii, *Bull. Volcanol.*, *29*, 629–656, 1966.

Peck, D. L., and T. Minakami, The formation of columnar joints in the upper part of the Kilauean Lava Lakes, Hawaii, *Geol. Soc. Amer. Bull.*, *79*, 1151–1166, 1968.

Rex, R. W., X-ray mineralogy studies: Leg 2 Deep Sea Drilling Project, 1970, in *Initial Reports of the Deep Sea Drilling Project, vol. 3*, edited by M. N. A. Peterson, U.S. Government Printing Office, Washington, 509–582, 1970b.

Skinner, J. J., and D. L. Peck, An immiscible sulfide melt from Hawaii, *Econ. Geol. Mono. 4*, 310–322, 1969.

Skornyakova, N. S., P. F. Andruschchenko, and L. S. Fomina, The chemical composition of iron-manganese concentrations of the Pacific Ocean, *Okeanologiya Akad. Sci.*, *2*, 2, 1962.

Taylor, H. P., Oxygen isotope evidence for large-scale interaction between meteoric ground waters and Tertiary Diorite inclusions, Western Cascade Range Oregon (abstract), *Eos, Trans. Amer. Geophys. Union*, *51*, 453, 1970.

van Andel, Tj. H., and C. O. Bowin, Mid-Atlantic ridge between 22° and 23°N latitude and the tectonics of mid-ocean rises, *J. Geophys. Res.*, *73*, 1279–1298, 1968.

von der Borch, C. C., and R. W. Rex, Amorphous iron oxide precipitates in sediments cored during leg 5, Deep Sea Drilling Project, 1970, in *Initial Reports of the Deep Sea Drilling Project, vol. 5.*, edited by D. A. McManus, U.S. Government Printing Office, Washington, 1970.

von der Borch, C. C., W. D. Nesteroff, and J. Galehouse, 1970, Iron rich sediments cored during Leg VIII of the Deep Sea Drilling Project, in *Initial Reports of the Deep Sea Drilling Project, vol. 8*, edited by J. Tracey, U.S. Government Printing Office, Washington, 1971.

White, D. E., Environments of generation of some base-metal ore deposits, *Econ. Geol.*, *63*, 301–335, 1968.

Yoder, H. S., Jr., and C. E. Tilley, Origin of basalt magmas: an experimental study of natural and synthetic rock systems, *J. Petrol.* *3*, 342, 1962.

Zelenov, K. K., Iron and manganese in exhalations of the submarine Banu Wuhu Volcano, Indonesia, *Doklady Akademii Nauk SSR*, Engl. Transl., *155*, 94–96, 1964.

(Received June 18, 1971;
revised August 27, 1971.)

12

Reprinted from *Internat. Geology Rev.* 7:2161-2174 (1965)

Dispersed iron and manganese in Pacific Ocean sediments

I.S. Skornyakova

UDC 552.56+553.32(265/266)=82=20

ABSTRACT: The article considers the areal distribution of dispersed iron and manganese on the bottom of the Pacific Ocean. -Author.

* * *

PROBLEMS AND MATERIAL FOR INVESTIGATION

Iron and manganese are important and geochemically very interesting components of oceanic sediments. Their investigation in Pacific Ocean sediments has until recently been limited only to partial determinations either in isolated samples or in groups of them, giving mean values (Murray and Renard, 1891; Revelle, 1944; Goldberg and Arrhenius, 1958; El-Wakeel and Riley, 1961). The first attempt involving mapping them was made in 1961 (Skornyakova, 1961; Bezrukov, et al. 1961), but the map covered only the northern half of the Pacific Ocean.

It has now become possible to examine the distribution of iron and manganese and to compare them on an areal basis over almost the whole of the Pacific Ocean. The purpose of the article is to present this aspect of the problem.

The article is based on numerous analyses of iron and manganese in the surface sediments of the Pacific Ocean (about 600 determinations). Samples were collected in the Pacific Ocean by the "Vityaz" (476 analyses), the "Ob'" (17 analyses) and the American expedition "Downwind" (30 samples) kindly sent to us by U.R. Ridel of the Scripps Oceanographic Institute. Specimens collected by dredging and direct-flow tubes were investigated.

Analyses of the material from the "Vityaz" and the "Downwind" were carried out in the geochemical laboratory of the Institute of Oceanology of the Academy of Sciences of the U.S.S.R. by L.S. Fomina, N.V. Petrov and S.V. Tsikunova. Those from the "Ob'" were analyzed in the Marine Geological and Chemical Laboratory of the All-Union Marine Research Institute. Total iron was determined colorimetrically with sulfasalicylic acid in ammoniacal solution. High iron contents were estimated volumetrically.

Managanese was determined mainly colorimetrically after breakdown in hydrofluoric and sulfuric acids and fusion with potassium pyrosulfate; high managanese was estimated volumetrically.

Analyses by Revelle (1944) and Goldberg and Arrhenius (1958) have also been incorporated.

DISTRIBUTION OF IRON AND MANGANESE BY SEDIMENT TYPE

The distribution of total iron in sediments, classified on grain size and independent of their composition, is given in Table 1.

TABLE 1. Mean Fe content by grain size in Pacific Ocean sediments, percent

Sediment type	Number of samples	Range	In carbonate and silica free material
Sands	•63	1.39-13.28	5.99
Coarse silts	29	1.68-10.6	5.05
Fine silts	62	1.23-10.0	5.62
Silt-clay muds	156	1.76- 9.88	4.80
Clay muds	225	2.35-13.84	5.33

In Table 2 the distribution of iron is related to the grain size ranges of sediments of different composition. The surprising uniformity of iron content through the whole range of grain sizes, from the very coarsest to the very finest, stands out immediately. Variations are small and irregular. Iron and manganese rich iron-manganese-carbonates are the only exception. In the fine-grained deposits - silt-clay and clay muds the iron content is essentially about or slightly higher than the clarke, while in the coarse sediments - sands and coarse and fine silts - it is much higher than the clarke of these sediments. There is in other words a definite concentration of iron in the littoral sand-silt deposits of the Pacific Ocean. It does not however occur as any form of chemically precipitated iron but by the prolific distribution of clastic and volcanic particles; magnetite, pyroxene, brown volcanic glass, rock fragments and andesite-basalt and basalt ash, etc. All these minerals and detritus are associated either with the mechanical breakdown of older and younger

Translated from Rasseyannyye zhelezo i marganets v osadkakh Tikhogo Okeana, Lithology and Mineral Resources, 1964, no. 5, p. 3-20. The author is with the Institute of Oceanology, Moscow.

TABLE 2. Fe content of different Pacific Ocean sediments, percent

Sediment Type	Fe in Natural Dry Sediment			Fe in Carbonate and Silica-Free Material		
	Number of analyses	Range	Mean contents	Number of analyses	Range	Mean contents
Terrigenous						
Sands	31	2.53—11.93	5.46	23	2.59—13.28	5.8
Coarse silts	18	2.12— 8.80	4.52	14	2.50— 8.82	4.86
Fine silty muds	26	2.73— 6.85	5.05	21	2.92— 6.36	5.5
Silt-clay muds	77	2.21— 9.25	4.62	69	2.25— 9.46	5.0
Clay muds	50	2.99— 6.11	4.71	45	3.26— 6.69	5.2
Terrigenous - low silica (diatomaceous)						
Coarse silts	3	3.51— 5.26	4.6	3	4.01— 6.11	5.3
Fine silty muds	9	3.40— 4.91	4.1	8	4.04— 5.64	4.9
Silt-clay muds	35	1.84— 6.57	3.7	29	2.64— 7.82	4.4
Clay muds	22	2.00— 6.88	3.8	19	2.35— 8.02	4.5
Diatomaceous muds	11	0.89— 2.21	1.64	11	2.16— 4.64	3.47
Radiolarian muds	18	3.06— 5.53	4.22	14	3.49— 5.84	4.66
Carbonate (foraminiferal)						
Sands	25	0.08— 5.92	1.3	25	1.39—11.28	5.63
Coarse silts	12	0.14— 4.35	1.5	12	1.68—10.6	5.22
Fine silts muds	19	0.12— 4.22	1.7	19	1.23—10.00	5.21
Silt-clay muds	39	0.16— 3.96	1.4	37	1.76— 9.88	5.05
Clay muds	25	0.50— 4.58	2,2	25	2.48—13.84	5.7
Carbonate (coral-shelly, bryozoan and others)						
Sand-gravel deposits	13	0.10— 5.92	1.64	13	1.63—11.28	6.68
Iron-manganese-carbonate	24	0.54—13.77	4.93	23	10.31—33.0	18.5
Red clays						
Silt-clay muds	17	3.92— 7.34	5.10	17	4.16— 7.53	5.29
Clay muds	116	3.60— 8.60	5.52	45	3.77— 7.79	5.64
Red clays, over all	143	3.28— 9.53	5.44	141	3.50—11.79	5.65
Vulcanogenic sediments	16	2.53—10.22	6.06	16	2.76—10.76	6.94

Note: Tables 2 and 6 are compiled from chemical analyses of samples collected by the "Vityaz" (476 samples) the "Ob" (50 samples) the "Carnegie" (38 samples, Revelle, 1944). Analytical results (35 samples) by Goldberg and Arrhenius (1958) and material from the "Downwind" were also used to determine the mean manganese and iron content in the red clays.

igneous rocks or with lithoclastic or vitroclastic material. Since magmatic rocks are widespread in the areas contributing detritus to the Pacific Ocean and as there are a great number (more than 330) of active volcanoes and many extinct volcanic centers within it, it is not surprising that the coarse littoral sediments carry significant concentrations of iron.

Comparison of the iron contents of similar types of littoral terrigenous and pelagic sediments and volcanic muds are of some interest (table 3).

The figures reveal two important features. First, the volcanic sand-silt sediments are notably richer in iron. This clearly demonstrates the deposition of iron in the coarse deposits as solid mineral phases and the effect of vulcanicity noted above. Second, the fine pelagic deposits are somewhat richer in iron than the corresponding terrigenous ones. In this instance concentration can only take place when the iron is carried in the very finest, colloidal fraction of the sediment, which is greater in quantity in

Table 3. Mean Fe contents of terrigenous, pelagic and volcanic sediments (percentage of carbonate and silica free material)

Sediment type	Number of Analyses	Range	Mean
Terrigenous deposits (from near the continents)			
Sands	31	2.59-13.28	5.8
Coarse silts	18	2.50- 8.82	4.86
Fine Silts	26	2.92- 7.36	5.5
Silt-clay muds	77	2.25- 9.46	5.0
Clay muds	50	3.26- 6.69	5.2
Pelagic red clays			
Silt-clay muds	17	4.16- 7.53	5.29
Clay muds	116	3.77- 7.79	5.64
Volcanic sand-silt deposits.	16	2.76-10.76	6.94

TABLE 4. Fe content of the colloidal fraction of some Pacific Ocean sediments (after Revelle, 1944) percent

Station Number	Sediment types	$CaCO_3$ content	Content of colloidal fraction	Over-all Fe content of sediment	Fe in colloidal fraction	Percent total Fe in colloidal fraction
149	Red clay	0.72	49.0	5.51	7.09	61.0
145	Red clay	0.72	46.0	4.92	6.38	59.0
142	Red clay	0.57	42.3	4.75	8.19	74.0
72	Green clay mud	0.25	50.7	4.42	6.82	79.0
69	Red clay	20.0	3.86	3.58	5.62	63.0
156	Globigerina silt-clay mud	40.0	27.7	2.04	2.84	39.0
60	Globigerina silt-clay mud	75.0	18.9	3.27	7.33	42.4
57	Ferruginous fine silty globigerina mud	84.0	14.3	3.69	6.48	24.8
52	Ferruginous fine silt globigerina mud	86.0	13.5	3.07	1.52	6.45

the pelagic than in the terrigenous deposits. This is clearly illustrated by other data (table 4).

Table 4 shows that the minimum content of dispersed colloidal iron occurs in the fine silty muds (6-24% Fe); it increases in the silt-clay muds to 40% and reaches a maximum in clayey muds at 63-79%.

The data in Tables 3 and 4 suggest that the distribution of the iron in the Pacific Ocean sediments results from cumulative action of the arrival of Fe in clastic and pyroclastic detritus and in the colloidal fraction; the importance of the first factor decreases and that of the second increases with reducing grain size of the deposit (fig. 1).

Concerning the mode of occurrence of the iron in the oceanic sediments, its carrier in the sand silt-fraction is, as mentioned above, clastic and pyroclastic detritus, (magnetite, ilmenite, pyroxene, dark volcanic glass, etc.); the carrier in the subcolloidal and colloidal fractions is free iron oxides, occasionally sulfides, chlorites and clay minerals (montmorillonite, hydromicas and others).

The distribution of manganese in the Pacific Ocean sediments is however substantially different (tables 5 and 6).

In the first instance, the manganese content of all grain size classifications of sediment is higher than the clarke, this excess being particularly prominent in the finer grain; it is greater by a factor of 7-8, whereas in the sandy members it is 3-4 times as much. In the second case, increase in the manganese content from coarse to fine sediments takes place regularly throughout the range of grain sizes. Intermediate maxima and minima which occur in the distribution

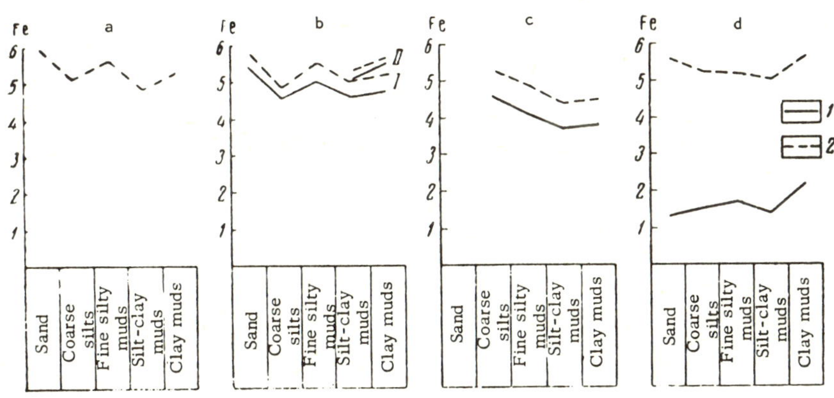

FIGURE 1. Distribution of iron by sediment grain size. Fe content, percent

1 - in natural dry sediment
2 - in carbonate and silica free sediment

a - in carbonate and silica free sediment
b - terrigenous sediment
c - interrigenous low silica sediments
d - carbonate sediments

TABLE 5. Mean Mn content by grain size in Pacific Ocean sediments, percent

Sediment Type	Number of Samples	Range	In carbonate and silica-free material
Sands	54	0.07—0.63	0.16
Coarse silts	32	0,05—0.65	0.17
Fine silts	47	0,04—2.66	0.32
Silt-clay muds	157	0.02—2.55	0.49
Clay muds	212	0.07—3.49	0.55

TABLE 6. Mn content of different Pacific Ocean sediments, percent

Sediment Type	Mn in natural, dry sediments			Mn in carbonate and silica-free material		
	Number of analyses	Range	Mean content	Number of analyses	Range	Mean content
Terrigenous						
Sands	30	0.075 —0.23	0.102	22	0.084—0.15	0,103
Coarse silts	17	0.052 —0.19	0.085	12	0.06 —0.19	0,091
Fine silty muds	27	0.01 —0.81	0.193	21	0.04 —1.01	0.225
Silt-clay muds	17	0.02 —2.04	0.30	69	0.02 —2.16	0,33
Clay muds	52	0.05 —1.89	0.50	45	0.07 —2,32	0.53
Terrigenous - low silica (diatomaceous)						
Coarse silts	3	0.04 —0.065	0.657	3	0.05 —0.08	0.066
Fine silty mud	10	0.06 —0.95	0.23	9	0.058—1.04	0.29
Silt-clay muds	37	0.032 —1.49	0.41	30	0.04 —1.74	0.43
Clay muds	21	0.05 —1.27	0.44	18	0.06 —1.60	0.50
Diatomaceous muds	11	0.04 —0.44	0.19	11	0.06 —0.75	0.40
Radiolarian muds	15	0.11 —1.13	0.45	15	0.12 —1.33	0.54
Carbonate (foraminiferal)						
Sands	11	0.0048—0.075	0.037	11	0.07 —0.63	0.27
Coarse silts	12	0.01 —0.20	0.065	12	0.07 —0.65	0.29
Fine silty muds	18	Trace —0.43	0.091	17	0.08 —2.66	0.53
Silt-clay muds	38	0.03 —0.43	0.16	38	0.14 —2,55	0.68
Clay muds	25	0.04 —1.15	0.26	25	0.097—1.85	0.67
Carbonate (coral-shelly, bryozoan and others)						
Sand-gravel deposits	13	Trace —0.09	0.022	10	0.067—0.50	0.20
Iron-manganese carbonate	24	0.21 —3.76	1.50	23	2.38 —12.9	5.89
Red clays						
Silt-clay muds	20	0.16 —1.55	0.49	20	0.24 —1,57	0.52
Clay muds	98	0.18 —1.68	0.53	98	0.18 —1,68	0.55
Red clay, over-all	143	0.16 —3.00	0.65	142	0.18 —3,49	0.67
Vulcanogenic sediments	16	0.06 —0.52	0.25	16	0.07 —0,63	0.29

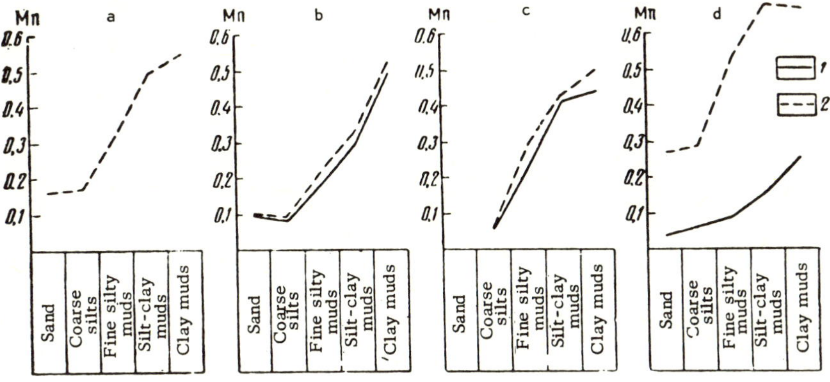

FIGURE 2. Distribution of manganese by sediment grain size. Mn content, percent

1 - in natural sediment
2 - in carbonate and silica free sediment

a - Pacific Ocean sediments overall
b - terrigenous sediments
c - terriaenous low silica sediments
d - carbonate sediments

of iron (fig. 2) are absent. Both these features clearly distinguish the distribution of manganese from that of iron.

The question then arises: what causes these differences? The answer lies first in comparison of the distribution of manganese in the terrigenous, pelagic and volcanic deposits (table 7).

The significant manganese content of the volcanic sand-silt deposits suggests that it was deposited in the sediments, especially in the littoral areas of the ocean, as a solid mineral phase. The clear increase in the manganese of the pelagic sediments relative to the terrigenous ones, in the same way as the rapid increase from sands to clayey muds, demonstrates the greater importance of fine suspension and solution as carriers of manganese.

The accumulation of manganese in the Pacific Ocean sediments thus takes place largely in a manner identical to the iron, although the quantitative relationships between the first and second forms of transport seem to be quite different. In particular, the contribution by fine suspension and solution is much greater for manganese than for iron. In the transition from coarse sandy to fine-grained sediments, the lack of manganese in clastic form is supplemented more than adequately by that in fine suspension and chemical precipitation from solution; the total content of manganese in the fine pelagic muds is therefore not only maintained but increases progressively, reaching a maximum in the clayey muds.

It must be borne in mind at the same time that the real primary contrasts in the manganese content of the fine terrigenous and pelagic sediments are probably much greater than shown in Table 7.

TABLE 7. Mean Mn contents of terrigenous, pelagic and volcanic sediments (percentage of carbonate and silica-free material)

Sediment type	Number of Analyses	Range	Mean
Terrigenous deposits (from near the continents)			
Sands	30	0.084-0.15	0.103
Coarse silts	17	0.06 -0.19	0.091
Fine silts	27	0.04 -1.01	0.225
Silt-clay muds	17	0.02 -2.16	0.33
Clay muds	52	0.07 -2.32	0.53
Pelagic red clays			
Silt-clay muds	20	0.24 -1.57	0.52
Clay muds	98	0.18 -1.68	0.55
Volcanic sand-silt deposits.	16	0.07 -0.63	0.29

The marked chemical mobility of manganese in reducing conditions (N.M. Strakhov, 1954) procures its redistribution in the sedimentary profile, its migration into the upper layers and, in this way, substantial enrichment of the superficial sediments.

In the peripheral areas of the Pacific Ocean where terrigenous muds are developed and the organic content of the sediments rises, there is a definite maximum in the manganese content of

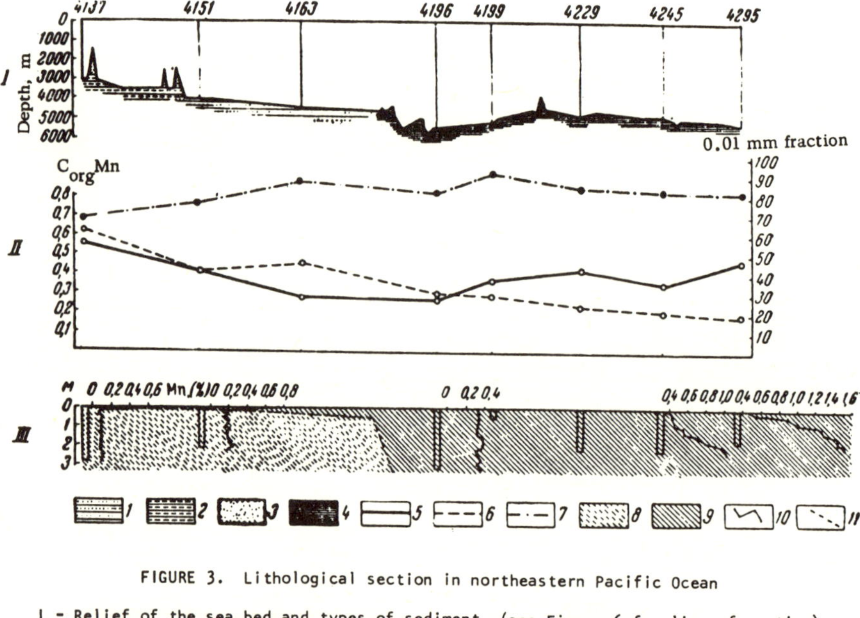

FIGURE 3. Lithological section in northeastern Pacific Ocean

I - Relief of the sea bed and types of sediment (see Figure 6 for line of section)
1 - terrigenous
2 - diatomaceous
3 - carbonate
4 - red clay

II - Manganese content, C_{org} and fraction < 0.01 mm in superficial layer of sediment
5 - manganese
6 - organic carbon
7 - fraction 0.01

III - Thickness of oxidized layer and manganese content from cores
8 - reduced sediments
9 - oxidized sediments
10 - Mn content, wt percent
11 - boundary of the oxidized layer

TABLE 8. Mean Mn content of the collodial fraction of Pacific Ocean sediments (after Revelle, 1944), percent)

Station number	Sediment Types	$CaCO_3$ content	Content of colloidal fraction	Over-all Mn content	Mn in colloidal fraction	Percent total Mn in colloidal fraction
52	Ferruginous fine silty globigerina mud	86	13.5	0.78	1.34	23.0
57	Ferruginous fine silty globigerina mud	84	14.3	1.12	1.02	13.0
60	Globigerina silt-clay mud	75	18.9	1.25	1.79	27.0
156	Silica-carbonate silt-clay mud	40	27.7	0.49	0.58	30.0
142	Red clay	0.57	42.3	0.18	0.30	66.0
145	Red clay	0.72	46.0	0.28	0.56	92.0
149	Red clay	0.72	49.0	0.38	0.61	78.9
69	Red clay	20	38.6	0.14	0.39	100.0
72	Green clay mud	0.25	50.7	0.37	0.86	100.0

the surface sediments associated with its diagenetic redistribution and migration from the lower to the upper horizons of the sediment (fig. 3).

Where an oxidized layer is lacking the manganese content of the clays does not rise above 0.15-0.18 percent and the sediments themselves are greenish gray in color.

It can therefore be assumed that the primary manganese content of the fine terrigenous sediments was about 0.2 to 0.3 percent, rising to 0.5 percent by diagenetic migration from below to above. If this is so, then the difference between the primary manganese contents of the fine terrigenous and pelagic sediments will actually be considerable and the importance of solution and fine suspension in its movement will become particularly clear. But if this does in fact take place, a still more definite association of manganese with the collodial fraction of the sediment might have been expected than has been observed for iron. Table 8 confirms this: the manganese gravitates to the colloidal fraction in the pelagic sediments to a greater extent than does the iron.

The following fundamentals can therefore be regarded as established:

1) The distribution of iron according to grain size is essentially different from that of manganese. The iron content of carbonate and silica-free material is virtually independent of grain size, indicating an inconsistent fall or rise on passing from sands to clayey muds. The manganese content on the other hand is clearly related to grain size, increasing steadily from sands to clayey muds.

2) The contrast in the distribution of iron and manganese by grain size of the sediment is caused by the differential importance of fine suspension and solution in the migration of these elements. In particular, the part played by suspension and solution in the migration of manganese is much greater than in the movement of iron.

AREAL DISTRIBUTION OF IRON ON PACIFIC OCEAN BED

In order to clarify the areal distribution of iron on the bed of the Pacific Ocean maps were drawn of its percentage content in the surface layer of natural sediments (fig. 4) and of iron content of carbonate and silica-free deposits (fig. 5). Since the diluting effect of $CaCO_3$ and authigenic SiO_2 has been cancelled in the latter, it follows that it is the most suitable for analysis of the mechanism of distribution of the iron deposited in the sediments. We will therefore consider this alone in the future and turn to the map of iron in naturally occurring sediment only when required.

A broad zoning, corresponding entirely to that of climatic sediment formation in the Pacific Ocean, is seen in the distribution of iron in the surface layer of sediments.

There are five broad zones with contrasting iron contents: northern and southern subtropical zones with higher iron contents and northern, equatorial and subantarctic zones with lower concentrations.

The northern subtropical area lies roughly between 30° and 10-14° north. It runs from California west to the Phillipine Trench. On the east of the zone a rather wide belt runs along the North American continent north to the Queen Charlotte Islands. Mainly fine clay sediments with a pelitic fraction amounting to 80-90 percent and sometimes more (mainly red clays) occur in this zone. Their iron content ranges from 5 to 6.5, averaging 5.6 percent. In areas of contemporary vulcanicity near Hawaii and the Marianas it rises to 8-9 percent. At the same time, the iron content of the fine silty sediments in this area is usually about 8 percent (7-8 percent in the clayey muds).

Areas of high iron content are apparently localized by two basic factors: first, with the inflow of iron in the fine fraction of the terrigenous suspension from the coast of America whence it is dispersed by the Californian and north trade wind currents and then deposited along its course and in the halistatic zone of anticyclonic circulation; second, by the peculiarities of sediment formation in the tropical part of the ocean. The tropical and subtropical areas of the ocean have a low sedimentation rate and minimal planktonic activity. This promotes coagulation and deposition of colloidal iron hydroxides and organic iron compounds. The enrichment in iron of the silty and clayey sediments near Hawaii and the Marianas is due to the supply of pyroclastic material.

The second east-west zone of high iron content takes in a considerable part of the southern half of the Pacific Ocean, stretching from 2-3° south in the west and 7-10° south in the east, almost to 45° south. This region has the maximum content of iron in the Pacific Ocean, from 5 to 13 percent (recalculated on carbonate free material) and coincides with carbonate sediments and red clays.

The iron content of the red clay in this area is from 5 to 11 percent, with a maximum of 9 to 11 percent near the Polynesian Islands. The mean concentration of iron in the red clays is 7.1 percent (6.94 on average). The highest contents occur in the region of the Tabuai, Society and Tuamotu Islands. The iron in the sand-silt sediments amounts to 6-9 percent; 7.5-11 percent in the clayey muds.

Along the western part of this region high

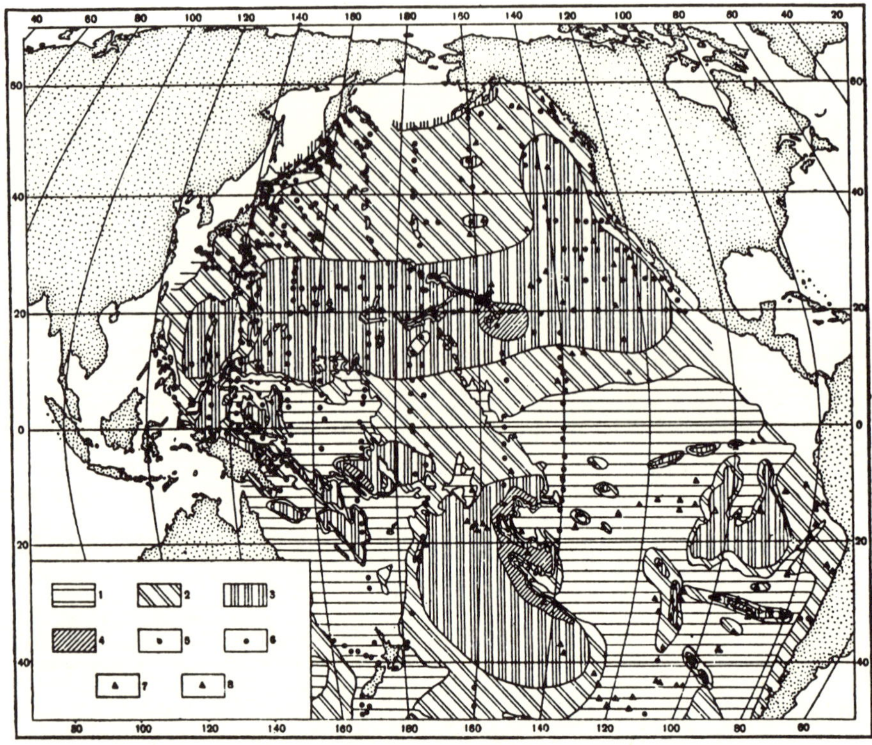

FIGURE 4. Distribution of iron in surface layer of Pacific Ocean sediment (wt. percent of natural sediment).

1 - < 3; 2 - 3-5; 3 - 5-7; 4 - 7-10

Stations on which iron determinations were carried out:

5 - "Vityaz"; 6 - "Ob'"; 7 - foreign expeditions; 8 - foreign expeditions with no iron determinations.

iron (6 to 13 percent) is confined to island arc sediments and deep trenches (Tonga, Samoa, New Hebrides and others).

The extensive region of iron concentration in the southern part of the Pacific Ocean owes its existence to identical factors which operate in the corresponding northern subtropical zone: low terrigenous contribution and low biological activity.

The iron enrichment of the sediments of the Tabuai, Society and Tuamotu Islands region and of the western part of the region around Tonga, Samoa, the New Hebrides, Ellice Islands and others has been produced by modern and recent vulcanicity and by the supply of iron in pyroclastic material. There is also a very interesting fact about the distribution of the iron in these regions: namely, the displacement in the maximum concentration of iron to the clayey sediments. This suggests the presence of supplementary sources of iron associated with the ingress of the element from volcanic exhalations.

In the eastern part of the subtropical zone of increased iron content there are iron-manganese carbonate sediments containing 10-25 percent iron in the carbonate-free material. With few exceptions, enrichment in pyroclastic material has not been recorded in these sediments.

El-Wakeel and Riley (1961) in their study of the composition of deep water sediments noted that certain carbonate deposits have a much higher iron content (in the carbonate-free base) than the clayey ones. They have put this down to the presence of oxides and hydroxides of iron liberated during the solution of globigerina and other organisms during their fall to the sea bottom. If marine foraminifera contain 1 percent Fe_2O_3 the authors admit that at least 85

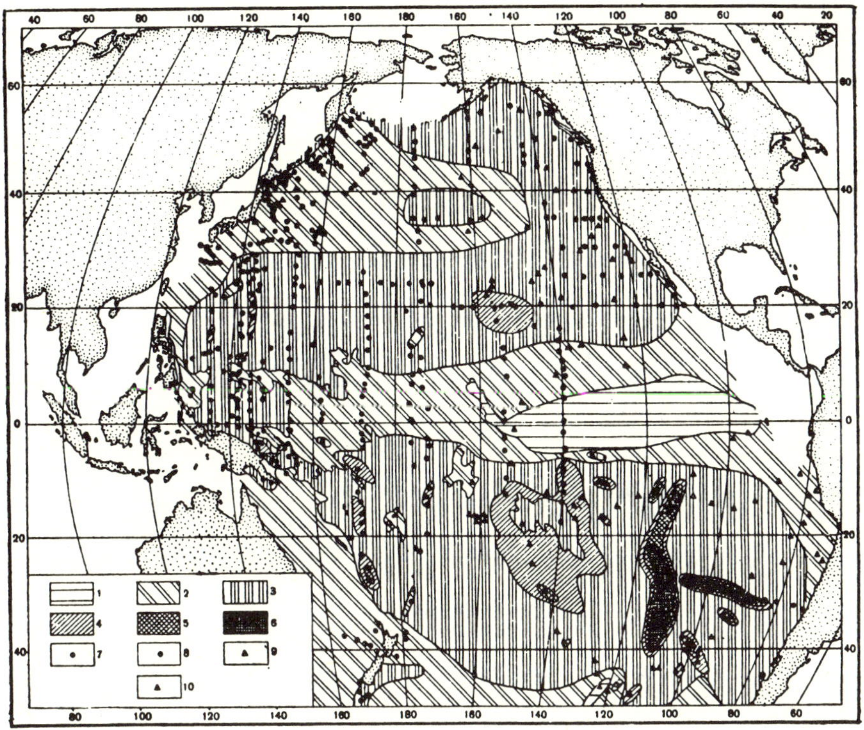

FIGURE 5. Distribution of iron in the surface layer of Pacific Ocean sediment (wt. percent of carbonate and silica free material)

1 - < 3; 2 - 3-5; 3 - 5-7; 4 - 7-10; 5 - 10-15; 6 - 15-20.

Stations on which iron determinations were carried out

7 - "Vityaz"; 8 - "Ob'"; 9 - foreign expeditions on which iron determinations were not carried out; [10 - omitted in original]

percent of the organic carbonate has to be dissolved to give 17 percent iron in the insoluble part of the sediment. This they think possible where there is a high level of CO_2 in the sea water.

But, as N.V. Belyayeva showed (1962) significant solution of foraminiferal tests does not take place during passage through the water, but only after they have lain for some time on the sea bed. She also observed traces of gradual solution of shells as the critical depths for the occurrence of carbonate sediments are found in red clays. It is just here that we might expect the maximum iron content. But in the red clays of the South Pacific and Bellingshausen trenches, which divide the East Pacific ridge, the iron content is 5-7 percent., i.e. roughly the same as in the natural iron-manganese high carbonate deposits of the Eastern Pacific Ridge. It is therefore impossible to account for the high concentration of iron in the east of this region by biogenic factors.

Petterson and Goldberg (1962) have recently carried out X-ray and microscopic investigations on the feldspars in the silty and pelitic fractions of the pelagic sediments of the South Pacific. They have demonstrated the volcanic origin of the bulk of the feldspars and the abundance of volcanic quartz. On this basis Petterson and Goldberg concluded that there is widespread contemporary vulcanism in the South Pacific.

This contemporary vulcanicity is primarily associated by them with the East Pacific Ridge. From this point of view, the occurrence of iron-manganese-carbonate sediments on the East Pacific Ridge and in local zones to the east of it, in an area of numerous submarine mountains, is very interesting. The formation of the iron-manganese-carbonate sediments seems to be associated with modern vulcanicity in the southeastern part of the Pacific Ocean and especially with post-volcanic phenomena, i.e hydrothermal sources.

Returning to the east-west zones of low iron content, the northermost of them stretches from Kamchatka, the Kurile Islands and Japan to approximately 140° west. It is situated in a zone of silt-clay diatomaceous and terrigenous muds and, in the eastern regions, in red clays. The limit of this zone is the 5 percent iron contour. The content of iron in natural sediments varies from 1.8 to 5 percent. Lower concentrations (less than 3 percent) are displaced to the southwest trough, where terrigenous low silica and siliceous muds are developed.

A second zone of low iron content - the equatorial minimum - occurs between the two subtropical zones of high iron content in the equatorial part of the Pacific Ocean. It extends in a rather wide belt from Central America to roughly 160° west, narrowing and disappearing in this direction.

The boundary of this zone coincides with the 5 percent iron isoline and lies roughly along 10-14° north and 2-7° south. In the eastern Pacific (east of 160° west) a belt of minimum iron content defined by the 3 percent isoline is distinguished in silica-carbonate muds within the zone. The iron content of radiolarian muds fluctuates from 3.5 to 5 percent, averaging 4.66 percent; it is 1 to 4.5, averaging 3.14 percent in carbonate sediments in the eastern equatorial zone and 3.3 to 6.1, averaging 4.69 percent in the western.

Finally, according to A.P. Lisitsyn, a zone of low iron content has been recognized in the subantarctic zone of the Pacific (south of 45 to 50° south) in an area of diatomaceous and diatomaceous-forminiferal sediments.

All three east-west zones of low iron content are found in regions where biogenic sediments are prevalent. These areas are at the same time characterized by higher biological activity. Their situation is determined by the fact that iron is an important biogenic element.

The iron is used in the biological cycle by the cells of planktonic organisms. When the cells and valves decay, one part of the iron is deposited along with the bodies of the organism and the other goes into the water in the form of organic iron compounds which are eventually hydrolyzed, coagulated and deposited on the sea bed. Part of the iron is returned to the upper layers of the water and utilized by the organisms.

In the upper active layer there is usually 2-3 times more suspended iron per gram than in the deep water. This distribution of iron in suspension in the waters of the open sea is only possible, as noted by Lisitsyn (1964), where there is some return of iron from depth to surface.

The formation of zones of low iron content is controlled by the active participation of the element in the biological cycle of organisms, its slow precipitation in areas of high biological activity and finally by the greater sedimentation rates in these parts of the ocean (greater by roughly 10 times or more than that in the subtropics).

The effect of active vulcanicity is superimposed on the east-west zonal distribution of finely dispersed iron in the Pacific Ocean. A zone of high iron content (in natural deposits) occurring only in the sand-silt sediments extends along the peripheral areas of the Pacific Ocean, from the Aleutian Islands in the north to New Zealand in the south. The iron content here is nearly always in the 5-7 percent range and more.

An iron content of 5-7 percent is recorded in the sand-silt sediments of the Aleutian Island arc through Kamchatka to the Kurile Islands. V.P. Petelin and E.A. Ostroumov (1958) have noted iron in excess of 6 percent in certain sands in the Kurile Islands. Its maximum concentration (up to 11 percent) is seen in the Chetvertiy Kuril'sk Bay.

Further south, a rise in the iron content takes place in sand-silt and silt-clay sediments in the southern part of the Idzu-Bonin and Mariana Island arcs. The maximum content here is 9.95 percent. Iron contents of 5 to 7 percent have been reported in sediments from near the Philippines, New Guinea, the Solomon Islands, New Hebrides and Tonga. The maximum content is on the coast of New Guinea.

Iron of volcanic origin also occurs near the Central Pacific islands (Hawaii, Easter Island, Tuamotu and others).

DISTRIBUTION OF MANGANESE IN SURFACE LAYERS OF PACIFIC OCEAN SEDIMENTS

Figure 6 indicates the distribution of manganese in the superficial 5 to 10 cm of the Pacific Ocean sediments (contents are quoted in wt. percent of carbonate and silica-free material). On drawing the high manganese isolines in the southeast region of the Pacific (corresponding to the distribution of the iron-manganese-carbonate sediments) the position of the "Carnegie" stations for which analyses were not made were taken into account; from Revelle's descriptions, the usual carbonate sediments occur there. In connection with the great mobility of manganese during diagenesis and the possibility of its redistribution throughout the sedimentary profile, the boundary of the area in which the thickness of the oxidized zone of the sediment is equal to

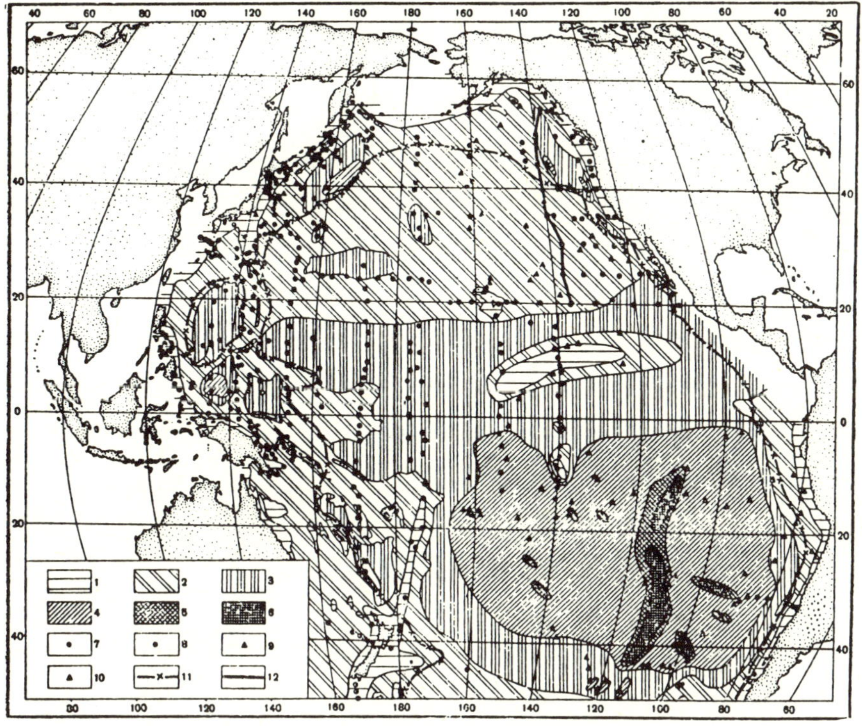

FIGURE 6. Distribution of manganese in surface layer of sediments (wt. percent of carbonate-free material)

1 - < 0.2; 2 - 0.2-0.5; 3 - 0.5-1.0; 4 - 1.0-3.0; 5 - 3.0-5; 6 - > 5.

Stations on which manganese determinations were carried out:

7 - "Vityaz"; 8 - "Ob'"; 9 - foreign expenditions;
10 - foreign expeditions, no determinations carried out;
11 - limits of the oxidized layer more than 1m thick.
12 - lithological section

or exceeds 1 to 2 m is drawn on the map and the vertical redistribution of the manganese reduced to a minimum. In the peripheral parts of the ocean outside this area the manganese content in the surface layer of sediments is largely determined by its vertical diagenetic redistribution.

A rapid glance at the map shows that the distribution of manganese in the superficial layers of the Pacific Ocean sediments is very different from that of iron. Without going into too much detail, the manganese content can be said to increase from the margin to the center of the ocean, reaching a maximum in the extensive, somewhat asymmetric region in the southern hemisphere between the equator and 40° south.

The manganese content is at a minimum in terrigenous sand-silt deposits in the peripheral areas of the Pacific (<0.2%). Seaward its concentration gradually increases in the clay and silt-clay muds. In the peripheral zone clays in the belt where vertical movement of manganese is possible during diagenesis, its content is usually in the range 0.2 to 0.5 percent. But in certain regions where these vertical migrations are apparently more marked the manganese content rises to 0.5-1.0 percent and sometimes higher (1.5-2 percent). Locally, however, concentrations of manganese in this belt fall to less than 0.2 percent giving its distribution along the margin of the Pacific a rather irregular aspect.

About one-third of the area of pelagic sediments nearer the center of the Pacific is covered with muds containing 0.2 to 0.5 percent manganese: these concentrations are localized almost everywhere in the external area of the pelagic region. At the same time the most extensive areas with manganese concentration of 0.2 to 0.5 percent are situated in the red clays of the North Pacific as far as 20° north. Concentrations of manganese ranging from 0.5 to 1.0

percent occur as a vast patch also (making up about one-third of the pelagic region), but it also contains a major zone where the manganese content lies between 1.0 and 3.0 percent. In carbonate sediments of this zone in the extreme eastern part of the ocean along the East Pacific Ridge and in isolated areas near it there are patches with manganese in excess of 5 percent.

In addition to the clear increases in manganese there are, however, in the pelagic sediments spots of very sparse accumulation of the element. They are usually small or very small and correspond to sand-silt deposits. The rather extensive patch of low manganese in clayey radiolarian muds north of the equator is however an exception, its content being as low as 0.2 percent.

Comparing the distributions of manganese and iron in the superficial sediments, it is easily seen that they are concentrated according to completely different schemes.

On the iron distribution map (figs. 4 and 6) the east-west zoning is the most characteristic feature. There is no such zoning on the manganese map, only traces of it remaining The east-west zoning is replaced by a concentric pattern of rising manganese content from the margin of the ocean to its center. The peripheral zones are deficient in manganese but the more central ones are clearly enriched in it. The greatest manganese contents are localized somewhat asymmetrically in the southeast between the equator and 40° south. This simple layout is complicated by numerous relatively small spots, in some of which the manganese content is very high while in others it is quite low.

The manganese distribution map differs from the iron one by the displacement of the main bulk of manganese from the edge of the oceanic area towards the center, particularly in the southeast.

There are two over-all features which nevertheless show through the distribution of iron and manganese: 1) high manganese (0.2 to 0.5 percent) in certain high iron sand-silt volcanic sediments in the peripheral areas of the ocean (Kurile, Idzu-Bonin, New Guinea, New Hebrides and other islands); 2) areas of abrupt increase of manganese in the southeast part of the ocean which almost exactly match areas of high iron.

Contrasting distribution of manganese and iron is not a specific property of the Pacific Ocean. The identical phenomenon has been noted by E.A. Ostroumov (1955) for the Sea of Okhotsk and by N.M. Strakhov (1960) for the Black Sea. Even the nature of the contrasts is the same: in each case the sediments at the edge of the basin are deficient in manganese relative to iron but manganese in the center is concentrated, i.e. in each case high manganese is displaced towards pelagic areas relative to iron.

As has been said above, the occurrence of iron and manganese by sediment grain size is also at variance, and the manganese profile differs from that of iron by a shift of the high concentrations towards the finer sediments. The contrasts in areal distribution of iron and manganese therefore agree well with variation in the disposition of sediment types.

What produces this pelagic displacement of the the manganese values relative to iron? It is apparently a single factor: manganese is geochemically more mobile than iron, which is expressed in the greater importance of solution and colloidal suspension in its migration, independent of whether the source is terrigenous or volcanic (Strakhov, 1960).

EFFECT OF VULCANICITY ON DISTRIBUTION OF MANGANESE AND IRON IN PACIFIC OCEAN SEDIMENTS

Iron and manganese appear to accumulate in the oceans as a result of inflow of terrigenous material from the continents and from vulcanicity. It is at present still difficult to evaluate the importance of volcanic products in the general balance of material participating in sedimentation, and in contemporary sediment formation in particular. Nevertheless, attempts to evaluate this factor have already been made. Of the total sediments deposited between D_1 and J_3 times, material of volcanic origin, according to A.B. Ronov (1959), constitutes some 24 percent. N.M. Strakhov gives a similar estimate (25 to 35 percent) for the proportion of volcanic material taking part in contemporary sedimentation. Lavas, agglomerate flows, lapilli, ejectimenta and ash of sand-silt grade form the bulk of the volcanic material in fact remains in the vicinity of the volcanic centers and only 10 to 12 percent is contributed to deep pelagic sediments (Strakhov, 1962).

Associated with the fact that some part of the iron and manganese, presently indeterminable, is certainly of volcanic origin, the problem naturally arises about the effect of volcanic activity on the distribution of these elements in the Pacific sediments. The solution requires elucidation of the trends and features of the distribution of iron and manganese which show a proven association with volcanic activity.

There seem to be two such features for iron. The first is a definite enrichment in the iron of coarse littoral deposits near centers of active volcanoes and, even though they are now extinct, young volcanic islands. This is seen along the Kurile, Aleutian and Mariana Island chains, at New Guinea, New Hebrides, Samoa, Tonga and other localities. Here iron is deposited as solids in the form of iron and iron-bearing volcanic minerals. Hydrothermal efflux of iron has no effect here.

The clear local concentrations of iron in the southeastern part of the ocean between the equator and 40° south are a further feature of its distribution in Pacific Ocean sediments. The greatest concentrations of iron here are confined to calcareous muds (iron - manganese-carbonate sediments) along the East Pacific Ridge. Pyroclastic material is almost typically absent from these sediments. Areas of calcareous mud enriched in iron form long narrow belts sometimes elongated north-south and sometimes east-west The thought unwillingly springs to mind that in this case the iron has been carried out in solution by underwater thermal springs rather than in the solid phase. Vulcanicity here influences the distribution of the iron through its liquid hydrothermal phase.

Local iron enrichment in the sediments about the Society Islands, Tuamotu and Tabuai may be connected with the ingress of dissolved exhalated iron. It is of interest to note that the highest iron concentrations are here displaced into the clayey deposits (red clays and clay-carbonate muds).

Volcanic efflux of manganese in pyroclastic material also takes place. The effect of this mechanism is very weak, as manganese content is low near the volcanic islands in the sand-silt deposits. Efflux of manganese is much more clearly demonstrated in thermal springs in those localities in the southeastern Pacific where iron is concentrated in calcareous muds. The problem arises as to whether, in general, the whole area of high manganese south of the equator (1 to 3 percent) is the result of hydrothermal action? It is not yet possible to give a definite answer.

As seen from the mechanical denudation process on the modern continents (Strakhov, 1960) the terrigenous material in the North Pacific is fundamentally derived from North America, Asia and humid tropical areas. In the southern, and particularly in the southeastern part, the supply of terrigenous material from South America is negligible (by virtue of the arid climate). The inconsiderable terrigenous inflow and the slow tempo of sediment formation in the southeast portion of the Pacific leads to an overall increase in the importance of the hydrogene components, especially manganese, in the constitution of the sediments and to a displacement of the highest manganese contents in this region.

Bramlette (1961), describing the pelagic clays of the North and South Pacific, remarked upon the high concentration of manganese in the latter. He associated this with the marked remoteness of the South Pacific from sources of terrigenous material and with extremely low sedimentation rates (0.5 to 0.3 mm/1000 years according to Goldberg and Koide, 1962).

The vast area of high manganese content in the southeast of the Pacific should thus apparently be related, not to particularly strong thermal activity in the region but to a definite reduction in the rate of sedimentation which, so to speak, has the effect of concentrating manganese (and iron) in the sediment not only by means of efflux but by the accumulation of Mn and Fe brought in by rivers far from the place of their fixation on the sea bed.

REFERENCES

Belyayeva N.V., 1962, Raspredeleniye planktonnykh foraminifer v tolshche vod Indiyskogo okeana [DISTRIBUTION OF PLANKTONIC FORAMINIFERA IN WATERS OF INDIAN OCEAN]: Byul. Mosk. o-va ispyt. prirody. Otd. geol., v. 37 (3).

Bezrukov, P.L., A.P. Lisitsyn, E.A. Romankevich and N.S. Skornyakova, 1961, Sovremennoye osadkoobrazovaniye v severnoy chasti Tikhogo Okeana. Sovremennyye osadki morey i okeanov [CONTEMPORARY SEDIMENT FORMATION IN THE NORTH PACIFIC MODERN SEDIMENTS OF THE SEAS AND OCEANS]: U.S.S.R. Acad. Sci. Press.

Bramlette M.N., 1961, PELAGIC SEDIMENTS: Oceanogr. publ. Amer. Asso. Advancement Science, Wash. D.C., no. 61.

El - Wakeel, S.K. and J.P. Riley, 1961, CHEMICAL AND MINERALOGICAL STUDIES OF DEEP-SEA SEDIMENTS. Geochim. et cosmochim. acta. v. 25, no. 2.

Goldberg, E.D., and G.O.S. Arrhenius, 1962, CHEMISTRY OF PACIFIC PELAGIC SEDIMENTS: Geochim. et cosmochim. acta., v. 13, no. 213.

Goldberg, E.D. and Koide, M., 1962, GEOCHRONOLOGICAL STUDIES OF DEEP-SEA SEDIMENTS: Geochim. et cosmochim. acta., v. 26.

Lisitsyn, A.P., 1964, Raspredeleniye i khimicheskiy sostav vzvesi v vodakh Indiyskogo okeana. Rezul'taty issledovania po programme MGG [DISTRIBUTION AND CHEMICAL COMPOSITION OF SUSPENSION IN WATERS OF INDIAN OCEAN. RESULTS OF INVESTIGATIONS IN THE MGG PROGRAMME]: Okeanologiya, no. 10.

Murray J. and A.F. Renard, 1891, DEEP-SEA DEPOSITS. Rep. on the scient. res. of the voyage of H.M.S. Challenger, 1873-1876.

Ostroumov, E.A., 1955, Raspredeleniye

margantsa v donnykh otlozheniyakh Okhotskogo morya [DISTRIBUTION OF MANGANESE IN BOTTOM SEDIMENTS OF SEA OF OKHOTSK]: Izv. AN SSSR, ser. geol. no. 5.

__________, 1955, Zhelezo v donnykh otlozheniyakh Okhotskogo morya [IRON IN BOTTOM SEDIMENTS OF THE SEA OF OKHOTSK]: Dokl. Akad. Nauk SSSR, v. 102, no. 1.

Petelin, V.P. and E.A.Ostroumov, 1958, O nekotorykh osobennostyakh raspredeleniya zheleza v osadkakh Okhotskogo morya [CERTAIN PROPERTIES OF DISTRIBUTION OF IRON IN SEDIMENTS OF SEA OF OKHOTSK]: Byul. Mosk. O-va ispyt. prirody. Otd. geol. no. 2.

Petterson, M.N.A. and Goldberg, E.D., 1962, FELDSPAR DISTRIBUTION IN SOUTH PACIFIC PELAGIC SEDIMENTS: J. Geophys. Res., v. 67. no. 9.

Revelle, R.R., 1944, MARINE BOTTOM SAMPLES COLLECTED IN THE PACIFIC OCEAN BY THE CARNEGIE ON ITS SEVENTH CRUISE. SCIENTIFIC RESULTS OF CRUISE VII OF THE CARNEGIE DURING 1928-1929: Carnegie Inst. Wash. Publ., no. 556.

Rovov, A.B., 1959, K posledokembriyskoy geokhimicheskoy istorii atmosfery i gidrosfery [POST-PRECAMBRIAN GEOCHEMICAL HISTORY OF THE ATMOSPHERE AND HYDROSPHERE]: Geokhimiya, no. 5.

Skornyakova, N.S., 1961, Donnyye otlozheniya severo-vostochnoy chasti Tikhogo okeana [BOTTOM SEDIMENTS OF NORTHEASTERN PART OF PACIFIC OCEAN]: Tr. in-ta okeanologii, v. 45.

Strakhov, N.M. et al., 1954, Obrazovaniye osadkov v sovremennykh vodoyemakh [SEDIMENT FORMATION IN MODERN BASINS]: U.S.S.R. Acad. Sci. Press.

__________, 1960, Osnovy teorii litogeneza [FUNDAMENTALS OF THEORY OF LITHOGENESIS]: v. 1 and 2. U.S.S.R. Acad. Sci. Press.

__________, 1962, K voprosu o znachenii vulkanicheskogo protsessa v osadochnym porodoobrazovanii [PROBLEM OF SIGNIFICANCE OF VULCANICITY IN SEDIMENTARY ROCK FORMATION]: Sovet. geol., no. 9.

Wedepohl, K.H., 1960, SPURENANALYTISCHE UNTERSUCHUNGEN AN TIEFSEETONEN AUS DEM ATLANTIK: Geochim. et cosmochim. acta., v. 18, no. 3 - 3 [sic].

13

Reprinted from *Econ. Geology* **61**:1258-1265 (1966)

PRECIPITATES FROM HYDROTHERMAL EXHALATIONS ON THE EAST PACIFIC RISE

K. BOSTROM AND M. N. A. PETERSON

ABSTRACT

On the very crest of the East Pacific Rise, in equatorial latitudes—particularly 12° to 16° South, the sediments are enriched in Fe, Mn, Cu, Cr, Ni, and Pb. The correlation of these areas of enrichment to areas of high heat flow is marked. It is believed that these precipitates are caused by ascending solutions of deep-seated origin, which are probably related to magmatic processes at depth. The Rise is considered to be a zone of exhalation from the mantle of the earth, and these emanations could serve as the original enrichment in certain ore-forming processes. Ba and Sr are enriched on the flanks of the Rise.

INTRODUCTION

VOLCANIC processes on the continents commonly release volatile emanations like H_2O, CO_2, H_2S, H_2, HF and various mineralizing aqueous solutions. Zies (17) and Rubey (12), among others, have estimated the amounts of subaerial volcanic emanations that are released into the atmosphere and hydrosphere. One may infer from Rubey (12) that the quantities of emanations released by submarine volcanic activity are of the same order of magnitude. In addition, many magmatic events are not manifested as surficial volcanism. The possibility that precipitates that originate from hydrothermal solutions might be accumulating in deep-sea sediments was suggested a long time ago by von Gümbel (14). The idea continues to attract attention: very recently Richards (11) has favored submarine exhalations as the source for the Broken Hill banded iron formations. Descriptions of solutions from hydrothermal wells near the Salton Sea, California (16) have added recent impetus to this general trend of thought.

THE EAST PACIFIC RISE SEDIMENTS

On the very crest of the East Pacific Rise in equatorial latitudes—6° North and 12°–16° South (where expedition RisePac of Scripps Institution crossed the Rise with two traverses), the sediments are different from most other sediments of the Pacific Ocean. They contain an abundant, brown, X-ray amorphous, metal oxide precipitate. Rather than being a coating or impregnation of other mineral grains, as is common for the normal red-brown sediments of the deep ocean, it is a separate, very finely disseminated or loosely aggregated material. It is possible to stabilize a suspension of this material that does not settle out for many months. Our attention was first attracted to the possibility that this material could be a precipitate related to hydrothermal activity by the observation that, during the treatment to remove calcium carbonate from the sediment, this very fine precipitate would not settle out after having been dispersed. Figure 1 shows a series of samples taken across the Rise (the southern traverse of RisePac Expedition—section AA of Fig. 1). The supernatant liquid of only those samples on the very crest of the Rise is colored a bright orange-brown by this finely dispersed material that did not settle from suspension after several weeks. This same effect, although not as marked, was clearly present for the northern traverse of RisePac expedition, which crossed the Rise at about 6° North Latitude. Where abundant, this precipitate stains the normal light, pure globigerina ooze a dark brown.

FIG. 1. Series of samples taken during southern traverse of RisePac expedition showing very finely disseminated metal oxide precipitate in suspension. Only those samples very near the crest of the Rise show this effect.

We therefore decided to study the distributions of various elements that could serve as tracers of hydrothermal processes.

The ideal tracer for this purpose should have a comparatively short residence time so that the element would precipitate as an aureole around the area of emanation. Too short a "residence time" would lead to deposition just around, or even deep inside, the orifice, and would make it difficult or impossible to locate regions of hydrothermal activity. Too long residence times would lead to mixing and transport over long distances and thus, not produce easily interpreted distribution patterns. The major constituents of sea water, of course, have very long residence times and are poor tracers of hydrothermal activity unless they were to occur in unusually large concentrations.

The locations of the analyzed sediment cores are shown in Figure 2. All samples were collected during RisePac expedition. Samples were taken from the upper several centimeters of the core. Two analytical procedures were used:

1) One set of samples, each consisting of 0.15 gm dry, ground unwashed sediment, was digested in boiling HCl, diluted and analyzed with a Perkin Elmer 303 Atomic Absorption spectrophotometer. All samples of the northern traverse of RisePac and that part of the southern traverse designated AA (Fig. 2) were analyzed in this way for Au, Ba, Co, Cr, Cu, Fe, Mn, Ni, Pb and Sr. Carbonate analyses were of this same mixed sample.

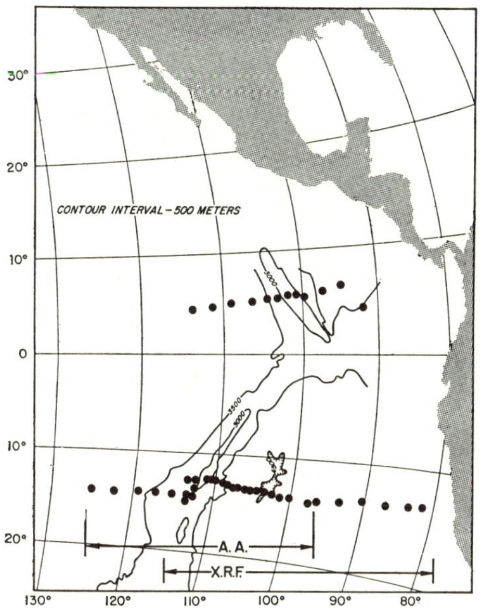

FIG. 2. Map of the East Pacific, showing the location of analyzed sediment samples.

2) The samples from the southern traverse designed XRF were pressed into pellets and analyzed on a Norelco X-ray fluorescence unit for Ba, Ca, Fe, Mn, Si, and Sr. The range of samples studied by XRF was purposely extended more to the east than those studied by method 1) for reasons given below.

Results of the analyses are in Figures 3, 4, and 5. Heat flow data and water depths for these two traverses across the Rise are also shown (15).

There is a marked enrichment of Fe, Mn, Cu, Cr, Pb, and Ni in sediments from the very crest of the Rise; in addition, a clear co-variation exists between these elements. There is also, however, especially on the southern traverse, a very pronounced maximum in the concentration of these elements to the

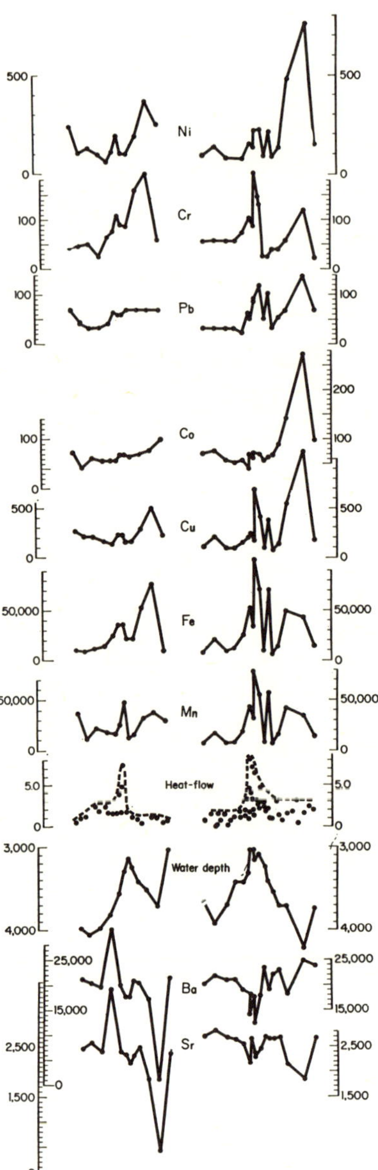

Fig. 3. Results of chemical analyses by atomic absorption. Analyses of samples from the northern traverse of the RisePac expedition are given to the left; to the right are given analyses of samples from the southern traverse of the RisePac expedition. Concentrations in ppm, water depths in m, heat flow in 10^{-6} cal cm^{-2} sec^{-1}.

east of the Rise, closely coinciding with a depression in the ocean floor between the Rise and a parallel ridge to the east. Because of the deep water, very little carbonate is present in these sediments of the depression. In order more clearly to determine the normal concentration values for Fe and Mn, and most abundant elements in the oxide precipitate, these elements were determined by X-ray fluorescence for samples farther to the east of the Rise. As can be seen, Figure 5, the high Fe and Mn values for the sediments of the depression mentioned above are not typical for most of the sediments on the sides of the Rise. On the contrary, it is evident that, calculated on a carbonate free basis, the total concentrations of Fe and Mn reach their highest concentrations on the very crest of the Rise (Figs. 4, 5).

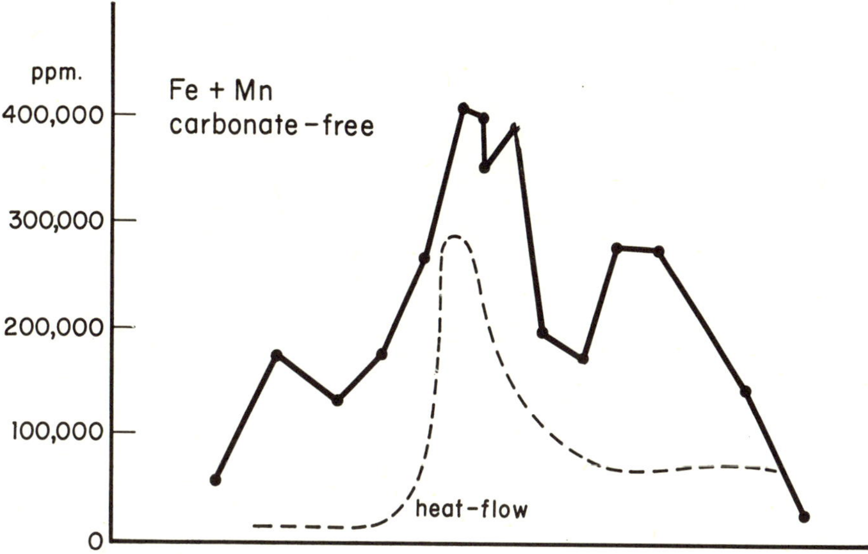

FIG. 4. Co-variation of heat flow and Fe and Mn (combined) on a carbonate free basis. Data from the southern traverse. Concentrations in ppm.

Ba and Sr are enriched on the flanks of the Rise, but not on its crest—a distribution pattern clearly different from that of Fe, Mn, and associated elements. A clear co-variation exists for Ba and Sr over the crest and flanks of the Rise; this co-variation is probably due to the fact that Ba and Sr occur as a celestobarite. Farther to the east, however, the distribution of Sr is that more commonly found in pelagic sediments, that is, Sr co-varies with Ca, presumably in a carbonate phase. In the depression Ba is probably present both in barite and a zeolite; both these minerals have been observed in sediments from this area (Bonatti, pers. comm.). Au showed no enrichment across the crest of the Rise. Summing up the evidence, Fe, Mn, Cu, Cr, Ni, and Pb are present in important quantities only on the crest of the

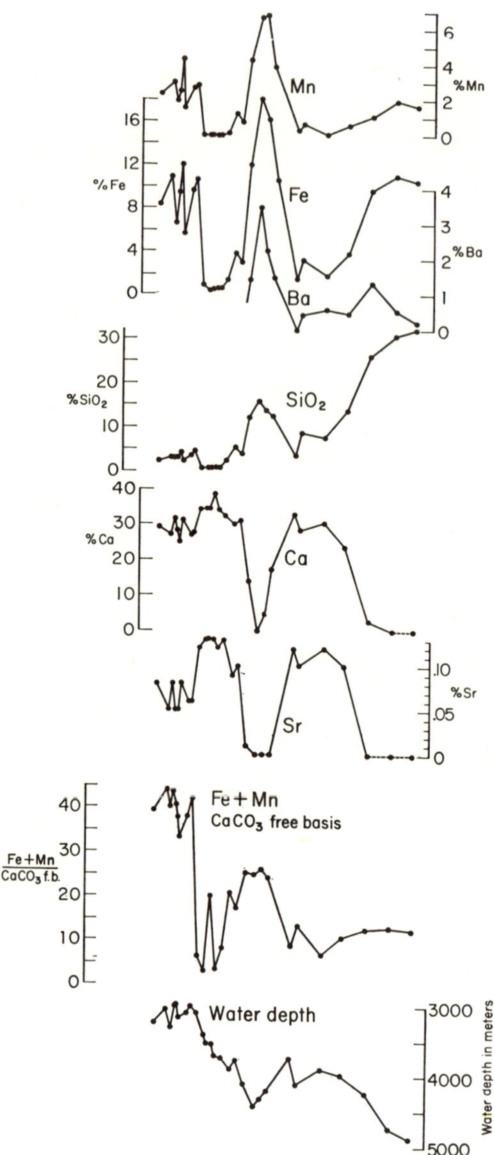

Fig. 5. Results of chemical analysis by X-ray fluorescence. Location of analyzed samples is indicated by the symbol XRF on Figure 1. Concentrations in weight percent.

Rise; on a carbonate free basis, these sediments may exceed 60 percent, by weight, metal oxide precipitate. Similar results for Fe and Mn have been reported by Skornyakova (13) to be due to hydrothermal activity. This is the only explanation that seems reasonable also to the present authors.

DISCUSSION

The crest of the East Pacific Rise appears to be a more or less persistent zone of magmatic activity. High heat flow anomalies on the East Pacific Rise (15) suggest that some magmatic event may exist under its crest (4, 6). This conclusion is supported by the evidence for volcanic activity that has been observed along the trend of the East Pacific Rise (9, 10). Silica cementation (1) and barite distribution in sediments in the vicinity of the Rise has also been suggested as due to hydrothermal processes (2). The hot springs of the western United States and the high heat flow areas of the top of the East Pacific Rise appear to be part of the same main geological feature (7). The presence of the Rise appears to be related to rifting of the Gulf of California, where the Rise extends under the continent; similar relationships are to be found in the Red Sea, in which hot pools of metalliferous brines have recently been discovered (8).

It is believed that these precipitates are caused by ascending solutions, of deep-seated origin, which are probably related to magmatic processes at depth. It is futile to speculate on the exact nature of the magmatic events, whether interflow, ascending magma chamber, rift filling or other. Drake and Girdler (5) interpret the data of the Red Sea rift to indicate the presence of a long, narrow igneous intrusive at shallow depth extending along the middle of the rift. Judging from the suite of elements that debouch at the surface, it would appear likely that the magma type would be basic. These ascending solutions, which would be expected to be acidic and reducing, would transport these elements upward, and precipitate them upon reaching alkaline, oxidizing environments associated with sea water and ocean bottom conditions. Such solutions would also surely mobilize elements such as Mn while traversing deeper sections of the sediment (3).

The major Rise-Rift systems of the world are exceedingly large features, and the reasons for their existence must most likely lie within the mantle of the earth. Whatever else the Rise-Rift systems may be, they appear also to be zones of emanation from the upper mantle. Additional evidence for this statement may be the anomalously high quantities of He that have been found in deep waters close to the crest of the East Pacific Rise (Bieri, Koide, and Goldberg, pers. comm.). The well known fact that high heat flow values are found on the crests of oceanic rises indeed suggests that the same deep seated processes that cause the formation of rises and their high heat flows also trigger the formation of ascending mineralizing solutions. It seems likely that we are seeing here an example of a process that could serve as the basic enrichment mechanism and source for the elements in some ore formations. If such metals in hydrothermal solution could find their way into the proper depositional situation and become deposited there, and later perhaps reworked in various ways or metamorphosed, these later processes would leave their imprint on the geologic history of the deposit, and leave in doubt the original concentration mechanism.

ACKNOWLEDGMENTS

We thank R. Bieri and E. Bonatti for valuable comments. Carbonate analyses were made by Miss B. Meinke in the laboratory of Dr. E. D. Goldberg. Mr. E. Jarosevich kindly helped us with the X-ray fluorescence analyses. Support by NSF grants GP-489, GP-5112 and the ONR is gratefully acknowledged.

SCRIPPS INSTITUTION OF OCEANOGRAPHY,
UCSD, LA JOLLA, CALIFORNIA,
August 16, 1966

REFERENCES

1. Arrhenius, G., 1952, Sediment cores from the east Pacific: Reports of the Swedish deep sea expedition, 1947–48, Vol. V., Goteborg.
2. Arrhenius, G., and Bonatti, E., 1965, Neptunism and volcanism in the ocean: in Progress in Oceanography, v. 3.
3. Bostrom, K. (in prep.), The problems of excess manganese in pelagic sediments.
4. Bullard, E. C., 1963, The flow of heat through the floor of the ocean: in The Sea, ed. M. N. Hill, Interscience Publishers, J. Wiley and Sons, N. Y.
5. Drake, C. L., and Girdler, R. W., 1964, A geophysical study of the Red Sea: Geophys. Jour. Royal Astro. Soc., v. 8, p. 473.
6. McBirney, A. R., 1963, Conductivity variations and terrestrial heat-flow distribution: J. Geophys. Research, v. 68, p. 6323–6329.
7. Menard, H. W., 1964, Marine Geology of the Pacific: McGraw-Hill, N. Y.
8. Miller, A. R., Densmore, C. D., Dogens, E. T., Hathaway, J. C., Manheim, F. T., McFarlin, P. F., Pocklington, R., and Jokela, A., 1966, Hot brines and recent iron deposits in deeps of the Red Sea: Geoch. et Cosmochim. Acta, v. 30, p. 341–359.
9. Peterson, M. N. A., and Goldberg, E. D., 1962, Feldspar distributions in South Pacific pelagic sediments: J. Geophys. Research, v. 67, p. 3477–3492.
10. Peterson, M. N. A., and Griffin, J., 1964, Volcanism and clay minerals in the southeastern Pacific: J. Marine Research, v. 22, p. 13–21.
11. Richards, S. M., 1966, The banded iron formations at Broken Hill, Australia, and their relationship to the lead-zinc ore bodies: ECON. GEOL., v. 61, p. 257–274.
12. Rubey, W. W., 1951, Geologic history of sea water: Geol. Soc. America Bull. v. 62, p. 1111.
13. Skornyakova, I. S., 1964, Dispersed iron and manganese in Pacific Ocean sediments: Lithology and Mineral Resources, v. 5, p. 3–20 (In Russian). Internat. Geol. Rev., v. 7, p. 2161–2174 (In English).
14. von Gümbel, G., 1878, Ueber die im Stillen Ocean auf dem meeresgrunde vorkommenden Manganknollen, Sitz. Berichte d. K. Bayerischen Akademie d. Wissenschaften Munchen, *Matem-Physik Klasse,* p. 189–209.
15. von Herzen, R. P., and Uyeda, S., 1963, Heat flow through the eastern Pacific Ocean floor: J. Geophysical Research, v. 68, p. 4219–4250.
16. White, D. E., Anderson, E. T., and Grubbs, D. K., 1963, Geothermal brine well: mile-deep drill hole may tap ore-bearing magmatic water and rocks undergoing metamorphism: Science, v. 139, p. 919–922.
17. Zies, E. G., 1929, The valley of ten thousand smokes. II The acid gases contributed to the sea during volcanic activity: Nat. Geogr. Soc. Contrib. Tech. Papers, Katmai Series 1, v. 4, p. 61–79.

14

Reprinted from *Earth and Planetary Sci. Letters* **12**:425 (1971)

GEOCHEMISTRY OF THREE CORES FROM THE EAST PACIFIC RISE*

Michael BENDER, Wallace BROECKER, Vivian GORNITZ**, Ursula MIDDEL,
Robert KAY, Shine-Soon SUN and Pierre BISCAYE
Lamont-Doherty Geological Observatory of Columbia University, Palisades, New York 10964, USA

Received 4 June 1971
Revised version received 30 August 1971

We have determined sedimentation rates of a core from the crest and two cores from the flank of the East Pacific Rise from the depth variations of their excess ^{230}Th content. We have also analyzed composite strip samples from these three cores for [Mn], [Fe], [Ni], SiO_2], [Al_2O_3], [NaCl], and mineralogy, and determined the rare earth contents and isotopic composition of lead, strontium, and uranium in the crest core. The Mn accumulation rate of the crest core is 25 times higher than the world average, supporting the conclusion of Bostrom and Peterson [1] that at least some of the elements in these sediments derive from local volcanism. The relative concentrations of the rare earth elements and the isotopic composition of U and Sr in the crest core indicate that these elements derive from sea water. On the other hand, thc isotopic composition of lead in the crest core indicates that the lead is mostly of local volcanic origin.

[*Editors' Note:* In the original, material follows this abstract.]

* Lamont-Doherty Geological Observatory of Columbia University contribution no. 1584.

** Present address: Institute for Space Studies, 112 St. and Broadway, New York City.

15

Reprinted from pages 541-544 of *Initial Reports of the Deep Sea Drilling Project,* Vol. V, D. A. McManus et al., eds., U.S. Government Printing Office, Washington, D.C., 1970, 827 pp.

AMORPHOUS IRON OXIDE PRECIPITATES IN SEDIMENTS CORED DURING LEG 5, DEEP SEA DRILLING PROJECT

C. C. von der Borch, Scripps Institution of Oceanography, La Jolla, California
and
R. W. Rex, University of California, Riverside, California

INTRODUCTION

The sediment column at Sites 37, 38 and 39, Leg 5, of the Deep Sea Drilling Project (locality map, Figure 1) is dominantly a dark brown unconsolidated "mud", composed in part of a non-crystalline iron oxide mineral. X-ray diffraction studies by Rex (this volume) verify its non-diffracting, amorphous nature. This ferruginous material appears identical to metal oxide precipitates described by Bostrom and Peterson (1966), Bostrom *et al.* (1969), and Bostrom and Peterson (1969) which were associated with areas of high heat flow near the crest of the mid-ocean ridge system, particularly the East Pacific Rise. It is also similar, at least in textural detail, to the so-called Amorphous Goethite Facies of the Red Sea metalliferous sediments, described by Bischoff (1969). In all documented cases, these unusual sediments appear to be associated with the mid-ocean ridge system or with related rift structures. If the hypothesis of sea-floor spreading is correct, then such a metal-rich facies would be expected to occur as a basal deposit immediately overlying basaltic "oceanic basement" in many areas of the ocean basins. Site 9, Leg 2, of the Deep Sea Drilling Project encountered hematite-rich sediments just above igneous rock, and this occurrence was interpreted by Peterson *et al.*, 1970, as being due to precipitation near hydrothermal sources at a prior oceanic ridge crest. The fact that basal iron-rich sediments were also encountered at Sites 37, 38 and 39, Leg 5, of the Deep Sea Drilling Project in areas remote from a present-day ridge system calls for a detailed description of the mineralogical and sedimentological associations occurring at these sites.

STRATIGRAPHIC RELATIONSHIPS

Generalized stratigraphic columns of Sites 37, 38 and 39 are shown in Figure 1. Basalt was reached at Sites 37 and 39 with no obvious indications in the recovered sediment of baking at the contact. No igneous rock was sampled at Site 38, although a hard layer, probably basalt, terminated the drilling. In all three holes, three distinct facies were recognized which are described in detail below. These facies are: Basal amorphous iron-manganese oxide facies; mixed amorphous iron-manganese oxide-detrital facies; and, detrital facies. The criteria by which these facies are recognized differ somewhat from those used in the earlier shipboard facies designations of McManus *et al.* (this volume).

Basal Amorphous Iron-Manganese Oxide Facies

Immediately overlying basalt at Sites 37 and 39, a section of 5 to 6 meters of dusky yellowish-brown amorphous goethite "mud" occurs. Small quantities of Lower Eocene calcareous nannoplankton occur near the base of the hole at Site 39. At Site 38, the basal 9-meter section of sediment is a mixture comprising sub-equal amounts of amorphous iron-manganese oxides and calcified Lower Eocene calcareous nannoplankton, overlain by 6 meters of the 100 per cent amorphous iron-manganese oxide sediment. In all three sites this sediment association is designated the "basal amorphous iron-manganese oxide facies". Apart from the local admixture of nannoplankton, an ash bed at Site 37 and scattered aggregates of phillipsite, this facies is essentially a pure iron-manganese oxide sediment, free from crystalline detritus. Bulk X-ray diffraction studies (Rex, this volume) and microscopic smear slide observations point to a relatively sharp boundary between this facies and the overlying unit described below.

Mixed Amorphous Iron-Manganese Oxide-Detritial Facies

Overlying the basal facies at all three sites is a dusky yellowish-brown sediment composed of a mixture of amorphous iron-manganese oxides and crystalline detrital components. This is termed the "mixed amorphous iron-manganese oxide-detrital facies". Microscopic examination shows that the proportion of amorphous iron-manganese oxides decreases in an up-hole direction along with a progressively increasing detrital admixture. Crystalline components, herein broadly characterized as "detrital", include silt-sized quartz and plagioclase, accompanied by kaolinite, mica and chlorite. These minerals may be largely of eolian origin, or they may represent nepheloid layer deposition. Other minerals detected by X-ray diffraction (Rex, this volume) are montmorillonite, phillipsite and localized barite, of authigenic or hydrothermal origin. Scattered manganese nodules occur within this facies at Site 37, and an ash bed is present at Site 38.

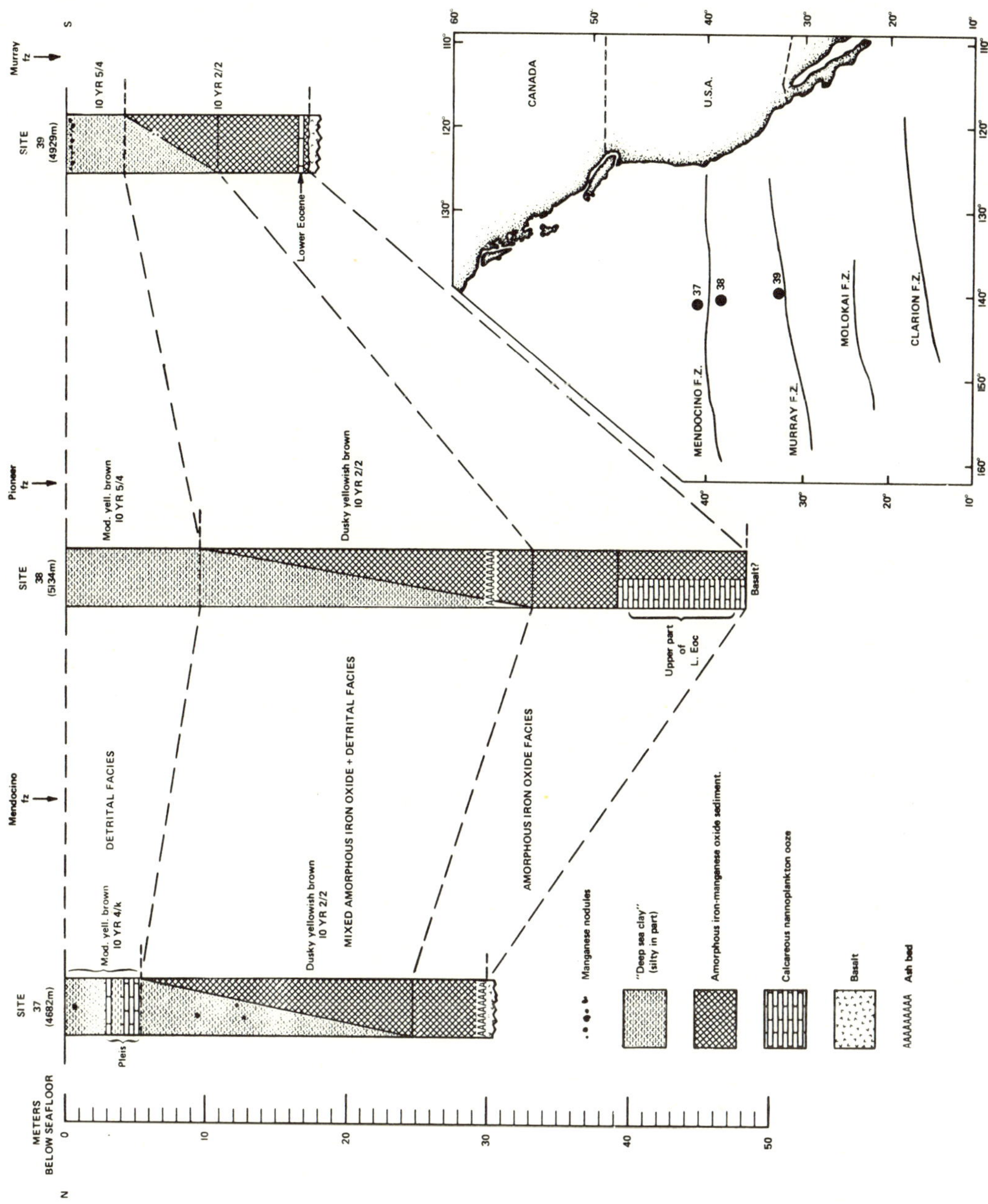

Figure 1. *Locality chart and stratigraphic columns of Sites 37, 38 and 39 of the Deep Sea Drilling Project. Color code is from the U. S. G. S. rock color chart.*

The upper level of this facies is not as well-defined as its lower boundary. However, the dashed line shown in the stratigraphic columns in Figure 1 approximates the zone above which essentially no amorphous iron-manganese oxides are observed in smear slides. Age relationships of this facies are uncertain, due to lack of fossil remains in all three sites.

Detrital Facies

The gradational upper boundary of the "mixed amorphous iron-manganese oxide-detrital facies" coincides with a color change from dusky yellowish-brown below to moderate yellowish-brown above. The lighter color persists to the sediment surface. The upper unit is termed the "detrital facies" and is largely composed of silt-sized quartz and plagioclase, along with kaolinite, mica, chlorite, barite (in Site 37) and montmorillonite (in Site 38). Generally speaking, this sediment type is equivalent to that described in "deep sea brown clay" by the majority of marine geologists. Several layers containing Pleistocene calcareous nannoplankton occur in this facies at Site 37. Manganese nodules and micronodules occur sporadically within the sediment at Sites 37 and 39.

MINERALOGY OF THE AMORPHOUS IRON-MANGANESE OXIDE SEDIMENT

Smear slide observations show the iron oxide sediment collected on Leg 5 to be composed of a mixture of colloidal material and microscopic sub-spherical grains. When dispersed in distilled water, a reddish-brown colloidal suspension persists for many days before settling, a property similar to that observed in equivalent sediments from the crest of the East Pacific Rise (Bostrom and Peterson, 1966). The actual iron oxide spherules are translucent and amorphous and vary in color from light to dark yellowish-brown in transmitted light. Darker hues may be indicative of admixtures of manganese oxides, in a manner comparable to East Pacific Rise analogues (Bostrom and Peterson, 1969).

The ferruginous spherules range in size from 0.5 microns to 25 microns, being dominantly about 5 microns in section. Larger particles appear to be aggregates that survived treatment with an ultrasonic probe. In general, the size range is equivalent to that observed by Bischoff (1969) in sediment of the Red Sea Amorphous Goethite Facies.

Amorphous iron-manganese oxides described by Bostrom and Peterson (1969) from the East Pacific Rise surface sediments show low aluminum and titanium values and significant enrichments of iron, manganese, boron, arsenic, cadmium, vanadium and chromium, compared with average pelagic sediments. Neutron activation analyses for major elements in selected Leg 5 samples (Kuykendall, this volume) show comparable enrichments in iron, up to 27.85 per cent of the sediment, and relatively low aluminum values of 0.86 to 2.82 per cent.

DISCUSSIONS

The occurrence at Sites 37, 38 and 39 of a basal amorphous iron-manganese oxide sediment suggests that this material is a primary precipitate. The necessary localized enrichment of iron, along with possible manganese and trace metals, is best explained by hypothesizing the presence of submarine hydrothermal exhalations synchronous with deposition. According to the present state of our knowledge, the observed stratigraphic and mineralogic relationships are best understood by invoking the sea-floor spreading hypothesis. With such a model, the volcanic "basement" at Sites 37, 38 and 39 would have originated in the vicinity of a ridge crest. In this particular case, the ridge may have been subsequently over-ridden by the southwestern portion of North America. Hydrothermal exhalations along the crestal zone of the ridge could cause precipitation of the metal oxides, leading to the trapping of a relatively pure deposit in possible topographic basement lows. Precipitation at the crestal zone may have been rapid enough that virtually no admixtures of eolian or other extraneous material occurred, with the exception of penecontemporaneous influxes of calcareous pelagic sediments at Site 38 and to a minor extent at Site 39. As sea-floor spreading subsequently moved the area away from the hydrothermal zone of the ridge crest, precipitation of metal oxides in any one area would have become progressively less, and sedimentation would have been increasingly dominated by detrital processes. This model, suggested by Bostrom and Peterson (1966) from observations of surface sediments across the East Pacific Rise (Figure 2), adequately explains the vertical sequence cored at Deep Sea Drilling Sites 37, 38 and 39.

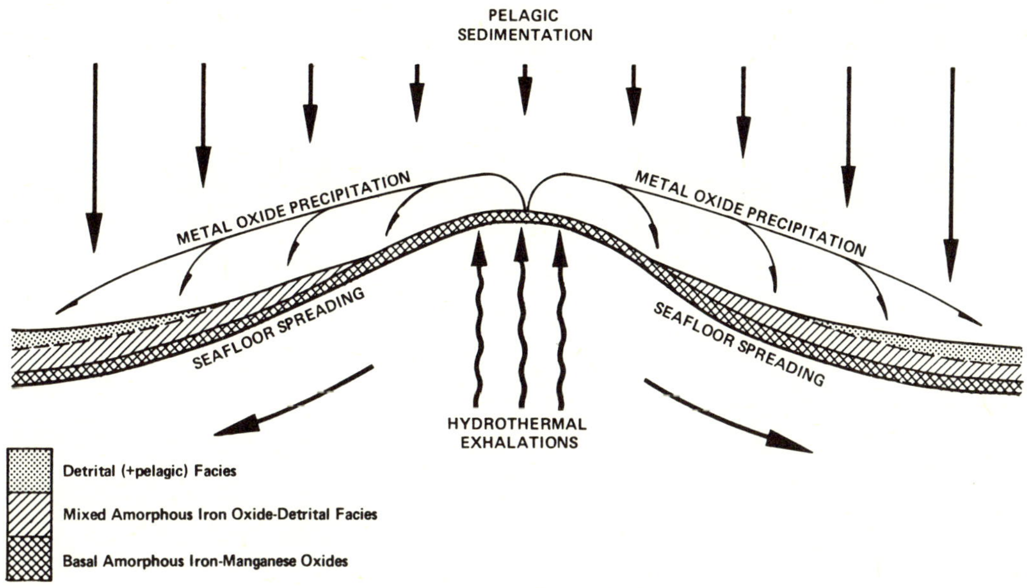

Figure 2. *Schematic representation of sedimentation processes on the East Pacific Rise. (Adapted and modified from Bostrom and Peterson, 1969.)*

REFERENCES

Bischoff, J. L., 1969. Red Sea geothermal brine deposits: Their mineralogy, chemistry and genesis. In *Hot Brines and Recent Heavy Metal Deposits in the Red Sea.* Degens and Ross (Eds.). New York (Springer-Verlag) 368.

Bostrom, K. and Peterson, M. N. A., 1966. Precipitates from hydrothermal exhalations on the East Pacific Rise. *Econ. Geol.* **61**, 1258.

———, 1969. The origin of aluminum-poor ferromanganoan sediments in areas of high heat flow on the East Pacific Rise. *Marine Geol.* **7**, 427.

Bostrom, K., Peterson, M. N. A., Joensuu, O. and Fisher, D. E., 1969. Aluminum-poor ferromanganoan sediments on active oceanic ridges. *J. Geophys. Res.* **74**, 3261.

Kuykendall, W. E., Hoffman, B. W. and Wainerdi, R. E., 1970. 14-MeV neutron activation analysis of selected Leg 5 core samples. In McManus, D. A. *et al.,* 1970. Initial Reports of the Deep Sea Drilling Project, Volume V. Washington (U.S. Government Printing Office).

Peterson, M. N. A., Edgar, N. T., von der Borch, C. C. and Rex, R. W., 1970. Cruise leg summary and discussion (Leg 2). In Peterson, M. N. A. *et al.,* 1970. Initial Reports of the Deep Sea Drilling Project, Volume II. Washington (U.S. Government Printing Office) 413.

Rex, R. W., 1970. Bulk X-ray diffraction results, Leg 5. In McManus, D. A. *et al.,* 1970. Initial Reports of the Deep Sea Drilling Project, Volume V. Washington (U.S. Government Printing Office).

16

Reprinted from pages 368-401 of *Hot Brines and Recent Heavy Metal Deposits in the Red Sea*, E. T. Degens and D. A. Ross, eds., Springer-Verlag, New York, 1969, 600 pp.

Red Sea Geothermal Brine Deposits: Their Mineralogy, Chemistry, and Genesis *

JAMES L. BISCHOFF †
Woods Hole Oceanographic Institution
Woods Hole, Massachusetts

Abstract

Chemical and mineralogical analyses were performed on samples from ten specially selected cores from the Red Sea geothermal deposit. The deposit was divided into seven bedded and laterally correlative facies as follows:

(1) detrital
(2) iron-montmorillonite
(3) goethite-amorphous
(4) sulfide
(5) manganosiderite
(6) anhydrite
(7) manganite

Distribution of the facies, their unconsolidated nature and age relations indicate that the solids were precipitated out of the overlying brine column, that the area of brine discharge is very local within the Atlantis II Deep, and that the chemistry of the brine has changed considerably with time.

Mechanisms of precipitation include simple cooling of subterranean brine as it discharges into the bottom of the Atlantis II Deep and mixing of the brine with the overlying sea water.

Introduction

The subjects of fundamental geochemical interest in the Red Sea geothermal system are the nature and source of the brine, the nature and distribution of the brine-derived deposits and the processes of precipitation of the deposits. The system appears to be an ore body in the process of formation (White, 1968), and the above subjects represent the major gaps in the understanding of general ore-forming processes. The Red Sea geothermal system is one of three remarkable similar systems: the others are the Salton Sea geothermal system (White *et al.*, 1963) and the recently discovered Cheleken geothermal system in the Soviet Union (Lebedev, 1967a).

This chapter is concerned with the Red Sea brine deposits themselves, their distribution, mineralogy, chemistry and mode of precipitation. Conclusions regarding the mode of precipitation hopefully will have application to the origin of ancient ore bodies. Preliminary investigations on the mineralogy and chemistry of these deposits are reported in Miller *et al.* (1966).

Methods

Sampling

Ten cores were chosen for detailed study as representative of the various areas in the geothermal region (Fig. 1). Cores 84K, 120K, 126P, 127P, 128P (K refers to Kasten cores, and P to piston cores) are from the floor of the Atlantis II Deep below the brine-sea water interface. Core 95K is from within the Atlantis II Deep but atop a knoll above the brine-sea water interface. 85K and 161K are from near the brine-sea water interface in the sill area separating the Atlantis II and Discovery Deeps. 119K is from the floor of the Discovery Deep

* Woods Hole Oceanographic Institution Contribution No. 2198.

† Present address: Dept. of Geological Sciences, University of Southern California, Los Angeles, California.

and 118K from about 4 miles southeast of the Discovery Deep.

The sediments in these cores are well bedded on a scale ranging from a few millimeters to several meters. The sediments from below the brine-sea water interface are characterized by striking colors of black, buff, ochre, orange, blue, green and brown, while those above or outside the Atlantis II Deep, *i.e.*, 95K, 118K and 119K, have more subdued colors of grey, buff and orange.

Samples were taken immediately upon opening the refrigerated cores. Samples were collected from each lithologic unit in each core, and in case of massive units, were collected every 50cm or so to insure proper representation. The samples were preserved at room temperature in air-tight, plastic-capped glass jars and were

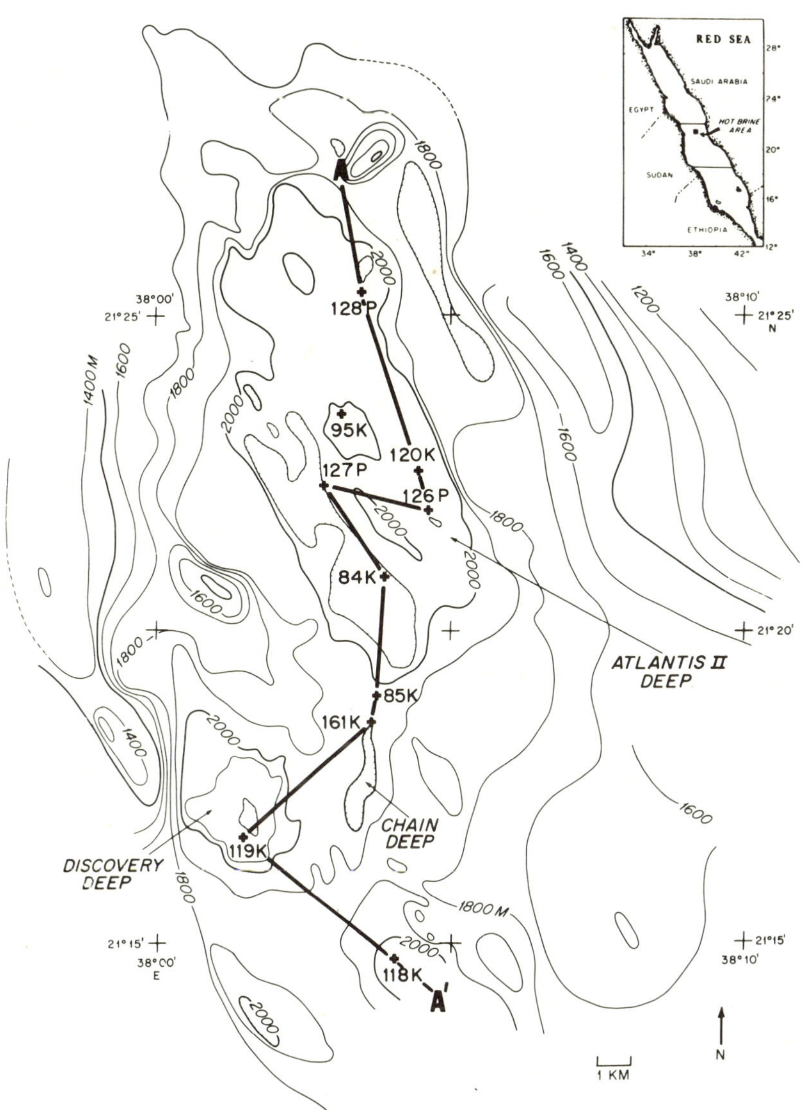

Fig. 1. Bathymetric map of geothermal area (Ross and Degens, 1969) showing core locations. Brine pools are limited to approximately below the 2,000m contour. Line A-A′ refers to a section shown in Fig. 2.

processed within one week to minimize oxidation and dehydration. Control studies on immediately processed samples indicated no such changes for the stored samples. In all, 133 samples were studied.

Analytical Techniques

Each sample was separated into three size fractions, >62μ, 2-62μ and <2μ, by first wet sieving through a 62μ screen with distilled water and then separating the <2μ fraction from the remainder by centrifuging, a process developed by Hathaway (1956). Each size fraction, now completely leached with distilled water, was dried at 50°C and weighed. The weight of the original wet sample and the individual dry separations were used for calculation of the weight per cent interstitial brine and the dry weight per cent of each size fraction (Table 1). Most of the brine-derived material consists of amorphous to microcrystalline spherules of various size, and it is the sizes of these spherules and spherule aggregates that largely constitute the size fractions (see Strangway *et al.*, 1969 for a discussion of crystal size in this material).

X-ray diffraction patterns were obtained for the <2μ and 2-62μ fractions, and grain mount slides for petrographic microscope examination were prepared from the 2-62μ and >62μ fractions (Table 1). A cured crystal monochrometer was used on the x-ray unit to discriminate against the high levels of iron radiation.

Fifty-five of the original samples were selected for chemical analysis, using the 2-62μ fractions for which x-ray patterns had been obtained. This fraction was chosen because it generally accounts for more than 90 per cent of the dry material and because the >62μ fraction consists predominantly of detrital foraminiferal debris. Total carbon, analyzed as CO_2, was determined by an induction furnace-gasometric technique (Laboratory Equipment Corporation, Saint Joseph, Michigan). Total sulfur, including sulfate and sulfide, was volatilized to SO_2 in an induction furnace, and determined by an iodometric titration (Laboratory Equipment Corporation; see also Conrad *et al.*, 1959). Ferrous iron was determined by the classic dichromate titration (Kolthoff and Sandell, 1952). Zn, K and Mn were determined by atomic absorption spectroscopy. K and Mn were additionally determined by spark emission spectrometry as was Si, Ti, Al, Fe (total), Mg, Ca, Sr, Ba, Na, Cu, Pb and P. The emission spectrometer technique is that of Landergren *et al.* (1964) as modified by Manheim (in preparation) and consists of fusing samples in a borate flux with internal standards of In and Be. The total composition of the samples were expressed as oxides and summarized (Table 2).

There appears to be some influence on Pb values by an Fe line on the spectrochemical analyses. This makes some of the low-heavy-metal, high Fe samples have erroneously high Pb (*e.g.* 118K-260, Table 2). It appears that about 50 per cent Fe_2O_3 gives about 0.03–0.04 per cent PbO. This is not a significant effect in the high heavy-metal samples (*e.g.* 85K-85, Table 2). Replicate analyses of the same sample by different analysts and laboratories are compared in Bischoff and Manheim (1969) and indicate the general reliability of the data. Analyses of most of the samples, moreover, sum close to 100 per cent, indicating that the above list of elements includes all major components.

Analytical Data: Facies Description

General Statement

Several facies, apparently correlative from core to core, were defined on the basis of physical appearance and mineralogy. Correlation was subsequently confirmed by age relations (Ku *et al.*, 1969). Distribution and lateral extent are shown in the cross section in Fig. 2, and averaged chemical composition in Table 3. Well defined bedding, inmixing of detrital material and age sequences leave little doubt that solids precipitated out of the overlaying water column and settled to their present position. As will be discussed later, various minerals appear to be forming at differing levels of the water column and settling to the same horizon on the sediment surface, resulting in a certain amount of mixing of the various facies; thus, considerable generalization has been necessary in defining and discussing the facies. In general, however, each facies represents a dominance of one process of mineral precipitation over another, and a description of each facies follows.

Detrital Facies. Buff-colored pelagic sedimentary material occurs throughout the geothermal area, both as discrete beds and as inmixtures with other facies. This material, essentially the same as described by

Table 1 Size Fraction, Brine Content and Mineralogy Data

Size fractions in terms of dry salt-free sediment and brine content in terms of weight per cent wet sediment. X's for mineral content refers to relative intensity of total x-ray pattern compared to standard of pure mineral. 4 X's indicates pattern is of equal intensity to pure standard, 2 X's half as strong and so on. Amorphous material content estimated by difference. For example, if an x-ray pattern of a given sample had sufficient peaks for a 2 X designation of one mineral and no other peaks were present, the 2 X's were assigned for amorphous material.

The cores are grouped by location in this table and in Table 2 and numerically within each group. For example, 95K, 118K and 119K are located above or outside the Atlantis II brine, 85K and 161K are located in the sill area, and the remaining, in numerical order, within the Atlantis II Deep.

SAMPLE (depth, cm)	interstitial brine content (%)	size distribution, μ	percent in size	amorphous	goethite	hematite	lepidocrocite	Fe montmorillonite	manganosiderite	anhydrite	sphalerite	chalcopyrite	pyrite	manganite	todorokite	detritals	miscellaneous and remarks
95K																	
	78.2	>62	21.7													XXXX	
-85		2-62	77.3	XX												XX	
		<2	1.0	XX												XX	
	71.1	>62	9.1													XXXX	
-110		2-62	89.8	X									tr.			XXX	
		<2	1.1													XXXX	
	72.8	>62	0.2													XXXX	
-150		2-62	99.2		XXXX												
		<2	0.6	X	XX											X	
	63.5	>62	5.2										X			XXX	
-180		2-62	92.0										XX			XX	
		<2	2.8	XX	X											X	
	56.2	>62	5.5		X								X			XX	
-185		2-62	89.5	X	X						tr.		tr.			XX	
		<2	5.0										X			XXX	
	---	>62	11.5													XXXX	tr. gypsum
-190		2-62	85.2		X											XXX	
		<2	3.3	X												XXX	
	80.8	>62	2.3													XXXX	
-340		2-62	95.6	XX	X								tr.			X	
		<2	2.1	XXX												X	
	23.5	>62	91.5													XXXX	
-395		2-62	8.1	X												XXX	
		<2	0.4	XX												XX	
118K																	
	55.5	>62	5.1													XXXX	
-40		2-62	93.7	X												XXX	
		<2	1.2													XXXX	
	79.5	>62	3.3	X	X											XX	
-90		2-62	93.9	XX	XX												
		<2	2.8		XX											XX	
	80.0	>62	19.8													XXXX	
-120		2-62	74.4	X				X								XX	
		<2	5.8					XX								XX	
	50.8	>62	17.7													XXXX	
-160		2-62	78.9										tr			XXXX	
		<2	3.4					XX								XX	
	78.1	>62	1.7	X	XX											X	
-200		2-62	93.5	XX	XX												
		<2	4.8	XX	XX												
	80.8	>62	2.2	X	XX											X	
-260		2-62	85.5	XX	XX												
		<2	12.3	XX	XX												
	75.1	>62	1.8				XXX									X	
-320		2-62	93.0				XXX									X	
		<2	5.2				XXXX										
	49.1	>62	2.8													X	pyrrhotite XXX
-385		2-62	87.6	X									X			XX	
		<2	9.6					XX			tr.					XX	

Table 1 (Continued)

SAMPLE (depth, cm)	interstitial brine content (%)	size distribution, μ	percent in size	amorphous	goethite	hematite	lepidocrocite	Fe montmorillonite	manganosiderite	anhydrite	sphalerite	chalcopyrite	pyrite	manganite	todorokite	detritals	miscellaneous and remarks
119K	94.3	>62	5.2													XXXX	
-50		2-62	80.9	X				XXX									
		<2	13.9	X				XXX									
-90	68.2	>62	12.4													XX	Greigite-XX
		2-62	80.2	X	X											XX	
		<2	7.4	X												XXX	
-110	71.2	>62	11.2													XXX	Greigite-X
		2-62	85.2	X	X											XX	
		<2	3.6													XXXX	
-130	67.5	>62	20.6													XXXX	
		2-62	64.1	X	tr.											XXX	
		<2	15.3	X												XXX	
-139	55.8	>62	18.7													XXXX	
		2-62	67.4	XX												XX	
		<2	13.9					XX								XX	
-151	70.5	>62	28.2													XXXX	
		2-62	66.3		X											XXX	
		<2	5.5	X				XXX									
-160	68.3	>62	4.9						XXXX								
		2-62	84.2	XXX	X												
		<2	10.4		X			XX	tr.							X	
-205	67.3	>62	6.2													XXXX	
		2-62	79.3	X	X											XX	
		<2	14.5		X			XX								X	
-231	55.4	>62	9.7													XXXX	Greigite tr.
		2-62	84.5	XX	X											X	
		<2	5.8													XXXX	
-239	62.3	>62	36.2													XXXX	
		2-62	56.3	X												XXX	
		<2	7.5													XXX	
-262	84.9	>62	1.0		XX											XX	
		2-62	97.3	XX	XX												
		<2	1.7	XXXX													
-270	89.5	>62	0.7													XXXX	
		2-62	96.0				XXX									X	
		<2	3.3	X												XXX	
-273	58.2	>62	2.6													XXXX	
		2-62	82.3													XXXX	
		<2	15.1													XXXX	
-282	52.3	>62	1.8													XXXX	
		2-62	95.6	X									XX				
		<2	2.6													XXXX	
-295	46.6	>62	54.6													XXXX	
		2-62	45.1													XXXX	
		<2	0.3													XXXX	
-310	52.6	>62	13.00													XXXX	
		2-62	45.6	XX												XX	
		<2	41.4													X	

Table 1 (Continued)

SAMPLE (depth, cm)	interstitial brine content (%)	size distribution, µ	percent in size	amorphous	goethite	hematite	lepidocrocite	Fe montmorillonite	manganosiderite	anhydrite	sphalerite	chalcopyrite	pyrite	manganite	todorokite	detritals	miscellaneous and remarks
119K Con't.																	
-324	80.3	>62	1.2	X	X											XX	
		2-62	81.2	XXX					tr.							X	
		<2	12.6					XXXX									
-329	58.1	>62	68.2					XXX	X								
		2-62	11.5		XX			XX									
		<2	20.3		X			XXX									
-355	50.7	>62	1.4													XXXX	
		2-62	88.4		X											XXX	
		<2	10.2	XXXX													
-380	52.3	>62	26.1	X												XXX	
		2-62	72.6	X												XXX	
		<2	1.3													XXXX	
-425	60.2	>62	12.6													XXXX	
		2-62	82.1													XXXX	
		<2	5.3													XXXX	
-450	55.8	>62	15.6													XXXX	
		2-62	80.3	X												XXX	
		<2	4.1	X												XXX	
<u>161K</u>																	
-135	84.8	>62	1.5													XXXX	
		2-62	50.5	XXX	X												
		<2	48.0	XXXX				tr.									
-200	87.0	>62	1.2													XXXX	
		2-62	50.0	XXX	X												
		<2	48.8	XXX	X												
-221	80.2	>62	1.3													XXXX	dolomite XX
		2-62	87.8	X	XXX											tr.	dolomite tr.
		<2	10.9	XXXX	tr.												
-231	82.9	>62	30.8													XXXX	dolomite XX
		2-62	57.1	XXX	X												
		<2	12.1	XXX	X												
-237	84.3	>62	3.6		X									X		XX	
		2-62	78.8	XXX													woodruffite? X
		<2	17.6	XXX													woodruffite? X
-265	86.2	>62	0.8													XXXX	
		2-62	89.4	XX	XX												
		<2	9.8	XXX	X												
-280	85.8	>62	2.1													XXXX	
		2-62	60.6	XX	XX												
		<2	37.3	XXXX													
-325	88.6	>62	5.3													XXXX	
		2-62	69.2	XXX	X												
		<2	25.5		XXXX												
-370	84.8	>62	1.6													XXXX	
		2-62	79.8	X	XXX												
		<2	18.6	X	XXX												
-389	83.7	>62	4.1													XXXX	
		2-62	88.3	X	XXX												
		<2	7.6	XX	XX												

Table 1 (Continued)

SAMPLE (depth, cm)	interstitial brine content (%)	size distribution, μ	percent in size	amorphous	goethite	hematite	lepidocrocite	Fe montmorillonite	manganosiderite	anhydrite	sphalerite	chalcopyrite	pyrite	manganite	todorokite	detritals	miscellaneous and remarks
161K Con't.																	
-392	88.2	>62	1.5													XXXX	
		2-62	60.3	X	XX											X	
		<2	38.2	XX												XX	
85K																	
-30	91.7	>62	0.3													XXXX	
		2-62	93.4	XX	XX												
		<2	6.3	XX	XX												
-85	92.9	>62	0.4													XXXX	
		2-62	95.7	X	XXX												
		<2	3.9	XXX	X												
-120	81.3	>62	5.3	XXXX													
		2-62	85.9	XXXX													
		<2	8.8	X				XXX									
-160	74.8	>62	37.5	XX	X											X	
		2-62	57.8	XXX	X												
		<2	4.7	X	XXX												
-180	88.0	>62	0.5	XX	X											X	
		2-62	80.8	XX	XX												
		<2	18.7		XXXX												
-210	89.2	>62	0.5	XX	X											X	
		2-62	72.3	X	XXX												
		<2	27.2	X	XXX												
-240	87.3	>62	3.9	XX	X											X	
		2-62	95.1	XXX	X												
		<2	1.0	XXXX													
-270	87.4	>62	5.4	X	XX											X	
		2-62	84.1	X	XXX												
		<2	10.5	X	XXX												
-300	88.0	>62	1.1	XX	X											X	
		2-62	95.6	XX	XX												
		<2	3.3	XXXX	tr.												
-320	77.7	>62	0.7	XX	X											X	
		2-62	82.9	XX	XX												
		<2	16.4	XXXX	tr.												
-340	80.9	>62	5.3	X	X											XX	
		2-62	85.4	X	X				X							X	
		<2	9.3	X	X											XX	
-380	58.6	>62	17.1								XXX	X					
		2-62	82.5	X							XX	X					
		<2	0.4	X							XX	X					
120K																	
-50	93.8	>62	9.6													XXXX	
		2-62	78.5	X				XXX	tr.							tr.	
		<2	11.9					XXXX									
-80	67.4	>62	28.9							XXXX							
		2-62	60.1							XXXX							
		<2	11.0							XXXX							
-100	68.9	>62	1.2	X				X								XX	
		2-62	85.3	X				XXX			tr.						
		<2	13.5	X				XXX		tr.	tr.						

Table 1 (Continued)

SAMPLE (depth, cm)	interstitial brine content (%)	size distribution, μ	percent in size	amorphous	goethite	hematite	lepidocrocite	Fe montmorillonite	maganosiderite	anhydrite	sphalerite	chalcopyrite	pyrite	manganite	todorokite	detritals	miscellaneous and remarks
120K Con't.																	
-125	79.6	>62	13.5	X					X							XX	
		2-62	83.2	XX					XX								
		<2	3.3	X	X											XX	
-140	93.2	>62	33.1													XXXX	
		2-62	48.0	X				X	XX								
		<2	18.9	X	X			XX	tr.								
-167	81.1	>62	18.2					X	X							XX	
		2-62	76.5					XX	X							X	
		<2	5.3					XX	X							X	
-200	93.4	>62	2.2													XXXX	
		2-62	73.9	X				XX	X								
		<2	23.9	XXX												X	
-282	95.2	>62	5.7							XX						XX	
		2-62	36.7	XXX				X									
		<2	57.6	XX				X								X	
-302	70.8	>62	39.8						XXXX								
		2-62	59.0						XXXX								
		<2	1.2	XXX				X									
-335	93.3	>62	1.6							XX						XX	
		2-62	39.3	XXXX													
		<2	59.1	XXX				X	tr.								
-360	74.0	>62	90.2						XXXX								
		2-62	6.5	XX				X	X								
		<2	3.3	XXXX													
-396	91.8	>62	6.7					XX								XX	
		2-62	61.4	XX				XX									
		<2	31.9	XX				XX									
-420	50.4	>62	48.2							XXXX							
		2-62	51.4							XX	X		X				
		<2	0.4	XXX							X						
84K																	
-120	91.6	>62	3.6							X	X					XX	
		2-62	85.9					XX			X		X				
		<2	10.5					XXX			X					tr.	
-190	93.0	>62	2.9													XXXX	
		2-62	85.8	X				XX			X						
		<2	11.3					XXX			X						
-235	---	>62	78.8							XX	XX		tr.				
		2-62	21.2							XX	XX		tr.				
		<2	0.0														
-260	94.3	>62	4.5							XXXX							
		2-62	46.8			XX		XX			tr.						
		<2	48.7			XX		XX			tr.						
-328	67.3	>62	45.7						X	XXX							
		2-62	49.5			X		XX		X							
		<2	4.8			X		XX		X							
-350	77.8	>62	2.0													XXXX	
		2-62	91.8			XXXX											
		<2	6.2			XXXX											

Table 1 (Continued)

SAMPLE (depth, cm)	interstitial brine content (%)	size distribution, μ	percent in size	amorphous	goethite	hematite	lepidocrocite	Fe montmorillonite	manganosiderite	anhydrite	sphalerite	chalcopyrite	pyrite	manganite	todorokite	detritals	miscellaneous and remarks
126P																	
-90	94.2	>62	2.7													XXXX	
		2-62	45.6	XX				XX									
		<2	51.7	X				XXX									
-280	93.7	>62	2.0													XXXX	
		2-62	85.3	XX				XX									
		<2	12.7	X				XXX									
-380	93.1	>62	3.0	X	X			X								X	
		2-62	81.7	X	X			X			X						
		<2	15.3					XXX			X						
-450	94.3	>62	1.5													XXXX	
		2-62	84.3	X				XXX			tr.						
		<2	14.2	X				XXX			tr.						
-650	90.2	>62	8.9	XXX												X	
		2-62	91.1	XXXX													
		<2	0.0														
-670	81.8	>62	47.8											XXX			Groutite X
		2-62	52.2	XXXX													
		<2	0.0														
-730	90.0	>62	1.2	X	XX											X	
		2-62	98.8	XX	XX												
		<2	0.0														
-744	55.6	>62	8.0													XXXX	
		2-62	61.8													XXXX	
		<2	30.2													XXXX	
-760	62.3	>62	9.2													XXXX	
		2-62	40.6	XX	X											X	
		<2	50.2	XX	X											X	
-800	89.8	>62	2.0	XXX												X	
		2-62	84.9	XXXX	tr.												
		<2	13.1		XX			XX									
127P																	
-80	95.7	>62	2.3													XXXX	
		2-62	86.4	XXXX													
		<2	11.3	XX				XX									
-150	94.9	>62	2.5													XXXX	
		2-62	84.2	X				XXX									
		<2	13.3					XXXX									
-237	97.3	>62	1.6													XXXX	
		2-62	91.8					XXX			X						
		<2	6.6					XXXX									
-260	96.0	>62	5.4													XXXX	
		2-62	81.7	XXX				X									
		<2	12.9	XX				XX									
-340	95.6	>62	9.8	X				X								XX	
		2-62	54.1	XX				XX									
		<2	36.1	X				XXX									
-380	96.9	>62	3.9													XXXX	
		2-62	86.3	X				XXX									
		<2	9.8	X				XXX									

Table 1 (Continued)

SAMPLE (depth, cm)	interstitial brine content (%)	size distribution, μ	percent in size	amorphous	goethite	hematite	lepidocrocite	Fe montmorillonite	manganosiderite	anhydrite	sphalerite	chalcopyrite	pyrite	manganite	todorokite	detritals	miscellaneous and remarks
127P Con't.																	
	92.6	>62	8.9														halite XXXX
-427		2-62	81.8	XX							X		X				
		<2	9.3	XXXX													
	82.6	>62	37.3		XXXX												
-455		2-62	58.2	X	XXX												
		<2	4.5	XX	XX												
	79.5	>62	33.6		XX				XX								
-460		2-62	63.6	X	XXX												
		<2	2.8	XX	XX												
	80.3	>62	3.8	XX	X											X	
-464		2-62	80.9	XX	XX												
		<2	15.3	XXX	X												
	92.6	>62	5.1						XXXX								
-495		2-62	92.8	XXXX	tr.			tr.									
		<2	2.1	XXXX													
	78.9	>62	9.4													XXXX	
-514		2-62	81.8	XX					XX								
		<2	8.8	X				XXX									
	93.4	>62	6.0	XX					XX								
-527		2-62	82.5	XX					XX								
		<2	11.5	XX												XX	
	80.6	>62	1.6						XX	XX							
-559		2-62	78.5	XX	X					X							
		<2	19.9	XXXX													
	59.3	>62	34.6						XX							XX	
-610		2-62	52.8						tr.		XXXX						
		<2	12.6								XXXX						
	89.7	>62	4.3						XXXX								
-645		2-62	79.4	XXX					X								
		<2	16.3	XXXX													
	85.1	>62	12.7													X	barite XXX
-710		2-62	86.4	X							X	X	X				
		<2	0.9	X							X	X	X				
	55.4	>62	70.4						XXX							X	
-750		2-62	28.5						XX							XX	
		<2	1.1	XX												XX	
	50.2	>62	2.3													XXXX	
-757		2-62	96.0	X					tr.							XXX	
		<2	1.7													XXXX	
	80.4	>62	1.4													XXXX	
-762		2-62	85.9	XX			XX									X	
		<2	12.7				XXXX										
	55.6	>62	24.1													XXXX	
-783		2-62	73.8	X												XXX	
		<2	2.1	XX												XX	

Table 1 (Continued)

SAMPLE (depth, cm)	interstitial brine content (%)	size distribution, μ	percent in size	amorphous	goethite	hematite	lepidocrocite	Fe montmorillonite	manganosiderite	anhydrite	sphalerite	chalcopyrite	pyrite	manganite	todorokite	detritals	miscellaneous and remarks
128P																	
	94.7	>62	6.5													XXXX	
-130		2-62	80.9	XXX				X									
		<2	12.6	XXX				X									
	95.7	>62	4.0													XXXX	
-162		2-62	85.8	XXX				X									
		<2	10.2	X				XXX									
	94.4	>62	5.5													XXXX	
-200		2-62	85.2					XXXX			tr.						
		<2	12.0					XXXX									
	91.7	>62	4.7													XXXX	
-226		2-62	90.6	XX				X			X						
		<2	4.7					XXX			X						
	95.2	>62	0.7													XXXX	
-244		2-62	69.2	X	X			XX									
		<2	30.1	X	X			XX									
	93.2	>62	1.4													XXXX	
-280		2-62	91.4		X			XXX			tr.						
		<2	7.2	X	X			XX	tr.								
	94.1	>62	2.1													XXXX	
-320		2-62	92.3	X				XXX			tr.						
		<2	5.2					XXX			X						
	95.0	>62	1.4													XXXX	
-357		2-62	86.1	XXX				X									
		<2	12.5	X				XXX			tr.						
	88.2	>62	6.7													XXXX	
-376		2-62	92.2	X				XX	X								
		<2	1.1	X				XXX									
	---	>62	100.0						XXXX								
-377		2-62	0.0						---								
		<2	0.0						---								
	95.2	>62	5.5													XXXX	
-420		2-62	93.6	XXX	X												
		<2	0.8	XX	XX												
	95.5	>62	3.8													XXXX	
-440		2-62	93.7	XXX	X												
		<2	2.5	XX	XX												
	72.6	>62	7.7													XXXX	
-470		2-62	92.0	X	X											XX	
		<2	0.3	XXXX													
	79.8	>62	60.6											XXXX			
-500		2-62	39.0												XXXX		
		<2	0.4												XXXX		
	81.0	>62	53.9											XXX	X		
-510		2-62	45.9												XXXX		
		<2	0.2												XXXX		
	81.4	>62	14.3													XXXX	
-540		2-62	83.7	XX	XX												
		<2	2.0	XXXX													

Table 1 (Continued)

SAMPLE (depth, cm)	interstitial brine content (%)	size distribution, µ	percent in size	amorphous	goethite	hematite	lepidocrocite	Fe montmorillonite	manganosiderite	anhydrite	sphalerite	chalcopyrite	pyrite	manganite	todorokite	detritals	miscellaneous and remarks
118K Con't.																	
-560	86.0	>62	9.4													XXXX	
		2-62	85.8	XX	XX												
		<2	4.8	XXXX													
-600	85.2	>62	10.6													XXXX	
		2-62	82.7	X				X								XX	
		<2	6.7					XXXX			tr.						
-640	92.0	>62	4.9													XXXX	
		2-62	88.9	XXXX													
		<2	6.2	XX	X			X									
-700	91.7	>62	1.0						XXXX								
		2-62	93.8	XXX					X								
		<2	5.2	XXXX					tr.								
-740	77.8	>62	8.5													XXXX	
		2-62	90.2	XX				XX									
		<2	1.3	XX				XX									
-795	89.2	>62	22.9						XXXX								
		2-62	70.9	XX							X	tr.				X	
		<2	6.2								X	X				XX	

Gevirtz and Friedman (1966) for the normal deep-sea sediments of the Red Sea, consists primarily of aragonitic pteropod shells, calcitic foraminiferal tests, coccoliths, clastic quartz, feldspar and clays. Detrital material predominates in cores 95K and 118K and occurs largely as separate beds while within the remaining cores it is largely mixed with the other facies. Detritus rarely exceeds 5 per cent in the Atlantis II deposits, whereas detrital material accounts for approximately 50 per cent of the Discovery sediments (core 119K).

Two samples from the sill area (161K-221 and 161K-231, Table 1) contain euhedral crystals of well-ordered dolomite ($\sim 70\mu$), which on the basis of its association is believed to be of detrital origin. No dolomite was found within the Atlantis II Deep.

The dominant chemical components are CaO, SiO_2, Al_2O_3, MgO and CO_2. Anomalous, but minor amounts of Fe_2O_3 are also present (see Table 3).

Iron Montmorillonite Facies. Dark brown, finely bedded, montmorillonitic mud comprises the uppermost facies throughout the Atlantis II Deep. Although approximately 70cm of this material occurs at the top of core 119K from the Discovery Deep, it is elsewhere primarily restricted to below the brine-sea water interface within the Atlantis II Deep where its thickness ranges from 4 to 6m. Discrepancies in thickness of this unit between piston and Kasten cores may be due to superpenetration of the latter. If a correction were possible, the correlation diagram (Fig. 2) could be simplified. Moreover, one cannot be sure that the iron montmorillonite facies is absent in the sill area since the cores from that area are of Kasten type (85K and 161K).

The iron montmorillonite facies, representing the most recent and perhaps presently occurring precipitation from the brine, is very liquid, generally containing from 90 to 96 weight per cent interstitial brine and consisting of various proportions of montmorillonite and amorphous material. The montmorillonite appears to contain ferrous and ferric iron in the octahedral position and is believed to be of inter-

Table 2 **Chemical analyses of Red Sea brine deposits** (Sequence of cores as in Table 1.)

CHEMICAL ANALYSES OF RED SEA BRINE DEPOSITS[a]

	95K-110[b]	95K-150	95K-185	95K-395	118K-90	118K-160	118K-260	118K-320
SiO_2	19.	<2.	30.	20.	8.0	34.	6.	8.3
TiO_2	<.01	<.01	0.81	0.46	<.01	0.96	0.01	0.17
Al_2O_3	7.8	<.2	10.2	7.6	0.94	9.4	<.2	2.9
Fe_2O_3 (tot.)	38.	85.	8.3	6.7	61.	6.5	78.	41.
(FeO)	(.84)	(.45)	(2.26)	(.62)	(.42)	(1.46)	(.30)	(.58)
Mn_3O_4	0.40	0.40	0.94	0.59	0.15	0.28	0.47	0.33
MgO	3.0	0.36	4.0	4.0	1.3	3.9	0.9	1.7
CaO	10.1	1.9	20.5	28.4	8.7	20.1	1.9	15.7
SrO	0.04	<.01	0.10	0.13	0.02	0.11	<.01	0.05
BaO	<.03	<.03	<.03	<.03	<.03	<.03	<.03	<.03
Na_2O	0.79	<.1	1.3	0.92	0.3	1.5	<.1	0.5
K_2O	0.56	0.08	0.93	0.62	0.14	0.68	0.11	0.35
ZnO	0.16	0.21	0.04	0.09	0.13	0.10	0.19	0.13
CuO	0.10	<.01	<.01	<.01	0.09	<.01	0.01	0.02
PbO	0.03	0.08	<.02	<.02	0.07	<.02	0.07	0.02
P_2O_5	<.4	<.4	<.4	<.4	<.4	<.4	<.4	<.4
Ig. loss[c]	19.1	13.4	23.6	29.3	18.4	21.6	15.2	21.8
sum	99.	101.	101.	99.	99.	99.	103.	101.
CO_2	11.9	1.98	20.5	26.69	8.29	20.09	2.49	14.23
S	0.10	0.06	0.21	0.11	0.15	0.60	0.07	0.09
Ag	<.001	<.001	<.001	<.001	<.001	<.001	<.001	<.001

	119K-90	119K-151	119K-262	119K-355	161K-200	161K-200 <2μ
SiO_2	18.	15.	7.2	25.	12.5	11.0
TiO_2	0.05	0.23	0.02	0.48	0.03	0.04
Al_2O_3	2.2	4.7	1.5	6.3	<.2	<.2
Fe_2O_3 (tot.)	38.	8.3	57.	4.5	63.	62.
(FeO)	(3.72)	(3.59)	(2.82)	(1.04)	(3.57)	-
Mn_3O_4	0.74	0.31	0.43	0.39	0.32	0.30
MgO	2.1	3.0	1.1	4.4	0.56	0.40
CaO	8.7	37.6	6.2	25.5	1.6	3.0
SrO	0.03	0.12	<.01	0.10	<.01	<.01
BaO	<.03	<.03	<.03	<.03	<.03	<.03
Na_2O	3.8	2.5	3.2	2.4	<.1	<.1
K_2O	0.40	0.69	0.18	0.74	0.13	0.76
ZnO	0.36	0.17	0.14	0.09	2.25	0.9
CuO	0.03	0.02	0.17	0.02	0.60	0.69
PbO	0.04	<.02	0.04	<.02	0.10	0.11
P_2O_5	<.4	<.4	<.4	<.4	<.4	0.9
Ig. loss	24.5	27.6	24.2	28.6	18.8	-
sum	99.	100.	101.	99.	100.	-
CO_2	9.13	32.89	6.01	24.86	1.83	2.79
S	0.18	2.98	0.34	0.14	0.50	0.56
Ag	<.001	<.001	<.001	<.001	.0043	.007

a. CO_2 and S determined by LECO combustometric technique, FeO by dichromate titration, ZnO by atomic absorption spectroscopy K_2O and Mn_3O_4 by both atomic absorption and emission spectroscopy, remainder done by direct reading spark emission spectroscopy.

b. Refers to core number and depth of sample in core in cm.

c. Includes loss of volatiles such as H_2O, CO_2, and S, and gain of oxygen. Elements are in oxide form during spark emission analysis.

	161K-221	161K-237	161K-280	161K-325	85K-85	85K-120	85K-160	85K-270
SiO_2	8.5	7.9	14.	13.5	7.0	16.	16.	5.5
TiO_2	0.16	<.01	0.07	0.05	<.01	<.01	0.08	<.01
Al_2O_3	2.1	<.2	1.0	1.5	<.2	1.0	1.7	<.2
Fe_2O_3 (tot.)	58.	36.	61.	53.	70.	42.	39.	70.
(FeO)	(.25)	(<.001)	(.87)	(3.42)	(.15)	(14.0)	(24.07)	(.46)
Mn_3O_4	5.55	26.	1.39	2.4	0.70	2.1	0.87	0.63
MgO	1.5	1.5	1.1	0.95	0.51	2.4	2.2	0.3
CaO	3.9	3.1	2.4	5.0	2.35	6.6	7.2	1.0
SrO	0.03	0.04	0.02	0.05	0.02	0.4	0.36	0.01
BaO	0.03	0.03	<.03	0.04	0.03	0.11	0.09	0.03
Na_2O	1.7	3.6	2.9	2.7	1.9	1.6	1.1	2.0
K_2O	0.14	0.16	0.10	0.19	0.10	0.17	0.14	0.08
ZnO	0.67	1.74	1.66	2.22	0.70	4.2	8.67	0.15
CuO	0.38	0.12	0.55	0.71	0.67	3.1	2.4	0.05
PbO	0.07	0.13	0.09	0.08	0.15	0.33	0.47	0.15
P_2O_5	<.4	<.4	<.4	<.4	<.4	0.5	<.4	<.4
Ig. loss	19.4	20.2	13.3	16.4	14.8	18.8	19.1	22.9
sum	102.	101.	100.	99.	99.	99.	99.	103.
CO_2	4.32	2.27	2.93	5.06	2.35	7.85	8.07	1.25
S	0.27	<.02	0.28	0.62	0.15	3.06	5.55	0.11
Ag	<.001	<.001	0.0049	0.005	0.0031	0.024	0.024	<.001

	85K-320	85K-340	85K-380	120K-50	120K-100	120K-125	120K-167	120K-282
SiO_2	14.	8.4	3.9	32.	21.	21.	24.	32.
TiO_2	0.30	0.05	<.01	0.17	0.09	0.14	0.14	0.03
Al_2O_3	5.1	1.1	<.2	1.2	1.5	3.4	3.3	<.2
Fe_2O_2 (tot.)	57.	63.	32.	38.	32.	42.	39.	46.
(FeO)	(.71)	(6.44)	(19.1)	(8.18)	(24.1)	(10.18)	(14.52)	(9.07)
Mn_3O_4	0.76	1.95	2.5	1.23	5.58	5.40	5.74	2.13
MgO	2.8	0.86	1.0	1.6	1.3	1.3	1.4	1.2
CaO	2.2	8.5	1.8	6.1	4.0	2.3	2.3	2.1
SrO	0.08	<.01	0.29	0.31	0.31	<.01	<.01	0.02
BaO	<.03	<.03	0.12	0.04	0.07	<.03	<.03	<.03
Na_2O	0.9	1.0	<.1	1.6	2.0	2.1	2.9	2.1
K_2O	0.08	0.16	0.10	0.35	0.46	0.60	1.1	0.73
ZnO	0.64	0.72	13.27	1.66	3.88	0.42	0.37	0.09
CuO	0.05	0.07	8.8	0.64	0.76	0.04	0.04	0.42
PbO	0.04	0.10	0.16	0.08	0.20	0.05	0.05	0.08
P_2O_5	<.4	<.4	<.4	<.4	<.4	<.4	<.4	<.4
Ig. loss	16.5	15.3	24.6	13.0	25.8	20.3	20.7	15.0
sum	100.	101.	98.	98.	99.	99.	101.	102.
CO_2	1.87	7.59	-	7.22	-	14.26	13.63	8.58
S	0.25	0.11	19.50	-	-	0.14	0.19	-
Ag	<.001	<.001	0.02	0.0048	0.0054	<.001	<.001	0.0022

Table 2 (Continued)

	120K-396	84K-120	84K-190	84K-260	84K-260 <2.	84K-328	84K-350	127P-80
SiO_2	24.	26.	18.	28.	27.	14.	5.0	25.
TiO_2	<.01	0.09	<.01	<.01	0.13	0.09	<.01	0.09
Al_2O_3	<.2	1.4	2.1	2.1	2.2	2.3	<.2	1.2
Fe_2O_3 (tot.)	31.	43.	28.	46.	47.	60.	85.	45.
(FeO)	(1.51)	(7.0)	(20.2)	(12.8)	(10.8)	(10.16)	(3.75)	(7.34)
Mn_3O_4	0.10	0.15	2.26	0.57	0.1	0.63	<.3	1.07
MgO	0.55	1.3	3.4	1.0	0.63	0.87	0.5	0.75
CaO	1.8	2.6	4.2	2.8	1.8	7.2	1.4	3.2
SrO	<.01	0.01	<.01	0.01	<.01	0.01	<.01	0.01
BaO	<.03	<.03	<.03	0.03	<.03	<.03	0.03	<.03
Na_2O	4.0	0.8	3.2	1.8	1.9	1.7	1.7	2.8
K_2O	0.73	0.17	0.16	0.17	0.14	0.18	0.10	0.40
ZnO	0.02	2.56	16.2	1.12	1.61	0.17	0.08	6.73
CuO	0.06	0.88	1.29	1.35	1.30	0.45	0.45	0.82
PbO	<.03	0.17	0.21	0.18	0.14	0.06	0.07	0.14
P_2O_5	<.4	<.4	<.4	<.4	<.4	<.4	<.4	<.4
Ig. loss	34.4	19.4	22.6	13.4	14.8	12.8	5.0	14.3
sum	97.	99.	102.	99.	99.	100.	99.	102.
CO_2	2.57	2.64	8.91	3.78	3.47	2.38	1.69	5.72
S	0.08	5.8	11.78	6.33	-	3.74	0.58	2.02
Ag	<.001	0.007	0.02	0.006	0.006	<.001	<.001	0.007

	127P-237	127P-427	127P-514	127P-610	127P-710	127P-762	128P-162	128P-200
SiO_2	22.	29.	28.	26.	40.	22.	28.	28.
TiO_2	0.04	0.11	0.40	<.01	0.06	0.23	0.11	0.11
Al_2O_3	1.0	1.5	6.7	2.7	1.5	3.8	1.3	1.7
Fe_2O_3 (tot.)	28.	27.	34.	17.	21.	44.	40.	38.
(FeO)	(8.24)	(14.6)	(22.7)	(8.4)	(11.6)	(1.06)	(6.02)	(7.43)
Mn_3O_4	0.57	0.43	2.77	0.56	0.83	1.55	0.89	0.70
MgO	1.0	0.21	1.9	0.4	0.39	1.2	1.6	1.2
CaO	6.9	5.1	3.3	1.2	1.7	7.0	7.8	7.8
SrO	0.02	0.05	0.01	<.01	0.02	0.02	0.03	0.03
BaO	0.03	0.06	<.03	<.03	0.04	<.03	<.03	<.03
Na_2O	4.8	4.0	1.0	2.9	1.7	2.0	3.5	3.5
K_2O	0.67	0.51	0.21	0.51	0.20	0.83	0.64	0.52
ZnO	2.54	6.59	0.48	21.0	7.85	0.39	0.86	2.20
CuO	0.33	1.5	0.36	3.7	4.0	0.15	0.61	0.38
PbO	0.04	0.21	0.06	0.27	0.31	0.06	0.06	0.06
P_2O_5	<.4	<.4	<.4	<.4	<.4	<.4	<.4	<.4
Ig. loss	31.6	21.0	21.7	25.6	19.5	16.9	16.4	16.4
sum	100.	97.	101.	102.	99.	100.	102.	101.
CO_2	6.27	4.77	21.5	5.39	6.86	6.78	8.91	7.48
S	-	13.1	-	18.2	16.3	0.75	0.50	1.10
Ag	0.0016	0.010	0.0031	0.016	0.017	<.001	0.0027	0.0027

	128P-226	128P-280	128P-320	128P-357	128P-376	128P-420	128P-470	128P-510	128P-795
SiO_2	25.	26.	20.	14.	31.	14.	9.0	7.1	17.
TiO_2	0.15	<.01	0.07	<.01	0.08	0.05	0.05	0.10	0.09
Al_2O_3	1.6	1.2	<.2	<.2	3.3	<.2	1.7	1.3	2.3
Fe_2O_3 (tot.)	30.	42.	52.	17.	27.	51.	57.	25.	48.
(FeO)	(10.68)	(8.4)	(19.6)	(3.74)	(4.98)	(6.74)	(1.94)	(.70)	(12.08)
Mn_3O_4	0.95	0.91	0.91	2.5	0.44	0.81	1.23	45	0.66
MgO	1.2	0.8	1.5	0.65	1.4	0.63	0.72	1.6	0.25
CaO	8.2	5.15	7.0	4.15	10.4	5.4	7.4	2.7	1.6
SrO	0.03	0.01	0.02	0.02	0.04	<.01	0.01	<.01	0.03
BaO	<.03	<.03	<.03	<.03	<.03	<.03	<.03	0.05	0.04
Na_2O	4.4	3.4	0.7	6.1	3.0	3.5	2.3	3.3	3.0
K_2O	0.90	1.01	0.45	1.2	0.81	0.42	0.33	0.35	0.84
ZnO	4.65	2.41	3.49	2.64	0.53	0.42	0.10	1.11	5.52
CuO	0.34	0.58	0.31	0.12	0.42	0.16	0.03	0.07	1.2
PbO	0.07	0.11	0.06	<.03	0.07	0.06	0.05	0.07	0.07
P_2O_5	<.4	<.4	0.5	<.4	<.4	<.4	<.4	<.4	<.4
Ig. loss	22.2	15.5	14.0	50.6	20.4	22.2	18.2	14.0	20.8
sum	100.	99.	101.	99.	99.	99.	98.	102.	101.
CO_2	7.77	4.95	-	4.14	9.75	5.06	6.64	2.20	14.04
S	2.50	1.79	-	1.63	1.34	1.59	-	1.14	6.79
Ag	0.0023	0.0058	0.0012	<.001	0.0018	<.001	<.001	0.0013	0.0012

Table 3 Averaged Chemical Analyses for Facies. Based on 43 "Selected" Typical Analyses from Table 2

	Detrital	Fe Mont-morillonite	Goethite-Amorphous	Sulfide	Manganite *
SiO_2	27.3	24.4	8.7	24.7	7.5
Al_2O_3	8.4	1.7	1.1	1.5	0.7
Fe_2O_3 (total)	6.5	37.1	64.2	24.3	30.5
FeO	1.4	11.7	2.7	13.4	0.4
Mn_3O_4	0.6	2.1	1.1	1.1	35.5
CaO	23.6	4.8	3.4	2.5	2.9
ZnO	0.08	3.2	0.7	12.2	1.4
CuO	<.01	0.8	0.3	4.5	0.1
CO_2	23.1	8.6	3.6	5.7	2.2
S	0.3	3.9	0.6	16.8	0.6

* Based on only 2 analyses.

mediate composition within the series between nontronite (dioctahedral) and the yet undescribed ferrous bearing montmorillonite (trioctahedral). Details of x-ray, chemical and Mössbauer analysis are now in process and will be presented later. The occurrence of montmorillonite in the Atlantis II Deep was noted in Miller *et al.* (1966).

Sphalerite, as a minor constituent, is usually present up to several per cent (Fig. 3), and probably accounts for the majority of the ZnO and S in the analyses (ZnO/S for sphalerite = 2.5). Other minor constituents include goethite and manganosiderite.

Averaged analyses of typical samples of this facies (Table 3) indicate the major components, FeO, Fe_2O_3 and SiO_2. Relatively high Na_2O values (1–5 per cent) are attributable to incomplete leaching of the interstitial brine and to adsorbed Na^+ at interlayer positions of the montmorillonite lattice.

Goethite-Amorphous Facies. Orange to yellow beds of the goethite-amorphous facies immediately underlie the iron montmorillonite facies within the Atlantis II Deep, usually with a rather abrupt transition zone of approximately 20cm, and extend considerably beyond the immediate area (Fig. 2). Similar material is found as a 1cm bed in core 153P located approximately 100 miles south of the Atlantis II

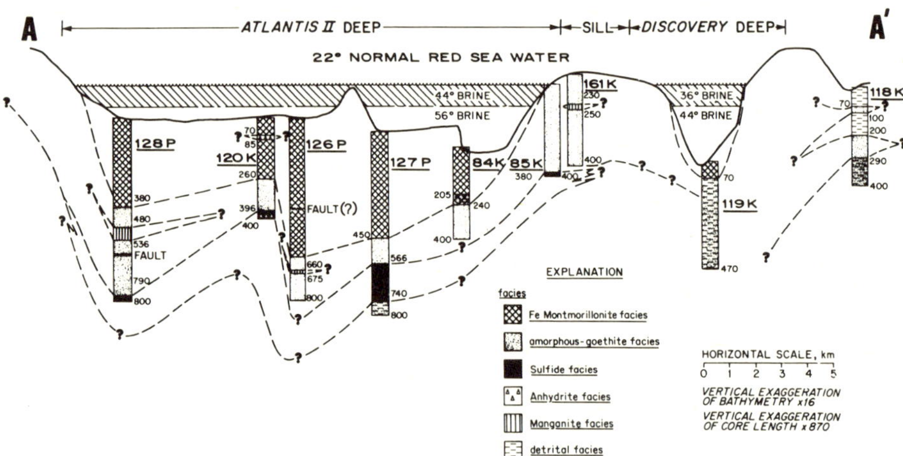

Fig. 2. Generalized cross-section through geothermal area projected along A-A′ in Fig. 1 showing facies distribution. Small numbers beside cores refer to depth in cm. Note superposition of facies patterns to denote mixtures.

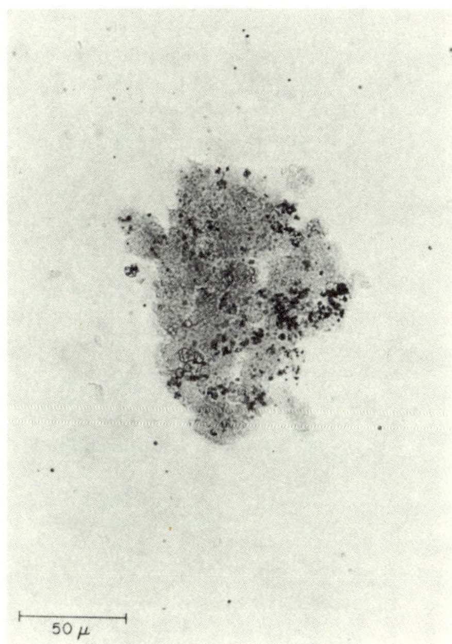

Fig. 3. Photomicrograph of iron-montmorillonite material. Dark spots are sphalerite. Refractive index of medium = 1.55.

Deep, but it is not clear whether its presence is due to the Atlantis II source or another, as yet undiscovered, "hot hole."

Facies thickness within the Atlantis II Deep is approximately one meter, with somewhat greater thickness in core 128P due to a clearly visible slump fault. The unit apparently thickens toward the sill area, where approximately four meters were penetrated by core 161K. Except for the upper 70cm of iron montmorillonite, cored sediments of the Discovery Deep are comprised entirely of this facies, although much diluted with detrital material. Core 118K, south of the deeps, contains several beds of goethite-amorphous material, relatively more crystalline than that in the Atlantis II Deep, and radiometric dating (Ku *et al.*, 1969) indicates that these beds are much older. Core 95K, from the "island" protruding above the brine in the Atlantis II Deep, contains similarly well crystallized material mixed with detrital material.

Interstitial brine contents in this facies are lower than in the iron montmorillonite muds, averaging approximately 80 per cent by wet weight. Poorly crystalline goethite and amorphous "limonite" of varying proportions are the major mineral components, again with minor proportions of other minerals. Microscopically, the samples are composed of 1-30μ spherules (Fig. 4) with the goethite-rich samples showing faint birefringence. Refractive index ranges from less than 1.55 to slightly greater than 2.0, indicating the extreme range of hydration. Pyrite, pyrrhotite and greigite are found associated with goethite in cores 95K, 118K and 119K.

Well crystallized lepidocrocite, a polymorph of goethite, occurs in samples 118K-320, 119K-270 and 127P-762 (Table 1) and may indicate a correlative bed from inside the Atlantis II Deep through the Discovery Deep and outside to the adjacent area. The lepidocrocite samples are physically indistinguishable from the goethite samples and may have been overlooked in the other cores.

The solids of this facies in core 84K from the deepest part of the Atlantis II Deep have transformed evidently to well crystallized hematite, indicating a temperature anomaly and the possibility of the proximity

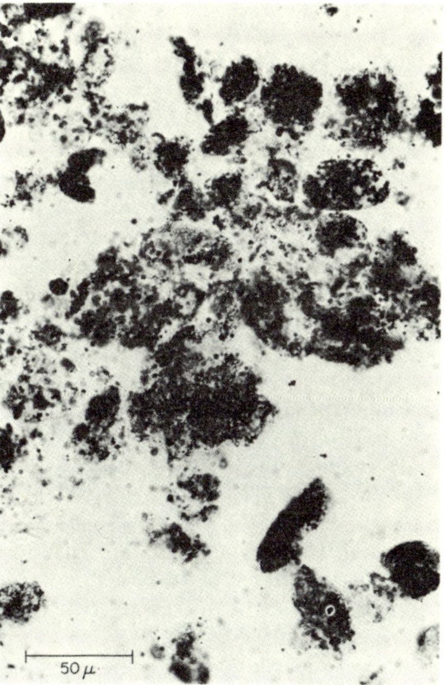

Fig. 4. Photomicrograph of goethite-amorphous material. Refractive index of medium = 1.55.

of the brine vent. Implications of this transformation are discussed elsewhere (Bischoff, 1969). Averaged analyses for this unit (Table 3) indicate the predominance of Fe_2O_3 and SiO_2. SiO_2 is somewhat lower than in the iron montmorillonite unit and evidently is amorphous, and FeO is markedly low. The somewhat high CaO values balance well with CO_2 and are attributable to detrital debris.

Sulfide Facies. The facies of most interest to those concerned with the genesis of syngenetic base-metal sulfide ores is the sulfide facies that occurs in an apparently continuous, homogeneously black bed throughout the bottom of the Atlantis II Deep (Fig. 2). It is the deepest horizon encountered and was penetrated by only four cores, 127P, 128P, 120K and 85K (Table 1). This unit was not observed beyond the brine-sea water interface of the Atlantis II Deep. 127P penetrated approximately a one-meter section, bottoming in detrital material, but it is not clear whether this detrital material represents the basement of the deposits, or a pause of brine activity such as is seen higher up in cores 126P and 128P. The remaining three cores barely penetrated the top 20cm of the sulfide unit, leaving a large uncertainty about its continuity and thickness. The absence of this facies in 126P may be that the unit pinched out locally, or that the core was not deep enough to penetrate it. Sphalerite, with lesser amounts of associate chalcopyrite and pyrite accounts for the bulk of the mineralogy (Plate 3). High FeO contents, excess S over requirements for Zn and Cu sulfides, and absence of the sensitive 32° x-ray peak of pyrite indicate that much x-ray amorphous iron-monosulfide must also be present. Coarsely crystalline barite is associated with this facies in one sample (127P-710, Table 1), being present in the 62μ fraction. Anhydrite is locally associated with this unit (120K-420 and 84K-235, Table 1). ZnO, FeO, CuO, SiO_2 and S are the major components (Table 3). The presence of radiolarian tests in the sulfide zones of 127P and 128P evidently account for a large part of the SiO_2.

Manganosiderite Facies. Thin, semi-lithified, buff-colored beds (1mm–2cm) of man-

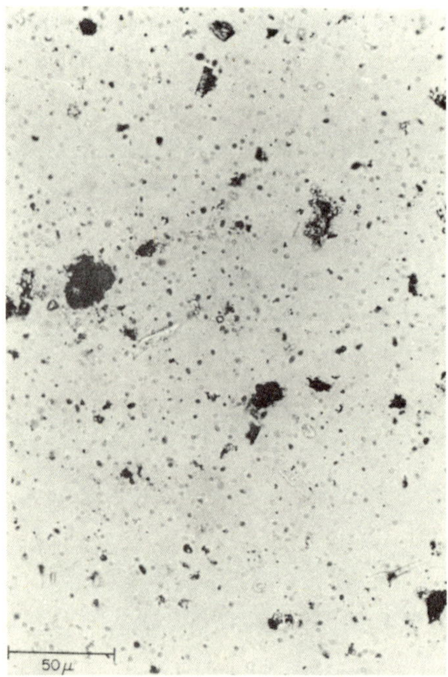

Fig. 5. Photomicrograph of sulfide material, predominantly sphalerite. Refractive index of medium = 1.55.

ganosiderite occur locally within the iron montmorillonite and goethite-amorphous facies within the bounds of the Atlantis II Deep, particularly in core 120K. The individual beds appear to be discontinuous and were too thin to show on the correlation diagram (Fig. 2). The mineral is well-crystallized manganosiderite (Fig. 6) and is of intermediate composition within the siderite-rhodochrosite isomorphous series. The beds also contain minor amounts of halite, and it is not clear whether the halite formed *in situ* at the same time as the manganosiderite or resulted from desiccation during storage. Halite was found only with manganosiderite, which argues for its formation *in situ*. The variation of the major diffraction peak position of manganosiderite for various samples indicates considerable range in the Fe:Mn ratio, and, as pointed out in Miller *et al.* (1966), some Ca may be present in the lattice.

Anhydrite Facies. White, massive beds (up to 20cm) of anhydrite were found locally in the Atlantis II Deep (cores 120K, 84K

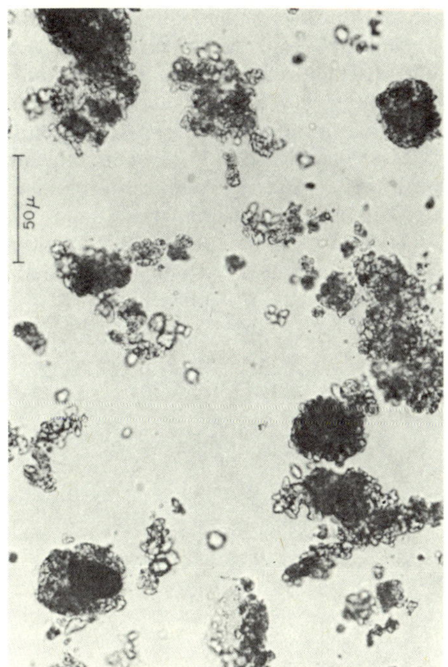

Fig. 6. Photomicrograph of manganosiderite. Refractive index of medium = 1.55.

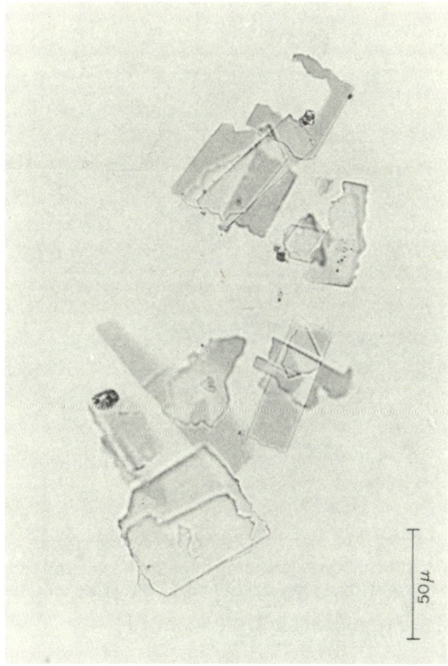

Fig. 7. Photomicrograph of anhydrite, crossed polarizers. Refractive index of medium = 1.55.

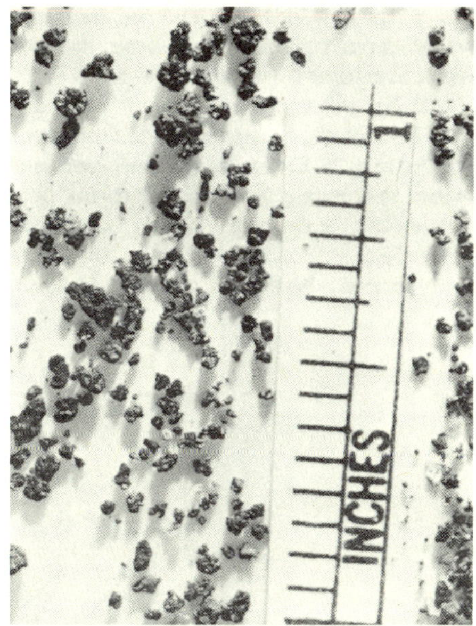

Fig. 8. Photomicrograph of manganite, reflected light.

and 127P, Table 1 and Fig. 2). The beds are composed of very pure, well-crystallized anhydrite (Fig. 7).

Manganite Facies. The manganite facies, also of local occurrence within the Atlantis II Deep, is found in cores 161K, 126P and 128P, in black 15- to 50-cm beds, not unlike the sulfide facies in physical appearance. Well-crystallized manganite is the characteristic mineral (Fig. 8), with locally associated todorokite, $(Mn, Mg, Ca, Ba, K, Na)_2Mn_5O_{12} \cdot 3H_2O$ (sample 128P-500), and groutite, MnOOH (126P-670), while woodruffite, $(Zn, Mn)_2Mn_5O_{12}$ $4H_2O$, represents this unit in sample 161K-237 (Tables 1 and 2). Mn_3O_4 and Fe_2O_3 are the major components (128P-510 and 161K-237, Table 2), with iron probably present mainly substituting for manganese and partly as amorphous limonite, and FeO markedly low.

Geochemical Relations

Chemical variations within individual facies, even within the same core, are very large. Plots of the concentrations of SiO_2, Fe_2O_3, FeO, Mn_3O_4, ZnO, CuO, Ag, BaO and S against depth for each of the cores reveal these variations and some interesting covariations (Figures 9a–9f). ZnO, CuO, Fe, Ag and BaO covary markedly with S, throughout the various facies within the Atlantis II Deep, suggesting these elements (excepting Ba) to be present predominantly as sulfides. Ba does not fit in a sulfide lattice, and because the barium sulfate mineral, barite, is commonly associated with Mississippi Valley type sphalerite deposits, Ba is most likely present as a sulfate. Outside the Atlantis II Deep only FeO varies with S while CuO and ZnO do not, a fact borne out mineralogically (cores 95K, 119K and 118K, Figs. 9a and b), and suggesting a process of sulfide precipitation different from that within the Atlantis II Deep. S and the metals of sulfide affinity, with the exception of Zn, which is present in

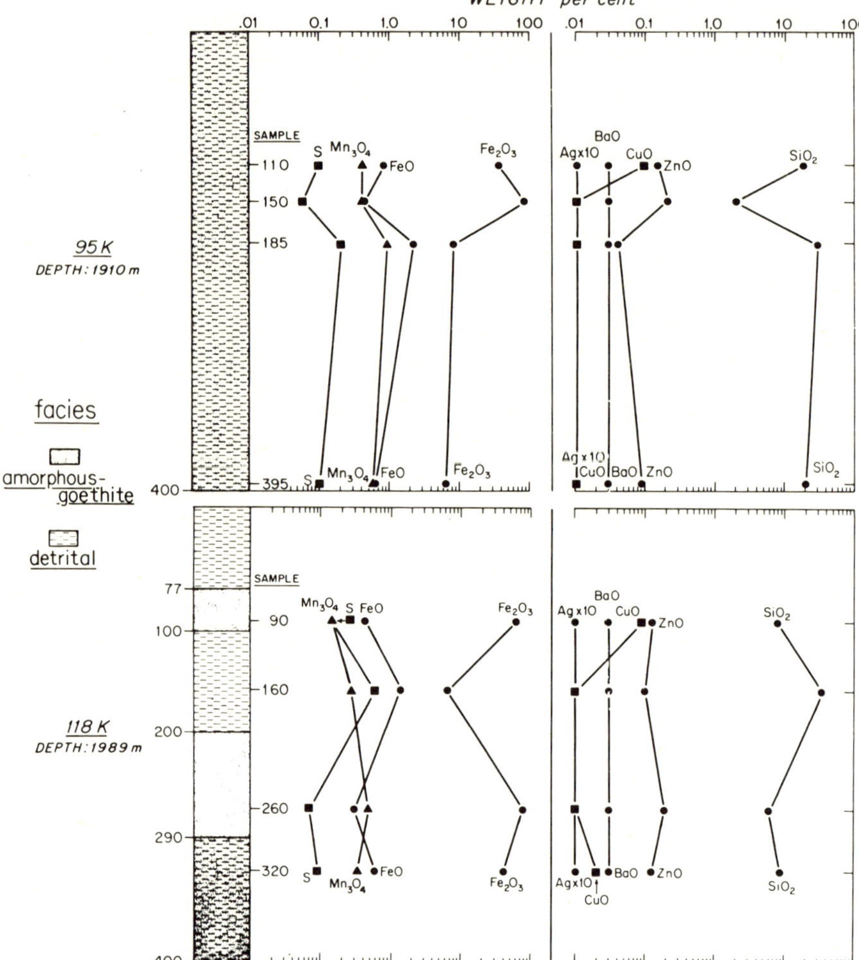

Fig. 9. Charts of cores showing facies and chemical variations with depth. Numbers to right of cores refer to sample locations in cm from top of core, numbers to left refer to facies boundaries. Chemical data on logarithmic scale.
a. Cores 95K * and 118K

* Discrepancy in core length and sample numbers of 95K between this paper and others in this book is because the core device did not completely penetrate. In this paper, samples were measured from top of core barrel, not top of sediment.

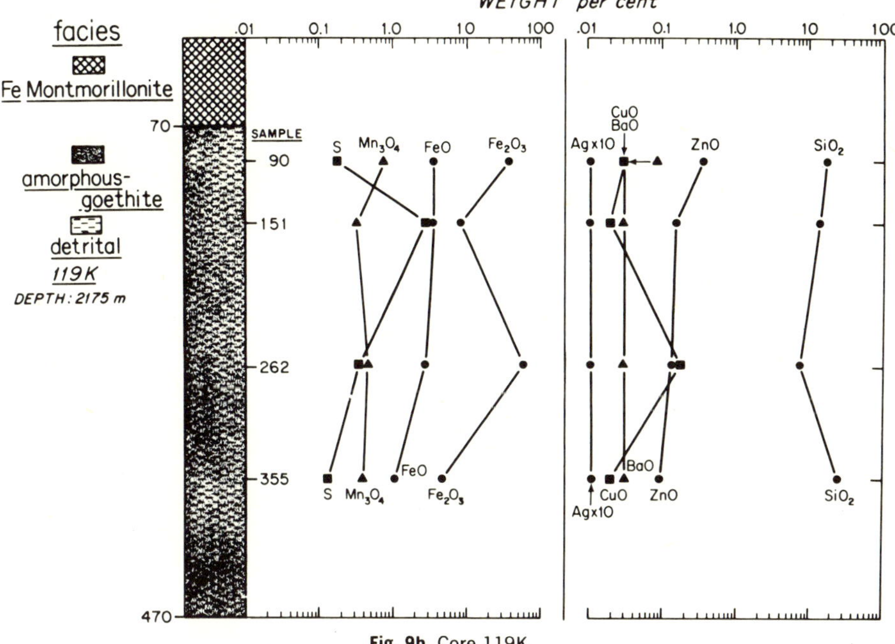

Fig. 9b. Core 119K

the lattice of woodruffite, are depleted in the manganite facies. CuO:ZnO ratios are rather low and constant in the sulfide facies but increase in the goethite-amorphous facies, suggesting co-precipitation of some Cu with colloidal ferric hydroxide. Such co-precipitation is expected at near neutral pH and appears to catalyze the oxidation process of ferrous iron (Hem and Skougstad, 1960). Within the iron montmorillonite facies (core 128P, Fig. 9f) ZnO is often in excess over that necessary to balance S. Electron probe work on the iron montmorillonite (G. Parks, personal communication) indicates occasional small areas of high Zn and no S. Thus, excess Zn is evidently neither in the montmorillonite lattice, nor in a sulfide.

CaO, MgO, Al_2O_3 and TiO_2 display strong correlation with the detrital facies, as expected, and were left off the plots in Fig. 9 to prevent undue complication. Mg:Ca ratios are very high throughout all facies (averaging about 1:5) and reflect the detrital component composition. There is insufficient dolomite or magnesian calcite present to account for this high ratio, so detrital chlorite may be the source of the Mg. High K values in the detrital facies are probably due to detrital illite. In the brine-derived facies, however, there is insufficient detrital material to account for the K, and it must be co-precipitated with the iron montmorillonite, amorphous or perhaps sulfate phases. Fe_2O_3 and SiO_2 are major components throughout all the brine-derived facies and show remarkably little variation. As iron montmorillonite is the only recognized SiO_2 bearing mineral, large amounts of amorphous SiO_2 must therefore be present throughout the other facies. The surface of colloidal ferric hydroxide evidently catalyzes the polymerization and precipitation of SiO_2 (Parks, 1967), explaining the large amounts of SiO_2 usually associated with natural limonite.

Although the brine contains somewhat more dissolved Mn than Fe, the brine-derived deposits rarely contain more than one per cent Mn_3O_4, except in the very local beds of the manganite facies. This is undoubtedly explained by the higher oxidation potential of divalent manganese with respect to divalent iron, such that at the

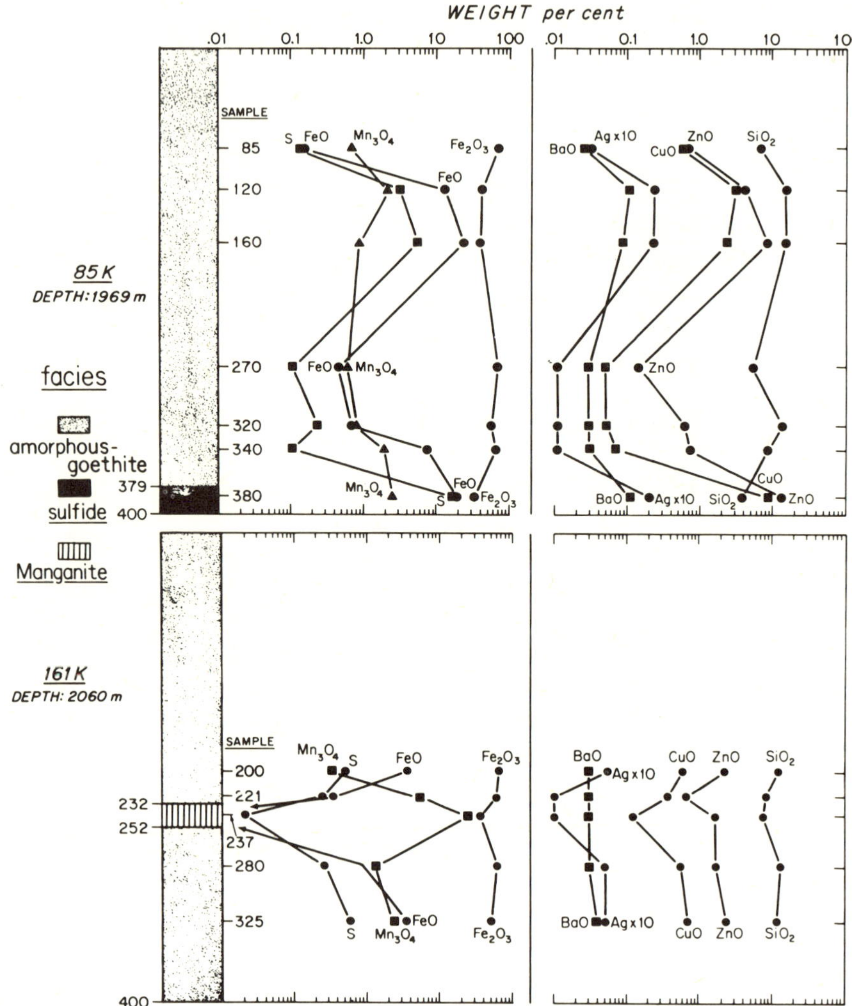

Fig. 9c. Cores 85K and 161K

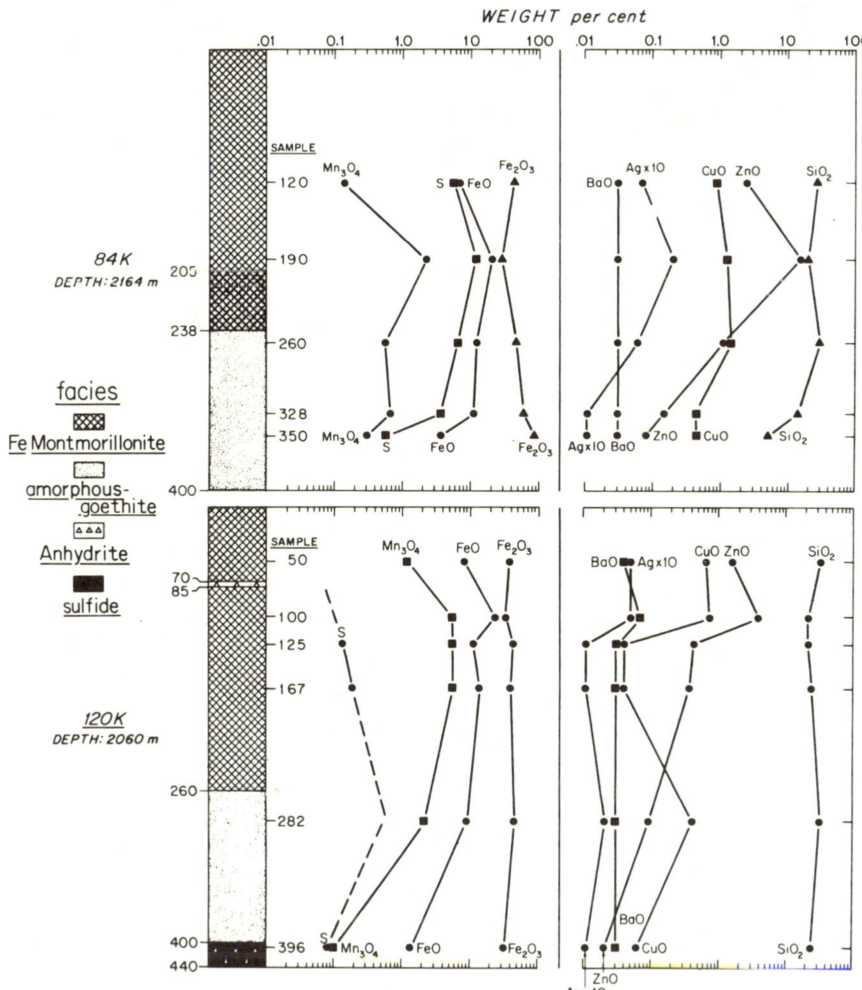

Fig. 9d. Cores 84K and 120K

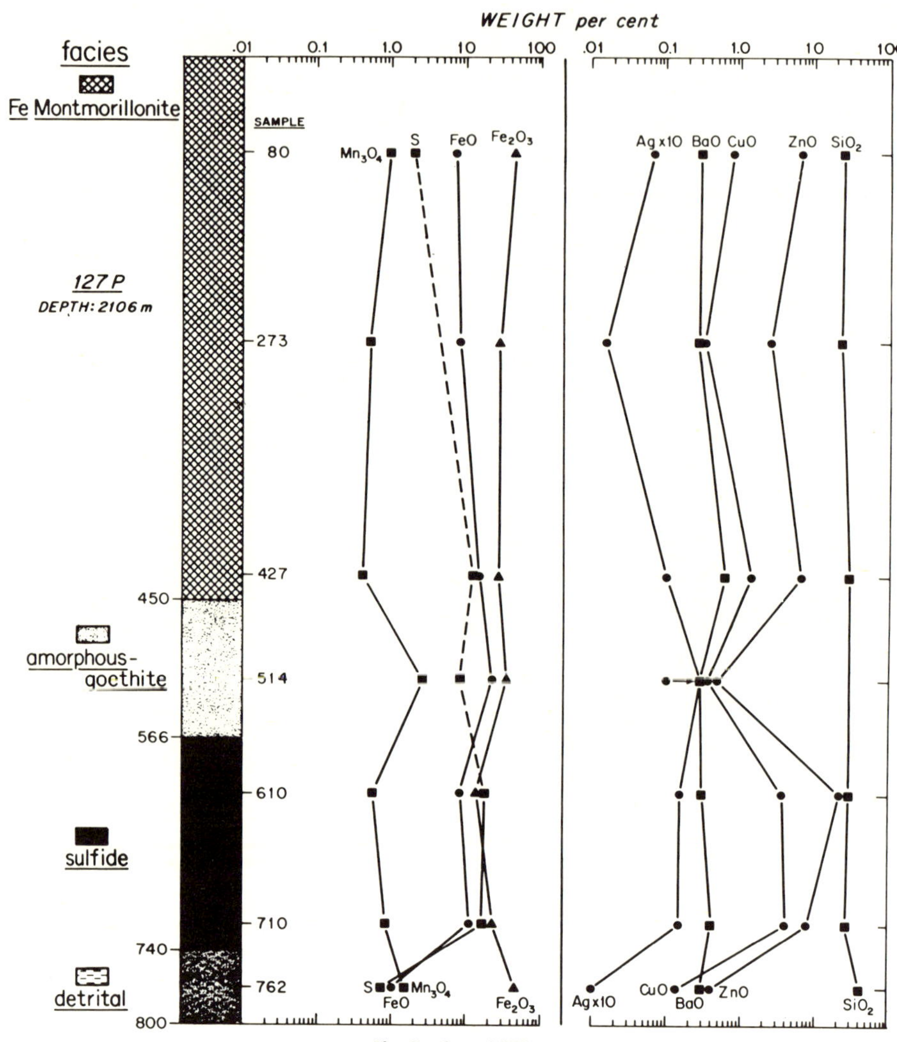

Fig. 9e. Core 127P

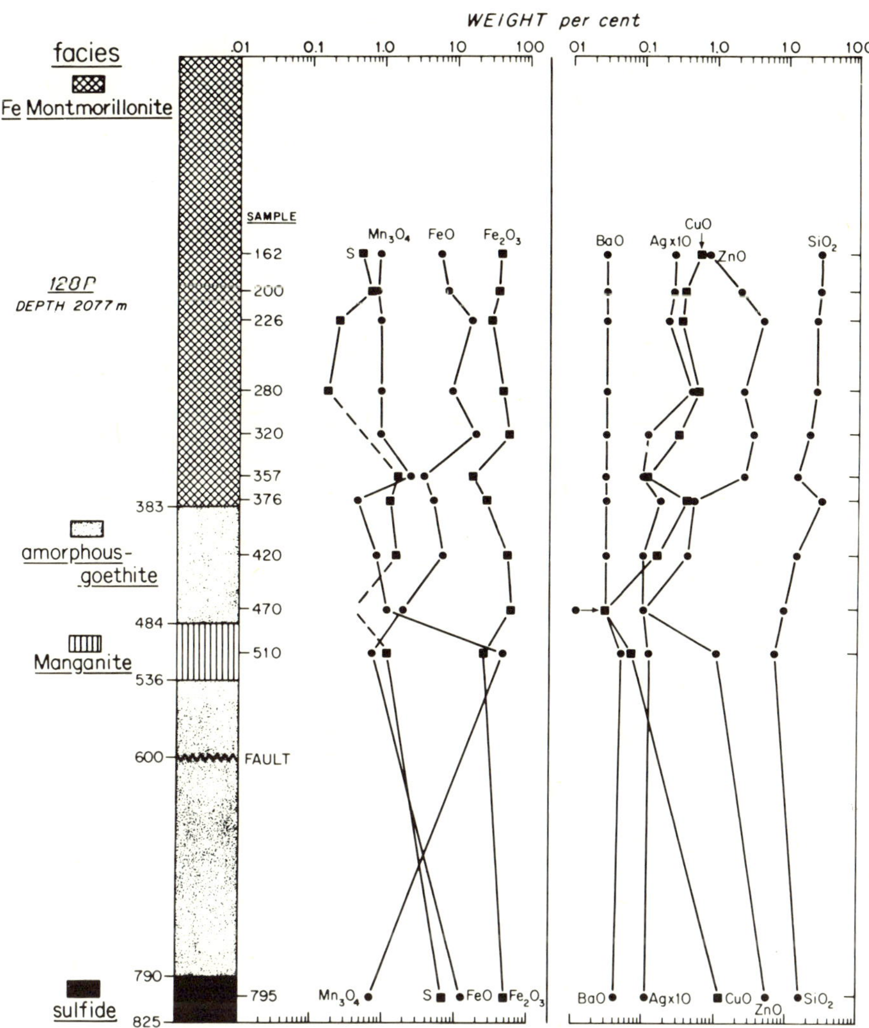

Fig. 9f. Core 128P

brine-sea water interface, the iron consumes the available oxygen and the manganese diffuses out of the system. The chemical data on the brine supports this model, the 56° brine being rich in both iron and manganese, while the 44° brine and the Discovery Deep brines are rich in manganese but markedly depleted in iron (Brewer and Spencer, 1969). Statistical treatment and further interpretations of the geochemical data will be discussed in Spencer and Bischoff (in preparation).

Processes

General Statement

The subject of most interest to those concerned with the genesis of ore bodies is the mode of concentration of the metals, that is, the mechanism of supersaturation and subsequent precipitation of the minerals from the brine. Although a detailed discussion of the origin of the brine is beyond the scope of this chapter, and would require additional data, the author prefers a model whereby Red Sea water circulates through the subterranean sediments of the Red Sea basin, exchanging various portions of original SO_4 for H_2S by early diagenetic bacterial activity, accumulating chloride salts from evaporite beds and heavy metals via chloride complexes from more shaley sequences, presumably fairly deep along channels of subsurface circulation. Encountering a high heat source while moving toward the central part of the Red Sea rift zone, the brine then percolates upward through fractures to the sea floor (see White, 1968, for a more complete discussion of possible origins of the Red Sea and Salton Sea geothermal brine systems). That the deposits of the Atlantis II Deep were precipitated from the overlying brine column and sedimented into their present position is amply demonstrated by the age sequences (Ku, 1969; Ku *et al.*, 1969), bedding relations and gel-like nature of the deposits. The point of discharge of the brine, moreover, must be within the Atlantis II Deep judging from the distribution of the deposits and on the basis of the brine chemistry and temperature relations (Brewer *et al.*, 1969). From the occurrence of hematite, the point of discharge is probably near station 84 (Bischoff, 1969). Thus, a picture emerges of a hot subterranean brine rich in chloride salts and heavy metals, slowly percolating into the bottom of the Atlantis II Deep and precipitating solids which then deposit at a rate of about 40cm per thousand years.

An order of magnitude approximation of the rate of brine discharge based on the amount of silica in the Atlantis II sediments can be made. The assumptions and approximations are as follows:

1. All SiO_2 in the Atlantis II sediments was precipitated from the brine (this ignores detrital contributions).
2. Origin of dissolved SiO_2 in brine is from subterranean equilibration with quartz, with reasonable equilibration temperatures taken as 110°C and 170°C, with respective solubilities of 90 and 226ppm SiO_2 (from the data of Morey *et al.*, 1962).
3. The amount of SiO_2 precipitated is represented by the difference between the high temperature quartz solubility and the present SiO_2 concentration in the Atlantis II 56° brine, 59ppm (Brewer and Spencer, 1969). This assumes no loss of SiO_2 during ascent of brine to discharge point and that no further precipitation below 59ppm takes place.

The mass of the dry salt-free sediment of the top 10m is taken as 19.4×10^7kg (Bischoff and Manheim, 1969). SiO_2 accounts for approximately 20 weight per cent of this material so that total precipitated SiO_2 in the top 10m is approximately 3.9×10^7kg.

Using the 110° quartz equilibrium value, each kilogram of brine precipitated 0.031 g of SiO_2, and 1.25×10^{12}kg of such brine was required to account for the deposit. Dividing by sediment age at the bottom of the 10m section (12,000 years, Ku *et al.*, 1969) yields a brine discharge rate of 10^8kg of brine per year. Similar calculations for the 170° quartz equilibrium value yields a discharge rate of 1.9×10^7kg of brine per year. Comparing these calculated rates with the amount of brine at present within the Atlantis II Deep (*ca.* 7.2×10^9kg) indicates that between 0.26 and 1.4 per cent of the total brine pool is being added yearly.

Although the most recent precipitation from the brine is dominantly iron-montmorillonite, small amounts of the other minerals appear to be precipitating concurrently, although contributing very little to the bulk deposits. Variation during the past 10,000 years in volume of mineral precipitates has given rise to the succession of facies as seen in the cores. A small part of this variation could, perhaps, be explained by fluctuating physical conditions such as temperature and level of brine-sea interface, but the major changes must have been due to a variation in the chemistry of the brine. Today the brine is an iron hydroxide-silicate producer as evidenced by the iron montmorillonite facies; in the past it has been an iron hydroxide, iron-manganese carbonate and base-metal sulfide producer as evidenced by the succession of facies. The length of time which the brine was a base-metal sulfide producer, *i.e.*, how deep the sulfide bed extends, is an interesting and economically pertinent question (see Bischoff and Manheim, 1969).

What causes this chemical variation? Has the brine chemistry changed cyclically, or has it changed gradually with time? At present these questions must remain open.

Precipitation of Goethite and Amorphous Iron Hydroxide Phases

Precipitation of the limonitic material requires oxidation of dissolved ferrous iron and subsequent precipitation of ferric hydroxide. This process appears to be occurring within the 44° brine of the Atlantis II Deep as indicated by the marked depletion of dissolved iron in this layer noted by Brewer and Spencer (1969). Further evidence is that initially clear samples of the 56° brine rapidly precipitated ferric hydroxide upon exposure to air (Miller *et al.*, 1966).

The pertinent half reactions for this process are,

$$Fe^{+2} + 3H_2O \leftrightharpoons Fe(OH)_3 + 3H^+ + e^-$$

$$O_2 + 2H_2O + 4e^- \leftrightharpoons 4OH^-$$

Combining,

$$2Fe^{+2} + \tfrac{1}{2}O_2 + 5H_2O \leftrightharpoons 2Fe(OH)_3 + 4H^+$$

with oxygen diffusing into the 44° brine from the overlying sea water. Note that this reaction releases hydrogen ions, lowers the pH and suggests that measured pH values from these brines are likely to be too low unless oxygen contamination was avoided. SiO_2 associated with the ferric hydroxide in the deposits was probably in a supersaturated state after cooling of the brine and simply polymerized on the catalyzing surfaces of the nascent ferric hydroxide colloids. Anomalous CuO concentrations associated with the goethite-amorphous facies probably resulted from similar surface adsorption during this period.

Such colloids, when first precipitated, are extremely small and strongly hydrated (Hem and Cropper, 1959) and would be expected to move laterally and up in the water column as well as down. Such mobility is offered as the explanation for the widespread occurrence of the goethite-amorphous facies outside the Atlantis II Deep. Enlargement of the brine pools and resultant elevation of the brine-sea water interface would also result in such a distribution, but this is unlikely, as the other facies are not so widely distributed.

After deposition, transformation of the amorphous ferric hydroxide to goethite, as a trend toward thermodynamic stability, would take place,

$$Fe(OH)_{3\,(limonite)} = FeOOH_{(goethite)} + H_2O$$

with resultant expulsion of adsorbed Cu and SiO_2. The goethite-amorphous facies in cores 95K and 118K is more goethitic than that of the Atlantis II Deep, and is evidently older (Ku *et al.*, 1969); CuO values are distinctly lower.

Transformation of goethite to hematite according to the reaction,

$$FeOOH_{(goethite)} = Fe_2O_{3\,(hematite)} + H_2O$$

a further dehydration, has evidently taken place in the sediments at station 84K, as mentioned earlier, and is discussed in Bischoff (1969).

The reasons for the occurrence of lepidocrocite instead of goethite in the several samples mentioned are not understood at present.

Precipitation of Iron-Montmorillonite

The somewhat limited distribution and high ferrous iron content of the iron-montmorillonite suggest precipitation in a zone of decreased oxygen availability, probably lower in the brine column than for the ferric hydroxide. Aside of the ferrous iron content, the iron-montmorillonite facies differs from the goethite-amorphous facies in having higher SiO_2, indicating a greater availability of SiO_2 from the brine at the time of precipitation. Thus, enough SiO_2 was available that when a portion of the iron was oxidized (below the 44° brine?), the montmorillonite became supersaturated and precipitated. Having an almost colloidal nature, portions moved over the sill and deposited in the Discovery Deep.

Precipitation of Anhydrite

Calcium sulfate solubility increases with decreasing temperature, and the brine, therefore, must become increasingly undersaturated with respect to anhydrite as it cools. Thus, the most likely mechanism of anhydrite precipitation is by mixing of SO_4^{--} from the overlying sea water with Ca^{++} of the brine. Sulfur isotope data of Hartmann and Nielsen (1966) and Kaplan *et al.* (1969) indicate the identity of the anhydrite SO_4 with that of normal Red Sea SO_4 and support the mixing conclusion. Recent studies on the stability fields of gypsum and anhydrite (Zen, 1965; Hardie, 1967) show anhydrite to be the stable phase at the temperature and salinity of the Atlantis II Deep.

It is puzzling why anhydrite has not precipitated continuously and more abundantly in the system, unless replenishment of SO_4 in the overlying sea water is extremely slow. Perhaps SO_4 has been depressed in the sea water immediately overlying the 44° brine by anaerobic bacteria. Watson and Waterbury (1969) and Trüper (1969) have shown that particulate organic matter is collected in this zone due to the density differential and that sulfate reducing bacteria are present.

Precipitation of Manganosiderite

Formation of the iron-manganese carbonate as a direct precipitate from the brine is difficult to explain for the same reason as for anhydrite precipitation; carbonates become increasingly soluble with decreasing temperature. An attractive possibility is that the manganosiderite forms by replacement of infalling detrital calcium carbonate according to the reaction,

$$CaCO_3 + Me^{++} = MeCO_3 + Ca^{++}$$

where Me^{++} refers to divalent iron and manganese in the brine. At equilibrium,

$$\frac{aCa^{++}}{aMe^{++}} = K_r$$

where K_r is the equilibrium constant of the above reaction and is calculated by dividing the solubility product constant of calcite by that of either siderite or rhodochrosite. For 25°C and one atmosphere pressure, the values of K_r are $10^{+2.35}$ and $10^{+1.73}$, respectively (K_{sp} values from Latimer, 1952).

If the reaction were rapid, replacement should take place until either the aCa^{++}/aMe^{++} ratio reaches the equilibrium value or all the calcium carbonate is consumed. Although it is not possible to evaluate the activities of Ca^{++}, Fe^{++} and Mn^{++} in the brine, molal ratios will give a qualitative indication of the nearness of the brine to equilibrium with both phases. Molal ratios of Ca^{++}/Fe^{++} and Ca^{++}/Mn^{++} calculated from the data in Miller *et al.* (1966) for the 56° brine yields $10^{+2.1}$ and $10^{+2.01}$, respectively, suggesting the brine at the present time is close to equilibrium.

It is suggested that periodic influx of brine was interrupted by periods of quiescence. At the beginning of such an influx, the brine was sufficiently concentrated in iron and manganese (and perhaps at a higher initial temperature) such that replacement of infalling detrital calcium carbonate took place rapidly; then, as the iron and manganese levels were reduced, additional infalling detrital calcium carbonate was unaffected. However, foraminiferal debris and manganosiderite were not found coexisting in the deposits, and petrographical evidence of partial replacement of foraminifera debris is lacking.

Another possibility is that CO_2 is produced from organic matter by bacterial activity in the Red Sea water above the 44° brine. Diffusion of such CO_2 into the brine could supply carbonate ion for manganosiderite precipitation.

Precipitation of Manganite

The occurrence of coarsely crystalline manganese oxide minerals presents an enigma. Certainly oxidation of divalent manganese is required, but if the process were the same as for the ferric hydroxide material, one would likewise expect a very poorly crystalline product. Marine manganese nodules are generally characterized by very poor crystallinity (F. T. Manheim, personal communication), and they generally have formation rates in terms of millions of years (Bender *et al.*, 1966). Bricker (1965) has pointed out, however, that coarsely crystalline manganese oxide minerals can be produced by oxidation of minerals of lower oxidation state such as rhodochrosite. Evidence of extreme oxidizing conditions in this facies is found in the extreme low level of FeO (Table 3). Thus, the Red Sea manganese oxides are possibly local "weathering" products of preexisting manganosiderite beds. Such a process could have occurred during a particularly long quiescent period of brine discharge during which the brine had diffused away and normal sea water was in contact with the surface deposits in the Atlantis II Deep.

Precipitation of Sulfide

The mechanism of base-metal sulfide precipitation in general continues to be a subject of controversy. In the present case, several alternative mechanisms have potential applicability.

The base-metals clearly came with the brine (Miller *et al.*, 1966) so that the central question concerns the source of sulfide. H_2S was not detected in the 56° brine (Brewer *et al.*, 1969) although small amounts must be present; the Atlantis II Deep brines and sediments are completely free of sulfate-reducing bacteria (Watson and Waterbury, 1969; Trüper, 1969). Sulfate-reducing bacteria are, however, abundant in the sea water immediately above the 44° brine of the Atlantis II Deep, evidently feeding on the organic matter which is collecting there (Watson and Waterbury, 1969). It is an attractive possibility that sufficient sulfide is produced in this area to precipitate the base-metals. Kaplan *et al.* (1969) have found that the δS^{34} content of the Atlantis II sulfides have a very narrow range, averaging about +6.0. Although bacterial reduction of sulfate in interstitial waters of marine sediments results in a wide range of sulfide δS^{34} values (Kaplan and Rittenberg, 1962), bacterial sulfate reduction in a water column with an infinite reservoir of available sulfate might be expected to have a narrow range of sulfide δS^{34}. Thus, the sulfur isotope data is not inconsistent with such a suggested mixing process for the precipitation of the base-metal sulfides. On the other hand, one might expect precipitation of the sulfides so high in the water column would result in a much wider distribution of these minerals than observed, and also more iron-sulfide.

Another possible mechanism is the mixing of two brines, one rich in H_2S, the other in base metals. This process appears to account for the precipitation of sphalerite in the storage tanks of the Chelekin geothermal brines (Lebedev, 1967b), where brines from several wells, each with a slightly different chemistry, are mixed. The chemical differences of the various Chelekin brines seem to be a function of depth, and overall they are remarkably similar to the Red Sea brines, particularly regarding temperature, total salinity, Ca/Mg ratio. The sulfides of the Atlantis II Deep may, therefore, have precipitated at a coalescing front between a metal-rich, sulfide-poor brine and a sulfide-rich, metal-poor brine, both coming along the same subterranean channel. Such a process, in fact, has been suggested by Beales and Jackson (1966) for the Pine Point deposits in Canada. This mechanism, however, fails to explain that sphalerite is forming at present within the iron-montmorillonite.

Barnes and Czamanske (1967) have suggested an oxidation process by which sphalerite might precipitate. Noting the common association of barite with sphalerite, they reason that, if the Zn is held in HS^- complexes, partial oxidation of sulfide results in barite precipitation and release of H^+ changing the sulfide distribution, and thereby releasing the Zn from HS^- and precipitation as sulfide. There is a strong correlation of Ba with the sulfides of the Atlantis II deposits, and even one occurrence of barite. However, the amount of Ba is very small compared to the heavy metals, and if HS^- complexes were responsible for metal transport, certainly H_2S would have been detected in the 56° brine. This process may operate in some geologic environments, but it appears unlikely for the Red Sea system.

The most simple and most probable process to the author is that of cooling. The association of epigenetic ore minerals with chloride brines has been known for a long time from fluid inclusion data (Sorby, 1858). Metal transport as chloride complexes in such solutions, first quantized by Garrels (1941), and later given a rigorous basis by Helgeson (1964), seems now, judging from the current literature, to be fairly well accepted. The formation of these metal chloride complexes involves an entropy increase and consequent increase in stability (and hence possible maximum concentration) with increasing temperature. If the brine approached equilibrium with sulfide minerals at depth and high temperature, most of the metals in solution would be as chloride complexes. During cooling, continuous release of metals from complexes would take place, with consequent supersaturation of the sulfide minerals. White (1968) has suggested a similar process for sulfides in the Salton Sea brine. Such a process for the Red Sea system would explain the narrow range of δS^{34} values, and the limited areal distribution of the sulfides as most precipitation would occur nearest the point of discharge in the Atlantis II Deep.

Barnes (1967) has pointed out that adequate Zn can be transported near 100°C as the $Zn(HS)_2$ complex ($S >> Zn$) and as $ZnCl_2$ ($Zn >> S$), but that it is not possible to transport just enough Zn and S to precipitate ZnS without a large excess of either.

The deposits as a whole and metals in the brine indicate that concentration of Zn and Cu is only slightly greater than S, making the bacterial reduction of SO_4 an attractive hypothesis. On the other hand, Hemley *et al.* (1967) have shown that ZnS is very soluble in silicate buffered chloride solutions at considerably elevated temperatures.

The iron sulfides within the goethite-amorphous facies outside the Atlantis II Deep (cores 118K, 119K) or above the brine-sea water interface (core 95K) require a different explanation. δS^{34} values scatter considerably around an approximate -25.0 average (Kaplan *et al.*, 1969) and strongly suggest a diagenetic origin from bacterial activity. A diagrammatic summary of some of the above discussed precipitation processes is shown in Figs. 10 and 11.

The Red Sea Deposit as an Ore Body

Besides being of scientific interest as an ore-forming system, the Red Sea brine deposits would certainly be economically exploited were they on land (Bischoff and Manheim, 1969). In its present form the deposit bears little resemblance to any ore body with which the author is familiar, particularly those of the sedimentary iron ores. Sedimentary iron ores are not generally associated with base metals, and, on the other hand, base metal deposits most commonly considered syngenetic seldom contain much iron as oxides or clay minerals. It is tempting to speculate about the appearance of the deposit after tectonism and metamorphism have operated, but there are so many variables such as oxidation, reduction, addition or subtraction of components, etc., that the attempt would be pointless.

That portion of the deposit which has been sampled by coring appears to fall into the "syngenetic" category of ore deposits, *i.e.*, bedded deposits forming at the same time as the surrounding sediments. It is, however, interesting to speculate about the

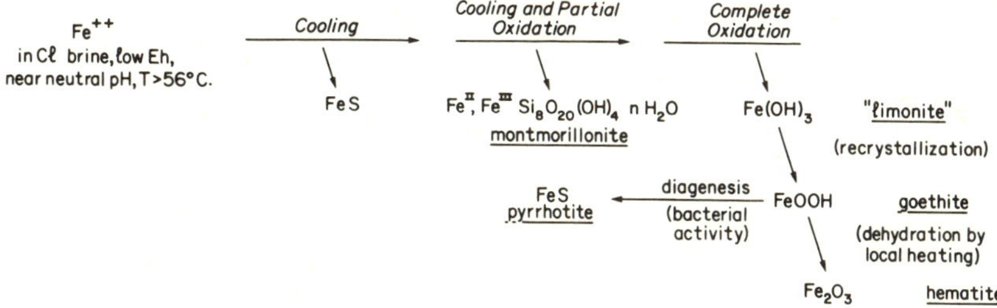

Fig. 10. Sequence of some suggested processes of formation of iron-bearing minerals from the Red Sea brine.

process in the country rock through which the brine must pass immediately prior to discharge into the Atlantis II Deep. If the model of base-metal sulfide precipitation by cooling and releasing of metal from chloride complexes is correct for the Atlantis II deposits, then such sulfides must be precipitating along channels through which the brine must pass, assuming continuously decreasing temperatures. The pathway of the brine must include fissures, faults, brecciated zones and bedding planes of the carbonate rocks underlying the present sea floor. Garrels (1941) has shown that in such a system, if Pb and Zn have comparable concentrations, PbS will precipitate earlier (at higher temperatures) because of the weaker affinity of Pb for chloride complexing. Little Pb was found in the Atlantis II deposits (Table 2), and this may be due to such a preferential removal process at depth.

Summary of fluid inclusion data (Yermakov, 1965) for desposits of this type indi-

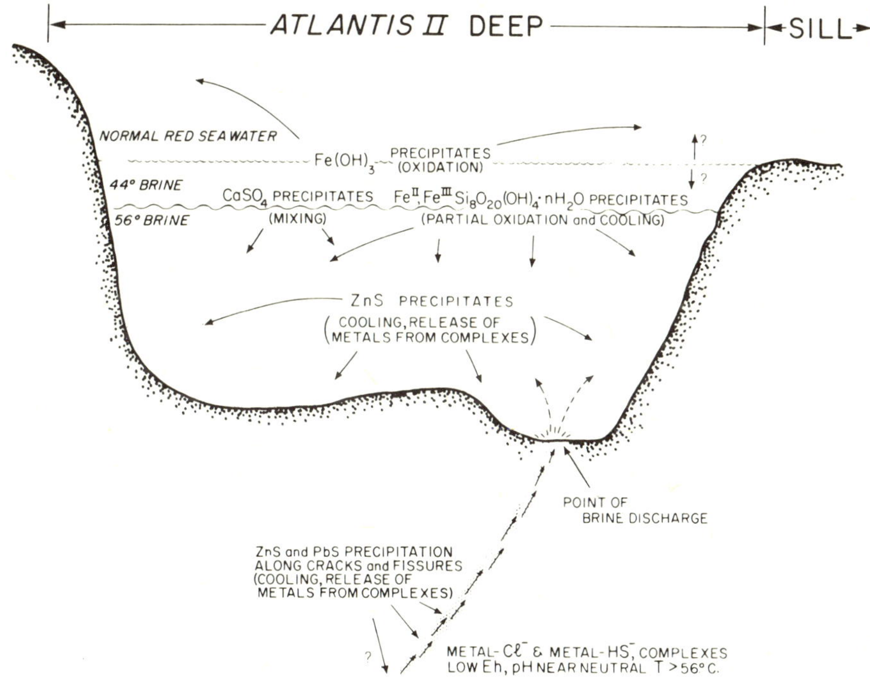

Fig. 11. Schematic representation of some of the suggested processes of mineral precipitation within the Atlantis II Deep.

cates chloride brines to be the mineralizing solutions and temperatures of deposition of around 150°C, conditions which one might expect deeper within the Red Sea geothermal system.

Thus, a model is suggested of epigenetic mineralization at depth within the Red Sea geothermal system with similarities to many of the Mississippi Valley type Pb-Zn deposits.

Acknowledgments

The author is indebted to K. O. Emery and the staff of the USGS-WHOI Continental Margin Program for generously providing laboratory space and equipment and to Mrs. S. McLeod and Mrs. H. Richards for assistance with the analytical work.

Sincere appreciation is due J. C. Hathaway and F. T. Manheim of the U.S. Geological Survey for advice and instruction on analytic techniques and for many hours of fruitful discussion. Completion of the project would have been impossible without their help.

D. E. White of the U.S. Geological Survey deserves special thanks for helpful suggestions as to organization and interpretation of the data.

This work was supported by National Science Foundation grant No. GA-584.

D. E. White, F. T. Manheim and K. O. Emery kindly read and improved the manuscript.

References

Barnes, H. L.: Sphalerite solubility in ore solutions of the Illinois-Wisconsin Districts. *In: Genesis of Strataform Lead-Zinc-Barite-Fluorite Deposits (Mississippi Valley type deposits), a Symposium.* J. S. Brown (ed.). Economic Geology Pub. Co., Monograph 3, New York, p. 326 (1967).

Barnes, H. L. and G. K. Czamanske: Solubilities and transport of ore minerals. *In: Geochemistry of Hydrothermal Ore Deposits,* H. L. Barnes (ed.). Holt, Rinehart and Winston, Inc., New York, 334–378 (1967).

Beales, F. W. and S. A. Jackson: Precipitation of lead-zinc ores in carbonate reservoirs as illustrated by Pine Point Ore Field, Canada. Trans. Canadian Inst. Mining and Metall. Appl. Earth Sci., **76,** B278 (1966).

Bender, M. L., T. L. Ku, and W. S. Broecker: Manganese nodules, their evolution. Science, **151,** 325 (1966).

Bischoff, J. L.: Goethite-hematite stability relations with relevance to sea water and the Red Sea brine system. *In: Hot brines and recent heavy metal deposits in the Red Sea,* E. T. Degens and D. A. Ross (eds.). Springer-Verlag New York Inc., 402–406 (1969).

——— and F. T. Manheim: Economic potential of the Red Sea heavy metal deposits. *In: Hot brines and recent heavy metal deposits in the Red Sea,* E. T. Degens and D. A. Ross (eds.). Springer-Verlag New York Inc., 535–541 (1969).

Brewer, P. G., C. D. Densmore, R. Munns, and R. J. Stanley: Hydrography of the Red Sea Brines. *In: Hot brines and recent heavy metal deposits in the Red Sea,* E. T. Degens and D. A. Ross (eds.). Springer-Verlag New York Inc., 138–147 (1969).

Brewer, P. G. and D. W. Spencer: A note on the chemical composition of the Red Sea brines. *In: Hot brines and recent heavy metal deposits in the Red Sea,* E. T. Degens and D. A. Ross (eds.). Springer-Verlag New York Inc., 174–179 (1969).

Bricker, Owen P.: Some stability relations in the system $Mn\text{-}O_2\text{-}H_2O$ at 25°C and one atmosphere total pressure. American Mineralogist, **50,** 1296 (1965).

Conrad, A. L., J. K. Evans, and V. F. Gaylor: Rapid determination of fluorine, sulfur, chlorine and bromine in catalysts with an induction furnace. Analytical Chemistry, **31,** 422 (1959).

Garrels, R. M.: The Mississippi Valley type Pb-Zn deposits and the problems of mineral zoning. Econ. Geology, **36,** 729 (1941).

Gevirtz, J. L. and G. M. Friedman: Deep-sea carbonate sediments of the Red Sea and their implications on marine lithification. Jour. Sed. Petrol., **36,** 143 (1966).

Hardie, L. A.: The gypsum-anhydrite equilibrium at one atmosphere pressure. The American Mineralogist, **52,** 171 (1967).

Hartmann, M. and H. Nielsen: Sulfur isotopes in the hot brine and sediment of Atlantis II Deep (Red Sea). Marine Geol., **4,** 305 (1966).

Hathaway, J. C.: Procedures for clay mineral analyses used in the sedimentary petrology laboratory of the U.S. Geological Survey. Clay Minerals Bull., **3,** 8 (1956).

Helgeson, H. C.: *Complexing and Hydrothermal Ore Deposition.* Pergamon Press, New York-London, 128 p. (1964).

Hem, J. D. and W. H. Cropper: Survey of ferrous-ferric chemical equilibria and redox potentials. U.S. Geol. Survey Water-Supply Paper, **1459-A** (1959).

Hem, J. D. and M. W. Skougstad: Coprecipitation effects in solutions containing ferrous, ferric, and cupric ions. Geological Survey Water-Supply Paper, **1459-E** (1960).

Hemley, J. J., C. Meyer, C. J. Hodgson, and A. B. Thatcher: Sulfide solubilities in alteration controlled systems. Science, **158,** 1580 (1967).

Kaplan, I. R. and S. C. Rittenberg: The microbiological fractionation of sulfur isotopes. *In: Biogeochemistry of Sulfur Isotopes,* M. L. Jensen (ed.). Proceedings of a National Science Foundation Symposium, Yale University, April (1962).

Kaplan, I. R., R. Sweeney, and A. Nissenbaum: Sulfur isotope studies on Red Sea brines and sediments.

In: Hot brines and recent heavy metal deposits in the Red Sea, E. T. Degens and D. A. Ross (eds.). Springer-Verlag New York Inc., 474–498 (1969).

Kolthoff, I. M. and E. B. Sandell: *Textbook of Quantitative Inorganic Analysis,* 3rd ed. The Macmillan Company, New York, 759 p. (1952).

Ku, T. L.: Uranium series isotopes in sediments from the Red Sea hot brine area. *In: Hot brines and recent heavy metal deposits in the Red Sea,* E. T. Degens and D. A. Ross (eds.). Springer-Verlag New York Inc., 512–524 (1969).

———, D. L. Thurber, and G. Mathieu: Radiocarbon chronology of Red Sea sediments. *In: Hot brines and recent heavy metal deposits in the Red Sea,* E. T. Degens and D. A. Ross (eds.). Springer-Verlag New York Inc., 348–359 (1969).

Landergren, S., M. William, and B. Rajandi. Analytical Methods. *In:* S. Landergren, *On the Geochemistry of Deep Sea Sediments,* Reports of the Swedish Deep Sea Expedition, **10,** Special Investigation No. 5 (1964).

Latimer, W. H.: *Oxidation Potentials,* 2nd ed. Prentice-Hall, Englewood Cliffs, N.J., 392 p. (1952).

Lebedev, L. M.: O sovremennom otlozhenii samarodnovo svintza iz termal'nykh rassolov Chelekena (On contemporary deposits of native lead from the thermal brines of Cheleken). Dokl. Akad. Nauk SSSR, **174,** 197 (1967a).

———: Sovreminnoye obrazovaniye sfalerita v proizvodstvennykh sooruzheniyakh Chelekenskovo mestorozhdeniya. Dokl. Akad. Nauk SSSR, **175,** 920 (1967b).

Miller, A. R., C. D. Densmore, E. T. Degens, J. C. Hathaway, F. T. Manheim, P. F. McFarlin, R. Pocklington, and A. Jokela: Hot brines and recent iron deposits in deeps of the Red Sea. Geochim. et Cosmochim. Acta, **30,** 341 (1966).

Morey, G. W., R. O. Fournier, and J. J. Rowe: The solubility of quartz in water in the temperature interval from 25°C to 300°C. Geochim. Cosmochim. Acta, **26,** 1029 (1962).

Parks, G. A.: Aqueous surface chemistry of oxides and complex oxide minerals, isoelectric point and zero point of charge. *In:* W. Stumm, Symposium chairman (chapter 6), *Equilibrium Concepts in Natural Water Systems,* American Chemical Society, Washington, D.C. (1967).

Ross, D. A. and E. T. Degens: Shipboard collection and preservation of sediment samples collected during CHAIN 61 from the Red Sea. *In: Hot brines and recent heavy metal deposits in the Red Sea,* E. T. Degens and D. A. Ross (eds.). Springer-Verlag New York Inc., 363–367 (1969).

Sorby, H. C.: On the microscopical structure of crystals, indicating the origin of minerals and rocks. Quart. Jour. Geol. Soc. London, **14,** 443 (1858).

Strangway, D. W., B. E. McMahon, and J. L. Bischoff: Magnetic properties of minerals from the Red Sea thermal brines. *In: Hot brines and recent heavy metal deposits in the Red Sea,* E. T. Degens and D. A. Ross (eds.). Springer-Verlag New York Inc., 460–473 (1969).

Trüper, H. G.: Bacterial sulfate reduction in the Red Sea hot brines. *In: Hot brines and recent heavy metal deposits in the Red Sea,* E. T. Degens and D. A. Ross (eds.). Springer-Verlag New York Inc., 263–271 (1969).

Watson, S. W. and J. B. Waterbury: Sterile hot brines of the Red Sea. *In: Hot brines and recent heavy metal deposits in the Red Sea,* E. T. Degens and D. A. Ross (eds.). Springer-Verlag New York Inc., 272–281 (1969).

White, D. E.: Environments of generation of some base-metal ore deposits. Econ. Geol., **63,** 301 (1968).

White, D. E., E. T. Anderson, and D. K. Grubbs: Geothermal brine well: mile deep drill hole may tap ore-bearing magmatic water and rocks undergoing metamorphism. Science, **139,** No. 3558, 919 (1963).

Yermakov, N. P. *Research on the Nature of Mineral Forming Solutions.* Trans. by V. P. Sokoloff. Edwin Roedeer (ed.). Pergamon Press, N.Y. (1965).

Zen, E-An: Solubility measurements in the system $CaSO_4$-$NaCl$-H_2O at 35°, 50°, and 70°C and one atmosphere pressure. Jour. Petrology, **6,** 124 (1965).

17

Reprinted from *Nature Phys. Sci.* **240**:153–158 (1972)

New Deeps with Brines and Metalliferous Sediments in the Red Sea

H. BACKER & M. SCHOELL

Preussag, Hannover & Bundesanstalt für Bodenforschung, Hannover

Investigation of Red Sea deeps has revealed 13 new brine pools and several areas covered by sediments of hydrothermal origin. There is a considerable variation of physical and chemical parameters within the sample of known Red Sea brines.

WATERS with anomalous salinity and temperature associated with metalliferous sediments have been known since 1964 near the central rift zone of the Red Sea between Djidda and Abu Shagara.

Following discoveries made from the Swedish vessel Albatross in 1948 (ref. 1), the brine area was intensively investigated from 1964 to 1966 (refs. 2–5). The result was the discovery and detailed description of three brine pools, the most important one being Atlantis II Deep[3]. Outside the Atlantis II Deep area only one indication (echo reflexions) has been reported from a deep near Brother Islands[5]. Tooms (personal communication) confirmed the existence of brines in that area and found sediments of hydrothermal origin in the Nereus Deep. These and additional observations called for a general survey of the whole central Red Sea trough. A general distribution of brine activity should have principal implications on the genesis of this type of ore deposit.

The whole central rift from Subair Islands to Abulkizaan (Daedalus Reef) has now been mapped using land-bound navigation systems (Shoran and Decca hi-fix) and a stabilized narrow-beam echo sounder (ELAC Schelfrandlot, 30 kc, beam width ±1.4°). For comparison, the Gulf of Aden rift system was included. During two Valdivia cruises, 13 new brine pools and several areas covered by sediments of hydrothermal origin were found. Compared with the well-known Atlantis II Deep area some of these findings show a remarkable variety of physical parameters and chemical components.

Some Aspects of the General Geological Setting

The bathymetric survey of the Red Sea rift system revealed considerable differences in the tectonic patterns between the southern and the central part of the Red Sea. In the area from Nereus Deep to the Suakin Deep the sea floor on both sides precipitates directly from the reefs down to a depth of 500–600 m, from where the bottom slopes only slightly over a distance of about 40 km to the edge of the main trough (at about 1,100 m), from which it reaches the final depth after a steep slope. Locally the central trough is flanked by high tectonic escarpments. In some cases there are two or three parallel basins with small ridges in between (Suakin Deep, Port Sudan/Volcano Deeps, Nereus Deep). In the Atlantis II Deep area five such parallel basins can be distinguished.

The rift system in the southern Red Sea from Suakin Deep to Djebel Tair is characterized by typical graben–horst structures. Long, flat-bottomed terraces and deeps are

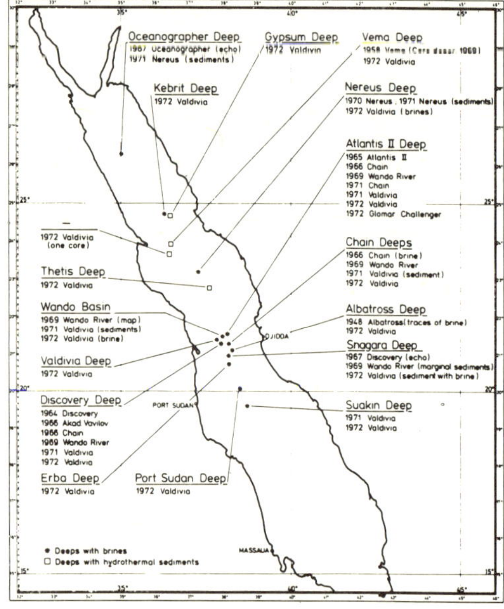

Fig. 1 Distribution of brine pools and brine derived sediments in the Red Sea. Principal investigations until May 1972.

accompanied by high ridges which are often delimited by steep scarps. Contrary to the northern Red Sea there is a central rough zone (similar to the Gulf of Aden) forming the deepest part of the sea. But the accompanying graben structures on both sides are only a few hundred metres higher.

This morphological situation is not very favourable for a transport of formation waters from the marginal evaporite bearing zones to the active central rough zone which offers the only significant trap structures with suitable low pelagic sedimentation rates. The different tectonic situation described is reflected by the geographic distribution of the brine pools discovered.

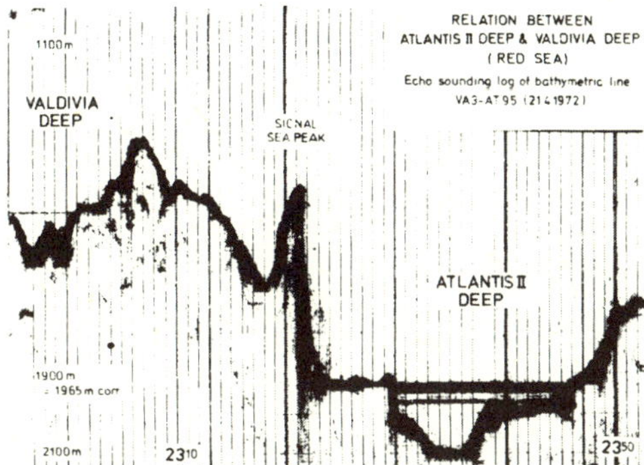

Fig. 2 Original echo sounding log of a profile in the Red Sea rift valley from Valdivia Deep (23.00: 21° 20.3′ N; 37° 56.9′ E) to Atlantis II Deep (23.50: 21° 21.2′ N; 38° 08.0′ E). ELAC narrow beam echo sounder 30 kHz, ± 1.4° beam width. Direction of the section 86°. The horizontal echoes correspond to the surface of brine layers.

A further new aspect of the general geologic situation in the Red Sea rift system is the proof of an extensive volcanic activity throughout the whole area. During our investigations 53 samples of tholeiitic basalts and basaltic glasses were collected along the main trough of the Red Sea between Hanish Islands and 23° N.

Besides the known volcanic surface structures of the central rift (Hanish and Subair Island, Djebel Tair and Suburged) a submarine volcano ("Ramad Seamount") has been found at 17° 00′ N and 40° 40′ E. Its flanks are covered with black ashes. Similar pyroclastites were found in a deep ("Volcano Deep" near Port Sudan Deep) at a depth of 2,440 m at 20° 01.3′ N and 38° 27.0′ E. Evidence of volcanic activity in the Atlantis II Deep is given by several lava flows which are intercalated in the hydrothermal sediments.

Tectonic and volcanic activity is reflected by the recently observed earthquake intensities. Epicentres have been registered during the past few years mainly in the Gulf of Aden and the southern Red Sea[6]. Two important groups are situated around Port Sudan Deep and Suakin Deep. The detailed analysis of one earthquake (March 31, 1969) in the northern Red Sea revealed the extent of important dislocations[7].

Many new heat flow measurements in the Red Sea and Gulf of Aden show in general high values and are in agreement with previously published data[8]. They range within values typical of mid-ocean ridges. Extreme values are connected with Atlantis II Deep, Suakin Deep and Nereus Deep, where they are found in marginal zones (J. Scheuch and R. Hänel, personal communication).

Evaporites seem to be one prerequisite for the formation of the Red Sea brines[5]. Miocene evaporites including large quantities of rock salt are described from wells and outcrops along the Red Sea from the Gulf of Suez to the Dakhlak Islands and even near Atlantis II Deep (R. D. Whitmarsh, personal communication). This sequence probably underlies large areas of the Red Sea basin and is covered by only a few hundred metres of Plio-Pleistocene calcareous clays and microfossiliferous oozes. Thus the evaporites can easily come into contact with the seawater during tectonic displacements near the surface.

The place of the different brine pools within the described geological setting is chiefly governed by recent tectonic lines. The brine basins generally occupy the deepest parts of local basins within the main central rift and are delimited by NW-striking faults. There are a few significant exceptions: Valdivia Deep (1,673 m) is a special depression on the slope of the central trough (2,200 m). Further, some of the deeps

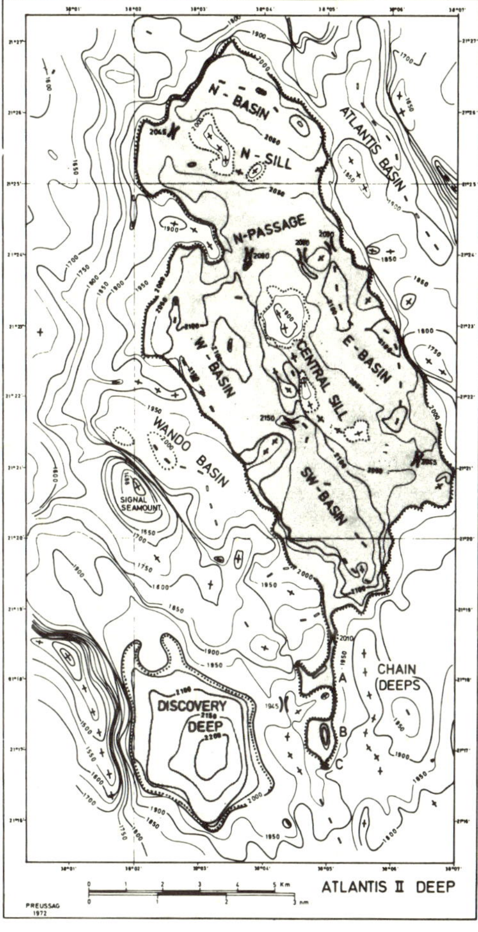

Fig. 3 Bathymetric chart of the Atlantis II Deep area, Red Sea. ELAC narrow beam echo sounder (30 kHz ± 1.4° beam width). Contours in metre, corrected after Matthews[13]. Decca hi-fix navigation. Track space 400 m, within the Atlantis II Deep and Chain Deeps 100 m. Dotted line uppermost echo reflector on top brine.

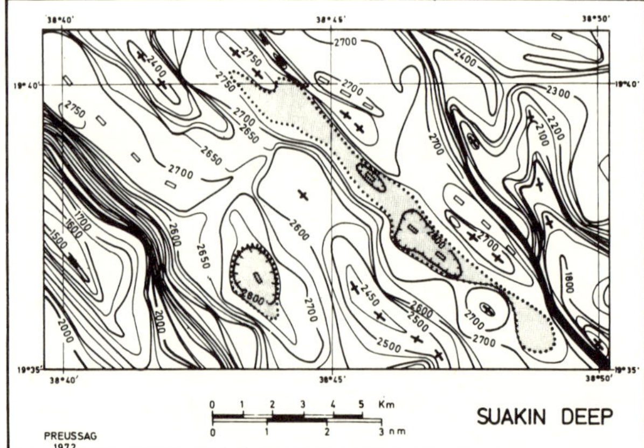

Fig. 4 Bathymetric chart of the central part of the Suakin Deep. Explanation see Fig. 3. Track space 1.4 km, locally less.

(Valdivia, Kebrit and Discovery Deeps) are more or less round structures not consistent with the general tectonic trends. The Albatross Deep has a unique position: any rise of the present brine level would cause an overflow into the Shagara Deep, the bottom of which is roughly 400 m below the Albatross Deep.

The Brines

Generally the deeps are filled with brines of different salt concentrations, the most concentrated one forming the principal and the lowermost brine layer. The development of the transition zone (that is, the mixing zone to the overlying seawater) is different. For some deeps a layered transition zone is characteristic (stepwise increase of temperature and chlorinity).

The most characteristic examples in this case are the Atlantis II Deep and nearby deeps. Similar stratifications have been detected in the Suakin Deep and the Port Sudan Deep. For all other deeps small undifferentiated transition zones are typical.

At least one of the brine pools now known is in an active state (Atlantis II Deep). Many observations since 1964 revealed a continuous rise of temperatures (1965–1972, 55.9 °C–60.1 °C). Convection currents are indicated by direct measurements (D. L. Dorson, personal communication) and have been confirmed by detailed temperature profiles[11]. As a consequence of this the 59–60° C brine which originates in the Atlantis II Deep SW basin is at the moment spreading to the other basins of the brine pool[11].

For the chemical characterization of the brines, their content in Cl, SO_4, Ca, Mg, Fe, Mn, Zn, Cu and the gases H_2S and CO_2 is most important. The Atlantis II Deep and neighbouring depressions have the highest chlorinities. In some areas with abundant hydrothermal sediments, however, only a very slight increase of the chlorinities could be detected in the connate water (Thetis, Vema and Gypsum Deeps). The Mg and SO_4 depletion observed in the Atlantis II Deep brine by several authors could not be found in the other deeps except those near the Atlantis II Deep and (for SO_4) in the Nereus Deep. Carbon dioxide was found in the Valdivia and Kebrit Deeps and in the Atlantis II Deep[9] in various concentrations. Hydrogen sulphide is present in remarkable quantities in the Kebrit Deep and gives rise to its name. The presence of H_2S in the incoming brine of Atlantis II Deep must also be concluded from the high amount of monosulphides in the recent sediments.

Some adjacent brine pools without "surface" connexions do not seem to be entirely separated. Evident conformity of the physical parameters and the chemical composition were detected suggesting subsurface connexions in the case of three groups of deeps: Atlantis II Deep with Discovery and Chain Deeps as well as Wando Basin; the two brine pools in Suakin Deep; the two basins of Nereus Deep.

The Sediments

Normal sediments in the central trough of the Red Sea are light-coloured marls with, in most cases, a significant content of nannofossils and pelagic microfossils, especially globigerinids and pteropods. Brownish coloured layers with

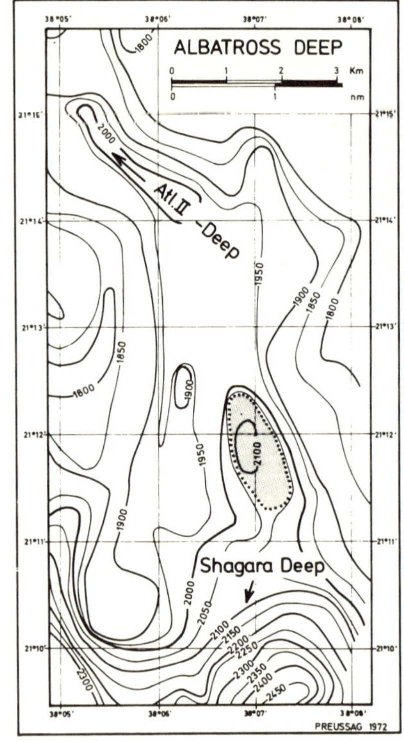

Fig. 5 Bathymetric chart of the Albatross Deep area. Explanation see Fig. 3. Track space about 1 km.

Table 1 Brine Filled Basins and Major Indications of Hydrothermal Sediments in the Red Sea Rift Zone

Name	Geographic coordinates of the centre of the brine covered areas or of the whole deep if no brine is present: N	E	free brine surface	important hydroth. sediment	Total depth (11)	Top brine (uppermost echo reflector) (11)(19)	Max. thickness of brine layer (11)	brine surface (approximative, km^2)	number of echo reflectors (thickness of reflector system)	sound velocity in the main brine (m/s)	saltcontent of most concentrated brine (chlorinity, ‰ Cl, (40))	Temperature of brine °C: max.	rising downward	± constant	decreasing downward	some variable brine components 10^{-3} g/kg: Mg	Ca	Zn	SO_4	main dissolved gases (++ major quantities): N_2	CO_2	H_2S	CH_4	C_2H_6
Suakin Deep SW Basin	19° 36,7'	38° 43,6'	+	+	2850 (29)	2776 (6)	74	2,5	1	1703 (1)	85,8	23,9 (3)(16)	+			1,4	2,1	0,05	3,2					
Suakin Deep NE Basin (30)	19° 38,0'	38° 46,3'	+	+	2830	2776 (6) (2743)(20)	54 (87) (20)	10	1 3 (20)	1704 (1)	85,9	24,6 (3)	+			1,4	2,2	0,04	3,2					
Port Sudan Deep	20° 3,8'	38° 30,8'	+	+	2800 2836 (6)	2514	286 322 (6)	5 (5)	3 (15m)	1795 (7)	125	36,2 (2)	+			1,5	1,1	0,17	4,0					
Erba Deep	20° 43,8'	38° 11,0'	+	(8) (+)	2395	2376	19	7 (5)	2-3 (15m)		86,5 (2386m)	27,9 (2) (2389m)	+			1,5	1,0	0,36	4,1					
Shagara Deep	21° 7,8'	38° 5,3'	+	+	2496	2488	8	1 (8)(9)	1 (10)		113 (12)													
Albatross Deep	21° 11,9'	38° 7,0'	+	+	2133	2051	72	1,5	1	1797 (1)	143,3	24,4 (1)			+									
Chain Deep C	21° 16,77'	38° 4,95'	+		2165 ?	1998 ?	167 ?	0,02		1820 (1)	154,6	44,5 (2)(16)	+											
Chain Deep B (18)	21° 17,17'	38° 4,95'	+	+	2130	1990	140	0,8	1 (2?)	1825 (1)	156,0	46,3 (2)(17) 46,7 (1)	+ (14)	+										
Chain Deep A (North) (21)	21° 18,0'	38° 4,9'	+	+	2072	1989 (2005)(20)	83 (67)	0,7	5 (6)(8)	1821 (1)	153,8 (2052m)	52,1 (2)(16) 53,2 (1)(16)	+		+ (14)									
Discovery Deep	21° 17,0'	38° 3,2'	+	+	2224	2015 (16) 2049 (22)	209 175 (23)	11,5	5-7 (34m)	1824 (1)	155,5	44,8 (16)(2)	+	+ (23)	+ (14)	0,8 (4)	5,1 (4)	0,8 (4)	0,6 (4)					
Wando Basin (24)	W: 21° 21,45' E: 21° 21,20'	38° 1,80' 38° 2,40'	+	+ +	2013 2007	1985 1985	28 22	0,2 0,5	1-2 2	1673 (1)	— 73,5	(16) 24,1 (1) (16) 29,3 (1)	+											
Atlantis II Deep (25)	21° 22,5'	38° 4,5'	+	+ +	2170 2194 (6)	1992 (26)	178 126 (27)	55	7-10 (52m)	1821 (1) 1830 (7)	156,5 (42)	(2) 60,0 (16) (1) 60,1 (16)	+	+ (42)		0,7 (4)	5,2 (4)	5,4 (4)	0,8 (4)	+ +	+	+? +	+	+
Valdivia Deep	21° 20,5'	37° 57,0'	+	–	1673	1550	123	4	1	1806 (1)	136,6 144,7 (43)	29,5 (3) 29,8 (43)	+			1,9	0,8	0,4	6,8	+ +	+		+	tr
Thetis Deep (36)	22° 43'	37° 36'	–	+ +	1970 1819 (37)						22,9 (12)	22,6 (2)(38)			+ (15)									
Nereus Deep E Basin	23° 11,5'	37° 15'	+	+	2458	2419	39	3 (31)	1	1675 (7)	129,5	30,2 (2)	+		+ (14)	1,5	7,9	0,5	0,9	+ +	tr		tr	
Nereus Deep W Basin	23° 11'	37° 12'	+	+	2432	2421	11	1	1															
Vema Deep (39)	23° 52,0'	36° 30,5'	–	+	1611						22,7 (12)													
Gypsum Deep (34)	24° 42,1'	36° 24,8'	–	+ +	1196						23,7 (35)													
Kebrit Deep	24° 43,35'	36° 16,6'	+	–	1558 1573 (6)	1466	92 107 (6)	2,5	1	1810 (1)	153,3	23,3 (2)		+		2,4	1,7		2,2	+	+ +	+ +	+	+
Oceanographer Deep (13)	26° 17,2'	35° 1,0'	+	–	>1446	~1364	>82		1															
Normal Red Sea water					2850 (29)					1570 (7)(32)	22,5	21,8 (3)		+		1,41 (33)	0,45 (33)		2,96 (33)					

1, Bathysonde; 2, reversal thermometers; 3, both reversal thermometers and bathysonde; 4, ref. 5; 5, uncertain; 6, corrected approximately with sound velocity in the brine; 7, calculated from distances of echoes of water samplers on the echosounder log, approximately[12]; 8, in the area now covered by brine little sediment; 9, the whole Shagara Deep is 12 km in diameter; 10, probably identical with that described in the Discovery account[5]; 11, meter, corrected after Matthews; 12, connate water; 13, information from ref. 5 and Tooms (personal communication); 14, lowermost part; 15, 1971; 16, 1972; 17, 1971 and 1972; 18, identical with that described by Ross and Hunt[4]; 19, two groups; 20, top of the second package of reflecting layers; water above without significantly higher salinity; 21, connexion by a channel with Atlantis II Deep; 22, top of the lowermost reflector, corresponding approximately to the top of the main brine layer; 23, main brine layer; 24, two small areas are now covered with brine which are possibly connected; 25, mainly data from the SW Basin 1972; 26, locally not very distinct reflecting layers are visible up to 1,971 m corrected; the top of the lowermost reflector, corresponding with the top of the 60° C brine is at 2,044 m (after Matthews[13], probably 2,049 m with consideration of higher velocity in the brine); 27, only the undiluted 60° C brine; thickness with consideration of sound velocity in brine about 150 m; 28, the brine pool has a length of 14 km and a maximum width of 1.5 km; 29, Suakin Deep SW Basin is probably the deepest point of the Red Sea; "corrected" depth values range up to 2,867 m, but the cable length was slightly less; 30, locally; 31, including a small brine pool NW from the main trough which might be separated by a low sill; the brine reflector appears about 2 m deeper; 32, around 2,000 m; 33, ref. 5: sea-water with 21.13 g kg^{-1} Cl; 34, the Gypsum Deep area is only a shallow depression but contains sediments with high content of hydrothermal material; 35, though there seems to be no free brine surface, a slightly elevated salinity was found in water from a piston core liner; 36, Thetis Deep, named after the nearby Thetis Reef, has a surface of about 20 by 30 km and consists of several more or less separated individual basins; the most important hydrothermal influences within the Thetis Deep are in the North Basin with a surface of approximately 3 by 10 km; 37, North-Basin; 38, at 1,800 m in the N-Basin; 39, Vema Deep measures about 24 by 6 km; sediments with hydrothermal components were first sampled by R/V Vema, ref. 5; 40, shipboard Mohr-Knudsen titrations; 41, the uppermost echo reflector corresponds generally with the first significant salinity increase[11]; 42, differentiated, for details see ref. 11; 43, in the main brine, two layers can be distinguished; the lower one (about 40 m) has a slightly higher temperature and a significantly higher chlorinity.

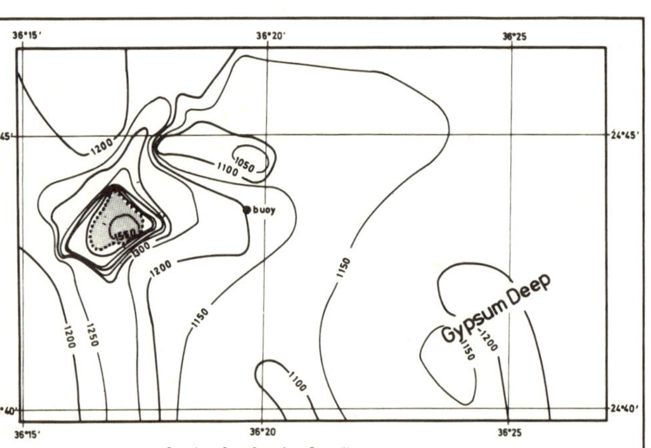

Fig. 6 Bathymetric chart of the Kebrit Deep area. ELAC narrow beam echo sounder (30 kHz, ±1.4° beam width). Contours in metre, corrected after Matthews[13]. Navigation with radar and bearing (buoy). Tracks irregular, radially from the centre of Kebrit Deep.

increased Fe and Mn content are often found in the sediments above the basalt along the central rift zone. This is consistent with observations made in the basal 40 m of Glomar Challenger cores near the Mid-Atlantic Ridge[10].

In the deeps with the strongest hydrothermal influence the bottom is covered by multicoloured sediments with extremely high sedimentation rates. Tectonic events cause repeated changes of the physico-chemical conditions and, therefore, different facies of the sediments. Normal oxidic conditions (especially during mixing with sea-water) lead to sediments with attributions of limonite, haematite and manganite facies. Sulphides are present only as pyrite and chalcopyrite.

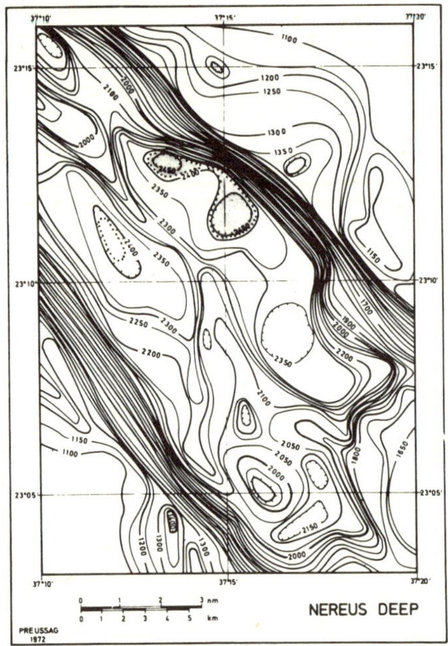

Fig. 7 Bathymetric chart of the central part of Nereus Deep. Explanation see Fig. 3. Track distance 1.8 km.

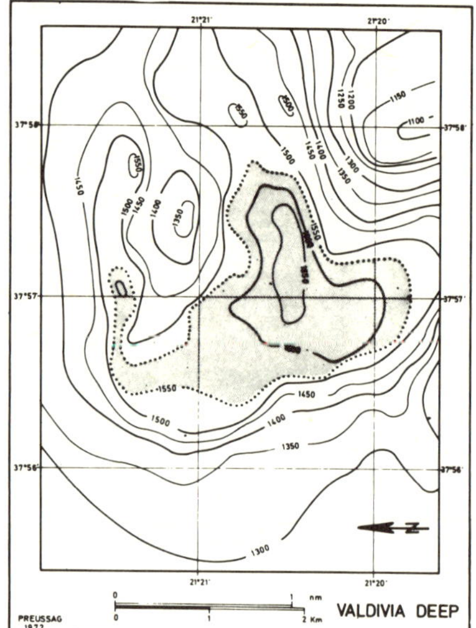

Fig. 8 Bathymetric chart of the Valdivia Deep. Explanation see Fig. 3. Track distance 500 m plus additional cross lines.

Under reduction conditions a sediment containing iron monosulphide and sphalerite together with Fe-montmorillonite and some manganosiderite is formed, but the presence of hydrogen sulphide is very important. There is also a high content of anhydrite in Atlantis II Deep and of gypsum in Gypsum Deep, well crystallized, dispersed in the other faces. The sediments in Valdivia Deep, Kebrit and probably Oceanographer Deep are predominantly carbonatic. The grain size of the hydrothermal part of the sediments is extremely small, mostly below 0.002 mm. The iron content in the limonite and the haematite facies is between 50 and 65%. In the Chain Deeps manganese contents show an average of between 30 and 40%. In Atlantis II Deep some sulphide layers have a zinc content of about 20%. But continuously changing facies conditions reduced the average content of the base metals to a few per cent.

The Valdivia cruises (VA 1 and VA 3) are sponsored by the Ministry of Education and Science, Bonn, and by Preussag AG, Hannover. The bathymetric mapping was done by H. Richter and K. Lange. D. Menz supplied the chemical analyses.

Received August 21; revised September 26, 1972.

[1] Bruneau, L., Jerlov, N. G., and Koczy, F., *Rep. Swedish Deep Sea Expedition*, **3**, 99 (1953).
[2] Swallow, J. C., and Crease, J., *Nature*, **205**, 165 (1965).
[3] Miller, A. R., Densmore, C. D., Degens, E. T., Hathaway, J. C., Manheim, F. T., McFarlin, P. F., Pocklington, R., and Jokela, A., *Geochim. Cosmochim. Acta*, **30**, 341 (1966).
[4] Ross, D. A., and Hunt, I. M., *Nature*, **213**, 687 (1967).
[5] *Hot Brines and Recent Heavy Metal Deposits in the Red Sea* (edit. by Degens, E. T., and Ross, D. A.) (Springer, New York, 1969).
[6] Fairhead, J. D., and Girdler, R. W., *Phil. Trans. Roy. Soc. Lond.*, A, **267**, 47 (1970).
[7] Ben-Menahem, A., and Aboodi, E., *J. Geophys. Res.*, **76**, 2674 (1971).
[8] Girdler, R. W., *Phil. Trans. Roy. Soc. Lond.*, A, **267**, 191 (1970).
[9] Schoell, M., and Stahl, W., *Earth Planet. Sci. Lett.*, **15**, 206 (1972).
[10] Boström, K., *Nature*, **227**, 104 (1970).
[11] Schoell, M., and Hartmann, M., *Marine Geol.* (in the press).
[12] Hartmann, M., *Marine Geol.*, **12**, M16 (1972).
[13] Matthews, D. J., *Publ. HD 282* (Hydrographic Department, Admiralty, 1939).

18

Reprinted from *Geophys. Res. Lett.* 1:355-358 (1974)

RAPIDLY ACCUMULATING MANGANESE DEPOSIT FROM THE MEDIAN VALLEY OF THE MID-ATLANTIC RIDGE

Martha R. Scott, Robert B. Scott, Peter A. Rona, Louis W. Butler, and Andrew J. Nalwalk

Abstract A manganese oxide crust from an extensive deposit in the median valley of the Mid-Atlantic Ridge was found to be unusually high in manganese (up to 39.4% Mn), low in Fe (as low as .01% Fe), low in trace metals and deficient in Th^{230} and Pa^{231} with respect to the parent uranium isotopes in the sample. The accumulation rate is 100 to 200 mm/10^6 y, or 2 orders of magnitude faster than the typical rate for deep-sea ferromanganese deposits. The rapid growth rate and unusual chemistry are consistent with a hydrothermal origin or with a diagenetic origin by manganese remobilized from reduced sediments. Because of the association with an active ridge, geophysical evidence indicative of hydrothermal activity, and a scarcity of sediment in the sampling area, we suggest that a submarine hot spring has created the deposit.

Introduction

Typical deep-sea ferromanganese nodules and crusts grow at a rate of about 1-10 mm/10^6 years and have an Fe to Mn ratio of .5 to 2. A manganese crust recently recovered from the median valley of the Mid-Atlantic Ridge has a growth rate two orders of magnitude faster than the typical rate and an extremely low iron concentration; it appears to have formed under conditions which are not typical of deep-sea manganese deposition. The location of the sampling site and the nature of the material immediately bring to mind a possible association between this manganese oxide deposit and the recently developed ideas concerning hydrothermal activity along active ridge crests [Anderson, 1972; Lister, 1972 and 1974; Deffeyes, 1970; Dymond et.al.,1973]. According to these theories seawater has percolated into the ocean crust and returned to the sea floor as hydrothermal solutions, heated and laden with dissolved metals; these fluids may be the source of the metal enrichment of sediments along ridge crests [Bostrom and Peterson, 1969; Bender et.al., 1971]. However, it has also been suggested [Turekian and Bertine, 1971] that organic-rich sediments may exist in ephemeral sediment ponds along ridge crests and cause enrichment of sediments in Mo and U. Conceivably remobilization of sedimentary manganese could also occur under such circumstances. The purpose of this paper is to examine the chemistry and physical surroundings of the sample to determine its probable origin.

Sample Description and Location

The rapidly growing MnO_2 crust (sample 13-21) described in this paper was recovered from Dredge site 13 in the Median Valley of the Mid-Atlantic Ridge at 26°N (Fig.1). It was collected during the 1972 Trans-Atlantic Geotraverse (TAG) program of the National Oceanic and Atmospheric Association [R.Scott et.al.,1974]. Dredge site 13 is a talus slope under 3.4 km of water at the base of the median valley scarp, only 5 km from the median valley axis. Sample 13-21 is a 42 mm thick manganese oxide deposit of birnessite and a trace of todorokite; the sea water interface is preserved and clearly identifiable. The material has a gray submetallic luster and conspicuous laminations 5 to 10 mm thick; some layers are very porous and exhibit bladed growth structures. Typical manganese crusts from sites 10G and 2B were also analyzed. Both are brownish-black with a dull earthy luster and uneven 1 to 5 mm laminations.

Analytical Methods

Radiochemical analyses were carried out by a method similar to that of Ku and Broecker (1969).

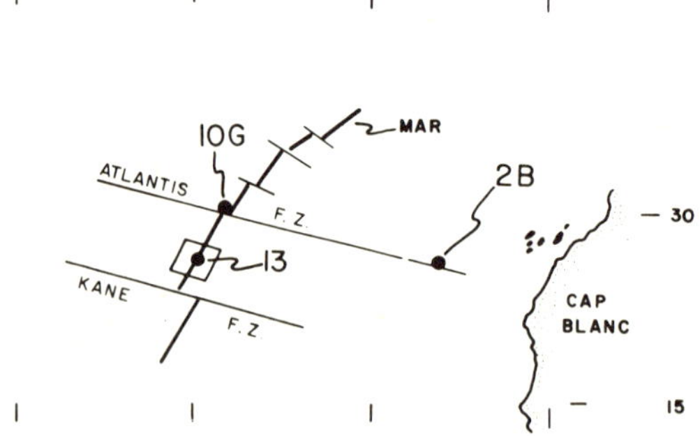

Figure 1. Sample locations for sample 13-21 and for typical manganese crusts from 10G and 2B.

Table 1a. Radiometric data for manganese crusts; typical crusts from the Atlantis Fracture Zone (Scott et al, 1972).

Sample and Interval (mm)		U (ppm)	Activity Ratio U^{234}/U^{238}	Th^{232} (ppm)	Th^{230} Excess (dpm/g)
2B2-1	0-0.33	13.76±.37	1.20±.04	218.54±6.65	417.7±10.5
2B2-2	.33-.67	18.10±.40	1.02±.03	146.39±20.66	6.46±.59
2B2-3	.67-.98	16.20±.42	1.15±.04	93.35±2.62	0.68±.59
10G-1	0-.46	12.78±.41	1.12±.04	159.77±4.52	530.86±11.92
10G-2	.46-.90	11.62±.29	1.22±.04	98.33±3.48	217.48±5.67
10G-3	.90-1.39	11.52±.38	1.02±.04	77.86±3.41	104.84±3.55

Thin layers were scraped from the samples, dissolved in HCl-HNO_3, and spiked with Th, U and Pa tracers. Isotopes were separated by ion exchange resins and deposited from organic scavenging solutions onto stainless steel planchets. Th and U isotopes were counted by alpha spectrometry using surface barrier solid state detectors and a multichannel pulse height analyzer; Pa isotopes were counted with a flow-type proportional counter. The results are listed in Table 1. Sample 13-21 dissolved completely; 10G and 2B yielded 5-10% insoluble residue.

Analyses for Fe, Mn, Co, Ni and Cu were made by atomic adsorption spectrophotometry. Using an HCl-HNO_3 sample solution procedure, the one sigma precision error is 2% of the Mn values, 3% of Fe, 5% of Co, 5% of Ni and 3% of Cu. Results are listed in Table 2.

Discussion

Radiochemical analysis of uranium series isotopes shows the site 13 manganese deposit to be very unusual in comparison to typical ferromanganese crusts such as those from nearby dredge sites 2B and 10G [Scott et.al.,1972], (Table 1; Figures 2 and 3). Manganese deposits found in the deep sea ordinarily contain amounts of Th^{230} and Pa^{231} far in excess of the amounts of equilibrium with the parent uranium isotopes [Ku and Broecker, 1969; Sackett, 1966], indicating that thorium and protactinium have been scavenged from sea water during the process of manganese accumulation. In contrast to the expected excess of Th^{230} and Pa^{231}, sample 13-21 was strikingly deficient in these isotopes relative to secular equilibrium with uranium. Commonly, manganese accumulation rates are measured by the decay of excess or unsupported Th^{230} and Pa^{231}. Because of the absence of these isotopes from the deposit, the accumulation rate was measured from the growth rate of Th^{230} and Pa^{231} toward secular equilibrium with their uranium parents. Growth of Th^{230} and Pa^{231} give independent measures of the accumulation rate, and the two determinations are in reasonably good agreement, Th^{230} yielding 130 mm/10^6y and Pa^{231} showing 250 mm/10^6 y (Fig.3). The Th data are clearly more precise.

The ages for layers of sample 13-21 (Table 1) are maximum ages calculated with the assumption that all the Th^{230} and Pa^{231} have grown in from the uranium in the sample. The apparent ages of surface material indicate that some small amounts of thorium and protactinium are scavenged by the rapidly accumulating manganese; the very low activity of these isotopes in the surface layer

Table 1b. Radiometric data from manganese crusts; sample 13-21 from the median valley of the Mid-Atlantic Ridge. One sigma errors from counting statistics are listed.

Sample and Interval (mm)	U (ppm)	Activity Ratio U^{234}/U^{238}	Th^{232} (ppm)	Activity Ratio Th^{230}/U^{234}	Activity Ratio Pa^{231}/U^{235}	Max. ages (x10^3y) from growth of Th^{230}	and Pa^{231}
13-21-1 0-.51	16.46±.39	1.06±.03	2.50±.28	.135±.010	.125±.034	15.7±1.2	6.3±1.7
13-21-2 .51-.91	13.21±.44	1.12±.04	3.27±.35	.160±.014	.131±.060	18.9±1.6	6.6±3.0
13-21-3 .91-1.41	13.61±.70	1.23±.07	4.54±.66	.225±.023	.188±.073	27.8±2.9	9.8±3.8
13-21-4 1.41-3.01	8.95±.28	1.16±.04	2.34±.58	.233±.033	.247±.093	28.8±4.1	13.3±5.0
13-21-5 3.01-4.61	9.25±.41	1.13±.06	3.95±.29	.336±.021	-	44.5±2.8	-

Table 2. Compositions of layers within the largest fragment of the Site 13 manganese deposit (sample 13-21); the typical ferromanganese crusts from dredge sites 10G and 2B, and from sites 19 and 15 within 50 km of site 13; and the average Pacific and average Atlantic nodules, (Bonatti et al 1972). The compositions listed for sites 10G and 2B are averages of four samples (Scott et al 1972).

Sample	Location	Composition				
		Mn(%)	Fe(%)	Co(ppm)	Ni(ppm)	Cu(ppm)
T3-72D 253-13-21	26°08'N,44°45'W					
Sequence of samples	a top 5 mm	39.2	0.011	18	100	12
through layers of 13-21	b 6 mm	38.5	0.078	25	790	119
	c 6 mm	38.6	0.106	25	660	93
	d 5 mm	39.4	0.070	20	200	23
	e 3 mm	39.2	0.072	18	400	23
	f 8 mm	39.1	0.038	16	270	19
	g 8 mm	39.3	0.036	14	50	11
T3-71D 160-10G	30°08'N, 42°29'W	9.8	18.1	2720	1280	880
T3-71D 148-2B	26°07'N, 25°21'W	14.1	16.1	7200	2200	750
T3-71D 254-15-2	26°33.9'N, 44°30'W	11.2	16.4	9480	900	300
T3-72D 255-19-3	26°16.9'N, 45°6.3'W	11.0	18.6	4950	1110	405
Average Pacific nodules		19.8	14.3	3810	7200	3660
Average Atlantic nodules		16.2	21.8	3090	3970	1090

make it difficult to determine accurately the ratio in which the Th and Pa were incorporated.

The growth rate of 13-21, about 200 mm/10^6 y, is two orders of magnitude faster than the typical growth rates for marine ferromanganese deposits. For example, the decay curves of excess Th^{230} with depth in the deposits from sites 2B and 10G yield growth rates of 1 and 5 mm/10^6 y, respectively, typical of rates found in other deep-sea manganese deposits (Ku and Broecker, 1969).

The major and trace element composition of this Mid-Atlantic Ridge manganese deposit is distinctly different from that of typical deep-sea ferromanganese nodules (Table 2); this sample is practically pure manganese oxide with as little as 0.01 weight % Fe. In addition the Cu, Ni, Co and Th contents are quite low compared to typical deep-sea hydrogenous deposits, including hydrogenous crusts collected within 50 km of site 13 (Tables 1 and 2). This appears to be characteristic of rapidly growing manganese deposits; the more slowly the material accumulates, the higher its concentration of certain trace metals removed from sea water [Bonatti et.al., 1972] regardless of the source of the manganese. The same effect may be seen in the low trace metal contents of the Loch Fyne and Jervis Inlet nodules [Ku and Glasby, 1972]. However, this generalization does not hold for all trace elements. Uranium, for example, does not appear to be affected by rapid growth rates (Table 2; Ku and Glasby, 1972), and its mechanism of incorporation into manganese deposits may differ from the other elements listed. The U^{234}/U^{238} activity ratios for all the layers of 13-21 are essentially the same as sea water, 1.15 within limits of counting errors (Table 1); the uranium in this deposit was probably incorporated from sea water.

The rapid growth rate, low trace metal content, extreme fractionation of Mn from Fe, and low surface ratios Th^{230}/U^{234} and Pa^{231}/U^{235} reported for sample 13-21 may typify either hydrothermal deposits such as the one described by Veeh and Bostrom (1971) from a Pacific seamount, or deposits formed by manganese remobilized from reduced sediments [Ku and Glasby, 1972; Manheim, 1965]. The overlapping chemical characteristics

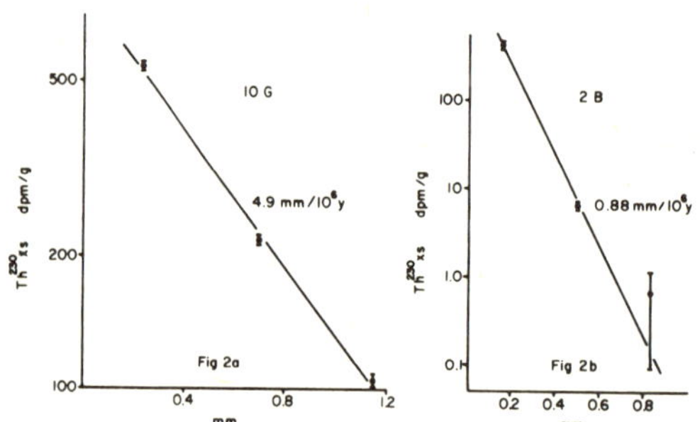

Figure 2. Rates of manganese crust accumulation based on decay of Th^{230} excess; a. 10G. b. 2B.

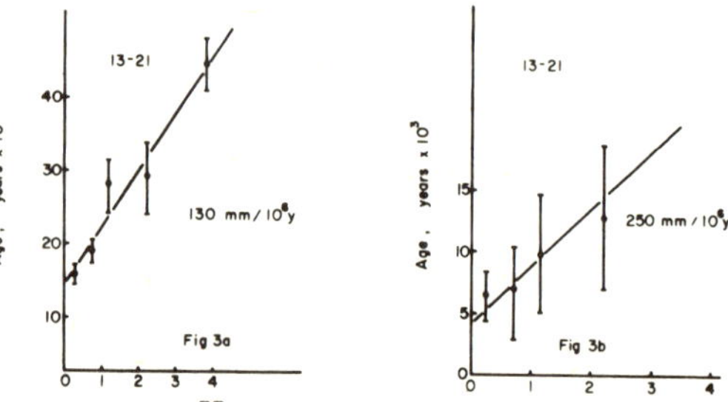

Figure 3. Rate of accumulation of site 13 manganese deposit. a. Rate from growth of Th^{230}; b. Rate from growth of Pa^{231}. Lines are best fit lines.

of these two types of deposits were pointed out by Bonatti et.al.,1972 ; but this Mid-Atlantic Ridge crust has much higher Mn/Fe ratios (360 to 3600) than values published for diagenetic continental margin nodules [Ku and Glasby, 1972; Manheim, 1965]

Bottom photographs show that the current-swept talus slope from which sample 13-21 was dredged is essentially sediment-free; the sample itself was recovered from a site at least 500 m above the median valley floor. Subsequent dredging of the site always yielded more thick MnO_2 encrustations; away from this site, pillow basalts were dredged but no thick MnO_2 crusts were found. Other work on the sampling site showed it to have a sharp 0.14°C increase in bottom water temperature over the site [Rona et.al.,1974] and very iron-rich suspended matter in the overlying water column [Betzer et.al.,1974]. Moreover, the Brunhes normal axial magnetic anomaly contains a localized magnetic low region which coincides with the sampling site, and which may reflect hydrothermal alteration of magnetic minerals in the pillow lavas [Watkins and Paster, 1971]. These features combined with other lines of evidence suggesting hydrothermal activity along active ridge crests [Anderson, 1972; Lister, 1972 and 1974; Dymond et.al.,1973] have led us to conclude that the MnO_2 crust described in this paper was deposited by a hydrothermal spring in the median valley of the Mid-Atlantic Ridge, rather than forming as a diagenetic deposit. The region itself has been described elsewhere as the TAG Hydrothermal Field [Scott et.al.,1974].

The extreme fractionation of Mn from Fe in sample 13-21 probably resulted from the differential solubilities of the two elements during oxidation by sea water oxygen [Krauskopf, 1958]. Iron in hydrothermal fluids may have been precipitated at lower levels within the pillow lavas or the talus as oxides or sulfides; but some iron apparently bypassed both the pillow lavas and the surface deposit to have become suspended in the water column [Betzer et.al.,1974].

An attempt to evaluate the magnitude of chemical effects due to hydrothermal effluents upon the composition of sea water would require estimates of the numbers of submarine hydrothermal orifices, the rate of flow from each, and the composition of the fluids of each. Obviously these data are absent at present. The nature and extent of possible hydrothermal deposits beneath the Atlantic and Indian Oceans, where slow spreading favors ocean water circulation in the crust, need to be evaluated.

Acknowledgements. Our research was supported by NSF grants GA-35456 and GA-29370 and by ONR Contract N00014-68-A-0308-0002.

References

Anderson, R.N., Petrological significance of low heat flow on the flanks of slow-spreading mid-ocean ridges, Geol.Soc.Amer.Bull.,83,2947-2956,1972.

Bender, M., W. Broecker, V. Gornitz, U. Middle, R. Kay, S.-S Sun and P. Biscaye, Geochemistry of three cores from the East Pacific Rise, Earth and Planet. Sci.Lett. 12, 424-433,1971.

Betzer, P.R., G.W. Bolger, B.A. McGregor and P.A. Rona, The Mid-Atlantic Ridge and its effect on the composition of particulate matter in the deep ocean (Abstract), Eos Trans. A.G.U.,55, 193,1974.

Bonatti, E., T. Kramer, and H.S. Rydell, Classification and genesis of submarine iron-manganese deposits, in Ferro-manganese Deposits on the Ocean Floor, pp 146-166, Lamont-Doherty of Columbia University, Palisades, N.Y. 1972 .

Bostrom, K., and M.N.A. Petersen, The origin of aluminum-poor ferromanganoan sediments in areas of high heat flow on the East Pacific Rise, Marine Geology, 7, 427-447, 1969.

Deffeyes, K.S., The axial valley: A steady state feature of the terrain, in Megatectonics of Continents and Oceans, pp 194-222, Rutgers University Press, New Brunswick, N.J., 1970.

Dymond, J.D., J.B. Corliss, G.R. Heath, C.W.Field, E.J. Dasch, H.H. Veeh, Origin of metalliferous sediments from the Pacific Ocean, Geol. Soc. Amer. Bull.,84, 3355-3372, 1973.

Krauskopf, K.B., Separation of manganese from iron in sedimentary processes, Geochim.Cosmochim.Acta, 12, 61-84,1958.

Ku, T.-L., and W.S. Broecker, Radiochemical studies on manganese nodules of deep sea origin, Deep-Sea Research, 16, 625-637, 1969.

Ku, T.-L, and G.P. Glasby, Radiometric evidence for rapid growth rate of shallow-water continental margin manganese nodules, Geochim. Cosmochim. Acta, 36, 699-704,1972.

Lister, C.R.B., On the thermal balance of a mid-ocean ridge, Geophys. J.R. Astr. Soc., 25, 515-535,1972.

Lister, C.R.B., Water Percolation in the Ocean Crust, Eos. Trans. A.G.U., 740-742, 1974.

Manheim, F.T., Manganese-iron accumulations in the shallow marine environment, Narragansett Mar. Lab. Publ. 3-1965, pp 217-276, University of Rhode Island, Kingston, Rhode Island,1965.

Rona, P.A., B.A. McGregor, P.R. Betzer, D.C. Krause, Anomalous water temperatures over Mid-Atlantic Ridge crest at 26°N (abstract), Eos Trans. AGU, 55, 193, 1974.

Sackett, W.M., Manganese nodules: thorium-230: protactinium-231 ratios, Science, 154,646-647, 1966.

Scott, R.B., P.A. Rona, L.W. Butler, A.J. Nalwalk, and M.R. Scott, Manganese crusts of the Atlantic fracture zone, Nature Phys. Sci.,239,1972. (see erraturm for this paper: Nature Phys.Sci. 242,95,1973.)

Scott, R.B., P.A. Rona, B.A. McGregor and M.R. Scott, The TAG Hydrothermal Field, Nature, 251, 301-302,1974.

Turekian, K.K., and K.K. Bertine, Deposition of molybdenum and uranium along the major ocean ridge systems, Nature, 229, 250, 1971.

Veeh, H.H., and K. Bostrom, Anomalous U^{234}/U^{238} on the East Pacific Rise, Earth Planet. Sci.Lett. 10, 372-374, 1971.

Watkins, N.D., and T.P. Paster, The magnetic properties of igneous rocks from the ocean floor, Phil. Trans. Roy. Soc. Lond. A., 268, 507-550, 1971.

(Received October 4, 1974;
accepted November 8, 1974.)

19

Reprinted from *Royal Astron. Soc. Geophys. Jour.* **38**:587-608 (1974)

The Galapagos Spreading Centre: Lithospheric Cooling and Hydrothermal Circulation*

David L. Williams,

R. P. Von Herzen, J. G. Sclater and R. N. Anderson

(Received 1974 March 15)†

Summary

The spatial pattern of cooling near a spreading ridge crest was investigated with a suite of 71 precisely-navigated heat-flow stations on the Galapagos spreading centre, East Pacific, near 86°W longitude. Stations are on crust less than 1·0 My old on which bathymetry and sediment distribution are well known. Values vary from near zero to greater than 30 HFU (10^{-6} cal cm^{-2} s^{-1}). The average over the entire region is significantly less than that predicted by theoretical conduction models of a cooling lithosphere. We observe a regular variation of the heat-flow pattern with a wavelength of 6 ± 1 km approximately normal to the ridge crest. Heat-flow maxima are characteristically located near faults and local topographic highs. The locations of fields of small sediment mounds, apparently hydrothermal vents, are also restricted to these faulted, elevated areas of high heat flow. Near-axis, bottom water temperature anomalies of several hundredths °C were detected. The low average, the low minima in the heat-flow pattern, and the water temperature anomalies suggest that hydrothermal circulation accounts for approximately 80 per cent of the geothermal heat released near the ridge crest. We conclude that the hydrothermal circulation pattern is controlled by one or more of the following physical properties of the system: highly developed cellular convection, discrete zones of high permeability, variation in the strength of heat sources near the base of the crust, or bottom topography. Our results imply that heat-flow studies near active oceanic ridges will be of most value if they are sufficiently detailed and well navigated to define the systematic small-scale variations that appear to be caused by hydrothermal circulation.

Introduction

In modern theories of sea-floor spreading, new oceanic lithosphere is created by the cooling of hot material as it moves away from the axis of a spreading centre. This process is one of the major mechanisms by which the Earth releases heat (McKenzie & Sclater 1969; Sleep 1969; Williams & Von Herzen 1974). Even before these new tectonic theories were established, ridges were recognized as regions of generally high but extremely variable heat flow (Von Herzen & Langseth 1966). Despite a tremendous increase in the number of heat-flow determinations, with few exceptions, this characterization of active ridge axis heat flow still holds (Langseth & Von Herzen 1971). With the evolution of the sea-floor spreading hypotheses came the

*Contribution No. 3302 of the Woods Hole Oceanographic Institution, and contribution of the Scripps Institution of Oceanography, new series.

†Received in original form 1974 January 14

development of several theoretical models of a conductively cooling lithosphere (Langseth, Le Pichon & Ewing 1966; McKenzie 1967; Sleep 1969; McKenzie & Sclater 1969; Sclater & Francheteau 1970; Parker & Oldenburg, 1973). With reasonable input parameters, they all predict heat-flow values near the ridge axes (within a few tens of kilometres) much higher than have generally been measured. Further, the models do not predict the scatter or general pattern of conductive heat flow that is normally observed at a spreading centre.

Heat-flow measurements on spreading ridges suffer from several limitations. First, with presently existing techniques, oceanic heat-flow measurements are possible only in sediments, whereas active spreading centres are ideally and actually characterized by little or no sediment cover. This fact limits measurements to sea floor at least old enough to have accumulated a sufficient thickness (a few metres) of sediments. Second, local environmental effects, which can disturb the near surface geothermal gradient (Langseth & Von Herzen 1971; Von Herzen & Uyeda 1963) tend to be more severe near mid-ocean ridges. Topographic variations are large and the sediment tends to accumulate unevenly, with greater thicknesses in the topographic lows. Finally, these techniques only measure the conductive component of the heat flux. There is a growing amount of evidence that the failure of conductive-cooling models to explain the observations is due to other heat transfer mechanisms, primarily hydrothermal convection in the crustal rocks (Elder 1965; Pálmason 1967; Erickson & Simmons 1969; Le Pichon & Langseth 1969; Sleep 1969; Deffeyes 1970; Talwani, Windisch & Langseth 1971; Langseth & Von Herzen 1971; Hyndman & Rankin 1972; Lister 1972; Anderson 1972; Sclater & Klitgord 1973).

Other indications of hydrothermal circulation near mid-ocean ridge crests, e.g. hydrothermally altered rocks (Aumento, Loncarevic & Ross 1971; Miyashiro, Shida & Ewing 1971; Nishimori & Anderson 1974, in press), deposits from hydrothermal emanations (Corliss 1971; Sayles & Bischoff 1973), and ophiolitic rocks (Spooner & Fyfe 1973) provide qualitative evidence that this mechanism may be an important mode of heat transfer. The disparity between measured conducted heat-flow values and the theoretical models might determine directly the quantitative significance of this circulation for the cooling of an ideal lithosphere. For this reason one of the major goals of our investigation was to determine as accurately as possible the magnitude and distribution of the conducted heat flux near an active spreading centre.

The Galapagos spreading centre and the adjacent regions of the eastern equatorial Pacific have been the subject of numerous geological and geophysical studies (Herron & Heirtzler 1967; Raff 1968; Grim 1970; Herron 1972; van Andel & Heath 1973; Sclater & Klitgord 1973). Our study included a small area immediately south of the spreading centre near 86°W longitude (Fig. 1), a region recently investigated in some detail by Sclater & Klitgord (1973). The results reported here are part of a subsequent and more detailed study obtained on legs 6 and 7 (June–August 1972) of Expedition South Tow from R/V Thomas Washington of the Scripps Institution of Oceanography. Detailed studies of bathymetry, magnetics, sediment distribution and near-bottom water temperature structure from both surface-ship and deeply towed instruments are described in Klitgord & Mudie (1974), and Detrick *et al.* (1974). In this paper we discuss results of heat-flow and near-bottom water temperature measurements, and their implications for the thermal structure and the cooling of the young lithosphere of this region.

The sea floor is spreading north and south of the E–W trending spreading centre located near 0°48′N. latitude, at a half rate of about 34.4 mm/yr (Klitgord & Mudie 1974). This sea-floor spreading system is particularly suitable for our study because (1) the spreading history is well known, (2) the bottom topography is extremely two-dimensional and relatively subdued (Fig. 2), and (3) the sedimentation rate is high due to proximity to the equatorial high productivity belt. The latter two factors allowed heat-flow measurements to within 5 km of the spreading axis (on crust as young as

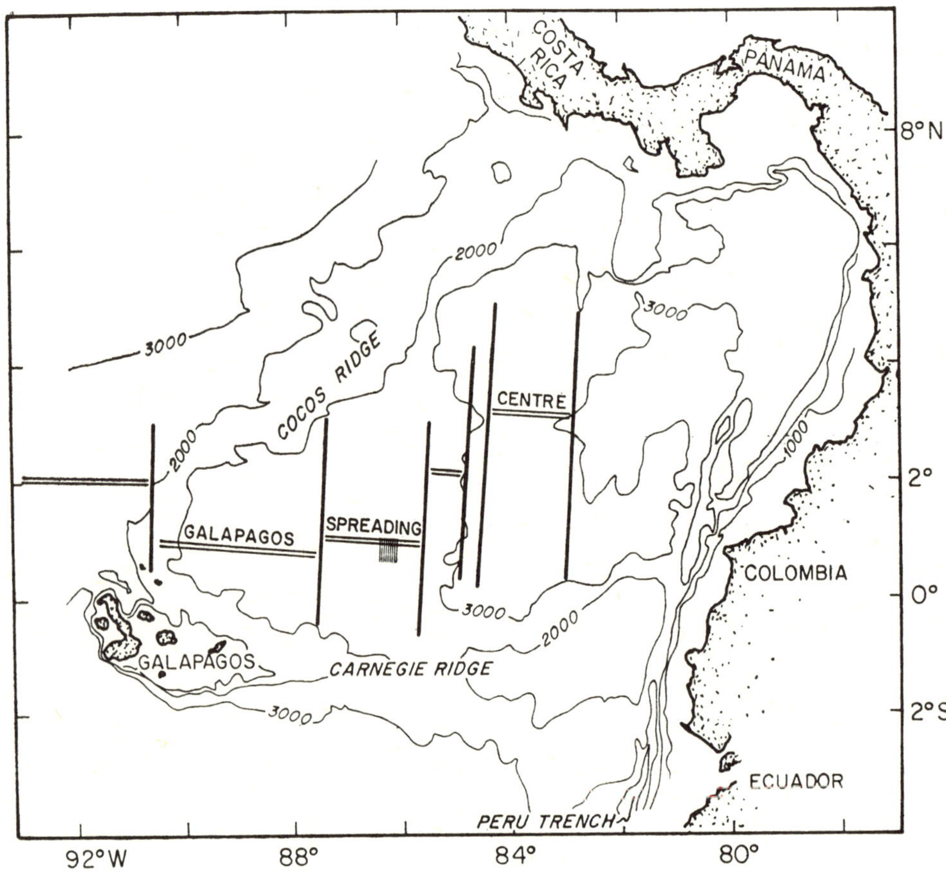

FIG. 1. Panama basin. Bathymetric contours are from van Andel *et al.* (1971) in uncorrected metres. Hatched lines mark area of detailed survey.

0·15 My), without significant bias in the location of each station due to the local sediment distribution.

Measurements and techniques

The 71 new heat-flow measurements (Table 1) and the bottom water temperature data reported in this paper were obtained on leg 7 of Expedition South Tow. Most of the temperature gradients in the sea floor were measured with probes designed for multiple penetrations of the sediments to depths up to 4m, with apparatus described previously by Von Herzen & Anderson (1972) and Corry, Dubois & Vacquier (1968). Two measurements of deeper penetration (greater than 10m) were made with thermistor probes attached to a piston corer. Thermal conductivity was determined by the needle probe method (Von Herzen & Maxwell 1959). The thermal conductivity of our piston cores averaged $1{\cdot}71 \times 10^{-3}$ cal cm^{-2} s^{-1} °C^{-1} for the upper two metres and $1{\cdot}80 \times 10^{-3}$ cal cm^{-2} s^{-1} °C^{-1} on the upper four metres. These values have been assumed, as appropriate, for the other stations (Table 1).

Navigation for both the leg 6 and leg 7 of Expedition South Tow surveys was accomplished utilizing, whenever possible, a net of 6 acoustic bottom transponders emplaced at the beginning of the leg 6 survey. On 12 of our heat-flow stations and both of the near-bottom horizontal water temperature profiles, the instrumentation was navigated inside the transponder net with near-bottom acoustic equipment attached to the hoisting cable (Boegeman *et al.* 1971); the positioning had an accuracy of ± 200 m relative to the bathymetry (Fig. 2). On an additional 10 heat-flow stations

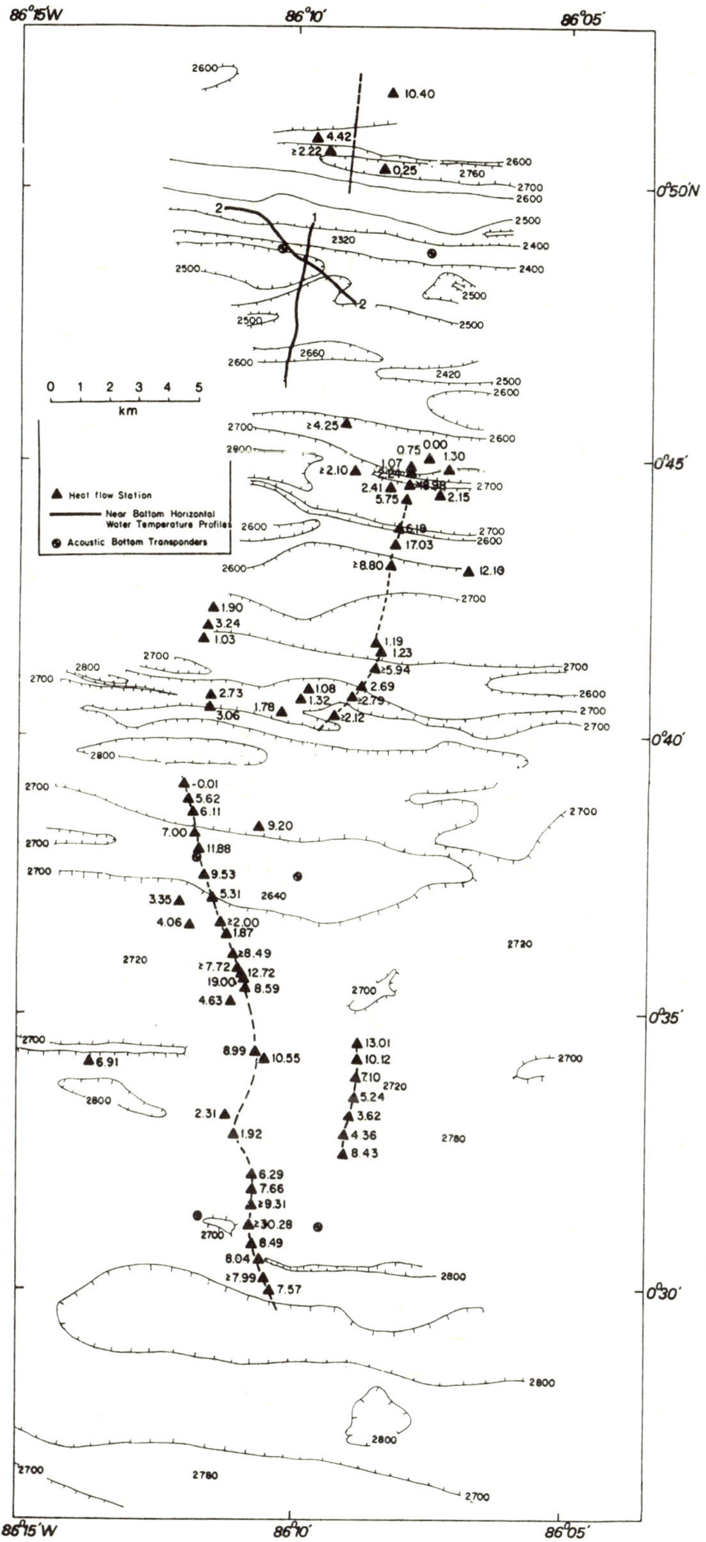

FIG. 2. Bathymetry and heat flow (HFU) in survey area. Dashed lines indicate location of topographic and heat-flow profiles illustrated in Fig. 8. Depths are in corrected metres at a 100 m contour interval.

Table 1.

Station No.	Lat. (°N)	Long. (°W)	Water depth	T	P	N	K	Q
43	0°50·3′	86°8·9′	2730		2·4	3	(1·71 (0·72))	0·25 (10) i
45(1)	0°50·9′	86°10·1′	2580		2·4	3	(1·71 (0·72))	4·42 (185) i
(2)	0°50·6′	86°9·9′	2630		0·7	1	(1·71 (0·72))	⩾ 2·22 (⩾92) i †
46	0°51·7′	86°8·8′	2620		2·4	3	(1·71 (0·72))	10·40 (436) v
50(1)	0°45·0′	86°7·4′	2690		2·4	3	(1·71 (0·72))	0·00 (0) v
(2)	0°44·8′	86°7·1′	2700	2·06	2·4	3	(1·71 (0·72))	1·30 (55) v
51	0°43·0′	86°6·7′	2600	2·05	2·4	3	(1·71 (0·72))	12·10 (507
52	0°44·5′	86°8·2′	2750	2·07	2·4	3	(1·71 (0·72))	2·41 (105) v
53(1)	0°45·7′	86°9·0′	2670	2·03	2·1	3	(1·71 (0·72))	⩾ 4·25 (⩾178) iw
(2)	0°44·9′	86°8·8′	2740	2·05	2·2	3	(1·71 (0·72))	⩾ 2·10 (⩾88) iw
54(1)	0°45·1′	86°7·8′	2690	2·04	2·5	3	(1·71 (0·72))	0·75 ± 0·03 (31 ± 1) w
(2)	0°44·9′	86°7·8′	2700	2·05	2·5	3	(1·71 (0·72)	1·07 ± 0·03 (45 ± 1) w
(3)	0°44·8′	86°7·8′	2680	2·04	2·9	3	(1·71 (0·72))	2·24 ± 0·15 (94 ± 6) w
(4)	0°44·5′	86°7·8′	2760	2·05	2·5	1	(1·71 (0·72))	⩾18·98 (⩾785) w
(5)	0°44·3′	86°7·9′	2760	2·05	2·9	3	(1·71 (0·72))	5·75 ± 0·47 (241 ± 20) w
(6)	0°43·8′	86°8·0′	2760	2·03	2·8	3	(1·71 (0·72))	6·18 ± 0·48 (259 ± 20) w
(7)	0°43·5′	86°8·0′	2590	2·03	2·5	2	(1·71 (0·72))	17·03 ± 1·00 (715 ± 22) w
(8)	0°43·1′	86°8·1′	2600	2·04	2·7	2	(1·71 (0·72))	⩾ 8·80 (⩾369) w
55(1)	0°42·2′	86°11·5′	2670		2·4	3	(1·71 (0·72))	1·90 (80)
(2)	0°42·0′	86°11·6′	2720		2·4	3	(1·71 (0·72))	3·24 (136)
(3)	0°41·7′	86°11·6′	2720	1·96	2·4	3	(1·71 (0·72))	1·03 (43)
56(1)	0°40·7′	86°11·5′	2770	2·06	2·4	3	(1·71 (0·72))	2·73 (104)
(2)	0°40·5′	86°11·5′	2730	2·06	2·4	3	(1·71 (0·72))	3·06 (128)
57(1)	0°36·6′	86°11·9′	2720		2·4	3	(1·71 (0·72))	3·35 (140) v
(2)	0°37·0′	86°12.1′	2720	2·06	2·4	3	(1·71 (0·72))	4·06 (170) v
58	0°44·3′	86°7·3′	2730	2·05	2·7	3	(1·71 (0·72))	2·15 ± 0·16 (90 ± 7) w
59(1)	0°41·7′	86°8·4′	2720	2·04	3·0	2	(1·71 (0·72))	1·19 ± 0·07 (50 ± 3) w
(2)	0°41·5′	86°8·3′	2720	2·04	2·8	2	(1·71 (0·72))	1·32 ± 0·07 (52 ± 3) w
(3)	0°41·2′	86°8·4′	2640	2·03	2·8	2	(1·71 (0·72))	⩾ 5·94 (⩾249) w
(4)	0°40·9′	86°8·7′	2680	2·04	3·0	2	(1·71 (0·72))	2·69 ± 0·11 (113 ± 5) w
(5)	0°40·6′	86°8·9′	2740	2·04	3·3	2	(1·71 (0·72))	⩾ 2·79 (⩾117) w
(6)	0°40·3′	86°9·2′	2690	2·04	3·0	2	(1·71 (0·72))	⩾ 2·12 (⩾89) w
60	0°38·3′	86°10·6′	2740	2·06	2·4	3	(1·71 (0·72))	9·20 (386)
61(1)	0°40·3′	86°10·2′	2740	2·06	2·4	3	(1·71 (0·72))	1·78 (75)
(2)	0°40·6′	86°9·9′	2750		2·4	3	(1·71 (0·72))	1·32 (55)
(3)	0°40·8′	86°9·6′	2730		2·4	3	(1·71 (0·72))	1·08 (45)
62	0°35·1′	86°11·1′	2720	2·04	*	2	1·73 (1·73)	4·63 ± 0·45 (194 ± 19) vw
63(1)	0°39·1′	86°12·0′	2740	2·04	2·5	3	(1·71 (0·72))	— 0·01 ± 0·03 (0 ± 1) vw
(2)	0°38·8′	86°12·0	2720	2·04	2·9	3	(1·71 (0·72))	5·62 ± 0·36 (236 ± 15) w

(3)	0°38·6′	86°11·9′	2720	2·04	3·0	3	(1·71 (0·72))	6·11 ± 0·36 (256 ± 15) iw
(4)	0°38·2′	86°11·9′	2650	2·04	2·9	3	(1·71 (0·72))	7·00 ± 0·22 (294 ± 9) w
(5)	0°37·9′	86°11·8′	2640	2·03	3·0	2	(1·71 (0·72))	11·88 ± 0·75 (497 ± 31) w
(6)	0°37·4′	86°11·7′	2660	2·04	2·9	3	(1·71 (0·72))	9·53 ± 0·52 (399 ± 22) w
(7)	0°37·0′	86°11·5′	2720	2·04	2·7	3	(1·71 (0·72))	5·31 ± 0·30 (223 ± 13) w
(8)	0°36·3′	86°11·3′	2720	2·04	2·6	3	(1·71 (0·72))	⩾ 2·00 (⩾84) w
(9)	0°36·4′	86°11·3′	2720	2·03	2·7	3	(1·71 (0·72))	1·87 ± ·06 (79 ± 1) w
(10)	0°36·0′	86°11·1′	2720	2·04	2·8	3	(1·71 (0·72))	⩾ 8·49 (⩾355) w
(11)	0°35·8′	86°11·0′	2720	2·03	2·8	3	(1·71 (0·72))	⩾ 7·72 (⩾324) w
64(1)	0°35·7′	86°10·9′	2720		2·4	3	(1·71 (0·72))	12·72 (533) v
(2)	0°35·6′	86°10·9′	2720		2·4	3	(1·71 (0·72))	19·00 (796) v
(3)	0°35·4′	86°10·9′	2720	2·05	2·4	3	(1·71 (0·72))	8·59 (360) v
65(1)	0°34·2′	86°10·6′	2720		2·4	3	(1·71 (0·72))	8·99 (376)
(2)	0°34·1′	86°10·5′	2720	2·05	2·4	3	(1·71 (0·72))	10·55 (442)
66(1)	0°33·1′	86°11·2′	2780	2·05	2·4	3	(1·71 (0·72))	2·31 (97) i
(2)	0°32·8′	86°11·1′	2780	2·05	2·4	3	(1.71 (0·72))	1·92 (80) i
67(1)	0°34·5′	86°8·7′	2720	2·04	2·7	2	(1·71 (0·72))	13·01 ± 0·68 (546 ± 29) w
(2)	0°34·1′	86°8·7′	2720	2·04	2·9	2	(1·71 (0·72))	10·12 ± 0·09 (421 ± 46) w
(3)	0°33·7′	86°8·8′	2700	2·03	2·9	3	(1·71 (0·72))	7·10 ± 0·44 (298 ± 46) w
(4)	0°33·4′	86°8·8′	2730	2·04	2·8	3	(1·71 (0·72))	5·24 ± 0·36 (215 ± 15) iw
(5)	0°33·0′	86°8·9′	2760	2·04	2·8	3	(1·71 (0·72))	3·62 ± 0·22 (152 ± 9) iw
(6)	0°32·8′	86°9·0′	2780	2·04	2·7	3	(1·71 (0·72))	4·36 ± 0·26 (183 ± 11) iw
(7)	0°32·4′	86°9·0′	2780	2·04	2·8	3	(1·71 (0·72))	8·43 ± 0·46 (353 ± 19) iw
68	0°34·0′	86°13·8′	2720	2·03	11·2	5	1·78 (0·75)	6·91 ± 0·31 (290 ± 13) w
69(1)	0°32·0′	86°10·7′	7770	2·03	2·9	3	(1·71 (0·72))	6·29 ± 0·17 (264 ± 7) w
(2)	0°31·8′	86°10·7′	2750	2·03	2·9	3	(1·71 (0·72))	7·66 ± 0·22 (322 ± 9) w
(3)	0°31·6′	86°10·7′	2750	2·03	2·9	3	(1·71 (0·72))	⩾ 9·31 (⩾390) w
(4)	0°31·1′	86°10·7′	2730	2·03	2·1	2	(1·71 (0·72))	⩾30·28 (⩾1270) w
(5)	0°30·8′	86°10·7′	2750	2·04	2·9	3	(1·71 (0·72))	8·49 ± 0·46 (356 ± 19) w
(6)	0°30·5′	86°10·6′	2790	2·04	2·9	3	(1·71 (0·72))	8·04 ± 0·44 (337 ± 18) w
(7)	0°30·2′	86°10·5′	2760	2·04	2·8	3	(1·71 (0·72))	⩾ 7·99 (⩾334) w
(8)	0°29·9′	86°10·4′	2780	2·04	2·9	3	(1·71 (0·72))	7·57 ± 0·22 (318 ± 9) w

T is bottom water temperature (°C).
P is the estimated sediment penetration (m) of lowermost probe used for temperature gradient measurements.
N is number of thermistors used for sediment temperature gradient measurements.
K is the thermal conductivity in 10^{-3} cal °C^{-1} cm^{-1} s^{-1} (W m^{-1} °K^{-1}). Outer parentheses indicate values assumed from measurements on nearby piston cores.
Q is heat flow in 10^{-6} cal cm^{-2} s^{-1} (10^{-3} W m^{-2}). Measurements made by Woods Hole Oceanographic Institution are indicated by a w with values and probable errors of these computed according to methods as in Von Herzen & Anderson (1972). An additonal uncertainty of ± 10 per cent in the heat flow values results when the thermal conductivity is assumed. Unless otherwise annotated, minimum values result from a tilt greater than our instruments can measure (> 30 degrees off vertical). i, v represent measurements in which the instrument or vessel respectively were acoustically navigated.
*Station 62 - some penetration beyond 10 m was indicated. The difficulty of fixing the actual penetration depth is discussed in the text.
† Only one thermistor on scale.
§ Water depth in corrected metres.

it was possible to navigate the ship only acoustically. The position of the heat-flow measurement was determined from an empirical relation between wire angle and horizontal distance to the relay transponder on stations where relay transponder navigation was available. We estimate the uncertainty of station positions determined by this method at ± 400 m. No accoustic navigation was available for the remaining 48 heat-flow stations. These stations were located by a combination of satellite navigation, bathymetry (Klitgord & Mudie 1974), ranges to a single bottom transponder, and computer-generated dead reckoning. The quality of these positions varies appreciably but the average is probably no better than ± 1 km. Our ability to locate all our stations is somewhat better in latitude than longitude due to the east–west grain of the bottom topography. On most stations multiple penetrations were made. The relative positions between individual measurements on these stations are generally determined better than ± 300 m.

As many as 11 heat-flow measurements over horizontal distance of greater than 7 km were made during a single station with the multipenetration probe described in Von Herzen & Anderson (1972).

In addition to our heat-flow measurements, two near-bottom horizontal water temperature profiles were obtained utilizing the apparatus depicted in Fig. 3. The lower thermistor was generally maintained within 10 m of the sea floor. The temperature of each of the lower two thermistors was recorded twice every 30 s at different sensitivities. The upper thermistor temperature was recorded at high sensitivity once

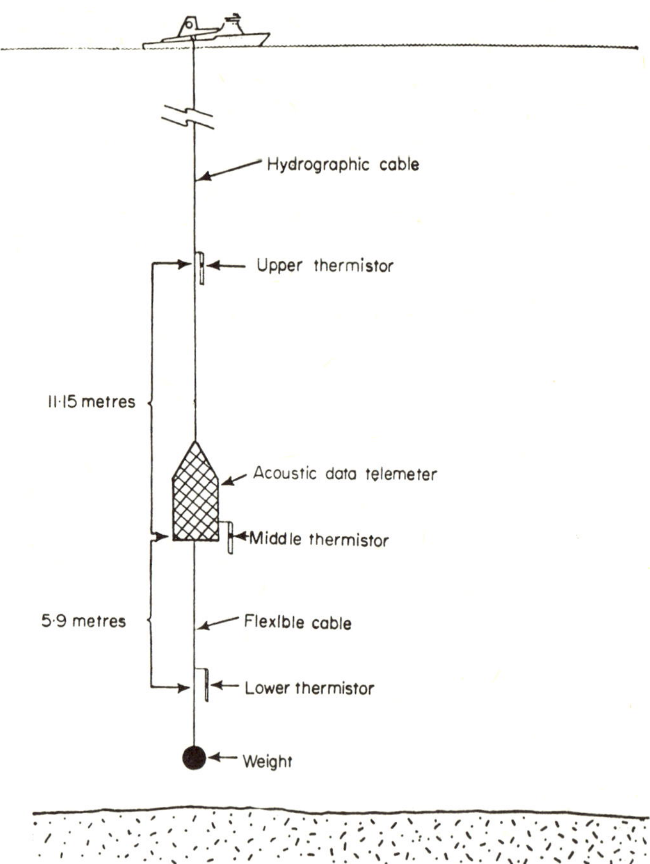

FIG. 3. Apparatus used in measuring horizontal water temperature profiles. Altitude above the bottom was maintained by echo sounding from the telemeter.

every 30 s. The potential temperatures have a precision of about ± 0·003°C at high sensitivity and ± 0·015°C at low sensitivity. Only the high sensitivity records are plotted in Fig. 8.

Observations

A synthesis of the heat-flow and water temperature data (Figs 2, 5, 6, 7, and 8) lead to the following observations: (1) conductive heat-flow measurements vary regularly from relatively low values averaging about 2×10^{-6} cal cm^{2-} s^{-1} (HFU) to high values averaging about 12 HFU, with extremes of individual measurements ranging from zero to greater than 30 HFU. A significant north–south variation in heat flow exists with a modulation wavelength of approximately 6 ± 1 km. (Fig. 5). It is clear from a close examination of Fig. 5, and particularly from the individual profiles in Fig. 6, that the heat-flow pattern is complicated. However, heat-flow maxima are marked by an absence of low values, and the heat-flow minima are almost as well defined. (2) The average heat flow gradually increases with distance south of the ridge

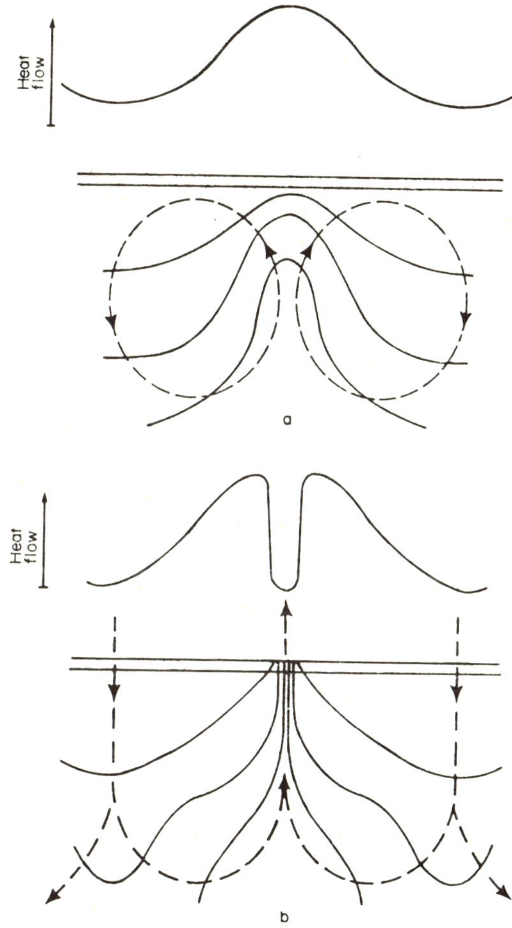

Fig. 4. Sketches illustrating the effect of hydrothermal circulation on geothermal gradients measured at the sea floor. Heavy solid lines represent isotherms, and dashed lines with arrows show the general pattern of convection. (a) Convection in crustal rocks below an impermeable sediment blanket. (b) Convection which penetrates the sediment cover allowing for exchange with the bottom waters. The effect on the conductive heat-flow pattern is shown above each sketch.

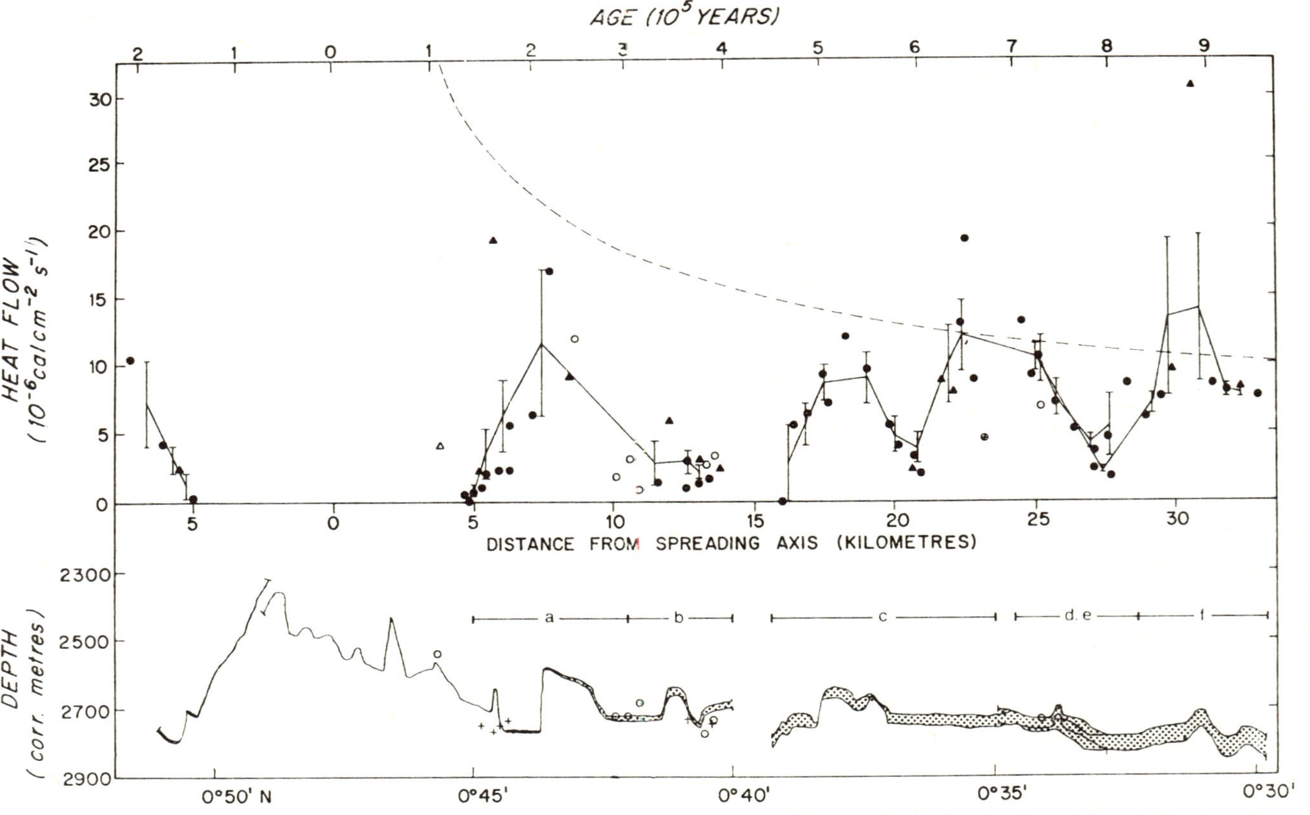

FIG. 5. Heat-flow data, topography and sediment thickness (stippled pattern) from the 4 profiles shown by dashed lines in Fig. 2 (projected north–south). The vertical exaggeration is 12·5:1. Closed circles are individual heat-flow values which were within 2 km of profile. Crosses represent depth of heat-flow measurement deviating 20 m or more from the plotted topography. Triangles indicate minimim values of heat flow. Open circles and triangles represent heat-flow stations more than 2 km away from the profile track. The circle with a cross in it is station 62. It is discussed in the text and is not used in the averages. The dashed line is the theoretical heat flow (from Sclater & Francheteau 1970) for an 85 km thick lithosphere, 1250°C on its lower boundary, an internal heat generation of $2{\cdot}0 \times 10^{-14}$ cal cm^{-3} s^{-1} and an average thermal conductivity of $6{\cdot}9 \times 10^{-3}$ cal cm^{-1} s^{-1}°C^{-1}. The solid lines connect averages of the measured heat flow values (vertical bars are ± standard error) averaged over 2 km intervals every 1 km along a profile.

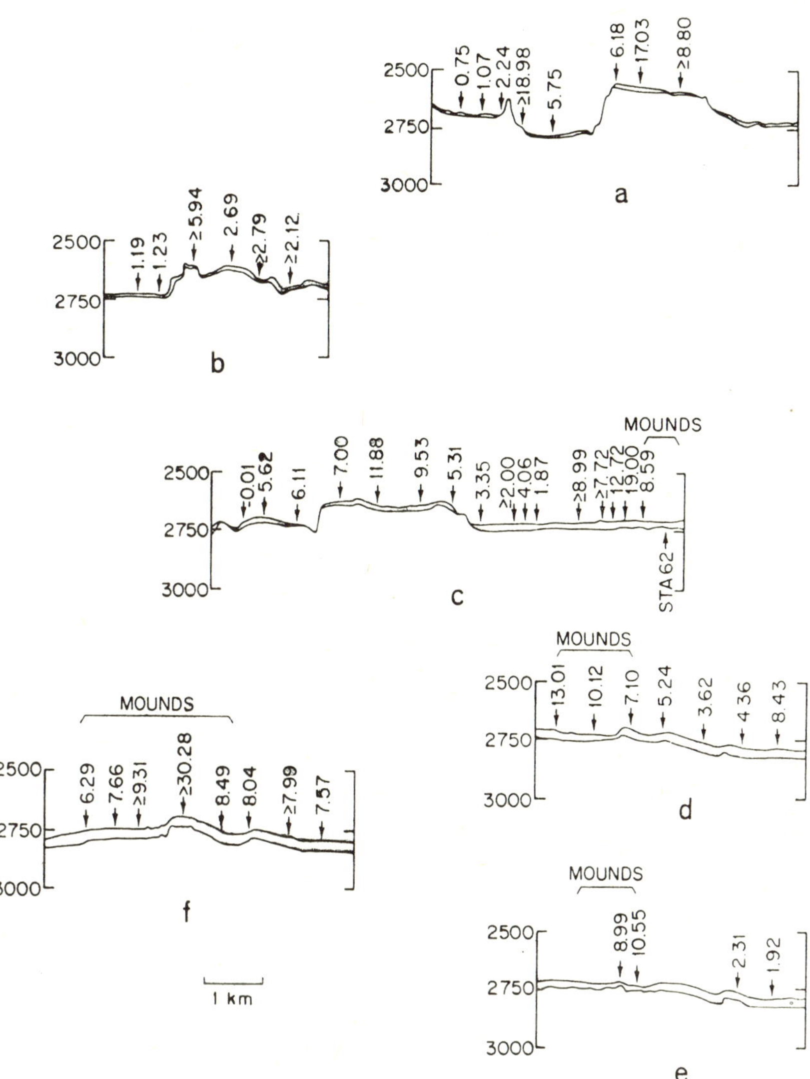

FIG. 6. Sediment and basement bathymetry from Klitgord & Mudie (1974). Each bathymetric profile is from an SIO deep-tow instrument package survey line that passes near the heat-flow station. The position of each profile is shown in Fig. 5. The location of mounds refers to the fields of sediment mounds discussed in the text. In general these mounds are too small to be seen at the vertical scale used in the figure. Numbers above the arrows are heat-flow values in 10^{-6} cal cm^{-2} s^{-1}. Depths are in corrected metres with a vertical exaggeration of 4:1. Since the heat flow stations were not located exactly on the bathymetric survey lines the relative geographical positions of the heat-flow stations have been adjusted to best reflect the actual bathymetry at the site of the heat-flow station. Only heat-flow measurements near each bathymetric profile are shown. As a result not all our measurements are illustrated.

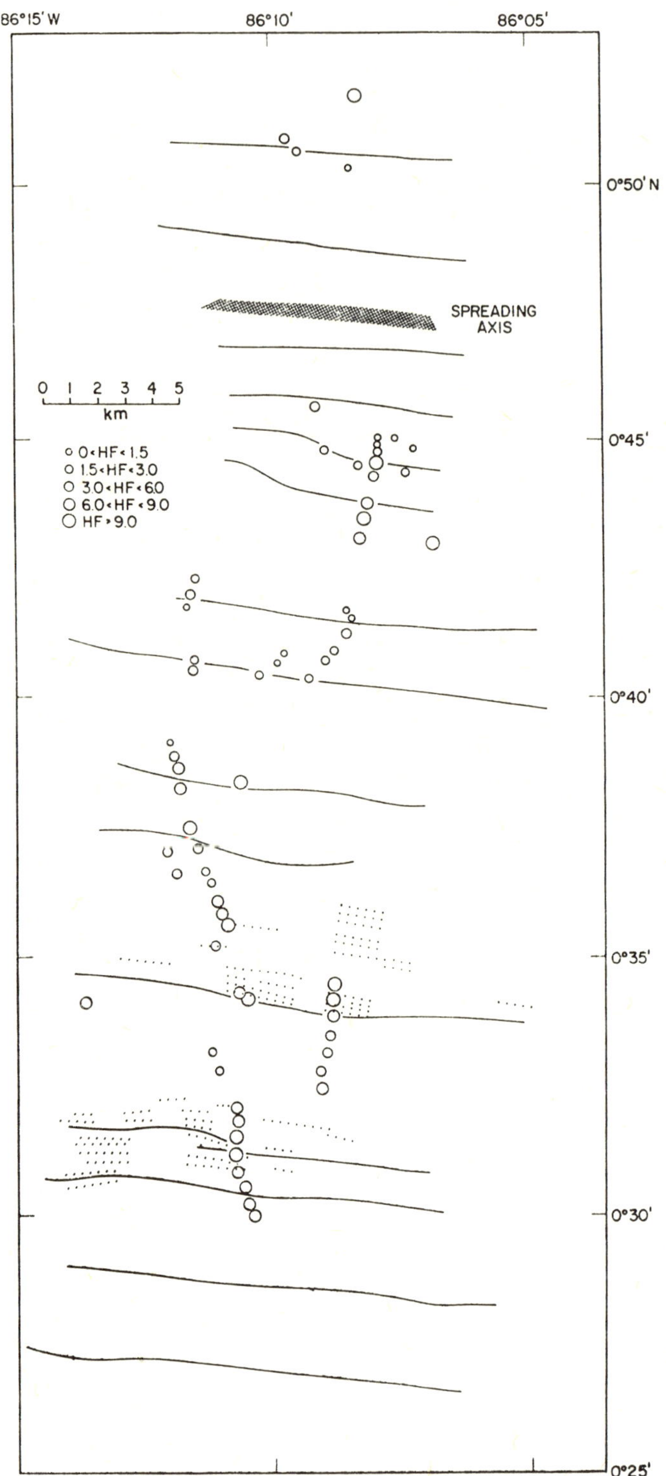

FIG. 7. A sketch map of our survey area. The heavy east–west trending dark lines mark the positions of fault scarps indentified by Klitgord & Mudie (1974). If these faults are zones of high permeability they could be important in the hydrothermal system. The small dots represent the general location of sediment mounds. Note the correlation of these mounds with areas of high heat flow.

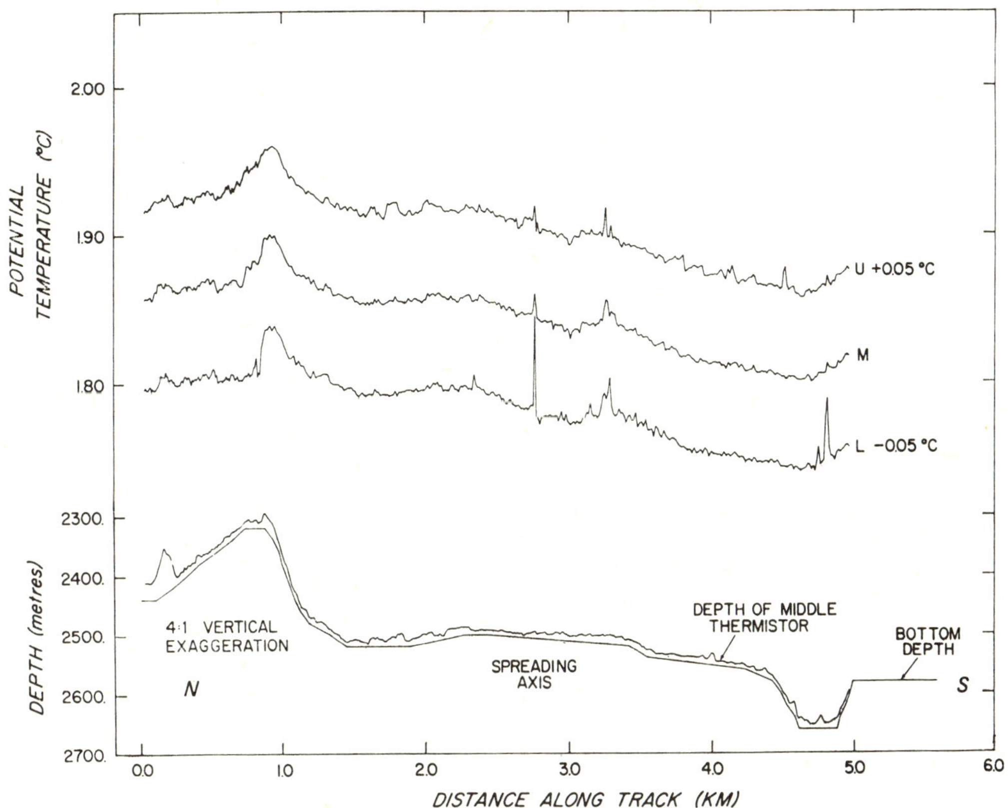

Fig. 8. Bottom water horizontal temperature profile, No. 1 (Fig. 2). U + 0·05°C is the potential temperature of the upper thermistor plus 0·05°C. M is the potential temperature of the middle thermistor and L −0·05°C is the lower thermistor potential temperature minus 0·05°C. The average horizontal speed was 1·28 km hr^{-1}. The thermistor depth trace shows much more roughness than the bottom because it is derived by subtracting a detailed thermistor altitude signal from less detailed bottom depth data. Therefore, much of the roughness in the thermistor depth is actually roughness of the bottom.

axis, whereas the theoretical heat flow, based on conductive cooling models (McKenzie 1967; Sclater & Francheteau 1970), decreases rapidly with distance from the axis (Fig. 5). (3) Near the spreading centre the average of the measured values is much less than the theoretical value but at the southern end of our survey this average clearly approaches the theoretical curve (Fig. 5). (4) The heat-flow minima tend to be on or near topographic lows and the heat-flow maxima are associated with either scarps, topographic highs, or fields of sediment mounds (Fig 5, 6 and 7). The mounds are found only in the southern half of our area an average about 5–10m high with a 25m radius, as determined from deep tow echo sounding and side-looking sonar (Klitgord & Mudie 1974). (5) On station 62 (Fig. 9) the observed thermal gradient extrapolated to the bottom water temperature yields an apparent penetration of 18·3 m for our 10 m piston core. Although there were other indications of some superpenetration, 8·3 m seems unlikely. (6) The average of all our measured heat-flow values is greater than 5·9 HFU. This is in good agreement with Sclater & Klitgord (1973). It is much higher than the world-wide average of conductive heat flow (1·5 HFU, Lee 1970). It is also significantly higher than has been measured at other spreading centres on crust of similar age (Langseth & Von Herzen 1971). (7) Several significant bottom water temperature anomalies can be seen in Fig. 8. The largest temperature anomaly is observed over the regional topographic high, an east–west ridge located at 0°49′N latitude. This ridge apparently represents a boundary between different water types above and below the ridge crest, as observed by Sclater & Klitgord (1973) and Detrick *et al.* (1974). The other water temperature anomalies cannot be explained by bottom temperature structure or currents. Anomalies located at 2·76, 3·26 and 4·80 km along our track are significantly above the noise level of our measurements and appear on more than one sensor.

Interpretation

Conductive heat-flow measurements in a hydrothermal area

(a) *General.* Since we will use hydrothermal circulation to explain our observations it is worthwhile to discuss what we should expect from conductive heat-flow measurements in a hydrothermal area. Thermal gradients can be radically distorted by hydrothermal circulation. The sketches in Fig. 4 illustrate the effect of a hypothesized water circulation pattern on sub-surface isotherms. When the circulating system is confined to permable rock which is overlain by an impermeable sediment cap (Fig. 4(a)), larger thermal gradients will be measured above the rising limbs than above descending limbs. Thermal gradients should grade evenly between the limbs. But if the circulating system has free discharge and re-charge (Fig. 4(b)) thermal gradients near the axis of both rising and descending limbs can be quite small. In this case, since the sediment temperatures near the rising limb are much higher, it is still possible to separate these two causes. The pattern of heat flow between the limbs depends on the character of the system but we should find a heat-flow maximum adjacent to the discharge vent. More generally, the net effect of hydrothermal waters venting through the sea floor is to lower the regional geothermal gradient.

(b) *Heat-flow minima.* Since the Galapagos spreading centre appears to be actively spreading and thus is a volcanically active region, high temperature rock and magma can be expected near the surface. If conduction is the dominant heat transfer mechanism, we should find a consistently high geothermal gradient; instead we find regularly spaced minima. Occasional low gradients might be produced by sediment slumping (Von Herzen & Uyeda 1963) but the sediment distribution on the spreading centre is fairly uniform. Most steep basement slopes (fault scarps) tend to have thinner sediment cover, as expected; the thin cover also extends to the base of these slopes, suggesting that these may be active fault scarps and that sediment slumping is minimal. Corrections for the effects of topography (e.g. Lachenbruch 1968), although small, would generally tend to accentuate the observed modulation. Similarly, thermal refraction caused by the conductivity contrast between rock and sediment is small because of the uniform sediment cover. These corrections were derived for regions in which thermal conduction is dominant and it would be inappropriate to apply them if thermal convection might be important. Therefore, no attempt has been made to modify our heat-flow values for these effects. We have seen no evidence in the bottom sediment gradients for significant temporal changes in bottom water temperature. Small variations in the temperature of the bottom water with periods shorter than a few days or longer than a few years might be difficult to detect but would not significantly affect our discussion.

The pattern of heat-flow variations, and the lack of evidence for other causes, seems to require hydrothermal circulation of sea water in the crustal rocks. Lacking any high sediment temperatures that could mark a hydrothermal vent, we conclude that the areas of low heat flow represent either (1) regions where the sea water is entering the crustal rocks, if the upper boundary of the system is still open, or (2) areas above the downgoing limb of a closed convective system.

(c) *Magnitude of hydrothermal cooling.* Using the same theoretical heat-flow model illustrated in Fig. 5 (Sclater & Francheteau 1970, equation (23)), lithospheric creation along the Galapagos spreading centre results in a continuous heat loss of 1100 cal s^{-1} cm^{-1} of ridge length. Of this total more than 330 cal s^{-1} cm^{-1} is released through crust less than 35 km from the axis. This is a minimum value because it is based on a conductive cooling model, and the combination of conduction and hydrothermal convection would remove heat more efficiently. The heat loss that would result if the upper 500 m of the crust were produced by extrusive lavas would account for only about 15 cal s^{-1} cm^{-1}. In this calculation we assume the lava is solidified causing each gramme to release 100 cal of latent heat and then cooled from 1250°C to 0°C which releases an additional 300 cal of sensible heat. The average of our conductive heat-flow measurements within 35 km of the axis gives 36 cal s^{-1} cm^{-1} of ridge length. This implies that more than 80 per cent of the geothermal heat released in our survey area escapes through hydrothermal vents.

(d) *Heat flow* vs. *age*. The increase of conductive heat flow southwards from the spreading centre (Fig. 5) suggests that the free exchange of water and therefore the magnitude of the hydrothermal component of heat loss decreases as sediment accumulates. Only after the sediment blanket is sufficiently thick that it begins to form an impermeable cap, as is apparently the case near the southern end of our survey, does thermal conduction become the dominant heat transfer mode at the sediment/water interface. The permeability of these ridge sediments is not known, but they apparently remain significantly penetrable to fluid flow to thicknesses of at least 50 m.

Increasing heat flux with age has been observed on other active ridges (Talwani *et al.* 1971; Hyndman & Rankin 1972). However, on the Galapagos spreading centre the measured heat flow approaches the theoretical value at a younger age. This is probably a result of the higher sedimentation rates and smoother topography, causing the upper boundary of the hydrothermal system to be sealed more quickly.

The observed thermal gradients imply the existence of large horizontal temperature variations and furthermore, that the average temperature of near surface rocks increases with age. This is contrary to what one might normally expect from a simple model.

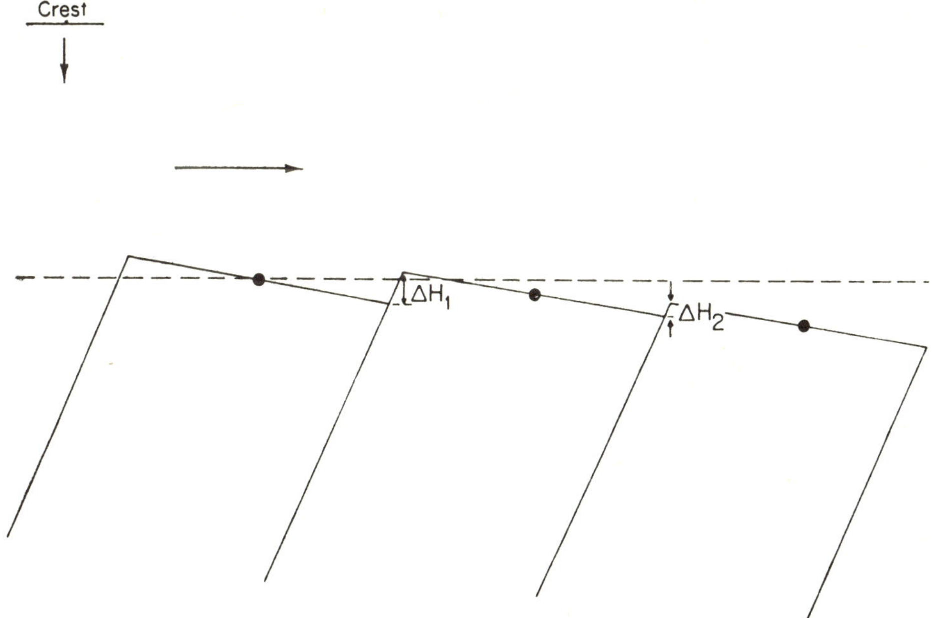

Fig. 9. Schematic representation of crestal block faulted topography. The average depth of each block increases with age whereas the scarp offset (ΔH) decreases. Dashed line indicates horizontal reference level.

(e) *Effect of hydrothermal circulation on crestal topography*. The crestal zone of the Galapagos spreading centre is dominated by faulted blocks tilted outward with scarps facing the crest (Fig. 9) (Klitgord and Mudie 1974). Scarps on blocks 0·25 My are offset (Δ H) an average of about 100 m, whereas those in 1·0 My old crust have a ΔH of only 50 m. Further, during this same period, the blocks increase in depth from approximately 2600–2700 m. Classical thermal contraction (Sclater & Francheteau 1970), lithospheric thickening (Williams & Poehls, in preparation) and the relaxation of dynamic uplift (Lachenbruch 1973) all predict this general decrease in elevation as the young crust ages.

Hydrothermal circulation provides us with an additional mechanism for generation of the crestal elevation changes. Circulation of sea water in the crustal blocks produces hydration of the oceanic crust. As the crustal blocks spread from the intrusion zone, portions of the blocks are reheated from below. This can produce metamorphic upgrading of the hydrous assemblage which may account for an additional overall elevation decrease. In general, low grade metamorphic rocks have larger volume changes (Δ V) of reaction than higher grade metamorphic rocks. This can be demonstrated by

Table 2.

Standard temperature and pressure, molar volumes for the reaction:

$$\underset{\text{Prehnite}}{5\,Ca_2Al_2Si_3O_{10}(OH)_2} + \underset{\text{Chlorite}}{Mg_5Al_2Si_3O_{10}(OH)_8} + \underset{\text{Quartz}}{2SiO_2} \rightleftharpoons \underset{\text{Epidote (Zoisite)}}{4Ca_2Al_3Si_3O_{12}(OH)} + \underset{\text{Tremolite}}{Ca_2Mg_5Si_8O_{22}(OH)_2} + \underset{\text{Water}}{6H_2O}$$

Component	Volume (cm³/mole)
Prehnite	142·20 §
Chlorite	213·01 ‡
Quartz	22·69 †
Epidote (Zoisite)	136·50 †
Tremolite	272·95 †
Water	18·07 †
Volume change of solids for the reaction *	= −42·05

* All volumes treated as constants.
† From Robie *et al.* (1966).
‡ Calculated from unit-cell dimensions of prochlorite from Steinfink (1958, 1961, 1962).
§ Calculated from unit-cell dimensions from Preisinger, (1965).

comparing volume changes from the upgrading of prehnite + chlorite + quartz to epidote + tremolite + water. This is a typical active ridge metamorphic suite and the reaction produces a volume change of −4 per cent (Table 2), at STP. Thermal expansion and compressibility coefficients for these universals are unknown but the low pressures and temperatures involved should make the STP calculation accurate to within ± 10 per cent.

Thus, if the hydrated crustal blocks are 5 km thick, less than 50 per cent of the crust undergoing this reaction would produce the entire observed elevation decrease assuming that all of the volume change goes into elevation change. Similarly, the scarp offsets may be maintained to some extent by the existence of large horizontal temperature differences within blocks and between adjacent blocks. We would expect the offset to be reduced as the crust ages and these temperature differences decrease. It is clear that hydrothermal circulation and the resulting chemical alterations can have a substantial effect on crestal topography. Until we have a better understanding of the thermal, petrologic and chemical properties of the oceanic lithosphere it will be difficult to separate these effects from those of classical thermal contraction, lithospheric thickening and dynamic uplift in controlling crestal topography.

(f) *Geographic distribution and magnitude of heat flow*. The geographical distribution of our data is insufficient to determine whether the observed variations in heat flow largely are two-dimensional, like the topography, or three-dimensional. Furthermore, a theoretical model with inherent uncertainties is required to estimate the magnitude of heat released by hydrothermal circulation on the spreading centre. We can only say that this circulation appears to control the near-axis conductive heat-flow pattern and that this pattern bears only a limited relationship to the actual heat flux. In regions where the hydrothermal system still has a free exchange with the bottom water, the theoretical models provide a more realistic means to estimate the heat flux than the measured values of conductive heat flow. Even where the sediment forms an impermeable upper layer, hydrothermal circulation may persist in the crustal rocks below the sediment, which clearly requires a large number of measurements spaced significantly closer than the scale of the circulation pattern to obtain a reliable value of the regional heat flux.

Character of the hydrothermal system

(a) *Hydrothermal vents.* Reasonably clear evidence for hydrothermal vents is presented from the bottom water temperature anomalies of Fig. 8. With expressions developed empirically by Rouse, Yih & Humphreys (1952), and discussed by Batchelor (1954) and Turner (1969), it is possible to make a rough estimate of the heat transfer by steady-state thermal plumes of this size. The rate of heat loss F represented by a two-dimensional plume is

$$F = \left[\frac{zg'}{2.6} \exp(41x^2/z^2)\right]^{3/2} \quad \text{cal s}^{-1}\,\text{cm}^{-1}$$

and for a three-dimensional plume

$$F = \left[\frac{z^{5/3}g'}{11} \exp(71r^2/z^2)\right]^{3/2} \quad \text{cal s}^{-1}$$

where $g' = \alpha g(T-T_0)$ with g the gravitational acceleration and T the anomalous temperature observed at distance z above the bottom and at distance x or radius r from the vertical axis of the plume. The coefficient of thermal expansion α is taken for the fluid medium at ambient temperature T_0. These equations assume that the composition of the plume water is not significantly different than the bottom water, that the plume is from a point or line source, and that is not significantly affected by bottom currents.

We estimate the anomaly near 2·76 km in Fig. 8 at 1600 $\pm$ 300 cal s^{-1} for a three-dimensional plume or 13·4 $\pm$ 3·6 cal s^{-1} cm^{-1} for a two-dimensional plume. The uncertainties derive from those associated with the temperature data. We have assumed that the profile was directly above the source, so that the values of F should be considered minimal.

Exrusive lava flows can be expected to heat the water that quenches and cools them. However, in an area as small as the one covered by our water temperature profile lava flows should be very infrequent with the lava cooling quickly. Therefore, we feel that the observed temperature anomalies are more probably caused by thermal plumes rising from hydrothermal vents.

The correlation between heat-flow maxima and the occurrence of small sediment mounds (Figs 5 and 6(d), (e) and (f) suggests that these mounds are hydrothermal vents. They seem to have no expression in the basement topography and they tend to be lineated as though they were above a hydrothermal fissure in the basement rock (Klitgord & Mudie 1974). We can only speculate about the character of the flow from these vents. It could be continuous or periodic. The fact that these mounds appear only in areas of high heat flow implies the elements of the hydrothermal system remain nearly fixed with respect to the underlying crust for long periods of time. At the observed sedimentation rates, it would take more than one hundred thousand years to bury one of these mounds after it became inactive. Interestingly, our station 62 has combined the qualities of a relatively low thermal gradient and high sediment temperatures (Fig.10) expected of a measurement near an active hydrothermal vent (Fig. 4(b)) and is located within one of these fields of sediment mounds. This explanation of the unusual temperature–depth relation seen in Fig. 10 may be more probable than excessive superpenetration of the coring apparatus.

The heat transfers we have calculated for some of the plumes, deduced from Fig. 8, represent a significant percentage of the total regional heat loss. Yet, just a few metres above the bottom, the predicted and observed temperature anomalies are less than 0·1 °C. Larger temperature anomalies would be expected closer to the vent. Also if the vented water were warmer and more dense (due to chemical differences) than the surrounding sea water, it might collect in pools such as those found in the Red Sea.

(b) *The deep hydrothermal system.* The circulation pattern could be controlled by one or more of the following physical properties of the system: (1) variation in the strength of heat sources near the base of the crust, (2) bottom topography, (3) discrete zones of

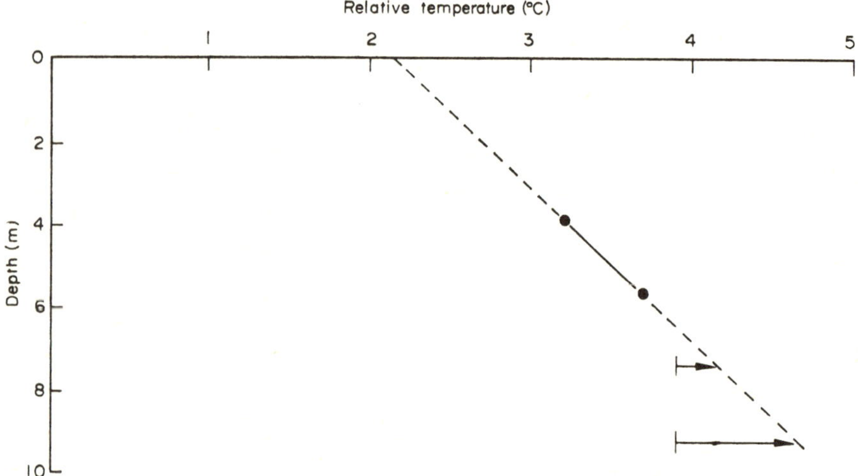

FIG. 10. Relative temperature versus depth for station 62. Depths are from the bottom of the piston corer weight stand. Any amount of superpenetration would increase these depths. Temperatures are relative to the bottom water temperature (2·04°C ± ·01°C). There were four thermistors mounted on the core barrel. The lower two, whose position are indicated by arrows, were sensing relative temperatures which were greater than 3·9°C, the highest our instrumentation was set to record.

high permeability, or (4) cellular convection. Only limited conclusions may be inferred from the heat-flow data. Hot intrusives rising to near the base of the oceanic crust, spaced at about 6 km along our track could force the entire system. The pattern of steep scarps suggests that the surface might have undergone north–south extension and block faulting which could be a by-product of deep intrusive east–west trending dikes associated with sea-floor spreading. The occurrence of heat-flow maxima near topographic highs and heat-flow minima near topographic lows may imply topographic control as proposed by Lister (1972). This correlation between topography and the heat-flow pattern, although approximate, appears to be more than a coincidence. The data seem to indicate that the convection is not entirely controlled by the major vertical faults. The heat-flow maxima and minima on the Galapagos spreading centre, are not always located near the steep scarps (faults?) (Fig. 7). This is in contrast to the pipe model of Bodvarsson & Lowell (1972) and to observations in the high temperature areas of Iceland (Bodvarsson 1961) and most continental geothermal areas (McNitt 1965) where the hydrothermal activity at the surface is in close proximity to dike contacts and vertical faults (zones of high vertical permeability). This could imply that the young oceanic crustal rocks on this spreading centre, and perhaps elsewhere, have a finer scale of permeability than rocks in continental geothermal areas.

(c) *Cellular convection.* Cellular convection through permeable rock was first suggested by Elder (1965) as being the mechanism by which large amounts of heat might be removed from active ocean ridges. The apparent regularity of the modulation in heat flow has led us to consider this mechanism.

Convection in a homogeneous medium is theoretically possible whenever the critical Rayleigh number is exceeded. This dimensionless parameter R is defined by

$$R = k\,\alpha\, g\,\Delta TH/\kappa\nu$$

where the properties of the fluid saturated medium are given by permeability k, thickness h, upper to lower surface temperature difference ΔT, and thermal diffusivity κ. The properties of fluid are defined by coefficient of thermal expansion α and kinematic viscosity ν. Erickson & Simmons (1969) and Lister (1972) discuss the magnitudes of these parameters and stress that the principal unknown is the permeability. It is clear that the Rayleigh number decreases away from the ridge crest because of the inevitable decrease in the temperature differences (ΔT) and average permeability (k). Local periodic changes in the strength of the heat source will also affect the ΔT and hence the Rayleigh number. In a closed system with a plane evenly-heated lower boundary and a plane isothermal upper boundary the critical Rayleigh number is near 40 (Lapwood 1948). However, when the upper boundary is not plane, any positive Rayleigh number is above the critical value. Despite this, low Rayleigh numbers would not be expected to produce large scale flows. Elder (1965, 1967) explains how problems can be studied utilizing computer-generated solutions or laboratory modelling (e.g. Hele-Shaw cell). Other studies have been reported by Lapwood (1948), Donaldson (1962), and Wooding (1959).

Hele-Shaw (1898) showed that cellular convective motion in a horizontal slab of homogenous saturated porous material of thickness H and lateral extent L is similar to that in a vertical cavity of width $a \ll H,L$ where the permeability $k = a^2/12$. The theory may be found in Lamb (1932, section 330). Although the motion in a Hele-Shaw cell is only two-dimensional most convective motion of this type perferentially forms two-dimensional rolls and there is the great advantage that the flow can be visualized.

Utilizing a Hele-Shaw cell similar to Elder's (1965), we have attempted to model the Galapagos spreading centre problem. From the few simple experiments so far conducted, only preliminary conclusions are possible (Willams 1974). (1) It appears the ridge topography has a strong influence on the position of the cellular convection, forcing the upgoing limbs under topographic highs as seen in Fig. 11 (a). This would produce a measured heat-flow pattern similar to Fig. 6(a) and (c). The downgoing limbs are forced under topographic lows (Fig. 11(b) and (c)). (2) In the Galapagos spreading centre model, the topography seems to have very little influence over the wavelength of the cellular convection. This may be because the amplitude of the topographic undulations is much less than the depth of circulation. (3) Most studies to date (Elder 1965; Donaldson 1962; Lapwood 1948), including ours, have found that Rayleigh cells are approximately equidimensional at Rayleigh numbers near the critical value. As the Rayleigh number increases the cells gradually become less constrained to the equidimensional shape (Fig. 12). The range of stable wavelengths in a porous medium has not yet been determined. With our observed modulation wavelength of 6 km, equidimensional Rayleigh cells would penetrate approximately 3 km and greater depths are certainly possible.

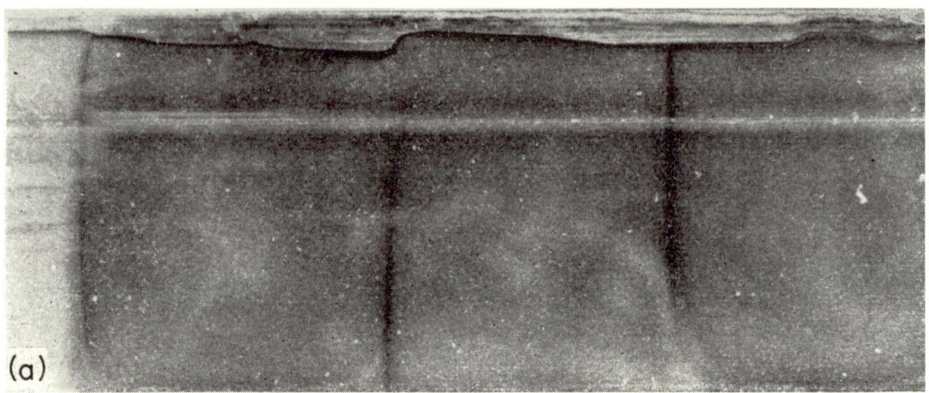

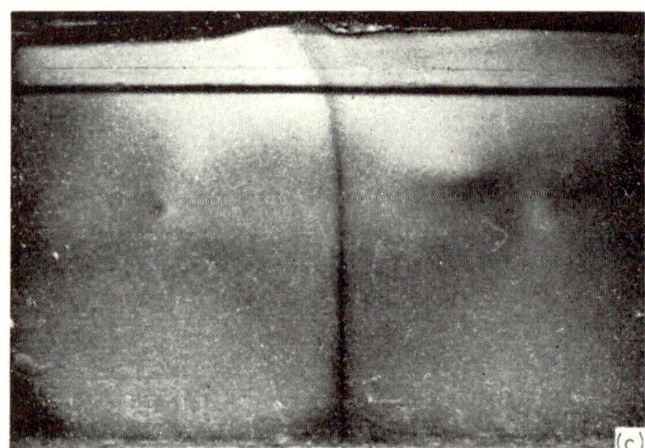

FIG. 11. Photographs illustrating the effects of topography on the motion in Hele-Shaw cell. The topography, simulating the Galapagos spreading centre, can be seen near the upper edge of the pictures and is scaled to a 3·75 km deep circulation system. The horizontal line just below this is part of the laboratory apparatus. The dark, nearly vertical lines are the rising and descending limbs of Rayleigh cells The three photographs show (a) a rising limb in the centre, flanked by two descending limbs. (b) A descending limb in a topographic low. (c) A rising limb under a topographic high. It can be seen that below the upper boundary the cells are not as rigidly constrained by the topography.

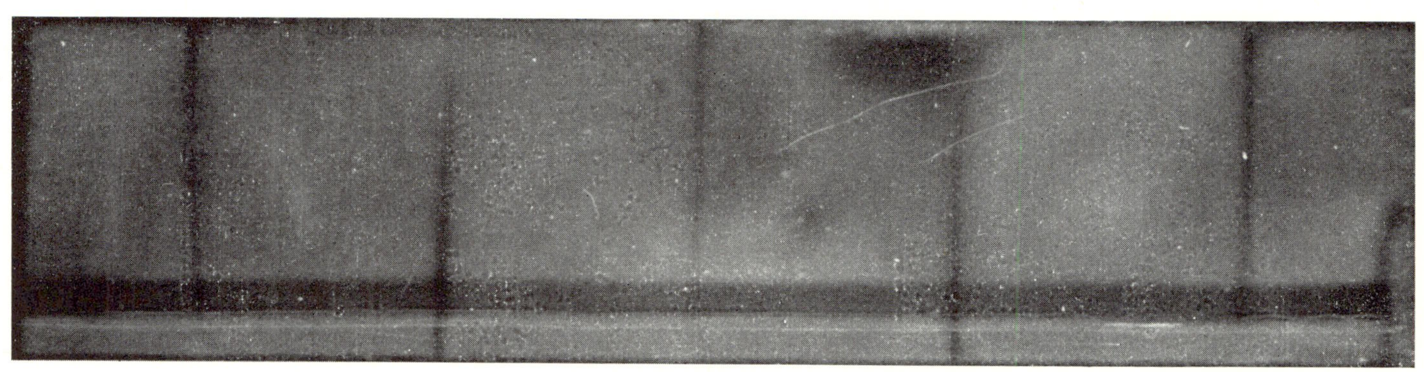

FIG. 12. Photograph of motion in a Hele-Shaw cell. There is no topography in this cell, i.e. the upper boundary is flat and closed. In this experiment the Rayleigh number was 360. From left or right the vertical dark columns are alternately rising (darkest and widest at bottom) and descending limbs.

These conclusions are primarily based on experiments with a Hele-Shaw cell modelling a completely closed and isothermal upper boundary. Preliminary experiments with a Hele-Shaw cell having an upper boundary open for free discharge and recharge have yielded similar results.

The motion in a Hele-Shaw cell is only two-dimensional. This is sufficient whenever the motion in the actual system is dominated by two-dimensional rolls. If, however, the motion were three-dimensional, the addition of another degree of freedom should tend to reduce the range of stable wavelengths but could also weaken the dependence between topography and wavelength.

This type of study, as mentioned earlier, cannot be conclusive in determining the controlling parameters of ocean ridge hydrothermal circulation. However, laboratory modelling and mathematical and numerical analysis can provide valuable insight. More work is obviously necessary in these areas before we can more accurately describe actual spreading ridge systems.

Conclusions

Lithospheric cooling along the Galapagos spreading centre at 86°W Longitude, as determined from surface heat-flow measurements, appears dominated by hydrothermal convection. The conductive heat-flow values alone do not provide an accurate estimate of the total heat flux, but if theoretical cooling models are accepted, hydrothermal convection accounts for approximately 80 per cent of the geothermal heat released along the Galapagos spreading centre through crust less than 1 My. As sediment accumulates, the rate of discharge from the hydrothermal system apparently decreases and the average of the observed heat-flow values begin to agree, in general, with theoretical conductive cooling models. The horizontal wavelength of the convective flow on the Galapagos spreading centre is 6 ± 1 km; water may penetrate several kilometers into the crust. The results of this study show the difficulties in resolving systematic patterns in the heat-flow distribution on spreading ridges from heat-flow surveys. If the Galapagos spreading centre is typical of these ridges, numerous, closely-spaced measurements with precise navigation combined with a relatively uniform sediment cover, appear to be necessary ingredients for recognition of the heat-flow pattern.

Acknowledgments

We would like to acknowledge the help of F. Dixon, J. Rogers, M. Hobart and V. Pavlicek who helped with, or took, many of the measurements presented in this paper. We are also indebted to Captain Bonham and the crew of the R/V Thomas Washington. We would also like to thank K. Klitgord, R. Detrick and J. Mudie for their invaluable assistance in gathering, reducing and interpreting the wealth of geophysical data gathered on legs 6 and 7 of 'South Tow'. Further, their critical review led to several improvements and additions to this paper. This research was supported by the National Science Foundation Grant GA-16078.

D. L. Williams and R. P. Von Herzen:
Woods Hole Oceanographic Institution
Woods Hole, Massachusetts 02543

J. G. Sclater:
Earth and Planetary Sciences,
Massachusetts Institute of Technology
Cambridge, Massachusetts 02139

R. N. Anderson:
University of California, San Diego
Marine Physical Laboratory of the
Scripps Institution of Oceanography,
La Jolla, California 92037

References

Anderson, R. N., 1972 Low heat flow on flanks of slow spreading mid-ocean ridges, *Geol. Soc. Am. Bull.*, **83,** 1947.

Aumento, F., Loncarevic, B. D. & Ross, D. I., 1971, Hudson geotraverse: geology of the Mid-Atlantic Ridge at 45°N, *Phil. Trans. Soc. Lond. A.*, **268,** 623.

Batchelor, G. K., 1954. Heat convection and buoyancy effects in fluids, *Meteoral. Soc.*, **80,** 339.

Bodvarsson, G., 1961. Physical characteristics of natural heat resources in Iceland, *Jökull*, **7,** 29.

Bodvarsson, G. & Lowell, R. P., 1972. Ocean-floor heat flow and the circulation of interstitial waters, *J. geophys. Res.*, **77,** 4472.

Boegeman, D. E. Miller, G. S., 1972. Precise positioning for near bottom equipment using a relay transponder, *Mar. geophys. Res.*, **7,** 381.

Corliss, J. B., 1971 The origin of metal-bearing submarine hydrothermal solutions, *J. geophys. Res.*, **76,** 8128.

Corry, C., Dubois C. & Vacquier, V., 1968. Instrument for measuring terrestrial heat flow through the ocean floor, *J. mar. Res.*, **26,** 165.

Deffeyes, K. S., 1970. The axial valley: a steady state feature of the terrain, in *Megatectonics of continents and oceans*, New Brunswick, Rutgers University Press.

Detrick, R. S., Williams, D. L. Mudie, J. D. & Sclater, J. G., 1974. The Galapagos Spreading Centre: bottom-water temperatures and the significance of geothermal heating *Geophys,. J.R. astr. Soc.*, **38,** 627–636.

Donaldson, I. G., 1962. Temperature gradients in the upper layers of the Earth's crust due to convective water flows, *J. geophys. Res.*, **67,** 3449.

Elder, J. W., 1965. Physical processes in geothermal areas, *Am. geophys. Un. Mono.*, **8,** 211.

Elder, J. W., 1967. Steady free convection in a porous medium heated from below, *J. fluid Mech.*, **27,** 29.

Erickson, A. J. & Simmons, G., 1969. Thermal measurements in the Red Sea hot brine pools, in *Hot brines and recent heavy metal deposits in the Red Sea*, eds E. T. Degens and D. A. Ross, Springer-Verlag, New York.

Grim, P. J., 1970. Connection of Panama fracture zone with the Galapagos rift zone, eastern tropical Pacific, *Mar. geophys. Res.*, **1,** 85.

Hele-Shaw, H. S. J., 1898. *Trans. Inst. Naval Architects*, **40,** 21.

Herron, E. M., 1972. Sea-floor spreading and the Cenozoic history of the east–central Pacific, *Geol. Soc. Am. Bull.*, **83,** 1671.

Herron, E. M. & Heirtzler, J. R., 1967, Sea-floor spreading near the Galapagos, *Science*, **158,** 775.

Hey, R. N., Deffreyes, K. S., Johnson, G. L. & Lowrie, A., 1972. The Galapagos triple junction and plate motions in the east Pacific, *Nature*, **237,** 20.

Hyndman, R. D. & Rankin, D. S., 1972. The Mid-Atlantic Ridge near 45°N XVIII: Heat-flow measurements, *Can. J. earth Sci.*, **9,** 664.

Klitgord, K. D. & Mudie, J. D., 1974. The Galapagos Spreading Centre: a near-bottom geophysical study, *Geophys. J. R. astr. Soc.* **38,** 563–586.

Lachenbruch, A. H., 1968. Rapid estimation of the topograhic disturbance to superficial thermal gradients, *Rev. Geophys.*, **6,** 365.

Lachenbruch, A. H., 1973. A simple mechanical model for oceanic spreading centers *J. geophys. Res.*, **78,** 3395.

Lamb, H., 1932. *Hydrodynamics*, Cambridge University Press, London.

Langseth, M. G., LePichon, X. & Ewing, M., 1966. Crustal structure of mid-ocean ridges. 5. Heat flow through the Atlantic Ocean floor and convection currents, *J. geophys. Res.*, **71,** 5321.

Langseth, M. G. & Von Herzen, R. P., 1971. Heat flow through the floor of the world oceans, in *The Sea*, Volume 4, Part 1, ed. A. E. Maxwell, Wiley-Interscience, 299.

Lapwood, E. R., 1948, Convection of a fluid in a porous medium, *Proc. Camb. phil. Soc.*, **44,** 508.

Lee, W. H. K., 1970, On the global variations of terrestrial heat flow, *Phys. Earth Planet. Int.*, **2,** 332.

Le Pichon, X. & Langseth, M. G., 1969. Heat flow from mid-ocean ridges and sea-floor spreading, *Tectonophys.*, **8,** 319.

Lister, C. R. B., 1972. On the thermal balance of a mid-ocean ridge, *Geophys. J. R. astr. Soc.*, **26,** 515.

McKenzie, D. P., 1967. Some remarks on heat flow and gravity anomalies, *J. geophys. Res.*, **72,** 6261.

McKenzie, D. P. & Sclater, J. G., 1969. Heat flow in the eastern Pacific and sea-floor spreading, *Bull. Volcanologique*, **33–1,** 101.

McNitt, J. R., 1965. Review of geothermal resources, *Am. geophys. Un. Mono.*, **9,** 240.

Nishimori, R. K. & Anderson, R. N., 1974. Gabbro, serpentinite, and mafic breccia from the East Pacific, *Earth planet. Sci. Lett.* in press.

Miyashiro, A., Shido, F. & Ewing, M., 1971. Metamorphism in the Mid-Atlantic Ridge near 24° and 30°N, *Phil. Trans. R. Soc. Lond. A.*, **268,** 589.

Pálmason, G., 1967. On heat flow in Iceland in relation to the Mid-Atlantic Ridge, *Visindafjelay Islendinaga*, Ret. 1967 Rep. of a symposium, Geoscience Soc., Iceland, Reykjavik, 111.

Parker, R. L. & Oldenburg, D. W., 1973. Thermal model of ocean ridges, *Nature*, **242,** 137

Preisinger, A., 1965. Prehnite—ein Schichtsilikatty, *Tschermaks Mineral Petrg. Mitt.*, **10,** 491–504.

Raff, A. D., 1968. Sea-floor spreading: Another rift, *J. geophys. Res.*, **73,** 3699.

Robie, R. A., Bethke, P. M., Toulmin, M. S. & Edwards, J. L., 1966, X-ray crystallographic data, densities, and molar volumes of minerals: *Geol. Soc. Am. Bull., Mem.* **97,** 27–73.

Rouse, H., Yih, C-S & Humphreys, H. W., 1952, Gravitational convection from a boundary source, *Tellus*, **4,** 201.

Sayles, F. L. & Bischoff, J. L., 1973, Ferromaganoan sediments in the equatorial East Pacific, *Earth Planet. Sci. Lett.*, **19,** 330.

Sclater, J. G. & Francheteau, J., 1970. The implications of terrestrial heat flow observations on current tectonic and geochemical models of the crust and upper mantle of the Earth, *Geophys. J. R. astr. Soc.*, **20,** 509.

Sclater, J. G. & Klitgord, K. D., 1973. A detailed heat flow, topographic and magnetic survey across the Galapagos spreading centre at 86°W, *J. geophys. Res.*, **78,** 6951.

Sleep, N. H., 1969. Sensitivity of heat flow and gravity to the mechanism of sea-floor spreading, *J. geophys. Res.*, **72,** 542.

Spooner, E. T. C. & Fyfe, W. S., 1973. Sub-sea-floor metamorphism, heat and mass transfer, *Contr. Mineral. Petrol.* **42,** 287.

Steinfink, H.,1958. The crystal structure of Chlorite. II. A triclinic polymorph: *Acta Cryst.*, **11,** 195–198.

Steinfink, H., 1961. Accuracy in structure analysis of layer silicates: Some further comments on the structure of prochlorite: *Acta Cryst.*, **14,** 198–199.

Steinfink, H., 1962 A correction—the crystal structure of Chlorite. I. A monoclinic polymorph: *Acta Cryst*, **15,** 1310.

Talwani, M., Windisch, C. C. & Langseth, M. G., 1971. Reykjanes ridge crest: a detailed geophysical study, *J. geophys. Res.*, **76,** 473.

Turner, J. S., 1969. Buoyant plumes and thermals, *Ann. Rev. fluid Mech.*, **1.** 29.

van Andel, T. H. & Heath, G. R., 1973. *Initial reports of the Deep Sea Drilling Project*, Vol. 16, Washington, D. C. (U.S. Government Printing Office).

van Andel, T. H., Heath, G. R., Malfait, B. T., Heinrichs, D. F. & Ewing, J. L., 1971, Tectonics of the Panama basin eastern equatorial Pacific, *Geol. Soc. Am. Bull.*, **82,** 1489.

Von Herzen, R. P. & Maxwell, A. E., 1959. The measurement of thermal conductivity of deep sea sediments by a needle-probe method. *J. geophys. Res.*, **64,** 1557.

Von Herzen, R. P. & Anderson, R. N., 1972, Implications of heat flow and bottom water temperature in the eastern equatorial Pacific, *Geophys. J. R. astr. Soc.*, **26,** 427.

Von Herzen, R. P. & Langseth, M. G., 1966, Present status of oceanic heat flow measurements, *Phys. Chem. Earth*, **6,** 365.

Von Herzen, R. P. & Uyeda, S., 1963, Heat flow through the eastern Pacific Ocean floor, *J. geophys. Res.*, **67,** 4219.

Williams, D. L. & Von Herzen, R. P., 1974. Heat loss from the Earth; New estimate, *Geology*, **2,** 327.

Williams, D. L., 1974. *Heat loss and hydrothermal circulation due to sea-floor spreading*, Ph.D. Thesis, Woods Hole Oceanographic Institution, Woods Hole, Massachusetts, 139p.

Wooding, R. A., 1960. Instability of a viscous liquid of variable density in a vertical Hele-Shaw cell, *J. fluid Mech.*, **7,** 501.

20

Reprinted from *Nature* **267**:600–603 (1977)

HYDROTHERMAL PLUMES IN THE GALAPAGOS RIFT

R. F. Weiss, P. Lonsdale, J. E. Lupton, A. E. Bainbridge, and H. Craig

ALTHOUGH there is indirect evidence that a major fraction of the heat loss from newly-created lithosphere occurs by convection of seawater through the porous crust[1–3], it has proved difficult to locate vents of deep-sea hydrothermal systems by direct measurement of the discharge fluid. Local increases in bottom water temperature up to 0.1 °C have been measured by towing arrays of thermistors a few metres above the axes of active oceanic spreading centres[4,5], but these data are ambiguous because small temperature anomalies may have a hydrographic explanation. We report here the first conclusive measurements of modified seawater discharging as buoyant hydrothermal plumes from fissures in young oceanic crust. We obtained samples of hydrothermal plumes in the Galapagos Rift[3], albeit after considerable dilution with surrounding bottom-water, and report the first results of the collection and analysis of these samples.

Our initial estimates of properties useful for detecting deep-ocean hydrothermal fluids were based on the geochemistry of the Red Sea brines[6,7], the only oceanic geothermal system previously sampled. In this system the basalt-interaction imprint is unfortunately superimposed on a saturated brine because the seawater flows through evaporite deposits before entering the basalts[6]. Predicted concentrations for constituents masked by the evaporite contribution must therefore be based on laboratory measurements[8]. Our estimates showed that helium isotopes were by far the most sensitive geochemical tracers: a mixture of one part of hydrothermal fluid in $\sim 10^5$ parts of seawater should be detectable in the $^3He/^4He$ isotope ratio, while a dilution of $\sim 2\times10^4$ should be detectable with total helium concentration. The latter figure is comparable with the detection limit for temperature differences, assuming a 50 °C fluid temperature. Transition metals such as Mn and Fe are also highly enriched in the Red Sea brines but the observed enrichments are consistent with an origin from evaporites[6] and ^3He is the only tracer which cannot be derived from such a source[7].

During the May 1976 Pleiades Expedition of the Scripps Institution, near-bottom hydrographic and geochemical surveys of sites on the East Pacific Rise (4°S, 102°W) and in the Galapagos Rift (1°N, 86°W) were made with a remote sensing and sampling system attached to the deep tow geophysical vehicle of the Scripps Marine Physical Laboratory[9]. The geochemical system included a high-precision CTD instrument to measure conductivity, temperature, and pressure, a light transmissometer, and an array of 8 remotely-triggered 9-litre sampling bottles de-

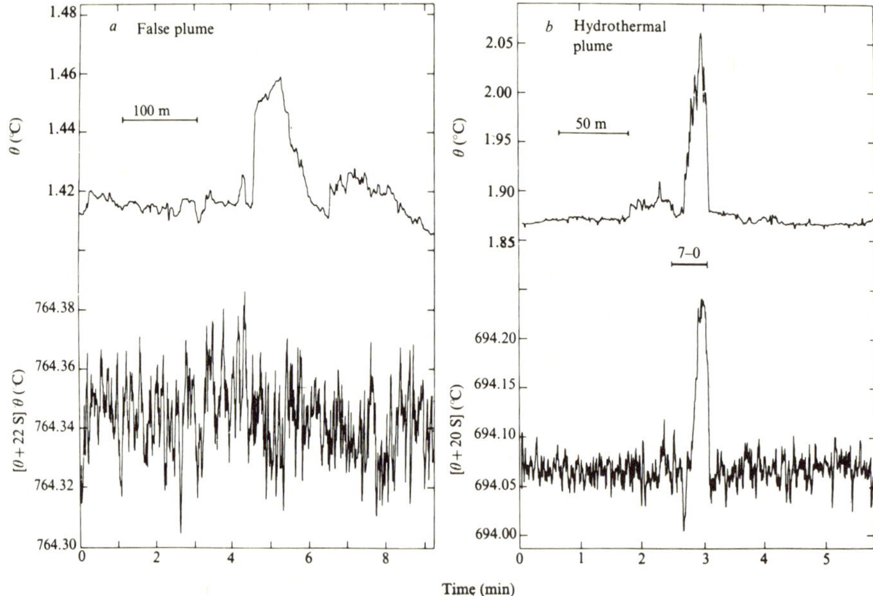

Fig. 1 Temperature records from an arbitrary time zero as the deep tow vehicle is towed ~ 10 m above bottom. Upper traces: potential temperature (θ) plotted at 10 points s^{-1}, corrected for a 200-ms time constant and smoothed by 10-point running average. Lower traces: the θ-salinity function ($\theta + XS$), which is invariant to mixing, with similar correction and smoothing. *a*. A record taken 1.9 km west of the East Pacific Rise axis, shows a 'false plume' in which the temperature spike is not seen in the θ–*S* function and is thus due to mixing. Tow speed was 50 m min^{-1}. *b*. The Galapagos Rift site data show a true hydrothermal spike which is reflected in both the temperature and $\theta + XS$ records. Tow speed was 43 m min^{-1} and sample 7–0 was collected during the interval shown beneath the θ record.

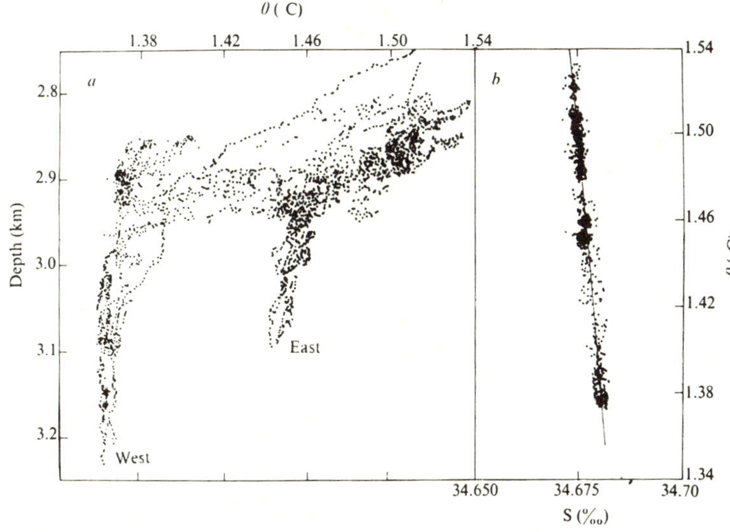

Fig. 2 Near-bottom (10–100 m) potential temperature data points measured over several hours of towing at the East Pacific Rise survey site, plotted against depth (Fig. 2*a*) and salinity (Fig. 2*b*). The θ–*S* diagram (2*b*) shows that the entire array of water types represents a single water mass; the θ–*S* slope here (= $-X$) is -22.

signed for rapid flushing. The data were telemetered to a shipboard computer which continuously monitored depth, potential temperature (θ), salinity (*S*), potential density at 4,000 decibars (σ_4), and light transmission in real time. Measurement precision was ±1 m in depth, ±0.0015 °C in θ, and ±0.002‰ in *S*. At each site the instrument package was towed ~8–100 m above bottom at speeds of ~1–2 knots (30–60 m min^{-1}) for about four days, using acoustic transponders for navigation and sonars and cameras for mapping bottom topography.

Near-bottom potential temperature excursions or 'spikes' were encountered in both survey areas; two examples are shown in Fig. 1 (upper traces). Hydrographic or mixing spikes ('false plumes') were characterised as such in the following way. The temperature–salinity (θ–*S*) relationship for the local water mass was first established by continuous soundings with the vehicle. The observed θ–*S* relationship (linear at the spreading axis depth) was then programmed into the computer and the appropriate function of θ,*S* which is invariant to mixing was monitored continuously. Deviations from this relationship were assumed to represent possible hydrothermal plumes and sampling was triggered manually when such deviations were observed.

Fig. 2 shows θ–*S* data points collected over several hours of towing during the East Pacific Rise survey. Bottom water temperatures west of the rise in this area are ~0.1 °C lower than at comparable depths east of the rise, while at the crest itself there is a transition region which is actually a sharp benthic thermocline (Fig. 2*a*). Fig. 2*b*, which contains all the rise-crest data from Fig. 2*a*, shows that only a single water mass, characterised by a linear θ–*S* relationship, is present over the entire region. The transition region at the rise crest thus simply reflects linear mixing between the water types on either side of the crest; as the deep tow vehicle navigates through the rough topography the θ record exhibits spikes caused by vertical excursions of the sensor and by encounters with various members of the continuous array of water types available in eddies and streams on the crest. Heated bottom water, however, will deviate from the ambient θ–*S* relationship characterised by $\theta + XS =$ constant, where X is the negative slope on the θ–*S* plot (Fig. 2*b*). Because of the possibility of conductive heating, the detection of such a 'θ–*S* spike' is a necessary, but not sufficient, condition for the existence of an actual hydrothermal plume.

The lower trace in Fig. 1*a* is the real time record of the θ–*S* function measured concurrently with the East Pacific Rise θ record shown above, and plotted to the same vertical scale. This spike is clearly due to mixing and is labelled a 'false plume'. It is apparent that a 'θ–*S* spike' requires an amplitude of >~0.02° for detection due to the scatter introduced by the conductivity measurement. Within this limitation, no hydrothermal spikes were observed during the East Pacific Rise survey.

Fig. 1*b* shows a potential temperature spike measured on the Galapagos Rift which shows a corresponding excursion in the θ–*S* function. In this area the near-bottom temperature structure reflects mixing between bottom water entering the Panama Basin through the Ecuador Trench[10] and warmer overlying water which spills across the Carnegie Ridge[11]. The θ–*S* excursion shows clearly that in this case the θ anomaly is not caused by mixing. Sample 7–0, collected somewhere in the indicated time interval, had a $^3He/^4He$ ratio equal to twice the atmospheric ratio, a spectacular anomaly relative to 'background' water which has a ratio only 30% greater than atmospheric[12]. This is by far the largest $^3He/^4He$ ratio yet observed in deep ocean water and there is thus no doubt that an actual hydrothermal plume was sampled.

Fig. 3 shows the bathymetry of the Galapagos Rift area as surveyed during the present study. The lava flows in the inner rift valley are fractured by many small faults and fissures along the rift margins, but near the central axis fracturing is confined to a few fissures about two or three metres wide extending along the central high[13,14]. Ten temperature spikes with corresponding 'θ–*S* spikes' were observed in this region during the shipboard survey; the locations of these ten spikes are plotted in Fig. 3 which shows that all ten lie on the axial fissure in the centre of the inner rift. Three of these spikes, 7–0 (the largest of the ten spikes) 8–3, and 8–6, were sampled and analysed; all three showed $^3He/^4He$ and radon anomalies relative to background water, verifying that the θ–*S* excursions in these ten locations represent true hydrothermal plumes.

A preliminary analysis of the remaining tow data has shown no evidence of hydrothermal activity off the central

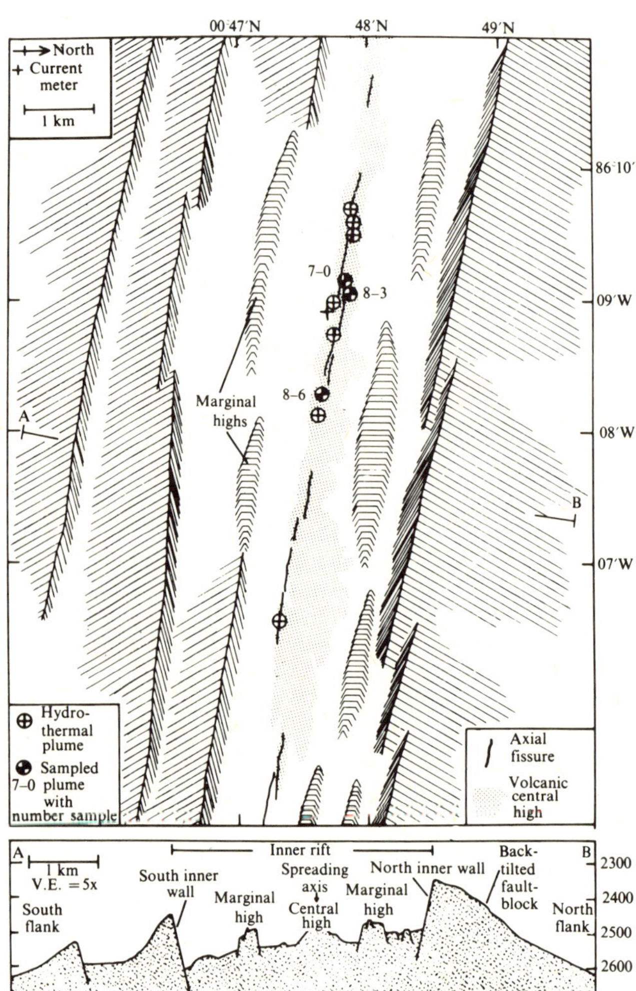

Fig. 3 Locations of hydrothermal plumes observed in the Galapagos Rift, with topography summarised from a deep tow structural map[14]. Numbers refer to plumes actually sampled. In addition to the axial fissures shown there are many fissures and minor faults along the minor rift margins; no plumes were detected in these marginal areas.

axis of the Galapagos Rift. One area of special interest in our survey was a region of 5–20 m high sediment mounds ~30 km south of the central axis which have been described as possible hydrothermal vents[13]. In this area the water structure is so uniform that hydrothermal spikes of ~0.002 °C should be observable, yet even with this sensitivity there was no evidence of hydrothermal plumes associated with the mounds.

Several of the hydrothermal plumes showed a small decrease (~0.1%) in light transmission; filtrates from the three sampled plumes and water have been distributed for a number of geochemical measurements. The helium and radon measurements are described in an accompanying paper[12]. Our results show that there is a one-to-one correspondence between the hydrographic identification of samples as heated water and the presence of excess quantities of ^{3}He/^{4}He and Rn in these samples. Together, these results constitute the first conclusive identification and sampling of hydrothermally circulating seawater in a deep-ocean spreading centre. Based on these findings, a detailed study of the Galapagos Rift hydrothermal vents using a manned submersible is now in progress.

We thank F. N. Spiess, J. Spiegelberg, J. Jain, and the entire Deep Tow technical group for their support, interest and assistance. Financial support for this research was provided by the US NSF.

1 Talwani, M., Windisch, C. C. & Langseth, M. G. *J. Geophys. Res.* **76,** 473–517 (1971).
2 Lister, C. R. B. *Geophys. J.R. astr. Soc.* **26,** 515–535 (1972).
3 Sclater, J. G. & Klitgord, K. D. *J. Geophys. Res.* **78,** 6951–6975 (1973).
4 Williams, D. L., Von Herzen, R. P., Sclater, J. G. & Anderson, R. N. *Geophys, J.R. astr. Soc.* **38,** 587–608 (1974).
5 Rona, P. A., McGregor, B. A., Betzer, P. R. & Krause, D. C. *Deep-Sea Res.* **22,** 611–618 (1975).
6 Craig, H. *Hot Brines and Recent Heavy Metal Deposits in the Red Sea* (eds Degens, E. T. & Ross, D. A.) Springer-Verlag, New York, 208–242 (1969).
7 Lupton, J. E., Weiss, R. F. & Craig, H. *Nature* **266,** 244–246 (1977).
8 Bischoff, J. L. & Dickson, F. W. *Earth Planet. Sci. Lett.* **25,** 385–397 (1975).
9 Spiess, F. N. & Mudie, J. D. *The Sea* (ed. Maxwell, A. E.) 205–220 (John Wiley and Sons, New York, 1970).
10 Laird, N. P. *J. Mar. Res.* **29,** 226–234 (1971).
11 Detrick, R. S., Williams, D. L., Mudie, J. D. & Sclater, J. G. *Geophys. J.R. astr. Soc.* **38,** 627–637 (1974).
12 Lupton, J. E., Weiss, R. F. & Craig, H. *Nature* **267,** 603–605 (1977).
13 Klitgord, K. D. & Mudie, J. D. *Geophys. J.R. astr. Soc.* **38,** 563–586 (1974).
14 Lonsdale, P. *Geology* **5,** 147–152 (1977).

21

Reprinted from *Nature* **267**:603-604 (1977)

MANTLE HELIUM IN HYDROTHERMAL PLUMES IN THE GALAPAGOS RIFT

J. E. Lupton, R. F. Weiss, and H. Craig

THE $^3He/^4He$ ratio in deep Pacific water is 20–30% higher than in atmospheric helium because of injection of primordial helium from the mantle[1,2]. The largest ^3He enrichments in the Pacific have been found in water on the crest of the East Pacific Rise where the isotopic ratios indicate[2] that the excess helium component has a $^3He/^4He$ ratio about ten times the atmospheric ratio, in agreement with the ratios measured in trapped helium in the glassy rims of oceanic tholeiites[3,4]. Recent measurements in this laboratory[5] have shown that the hot brines in the axial rift of the Red Sea are very highly enriched in mantle helium. ^3He and ^4He are respectively 3300 and 380 times supersaturated relative to atmospheric solubility equilibrium in seawater, with a $^3He/^4He$ ratio of 1.2×10^{-5}, or 8.6 times the ratio in atmospheric helium. Comparison of the enrichments of various elements in the Red Sea brines and in brines associated with salt domes[6] shows that helium is the only component in the Red Sea brines which unequivocally requires derivation from hydrothermal circulation of seawater in basalts. The helium isotopes are thus an extremely powerful and sensitive tracer for the detection and mapping of hydrothermal systems in oceanic spreading centres.

The Red Sea brines represent an extreme example of the most probable mechanism for the actual injection of mantle helium into the deep sea, namely the penetration and hydrothermal circulation of seawater in basalts on the crests of mid-ocean rises. In the Red Sea the circulating fluid is a brine because the downward penetrating seawater encounters evaporites before reaching the basalt[6], and thus the upward convecting fluid is stabilised in brine pools in depressions on the sea floor. In normal oceanic spreading centres, however, where evaporites are absent, the heated seawater should emerge as hydrothermal plumes which should be detectable by the association of temperature anomalies with the mantle helium signature. In this letter we report the first He^3 measurements on such plume samples, showing that hydrothermal circulation of seawater in basalts is an active process at spreading centres which transfers mantle helium to deep ocean water.

In May 1976 nine samples of bottom water were collected along the central axis of the Galapagos Spreading Center[7,8] using a newly-devised hydrographic sampling sled[9,10] attached to the Scripps Institution of Oceanography deep-tow instrument. The location of the samples relative to the local topography is described in the accompanying paper[10]. Three of these samples were collected during 'hydrothermal spike' temperature excursions whose origin could not be explained by turbulent mixing within the ambient water mass. The samples were taken about 10 m above the axial fissure opening, at the same level as the CTD sensor which recorded the temperature anomalies. When the sled was recovered, samples for helium analysis were sealed in copper tubing pinch-clamp containers and returned to the laboratory for analysis on a $^3He/^4He$ ratio mass spectrometer as described previously[3,5]. Radon activities were measured at sea on six of these samples, using our normal shipboard extraction and counting techniques[11] with the exception that the analyses were made on one, rather than twenty, litres of water.

Table 1 lists the ^3He results on the three samples taken during temperature spikes (7–0, 8–3, and 8–6) and on the samples of normal or 'background' bottom water where no temperature anomalies were observed on the continuous tow record. The measurements are tabulated as percent deviations of the $^3He/^4He$ ratio (R) from the atmospheric ratio (1.40×10^{-6}), that is

$$\delta(^3He) = 100\,[(R/R_{Atm})-1)]$$

The mean background water anomaly is $\delta(^3He) = 30.4\%$, in excellent agreement with values of 30.3 and 31.0% observed on the crest and at the mid-depth ^3He maximum on the eastern flank of the East Pacific Rise, south and west of the present area[2]. Relative to this background water, which is presumably the water entering the hydrothermal system, the $^3He/^4He$ ratio in sample 7–0 is enriched by 53% (1.99/1.30). The other two plume samples are also enriched in ^3He relative to ambient background, but by much smaller amounts. The $\delta(^3He)$ value of 99% measured on sample 7–0 is by far the largest value yet observed in ocean waters, and establishes for the first time a clear connection between the mantle helium in oceanic basalts and the excess ^3He observed in the deep oceans, by seawater penetration and circulation in the basalts.

Radon concentrations (Table 1) are reported as 'excess radon' relative to the activity of ^{226}Ra. The ^{226}Ra activity was assumed to be 33 d.p.m. per 100 kg from a previous study by this laboratory[11] on bottom waters 29 km south-west of the present location in approximately the same depth; rough measurements made on the present small samples were in agreement within the larger errors. The excess radon concentrations in the three plume samples (80–229 d.p.m. per 100 kg) are clearly much greater than the background value of about forty, although they do not scale with the ^3He anomalies. Because of the short radioactive mean life (5.5 d) the ^{222}Rn enrichments in plume waters are probably due to extraction of radon from the basalts, although some contribution from sediment material trapped within the axial fissure cannot be excluded.

Fig. 1 shows the absolute ^3He and ^4He concentrations in the deep tow samples, together with atmospheric solubility values and vectors for addition of atmospheric helium to air-saturated deep water, and for addition of radiogenic ^4He and pure 'mantle helium' (basalt values[3,4]) to background water

Table 1 ^3He and ^{222}Rn measurements on hydrothermal plume and background water on the Galapagos Rift (~0°48′N, 85°55′W)

Tow no., bottle no.	$\delta(^3He)$ (%)	Excess ^{222}Rn (d.p.m. per 100 kg)
a Samples with potential-temperature spikes		
7–0*	99.3	191
8–3	35.7	80
8–6	42.6	229
b 'Background water'		
7–1	27.8	39
7–2	—	49
8–1	28.6	—
8–2	30.2	33
8–5	33.4	—
8–7	32.2	—
Mean	30.4±2	40±8

Radon precision = ±3 d.p.m. per 100 kg (counting error); ^{226}Ra activity assumed = 33 d.p.m. per 100 kg. $^3He/^4He$ ratios are accurate to ±0.7% in $\delta(^3He)$.

*The temperature spike for sample 7–0 is shown in Fig. 1 of ref. 10; samples 8–3 and 8–6 were collected from plumes showing smaller temperature spikes.

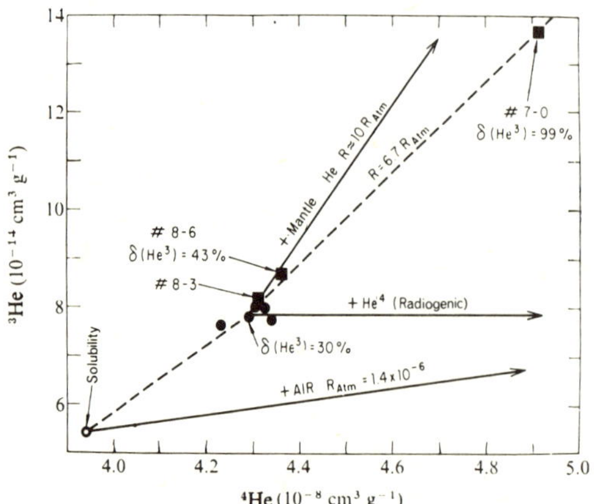

Fig. 1 ^{3}He and ^{4}He concentrations in water samples with temperature spikes (squares) and in ambient 'background water' (circles) on the axial fissure of the Galapagos Rift. The point labelled 'solubility' indicates the equilibrium air saturation concentrations for deep water. ● Normal bottom waters; ■ Hydrothermal plumes.

on the Galapagos spreading centre. (The 'absolute' concentrations, based on preliminary 'peak-height' determinations for ^{4}He, are accurate to about 5%; more accurate isotope dilution measurements are in progress.) ^{3}He and ^{4}He enrichments in the ambient background water relative to equilibrium solubility represent the net effect of air injection[12] and addition of mantle and radiogenic helium to deep water during its circulation. The absolute helium anomalies in sample 7–0 are very much larger: ^{3}He and ^{4}He concentrations are respectively 75% and 14% greater in this sample than in the mean Galapagos background water.

The added increment of ^{3}He relative to background water is 5.9×10^{-14} cm^3 STP per g in sample 7–0, an amount approximately equal to the initial solubility equilibrium (Fig. 1) and thus about 0.03% of the ^{3}He concentration observed in the Red Sea brines. The estimated temperature of inflowing Red Sea brines is ~100 °C (ref. 13); thus if the hydrothermal systems in Red Sea and Galapagos basalts are at all similar, a temperature spike of ~0.03 °C is predicted for sample 7–0, in reasonable agreement with the signal observed during sampling[10].

In Fig. 1 the vector from the mean background water to sample 7–0 corresponds to addition of helium with ^{3}He/^{4}He = 9.4×10^{-6} or $R = 6.7R_{Atm}$, significantly lower than the 'mantle helium' ratio ($R \sim 10R_{Atm}$) observed in ocean-ridge basalts[3,4], which implies that about 30% of the added helium is radiogenic ^{4}He generated in the oceanic crust. Helium with $R = 6.7\ R_{Atm}$ would be found in Pacific ridge-basalts after about 30 my of ^{4}He production (ref. 4), an age far too great for basalts in such close proximity to the rift axis. The holocrystalline portions of pillow basalts exchange trapped gases with seawater on a much shorter time scale than this[14]; thus the helium isotope ratios in basalt-seawater hydrothermal systems probably decrease coutinuously with time as the initial mantle helium component is progressively stripped out by the fluid phase and the radiogenic component grows in. It may therefore be possible to characterise such hydrothermal systems in time and space by detailed mapping of helium isotope ratios in plume samples.

The extreme sensitivity of helium and radon for plume detection is indicated by our estimate that sample 7–0 contained of the order of 0.03% of hydrothermal water. A sample with a 1 °C temperature spike would, on this basis, contain helium with $R \sim 5.5R_{Atm}$ and a ^{222}Rn activity of ~ 5,000 d.p.m. per 100 kg. Radon is of course a radiogenic isotope which is present in very high activity in marine sediments; it is therefore not a completely unamiguous tracer for seawater which has penetrated basalts. The ^{3}He signature, however, provided unequivocal evidence for the existence of a hydrothermal basalt–seawater system in the axial region of the Galapagos Rift.

We thank Dr F. N. Spiess for his strong interest and encouragement in this work, P. Lonsdale for discussion and for assistance at sea, A. Birket for assistance in the laboratory, and the Ocean Sciences section of the US NSF for financial support.

1 Clarke, W. B., Beg, M. A. & Craig, H. *Earth planet. Sci. Lett.* **6**, 213 (1969).
2 Craig, H., Clarke, W. B. & Beg, M. A. *Earth planet. Sci. Lett.* **26**, 125 (1975).
3 Lupton, J. E. & Craig, H. *Earth planet. Sci. Lett.* **26**, 133 (1975).
4 Craig, H. & Lupton, J. E. *Earth planet. Sci. Lett.* **31**, 369 (1976).
5 Lupton, J. E., Weiss, R. F. & Craig, H. *Nature* **266**, 244 (1977).
6 Craig, H., *Hot Brines and Recent Heavy Metal Deposits in the Red Sea*, (eds Degens, E. T. and Ross, D. A.) 208–242 (Springer-Verlag, New York, 1969).
7 Sclater, J. G. & Klitgord, K. D. *J. geophys. Res.* **78**, 6951 (1973).
8 Williams, D. L., Von Herzen, R. P., Sclater, J. G. & Anderson, R. N. *Geophys. J. R. Astr. Soc.* **38**, 587 (1974).
9 Weiss, R. F. *et al.*, *Trans. Am. Geophys. Union* **57**, 935 (1976).
10 Weiss, R. F. *et al. Nature* **267**, 600–603 (1977).
11 Chung, Y–C & Craig, H. *Earth planet. Sci. Lett.* **14**, 55 (1972).
12 Craig, H. & Weiss, R. F. *Earth planet. Sci. Lett.* **10**, 289 (1971).
13 Ross, D. A. *Science* **175**, 1455 (1972).
14 Dymond, J. & Hogan, L. *Earth planet. Sci. Lett.* **20**, 131 (1973).

22

Reprinted from *Nature* **269**:319-320 (1977)

HYDROTHERMAL MANGANESE IN THE GALAPAGOS RIFT

G. Klinkhammer, M. Bender, and R. F. Weiss

HYDROTHERMAL emanations originating at mid-ocean ridges have been thought[1-5] to provide a substantial source of manganese to the ocean but the evidence supporting this hypothesis has been indirect. Anomalous manganese concentrations have been measured in naturally occurring systems where seawater is in direct contact with lava flows[6-8]. Laboratory studies have shown that seawater tends to leach manganese from basalts at elevated temperatures and pressures[9-11]. Anomalously high manganese accumulation rates have also been determined for sediments adjacent to active ridge systems, most notably the East Pacific Rise[12-14]. No measurements of manganese concentrations in seawater near mid-ocean ridges, or in hydrothermal fluids emanating from these ridges, have yet been made, however. We report here the results of the first such direct measurements, which show that manganese is being injected into the deep sea by hydrothermal circulation of seawater through newly-formed oceanic crust.

We analysed water samples collected in the Galapagos Rift during two cruises of the RV Melville (Pleiades expedition, Legs 1 and 2). These samples were collected from hydrocasts and with a remote-controlled sampler attached to the Deep Tow vehicle of the Scripps Marine Physical Laboratory[15]. Unfiltered aliquots of these samples were acidified to pH 2 with redistilled 6M HCl. Manganese was determined by addition of radioactive ^{54}Mn, adjustment of the pH to 8, extraction of manganese into 8-hydroxyquinoline in chloroform, and back extraction into 3M HNO (ref. 1). Manganese concentration in the 3M HNO_3 solution was measured using an atomic absorption spectrophotometer with a graphite furnace and the yield was determined by counting the 835 keV ^{54}Mn γ ray. From these data and the volumes of the initial sample and concentrate, the manganese concentration of the initial sample was calculated. This value, which is believed to include all dissolved and particulate manganese except that fraction in the lattice of silicate minerals, is called 'total dissolvable manganese' (TDM)[1].

Figure 1 shows how TDM concentrations vary with depth at two separate locations. The shaded circles represent samples collected during the Pacific Geosecs program at station 343 (16° 31.6′ N, 123° 1.4′ W). Of the locations at which we have made these measurements, station 343 is the closest geographically to the Galapagos Rift. Below 1,000 m, the TDM profile at this station is typical, with concentrations ranging from 0.017 to 0.087 µg per kg. Also shown in Fig. 1 are the TDM concentrations found for the samples collected from hydrocasts over the Galapagos Rift. They include samples taken at one station during Leg 1 and seven stations during Leg 2. These samples show TDM concentrations ranging from 0.046 to 0.90 µg per kg, with a maximum at the depth of the ridge crest in the study area. We believe that these anomalously high concentrations are the result of the injection of hydrothermal plumes into this area. Furthermore, the fact that the TDM concentration increases rapidly with increasing depth implies scavenging of hydrothermal manganese from the water column.

In addition to the high manganese concentrations found in the hydrographic profiles, we have made direct measurements of manganese in hydrothermal plume samples collected in the Galapagos Rift by the Deep Tow sampling system. The hydrothermal origin of these plume samples has been unequivocally demonstrated by measurements of excess ^3He and ^{222}Rn (ref. 16) as well as by the hydrographic evidence[15]. The TDM results for the Deep Tow samples are listed in Table 1, together with the ^{222}Rn results[16] for the same samples. Sample 8-3, which had a ^{222}Rn anomaly of 80 d.p.m. per 100 kg, had a small positive δ^3He anomaly, and sample 7-0, which had a ^{222}Rn anomaly of 191 d.p.m. per 100 kg, had a large δ^3He anomaly. These data show that there is a one-to-one correspondence between the presence of very high TDM concentrations, the hydrographic identification of plumes, and the presence of excess quantities of helium and radon. The TDM concentration in plume sample 7-0 is 50 times the concentration

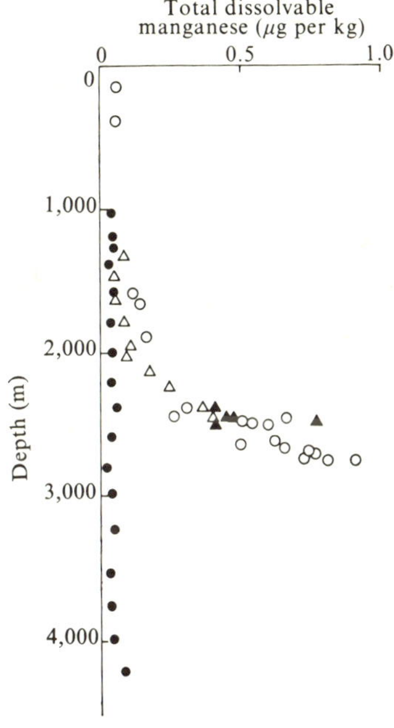

Fig. 1 The concentrations of TDM plotted against depth for samples collected from hydrocasts at two Pacific locations. ●, Samples collected during the Geosecs program at station 343 (16°31·6′N, 123°1.4′W). △ And ○, samples from eight Pleiades Expedition stations taken at the Galapagos Rift survey site (0°35.5′ to 0°48.0′N, 86°4.7′ to 86°10.2′W: see map ref. 15) △, Pleiades Leg 1; ○, Pleiades Leg 2. ▲, Deep Tow samples, although results for samples 7-0 and 8-3 are not plotted. The size of the symbols approximates the analytical uncertainty of the measurements.

Table 1 Total dissolvable manganese measurements on samples collected with the Deep Tow vehicle on the Galapagos Rift (0°48′N, 85°55′W), Pleiades Expedition Leg 1

Tow no.–bottle no.	TDM (μg per kg)	Excess ^{222}Rn[15] (d.p.m. per 100 kg)
Samples with potential temperature spikes		
7–0	24	191
8–3	2.1	80
Near-bottom samples with no temperature spikes		
7–1	0.41	39
7–2	0.41	49
8–1	0.77	—
8–5	0.48	—
8–7	0.48	—

in the surrounding bottom water. TDM is thus an exceptionally sensitive tracer of hydrothermal fluids: dilutions of the order of one part in 10^6 parts bottom water should be measurable.

Sample 7–0 was collected from a hydrothermal plume with a potential temperature anomaly of +0.20 °C relative to the surrounding bottom water[15]. Assuming that hydrothermal water in contact with basalt has a temperature of 100 °C (ref. 17), and that a negligible fraction of the TDM in sample 7–0 was scavenged before sampling, we calculate that the manganese concentration in the undiluted hydrothermal solution is 12 mg per kg. This value agrees well with predictions based on measurements of dissolved manganese in seawater basalt systems[6–11].

Our measurements provide the first direct evidence that manganese is injected into the deep water column as a result of leaching from newly-formed crust by hydrothermally circulating seawater. Although further measurements are required to establish the areal and temporal variability of this process, our data suggest that hydrothermal sources represent an important part of the manganese cycle in seawater.

We thank R. Collier, J. M. Edmond, J. Corliss, F. N. Spiess, P. F. Lonsdale, H. Craig, J. Dymond, and the participants of Pleiades Expedition Legs 1 and 2 of the RV Nelville who made the collection of these data possible. We also thank A. E. Bainbridge and the Geosecs Operations Group for collecting the Geosecs samples. This work was supported by the USNSF.

1 Bender, M., Klinkhammer, G. & Spencer, D. *Deep-Sea Res.* (in the press).
2 Seyfield, W. & Bischoff, J. J. *Earth planet Sci. Lett.* **34**, 71–77 (1977).
3 Elderfield, H. *Mar. Chem.* **4**, 103–132 (1976).
4 Lyle, M. *Geology* **4**, 733–736 (1976).
5 Corliss, J. B. *J. geophys. Res.* **76**, 8128–8138 (1971).
6 Elderfield, H. *Mar. Geol.* **13**, M1–M6 (1972).
7 Ferguson, J. & Lambert, I. B. *Econ. Geol.* **67**, 25–37 (1972).
8 Olaffson, J. *Nature* **255**, 138–141 (1975).
9 Bischoff, J. L. & Dickson, F. W. *Earth planet. Sci. Lett.* **25**, 385–397 (1975).
10 Mottl, M. J., Corr, R. F. & Holland, H. D. *Geol. Soc. Am. Abs. Program.* **6**, 879 (1974).
11 Hajash, A. *Contr. Miner. Petrol.* **53**, 205–226 (1975).
12 Bostrom, K. & Peterson, M. N. A. *Econ. Geol.* **61**, 1258–1265 (1966).
13 Bender, M. *et al. Earth planet Sci. Lett.* **12**, 425–433 (1971).
14 Dymond, J. & Veeh, H. H. *Earth planet Sci. Lett.* **28**, 13–22 (1975).
15 Weiss, R. F., Lonsdale, P. F., Lupton, J. E., Bainbridge, A. E. & Craig, H. *Nature* **267**, 600–603 (1977).
16 Lupton, J. E., Weiss, R. F. & Craig, H. *Nature* **267**, 603–604 (1977).
17 Ross, D. A. *Science* **175**, 1455–1457 (1972).

23

Reprinted from *Nature* 272:156–158 (1978)

EXCESS ^{3}He AND ^{4}He IN GALAPAGOS SUBMARINE HYDROTHERMAL WATERS

W. J. Jenkins, J. M. Edmond, and J. B. Corliss

SINCE the discovery of excess ^{3}He in Pacific deep waters[1], it has been argued that the source of this anomaly is the mid-depth injection of primordial helium from seafloor spreading centres[1,2]. This hypothesis is consistent with the spatial distribution of excess ^{3}He in the deep waters[3–5], its presence in the glassy (rapidly quenched) margins of extruded ocean basalts[6,7], and the detection of a large ^{3}He excess in a 'thermal plume' (thermal anomaly ~0.1 °C, sampled using a deep-tow sled) over the Galapagos Spreading Centre[8–10]. In this report, we summarise 37 measurements of excess helium in hydrothermal waters sampled using the Alvin deep submersible at the Galapagos Spreading Centre during February and March of 1977.

The samples were taken from a series of four distinct vent fields using a continuous flow (pumped) system on the submersible which flushed an array of eight 8.8 l Plexiglass sampling bottles. The water was monitored using a CTD probe and numerous chemical measurements have been made on these samples. The temperature of waters sampled ranged from ambient (2.05 °C) to 15.0 °C, with a maximum uncertainty in the temperature of about 0.5 °C.

The water samples were transferred on board ship to helium leak-tight copper tube samplers and returned for the shore-based helium analysis. The helium was extracted in a metal high vacuum system and stored in alumino-silicate glass sample tubes (Corning no. 1720). The isotopic ratio and amount of helium was measured by purifying and introducing a known fraction of the sample into a dual collection, statically operated helium isotope mass spectrometer[3,4]. The helium contents were thus determined to $\pm 1\%$ and the isotopic ratio to $\pm 0.5\%$.

The helium found in the water is as much as 11 times enriched in ^{4}He, and 60 times enriched in ^{3}He relative to ambient seawater concentrations. All the samples cluster quite tightly about a straight line on a plot of ^{3}He against ^{4}He (Fig. 1), with an isotopic ratio (^{3}He/^{4}He) nearly eight times atmospheric. The black square in Fig. 1 represents ambient bottom water, that is water with helium contents typical of the area. These ambient waters also show an excess of ^{3}He and ^{4}He (~40% and 10% of saturation, respectively) although much smaller than the effects observed here. The apparent scatter about the line (3.8%) is significant with respect to experimental error (0.5%); but the tightness of the correlation, despite the nearly fivefold range in absolute concentrations, suggests a common origin for the helium for all the vent fields. The hottest vents encountered, in the 'Garden of Eden' show the greatest scatter but with no significant bias from the general trend. This is significant in the light of the temperature data (discussed later). The isotopic ratio (^{3}He/^{4}He) of this excess helium ($1.08 \pm 0.02 \times 10^{-5}$) is in general agreement with the value $9.4 \pm 0.5 \times 10^{-6}$ observed by Lupton *et al.* using a deep-tow sled in a thermal plume ($\Delta T \sim 0.1$ °C) in this area (the uncertainty in their value we have estimated from their Fig. 1)[8–10].

The isotopic ratio is substantially lower than that found in tholeiitic glasses[6,7] (1.3–1.7×10^{-5}). It is not possible to explain the higher isotopic ratios observed in the glasses by *in situ* processes or partitioning between the water and rock. Diffusive loss from the glass after formation, partial equilibration with seawater, or *in situ* decay of U and Th all serve to lower the isotopic ratio of the glass helium. Further, this helium isotopic ratio is significantly lower than that observed in Red Sea brines[11] (1.2×10^{-5}), and distinctly lower than that of Kilauea gas, reported to be $2.09 \pm 0.11 \times 10^{-5}$ (ref. 4) and $1.94 \pm 0.04 \times 10^{-5}$ as measured in our laboratory. This suggests a significant degree of mantle heterogeneity (at least with regard to helium isotopes), which may be due either to a large scale variation in U and Th concentrations (and/or ^{4}He grow-in times) or to spatial inhomogeneity in the extent of mantle degassing. In either case, the measurement of helium isotopes should prove useful in studying mantle outgassing processes.

The ^{3}He concentrations measured correlate rather well with the temperature of the samples (Fig. 2). In fact, excluding the Garden of Eden samples, ^{3}He and temperature correlate linearly within errors ($\chi^2 \approx 1.9$), giving a slope of $7.6 \pm 0.5 \times 10^{-8}$ cal per atom. The Garden of Eden samples show consistent and significant bias from this correlation, appearing to be relatively rich in heat or poor in ^{3}He relative to the general trend. This is further mirrored by the isotopic ratio scatter (Fig. 1) and suggests that somewhat different processes occur in the vent field. Although not the only explanation, it is feasible that this effect may be due to an additional subsurface conductive transfer of heat to the circulating water, associated with little or no transfer of volatiles (that is, helium). It seems significant that this same vent shows anomalous behaviour (relative to the other vents) in a number of properties, including conductivity, alkalinity and *p*H. If the Garden of Eden data are included, the slope is changed to $9.5 \pm 0.7 \times 10^{-8}$ cal per atom, with a correspondingly increased χ^2 (6.2). In the light of other uncertainties in the calculations that follow, this difference is only marginally significant. Regardless of the detailed mechanism, such a variation weakens the generalisations that we are about to make. However, until a more comprehensive survey is performed, our best approach will be to use the correlation

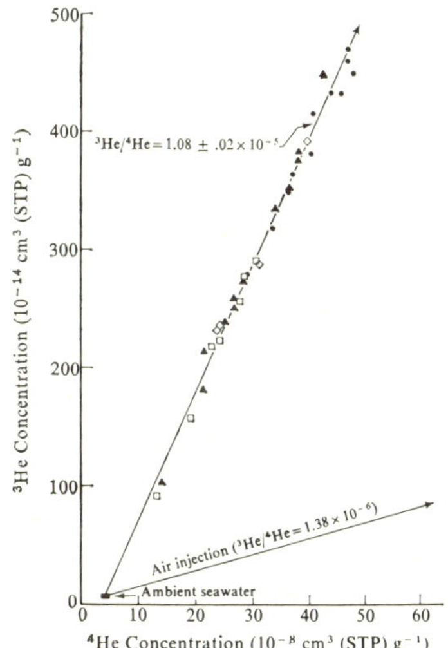

Fig. 1 The correlation between ^{3}He and ^{4}He for Galapagos hydrothermal waters. Ambient seawater (local bottom water, the black square) has ^{3}He and ^{4}He concentrations of 7.7×10^{-14} and 4.3×10^{-8} cm^3 (STP) g^{-1} respectively. The air injection line represents the trend in concentrations that would result from the addition of atmospheric helium. The linear correlation coefficient for the linear regression line is 0.994. Vent areas: ●, Garden of Eden; ▲, Clambake; □, Oyster bed; ◇, Dandelion.

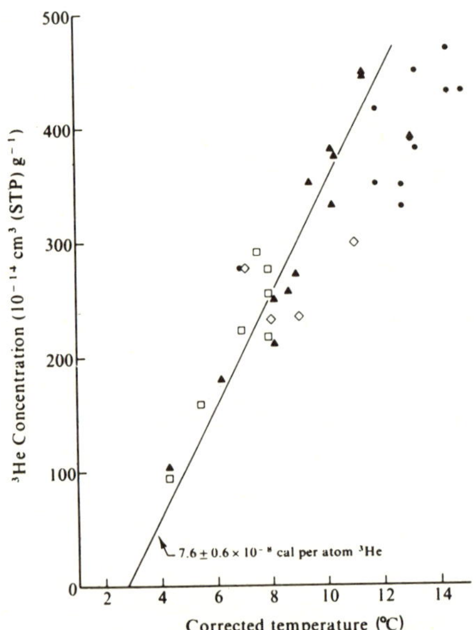

Fig. 2 The samples, exclusive of Garden of Eden, exhibit a tight correlation between excess ^3He and temperature ($\sigma(T) \approx 0.7$ °C), with a slope of 7.6×10^{-8} cal per atom. Exclusive of the Garden of Eden samples (see text). Note the substantial and systematic offset of the Garden of Eden samples. Vent areas as in Fig. 1.

obtained exclusive of the Garden samples, with the understanding that reinterpretation may be necessary later.

Because the ^3He anomaly has been detected throughout most of the world's ocean, and because (in the absence of tritium; as in the deep waters) helium is biologically and chemically 'conservative,' it should be possible to determine in a relatively unambiguous fashion its average oceanic flux. This estimate is further constrained by estimates of the ^3He loss rate in the upper atmosphere[12-13] which must balance this flux (in addition to crustal and atmospheric production rates which are relatively small). The most recent estimate of this oceanic flux[4] is 4 ± 1 atoms cm^{-2} s^{-1} corresponding to 6.5×10^{26} atoms yr^{-1}. Using the observed ^3He-heat correlation, we obtain a net 'convective hydrothermal heat flux' of $4.9 \pm 1.2 \times 10^{19}$ cal yr^{-1}. This value is in excellent agreement with the estimates of Wolery and Sleep[14] (4.0×10^{19} cal yr^{-1}) and Williams and Von Herzen[15] (6.4×10^{19} cal yr^{-1}) and represents the first independent confirmation of their calculations. Obviously, calculations of this sort may be extended to other (conservative) properties, thereby allowing an estimate of their ocean-mantle fluxes and enabling a more accurate balancing of their geochemical cycles.

It should be noted, however, that the ^3He-heat relationship gives a lower limit estimate to the hydrothermal flux as it is possible (in principle) to warm the circulating waters by heat conduction in the rock without exchange of helium, as perhaps suggested by the Garden of Eden results. Further, it must be established that the samples measured are indeed representative of all hydrothermal systems.

We have, therefore, detected a large component of excess helium in Galapagos hydrothermal waters with an isotopic ratio ($^3He/^4He$) substantially greater than atmospheric, but significantly smaller than primordial helium measured elsewhere, further substantiating the hypothesis of mantle outgassing of helium, and indicating mantle heterogeneity. We have used an observed correlation between excess ^3He and temperature of the waters to compute a global hydrothermal convective heat flux of $4.9 \pm 1.2 \times 10^{19}$ cal yr^{-1}, which is in excellent agreement with theoretical estimates.

We thank M. Clauson, A. Bainbridge and the Alvin engineering group for the design of the sampling system and excellent technical support. This work was funded under NSF grant IDE 76-00389. Woods Hole Oceanographic Institution contribution No. 4035.

1. Clarke, W. B., Beg, M. A. & Craig, H. *Earth planet. Sci. Lett.* **6**, 213 (1969).
2. Craig, H. & Clarke, W. B. *Earth planet. Sci. Lett.* **9**, 45 (1970).
3. Jenkins, W. J. thesis, McMaster Univ., Hamilton, (1974).
4. Craig, H., Clarke, W. B. & Beg, M. A. *Earth planet. Sci. Lett.* **26**, 125, (1975).
5. Jenkins, W. J. & Clarke, W. B. *Deep-Sea Res.* **23**, 481 (1975).
6. Lupton, J. E. & Craig, H. *Earth planet. Sci. Lett.* **26**, 133 (1975).
7. Craig, H. & Lupton, J. E. *Earth planet. Sci. Lett.* **31**, 369 (1976).
8. Lupton, J. E., Weiss, R. F. & Craig, H. *Nature* **267**, 603 (1977).
9. Weiss, R. F., Lonsdale, P., Lupton, J. E., Bainbridge, A. E. & Craig, H. *Nature*, **267**, 600 (1977).
10. Lonsdale, P. *Deep-Sea Res.* **24**, 857 (1977).
11. Lupton, J. E., Weiss, R. F. & Craig, H. *Nature* **266**, 244 (1977).
12. Kockarts, G. & Nicolet, M. *Ann. Geophys.* **18**, 269 (1962).
13. Johnson, H. E. & Axford, W. I. *J. geophys. Res.* **74**, 2433 (1969).
14. Wolery, T. J. & Sleep, N. H. *J. Geol.* **84**, 249 (1976).
15. Williams, D. L. & Von Herzen, R. P. *Geology* 32 (1974).

24

Reprinted from *Earth and Planetary Sci. Letters* **46**:1 (1979)

RIDGE CREST HYDROTHERMAL ACTIVITY AND THE BALANCES OF THE MAJOR AND MINOR ELEMENTS IN THE OCEAN: THE GALAPAGOS DATA

J.M. EDMOND, C. MEASURES, R.E. McDUFF, L.H. CHAN[1], R. COLLIER, B. GRANT

Department of Earth and Planetary Sciences, Massachusetts Institute of Technology, Cambridge, MA 02139 (U.S.A.)

L.I. GORDON and J.B. CORLISS

School of Oceanography, Oregon State University, Corvallis, OR 97331 (U.S.A.)

Received February 15, 1979
Revised version received August 20, 1979

Samples collected by the deep submersible "Alvin" from four hot spring fields (T = 3–13°C) on the crest of the Galapagos spreading ridge show pronounced and varied compositional anomalies. If it is assumed that these have a general significance, that they are associated with hydrothermal reactions between seawater and basalt wherever new oceanic crust is being produced then global fluxes can be computed. These are large. For Mg and SO_4 they balance the river input. For Li and Rb they exceed it by factors of between five and ten. Calcium is supplied at a rate equivalent to that of non-carbonate Ca from the continents. The additions of K, Ba and Si are between one third and two thirds of the river load. There are large positive and negative anomalies for Cl and Na indicating that substantial amounts of Cl may be taken up by the newly formed crust and transported deep into the subduction zones.

Where there are data in common, the field measurements agree with the experimental findings at low (<5) water/rock ratios.

[*Editors' Note:* In the original, material follows this abstract.]

[1] Permanent address: Department of Geology, Louisiana State University, Baton Rouge, LA 70803, U.S.A.

25

Reprinted from *Earth and Planetary Sci. Letters* **46**:19 (1979)

ON THE FORMATION OF METAL-RICH DEPOSITS AT RIDGE CRESTS

J.M. EDMOND, C. MEASURES, B. MANGUM, B. GRANT, F.R. SCLATER, R. COLLIER, A. HUDSON

Department of Earth and Planetary Sciences, Massachusetts Institute of Technology, Cambridge, MA 02139 (U.S.A.)

L.I. GORDON and J.B. CORLISS

School of Oceanography, Oregon State University, Corvallis, OR 97331 (U.S.A.)

Received March 2, 1979
Revised version received August 24, 1979

Data from the hot springs at the Galapagos spreading center (T = 3–13°C) show depletions of the exiting waters in Cu, Ni, Cd, Se, Cr and U relative to ambient seawater. Manganese is strongly enriched. Iron shows highly variable behavior between vent fields but is in general low. The data confirm the occurrence of extensive subsurface mixing between the primary high-temperature, acid, reducing hydrothermal fluids and "groundwater". The composition of the latter is indistinguishable from that of the free water column adjacent to the ridge axis. The final solutions are on the boundary between those forming MnO_2 crusts and those producing iron-manganese rich sediments. The suite of metal rich deposits observed at ridge crests – Mn-O, Fe-Mn-O, Fe-S – can be explained as the manifestation of the degree of subsurface mixing, decreasing from >100 : 1 to <1 : 1 across the series (assuming an end-member temperature of 350°C).

[*Editors' Note:* In the original, material follows this abstract.]

26

Reprinted from *Econ. Geology* 71:1515-1525 (1976)

Copper-Iron Sulfide Mineralizations from the Equatorial Mid-Atlantic Ridge

ENRICO BONATTI, BEATRIZ-MELBA GUERSTEIN-HONNOREZ, AND JOSE HONNOREZ

Abstract

Copper-iron sulfide mineralizations were found in metabasalts dredged from offsets of the axis of the Mid-Atlantic Ridge in the Romanche and Vema fracture zones. The basalts were probably originally emplaced at ridge axis. The Cu-Fe sulfides form "disseminated"- and "stockwork"-type deposits. Chalcopyrite is the main sulfide phase; pyrite and pyrrhotite are also present. Fe-hydroxides, containing as much as 13 percent Cu, are a common alteration product of the chalcopyrite; atacamite ($Cu_2Cl(OH)_3$), delafossite ($FeCuO_2$) or tenorite (CuO), and probably mooihoekite ($Cu_9Fe_9S_{16}$) or haycockite ($Cu_4Fe_5S_8$) were also observed as secondary phases. These findings are explained by a model of metallogenesis which shows that metal sulfide deposits are formed within the oceanic crust at spreading centers by deposition from hydrothermal systems and that, upon discharge on the sea floor, they give rise to metalliferous sediments. The main source of metals for the hydrothermal systems is the basaltic oceanic crust, though for sulfur reduction of seawater SO_4^{-2} is also important. Sulfide deposition below the sea floor causes fractionation of metals, particularly of Fe from Mn. These findings support the hypothesis that "massive" metal sulfide deposits are present in the oceanic crust and that "massive" sulfide ores of ophiolite complexes were formed in ancient spreading centers.

Introduction

IN the last few years it has been established that diverging plate margins (active oceanic ridges and other spreading centers) are the locus of intense hydrothermal activity. Metalliferous sediments, which are found along active oceanic ridges, are commonly ascribed to such hydrothermal activity, which causes metals to be extracted from the oceanic crust by circulating thermal waters. In this paper we describe Fe-Cu sulfide mineralizations in metabasalts from the equatorial Mid-Atlantic Ridge, which we believe offer direct evidence of metal mobilization within the basaltic oceanic upper crust at spreading centers. A model is discussed, which explains the formation of both metal sulfide deposits and metalliferous sediments at spreading centers. The presence of extensive metal sulfide mineralizations in ocean ridge basalts is also of interest in the context of understanding the origin of metal ore deposits in the geological record, especially those widespread deposits associated with emerged fragments of ancient oceanic crust, i.e., ophiolitic complexes.

Fe-Cu Sulfide Mineralizations in Metabasalt GS7309-93

The samples under study, consisting of five angular blocks (10 to 30 cm across) of a metabasalt were dredged from station GS7309-93 at 0°58.6′S, 24°30.8′W from 2,700 to 3,100 meters below sea level. This site is close to the intersection of the Romanche offset zone with the western axial segment of the Mid-Atlantic Ridge (Fig. 1). The Romanche Fracture Zone is morphologically complex in this area, being manifested by several parallel transverse (east-west) ridges and valleys. The dredging site is on the northern side of a valley situated at the southern termination of the Mid-Atlantic Ridge axial segment, about 30 km east of the Mid-Atlantic Ridge axis.

Description of the metabasalt

Megascopic: The metabasalt specimens are coated by a manganese-oxide crust up to 3 mm thick. Their crystallinity ranges from aphanitic to porphyritic. Upon sectioning one block displays color zonation due to alteration, from a yellowish outer rim to a darker core; this zonation is parallel to the contours of the block and to fracture zones crossing it. The specimens are criss-crossed by veins, generally less than 1 mm, but locally up to a few mm thick. The sulfide mineralizations are contained in these veins but also appear as isolated aggregates disseminated within the rock.

Microscopic: The metabasalt is made up of plagioclase An_{70} microlites (up to 0.7 mm long) partly replaced by albite and chlorite; fresh, zoned subeuhedral augite microphenocrysts (up to 0.3 mm in diameter); and chloritic aggregates in a fine groundmass.

The chloritic aggregates are of three types: (a) pseudomorphs after euhedral crystals, probably of olivine; (b) rounded aggregates, possibly representing infilling of former gas vesicles; and (c) intersertal patches as replacement of a devitrified glassy matrix.

The groundmass consists of plagioclase microlites (0.015 mm average length); pyroxene granules (0.006 mm average diameter); and sphene aggregates.

Veins: Two types of vein can be observed under the microscope. The first type has smooth boundaries, with matching walls and with sharp contact between vein minerals and host rock minerals. All the minerals in the veins are anhedral; chlorite minerals are the main components of these veins; toward the core of the veins a dark green pleochroic chlorite is present in fibroradial aggregates, while toward the margins a pale green pleochroic chlorite with a lower index of refraction forms parallel fibrous aggregates. Chlorites frequently grade into an orange, more birefringent fibrous phase, probably nontronite. The chlorite minerals are generally associated in these veins with zoned epidote, albite, calcite, and sulfides.

The second type of vein, also with matching walls, runs zig-zag and is thinner (from 0.01 to 0.05 mm thick) than the first type. These finer veinlets consist essentially of an unidentified fibrous zeolite, and they frequently intersect and displace or intrude the veins of the first type.

Chemistry of the metabasalt

The results of two chemical analyses of the basalt and their CIPW norms (Table 1) suggest that the rock was originally close to the olivine tholeiite-quartz tholeiite joint, using the normative basalt classification of Yoder and Tilley (1962). This is the magma type most commonly erupted along oceanic ridges. The relatively high water and Na_2O content of this rock indicates that it has been subjected to low-grade metamorphism with development of hydrated phases such as chlorites and zeolites.

Description of the sulfide mineralizations in the metabasalt

Metal sulfide phases are widespread in the basalt as disseminated aggregates and in the first type of vein mentioned above. The study of several polished thin sections by reflection microscopy, complemented by X-ray diffraction and electron microprobe analyses, indicates that the following two different types of sulfide phase paragenesis exist in these rocks.

Pyrrhotite-pyrite association: Small (0.35–0.40 mm in diameter) aggregates are scattered in the microcrystalline basaltic matrix. These aggregates consist of tabular idiomorphic (or idioblastic) sulfide crystals, each from 0.01 mm to 0.02 mm in size. The majority of these crystals are of pyrrhotite; some pyrite is also present (Figure 2). Frequently one observes only relicts of the original pyrrhotite-pyrite aggregates because they have been altered into micro- or crypto-crystalline limonite. No intermediate products of oxidation are observed.

Chalcopyrite-Fe hydroxide-Cu chloride association: The opaque minerals commonly associated with the first type of vein (Figs. 3 and 4) consist of grains ranging in size from about 0.5 to 2 mm. The major

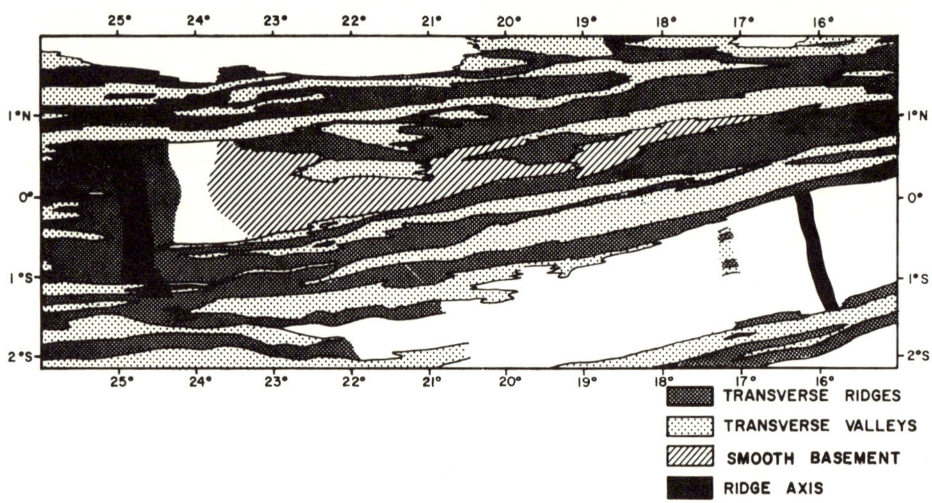

FIG. 1. Main morphological features of the basement in the Romanche Fracture Zone area, equatorial Atlantic (from Gorini, 1976). The location where the sulfide-containing metabasalts were recovered is shown by a triangle.

TABLE 1. Major Element Composition of Two Samples of Metabasalt GS7309-93F

	%	%		ppm
SiO_2	52.60	51.40	B	37
TiO_2	1.26	1.33	Ba	<50
Al_2O_3	14.63	14.82	Co	45
Fe_2O_3	2.04	1.97	Cr	160
FeO	7.04	7.04	Cu	160
MnO	0.17	0.17	Ni	80
MgO	6.00	6.23	Sc	40
CaO	10.49	10.69	V	250
Na_2O	3.84	3.82	Y	40
K_2O	0.062	0.063	Zr	110
P_2O_5	0.178	0.173	Zn	62
H_2O+	2.86	3.66		
Total	101.17	101.37		
	CIPW norms			
Q	0.15	—		
OR	0.359	0.363		
AB	33.30	33.08		
AN	22.54	23.66		
WO	13.12	12.31		
EN	15.11	13.94		
FS	9.59	8.46		
FO	—	1.36		
FA	—	0.91		
MT	2.99	2.92		
IL	2.42	2.58		
AP	0.43	0.41		

CIPW norms are indicated. Trace element composition of a sample of metabasalt GS7309-93F (obtained by optical emission spectroscopy) is also shown. O. Joensuu and M. Riera, analysts.

component of these grains is yellow under the microscope in reflected light and displays optical properties of chalcopyrite, although no anistropy was observed. The chalcopyrite is partially and centripetally replaced by a gray mineral with low reflectivity and with red and brown internal reflections, identified as Fe-hydroxide, goethite, and/or lepidocrocite (Figs. 5 and 6).

The presence of contraction cracks in the Fe-hydroxide aggregate suggests that the replacement of chalcopyrite is accompanied by a decrease in volume. No intermediate product of oxidation was observed between chalcopyrite and Fe-hydroxide. The alteration of chalcopyrite directly into Fe-hydroxide in the zone of oxidation is not unusual (Ramdohr, 1969). A Debye-Scherrer X-ray made on a sample of sulfide phases drilled in a vein from specimen GS7309-93H confirmed the presence of chalcopyrite and goethite.

In some instances a transparent green mineral was observed associated with the chalcopyrite-Fe-hydroxide masses, with optical properties coinciding with those of atacamite ($Cu_2Cl(OH)_3$) (Fig. 4). Atacamite and its closely related mineral malachite have been shown to be stable phases in seawater (Bianchi and Longhi, 1973). In one section a light brown, strongly anisotropic and bireflective mineral was observed as an alteration product of chalcopyrite, in addition to Fe-hydroxide. Its texture and optical properties are similar to those of delafossite ($FeCuO_2$) or tenorite (CuO) (Fig. 7).

One example of the chalcopyrite-Fe hydroxide-Cu hydroxide (atacamite) association is shown in Figure 8. The elemental distribution of S, Fe, Cu, and Si in this grain and in the adjacent area was determined by electron scanning and monitoring of the characteristic X-rays, using a Cambridge SEM instrument.

Chemical analyses of selected points on this grain were carried out with an ARL electron probe. The results (Table 2) essentially confirm the microscopic determinations. The analyses of chalcopyrite are close to the stoichiometric composition of this mineral. Notable are the relatively high Cu concentrations found in the region of the grain identified optically as Fe-hydroxide. The absence of S from this region of the grain excludes the presence here of Cu as a sulfide; possibly it is contained as impurities dispersed in the Fe-hydroxide. Notable is also the virtual absence of Mn both from the sulfide and Fe-hydroxide phases.

Fe-Cu Sulfide Mineralization in Metabasalt P7003–25N

Metal sulfide mineralizations were observed in metabasalts dredged at station P7003-25, 10°51'N, 41°48'W, at 2,950 to 3,650 meters below sea level. This station is located on the northern wall of the Vema Fracture Zone transverse valley (Fig. 9). Non-metamorphic basalts from several sites along the north wall of the valley have an olivine-tholeiite normative composition (Bonatti et al., in prep.).

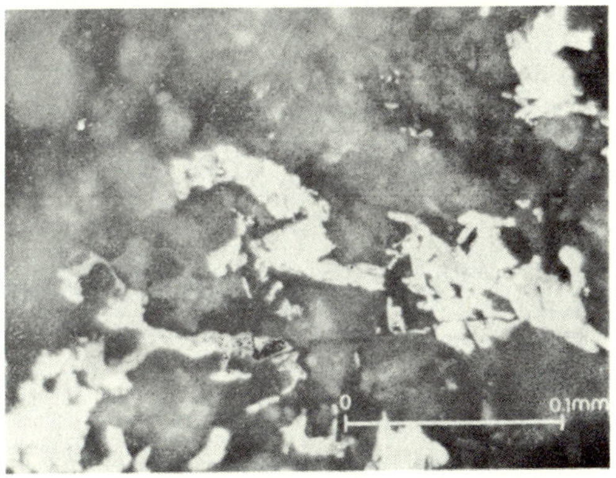

FIG. 2. Metabasalt GS7309-93F. Microphoto (in reflected light) of platy crystals of pyrrhotite with minor pyrite (both white). Fe-hydroxide (light gray) is present as replacement of pyrrhotite. Gangue silicates are dark and medium gray (plane polarized light; ×400).

The sulfide-containing sample consists of a fragment of pillow basalt the formerly glassy rim of which is plastered with a related hyaloclastite. Both pillow and hyaloclastite have undergone low-grade metamorphism transitional between zeolite and greenschist facies. The hyaloclastite and the old glassy rim of the pillow have essentially been replaced by a mixture of nontronite and chlorite. Minor amounts of quartz, albite, stilbite, analcite (?), sphene, and an actinolitic amphibole are also present. In the innermost portion of the pillow the metamorphism is evidenced by chlorite and nontronite which replaced the groundmass, whereas the few augite and plagioclase microphenocrysts and microliths are unchanged.

Numerous veins with matching walls criss-cross the whole sample; they reach a thickness of over 1 mm. The major minerals forming these veins are in decreasing order of abundance: penninite and nontronite, quartz, epidote, and calcite. The sulfide minerals are primarily contained in these veins.

The main sulfide phase is chalcopyrite. Its optical properties are identical to those described for samples GS7309-93. The chalcopyrite crystals are rimmed by a thin alteration zone of Fe-hydroxide (Fig. 10). Electron probe analyses of the chalcopyrite and Fe-hydroxide shown in Figure 11 are reported in Table 2. A film of lath- and needle-shaped crystals separates the chalcopyrite crystals from the outer Fe-

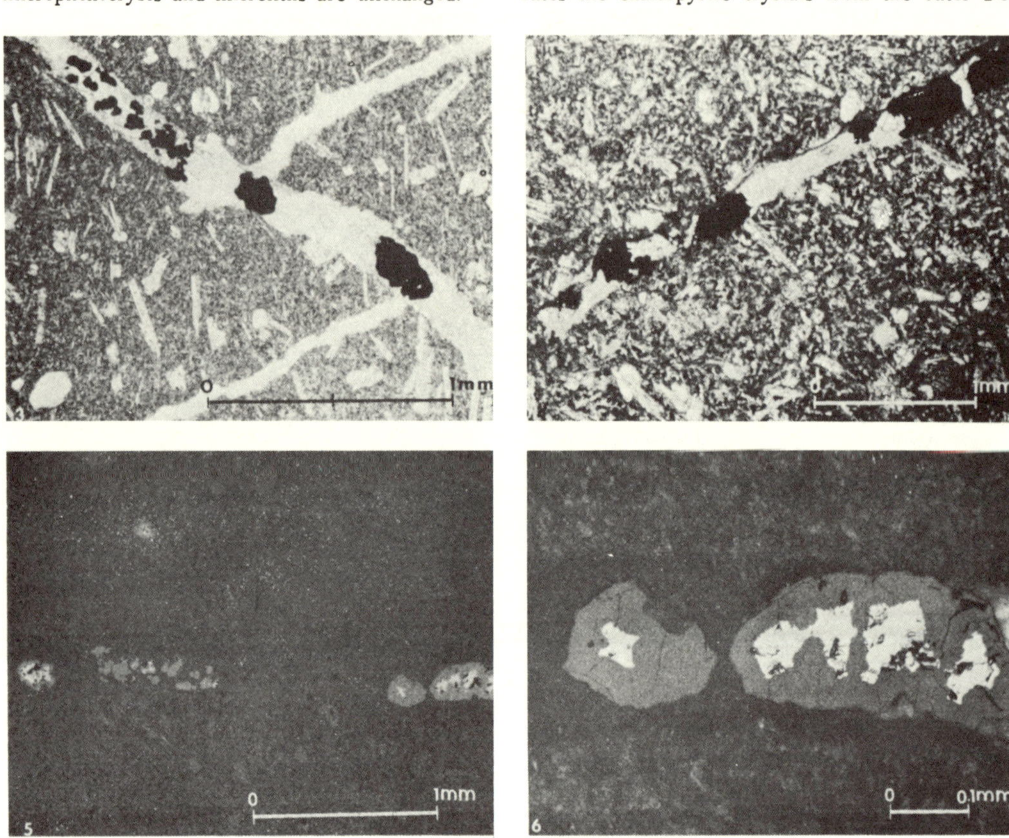

FIG. 3. Metabasalt GS7309-93F. Microphoto (in transmitted light) of a thin zeolitic vein, interesceted and displaced by a thicker vein containing sulfides (black), epidotes (gray, stronger relief), and chlorites (light gray, lower relief). The Ca-plagioclase microliths and the olivine microphenocrystals of the host basalt have been partly replaced by albite and chlorite (plane polarized light; × 40).

FIG. 4. Metabasalt GS7309-93F. Microphoto (in transmitted light) of a vein containing chalcopyrite (black), atacamite (white, very strong relief), epidote (gray, medium relief), and chlorites (light gray, low relief) (plane polarized light; × 40).

FIG. 5. Metabasalt GS7309-93F. Microphoto (in reflected light) of a vein containing chalcopyrite (white) replaced by goethite (gray) (plane polarized light; × 40).

FIG. 6. Detail of Figure 5 (× 160).

hydroxide rim; their optical properties are similar to those of chalcopyrite except for their moderate to strong anistropy. These minerals are probably mooihoekite ($Cu_9Fe_9S_{16}$) or haycockite ($Cu_4Fe_5S_8$), phases of the Cu-Fe-S system which are closely related to chalcopyrite (Cabri and Hall, 1972).

Genesis of the Sulfide Mineralizations in the Metabasalts

The Cu-Fe sulfide mineralizations in metabasalts GS7309-93 and P7003-25N resemble "stockwork"- and "disseminated"-type mineralizations of sulfide ore deposits on land. The host rocks in both cases have been subjected to a low-grade metamorphism transitional between zeolite and greenschist facies. However, these metabasalts do not seem to have reached an equilibrium since in both cases the clinopyroxene microphenocrysts remained essentially unaffected by the metamorphism, and the plagioclase microliths were only partly replaced by albite and chlorite in the sample from GS7309-93 or remained completely unchanged in sample P7003-25N. On the contrary the metamorphic mineral assemblages accompanying the sulfides in the veins (chlorite, quartz, albite, epidote, calcite) appear to be characteristic of a true greenschist facies except for the presence of the nontronite-chlorite association which indicates either a partial retromorphosis of the chlorite or an incomplete equilibrium under the conditions of the greenschist facies. We can, therefore, assume that the slight metamorphic process which affected the host basaltic rocks resulted or was initiated from the same mineralizing fluids which were responsible for the crystallization, in the fractures, of the more complex and higher grade metamorphic mineral assemblages, including the Cu-Fe sulfides.

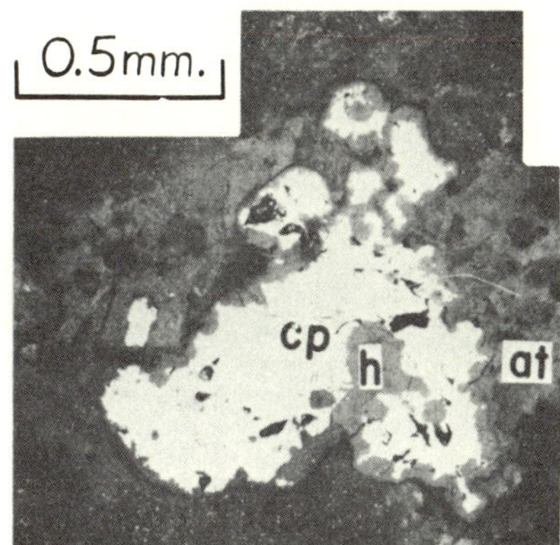

Fig. 8. Metabasalt GS7309-93I. Microphoto (in reflected light) of a Cu-Fe sulfide grain; cp = chalcopyrite, h = iron hydroxide, at = atacamite.

Fig. 7. Metabasalt GS7309-93F. Microphoto (in reflected light) of "flaky" delafossitelike minerals (white). They are altered along their margins to limonite (gray). Interstitial chalcopyrite (white) can be seen at the center of the picture. (Crossed nicols; plane polarized light; × 400.)

The latter metamorphic vein minerals, including the Cu-Fe sulfides, probably resulted from the action of hydrothermal fluids which circulated along the fractures of the host rocks and mobilized metals and sulfur from the oceanic crust.

It is probable that the hydrothermal activity, which resulted in the formation of the Cu-Fe sulfide deposits in metabasalts GS7309-93, took place close to the axial, active segment of the Mid-Atlantic Ridge, presently located about 30 km west of station GS7309-93. This assumes that the basalt was originally emplaced at the Mid-Atlantic Ridge and spread subsequently to its present location, in line with the sea-floor spreading theory. A similar assumption is probably valid for metabasalt P7003-25N, dredged from the north wall of the Vema Fracture Zone transverse valley about 210 km from the axial ridge segments. Tectonic, petrochemical, and textural relationships of basalts from the north wall of the Vema transverse valley indicate that they were originally emplaced in the active axial segment of the Mid-Atlantic Ridge (Bonatti et al., in prep.).

Hydrothermal Model of Metal Sulfide Genesis at Spreading Centers

Minor Fe-sulfide concentrations are fairly common in oceanic basalts, probably representing primary magmatic segregations. More extensive sulfide mineralizations of the type described in this paper have been reported in rocks from the Mid-Indian Ridge (Dmitriyev et al., 1970; Baturin, 1971). Massive-type sulfide ore deposits are probably formed at

TABLE 2. Results of Electron Probe Analyses of Sulfide and Associated Phases in Metabasalts GS7309-93 and P7003-25

		% Fe	% Cu	% S	% Si	% Ca	% Mn	% Cr
GS7309-93								
Chalcopyrite	1	33.0	31.2	34.5	—	—	<0.1	<0.1
	2	33.4	30.9	34.3	—	—	<0.1	<0.1
	3	32.0	30.8	34.6	—	—	<0.1	<0.1
Fe-hydroxide	4	40.0	12.1	0.3	1.8	0.5	<0.1	—
	5	37.9	13.5	0.4	2.3	0.6	<0.1	—
	6	37.2	9.1	0.2	3.6	1.3	<0.1	—
	7	36.3	11.5	2.7	4.0	1.1	<0.1	—
	8	37.3	5.7	0.2	4.6	1.4	<0.1	—
	9	40.1	4.8	0.2	5.2	1.8	<0.1	—
Atacamite	10	1.2	50.0	<0.2	—	—	<0.1	—
	11	0.8	49.5	<0.2	—	—	<0.1	—
	12	0.7	48.6	<0.2	—	—	<0.1	—
	13	1.6	51.1	<0.2	—	—	—	<0.1
P7003-25								
Chalcopyrite	14	33.2	30.8	34.8	—	—	—	<0.1
	15	31.8	30.7	34.0	—	—	—	<0.1
Fe-hydroxide	16	41.1	10.6	<0.2	—	—	—	<0.1

Analyses were obtained with ARL electron probes at the Smithsonian Institution, Washington, D. C., and at the Lamont-Doherty Geological Observatory. Sample GS7309-93 is described in the text and shown in Figures 8 and 9. Sample P7003-25 is shown in Figure 11. A number of points were analyzed in each phase. The optical microscopic identification of the analyzed phases is indicated.

spreading centers (even though they have not been sampled yet) as discussed later in this paper.

We will now discuss the genesis of metal sulfide deposits in the oceanic crust, within a model which assumes the presence of sub-sea-floor hydrothermal systems at active spreading centers. The model is schematized in Figure 11 (from Bonatti, 1975).

Hydrothermal systems at spreading centers

Several years ago it was proposed that sub-bottom hydrothermal systems operate along active oceanic ridges (Skornyakova, 1964; Arrhenius and Bonatti, 1965; Elder, 1965; Bonatti and Joensuu, 1966; Bostrom and Peterson, 1966). More recent work has confirmed those earlier suggestions (i.e., Spooner and Fyfe, 1973). In the convective-hydrothermal circulation model, seawater enters within the basaltic oceanic crust and is subsequently driven back up by thermal convection caused mainly by hot basaltic dikes and intrusions emplaced beneath the spreading center. A penetration depth of the convective circulation amounting to 2 to 3 km beneath the sea floor has been estimated for the Galapagos spreading center (Williams et al., 1975) and one of several kilometers for the Reykjanes spreading center in Iceland (Palmason and Semundsson, 1974). Estimates of the

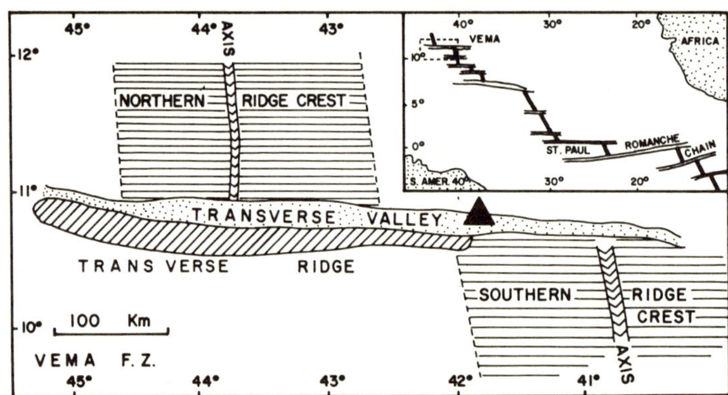

FIG. 9. Morphology of the Vema Fracture Zone, from Van Andel et al. (1967). Site where the sulfide-containing metabasalts were dredged (P7003-25) is shown with a triangle.

permeability of the oceanic crust are consistent with such circulation (Palmason, 1967; Deffeyes, 1970; Lister, 1972). Metalliferous sediments commonly found in the vicinity of spreading centers are probably formed by deposition of metals supplied by the hydrothermal solutions discharging through the sea floor. These same sub-bottom hydrothermal systems operating at spreading centers may also cause the creation of metal sulfide deposits within the basaltic oceanic crust, including the sulfide mineralizations of the type described in this paper.

Sources of metals and sulfur for the hydrothermal systems

The composition of the hydrothermal solutions upon discharge is determined by processes occurring during their sub-sea-floor circulation, namely by: (a) reactions with sediments, (b) reactions with the basaltic oceanic crust, and (c) contributions from deep-seated (mantle) volatile sources. Process (a), while important in situations such as the Red Sea geothermal system, is negligible in most oceanic spreading centers due to the lack or paucity of sediments close to a spreading axis.

That elements may be contributed to the hydrothermal solutions by direct degassing of the mantle at spreading centers and in deep fracture zones, process (c), is suggested by the discovery of excess ^3He in deep ocean water above the East Pacific Ridge (Clark et al., 1969; Craig et al., 1975) and above the Mid-Atlantic Ridge near the Gibbs Fracture Zone (Jenkins et al., 1972). The excess ^3He has been found in some cases to be associated with anomalous maxima in the concentrations of Fe, Cu, and other metals in seawater (Brewer et al., 1972).

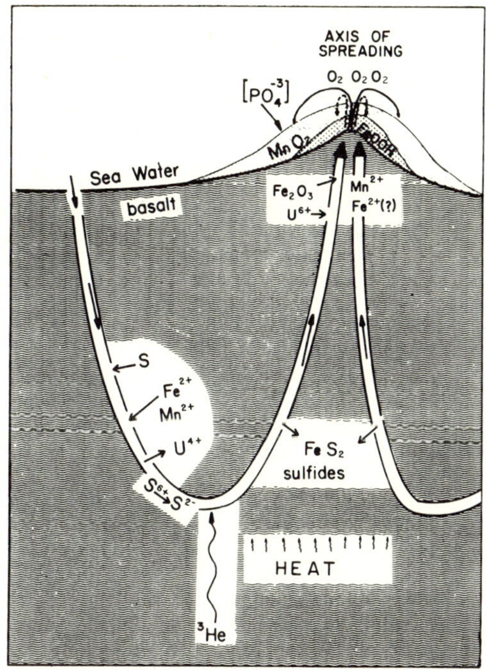

Fig. 11. Qualitative scheme illustrating the hydrothermal model of metallogenesis at spreading centers. Fe and Mn stand for metals leached from the basaltic crust. U stands for elements which may be lost by the hydrothermal waters and acquired by the basaltic crust. ^3He stands for volatile components supplied by the upper mantle. PO_4^{-3} stands for elements scavenged from seawater by FeOOH and MnO_2 particles. For further discussion of this model, see Bonatti, 1975.

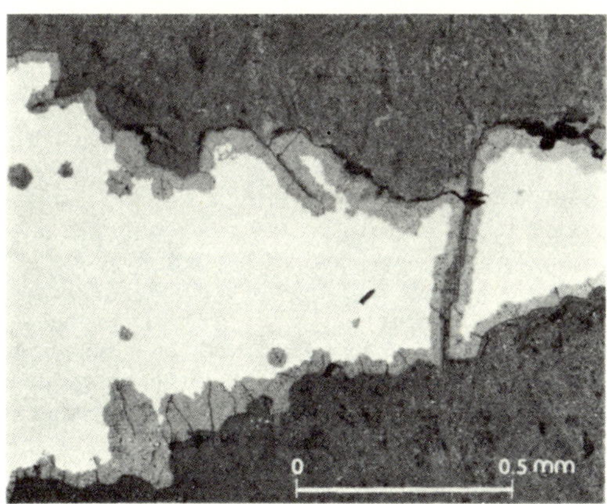

Fig. 10. Metabasalt P7003-25N. Microphoto (in reflected light) of a vein containing chalcopyrite (white) with a thin alteration rim of Fe-hydroxide (gray). Maximum thickness of this vein is about 2 mm.

According to Bostrom (1974) most of the elements enriched in metalliferous sediments from spreading centers are supplied by mantle-derived, CO_2-rich volatile phase. Whatever their relative importance in this context, mantle-derived volatile components probably travel to the sea floor largely via sub-bottom hydrothermal systems. Sulfur may be one of the elements contributed by this mechanism to the hydrothermal systems.

Reactions between seawater and basalt (and/or gabbro) occurring at temperatures of up to a few hundred degrees centigrade, process (b), result in leaching of metals and silica from the basalt (Bonatti and Nayudu, 1965; Corliss, 1971). This process, which probably constitutes the main source of metals for sub-sea-floor hydrothermal systems at spreading centers, has been verified experimentally in the laboratory (Krauskopf, 1956, 1957, 1967; Bischoff and Dickson, 1975; Hajash, 1975). The experiments by Hajash (1975) have shown that as a result of seawater-basalt reactions in the 200° to 500°C temperature range, the original mildly alkaline, oxygenated seawater becomes an acidic, reducing solution highly enriched in Fe, Mn, and Cu. A similar metal enrichment is noted in natural hydrothermal systems where seawater circulates within an igneous basic substratum (Bonatti, 1975, and references therein). Data from both natural and experimental hydrothermal systems indicate that some elements are lost by the solution and gained by the basalt; for instance Mg among the major elements (Hajash, 1975; Bischoff and Dickson, 1975) and U among the trace elements (Rydell and Bonatti, 1973).

Deposition of sulfides in the hydrothermal systems

Spooner and Fyfe (1973) point out that a number of redox reactions leading to increasingly reducing conditions are possible during the sub-bottom hydrothermal circulation, as exemplified by:

$$\underset{\text{(fayalite)}}{11Fe_2SiO_4} + SO_4^{-2} + 4H^+ = \underset{\text{(magnetite)}}{7Fe_3O_4} + \underset{\text{(pyrite)}}{FeS_2} + 11SiO_2 + 2H_2O$$

Precipitation of metal sulfides is likely under such conditions, provided metals and sulfur are present in the solution. The concept that metal sulfides are formed during high-temperature basalt-seawater reactions is strongly supported by laboratory experiments. Hajash (1975) produced chalcopyrite and pyrrhotite during seawater-basalt reactions at 400° and 500°C, while the original seawater solution became strongly depleted in SO_4^{-2}. The sulfide ion can be supplied by various sources, the main one probably being the reduction of seawater SO_4^{-2} during the sub-bottom circulation through a number of redox reactions exemplified by the reaction written above. It is significant in this context that in geothermal systems where seawater circulates within basalt the SO_4^{-2} content of the solutions at discharge is generally drastically lowered relative to seawater (see Bonatti, 1975, and references therein). A similar depletion of SO_4^{-2} is noted in seawater solutions which have reacted with basalt in laboratory experiments (Hajash, 1975; Bischoff and Dickson, 1975). While a portion of the missing SO_4^{-2} may precipitate in the form of anhydrite, most of it is probably reduced and taken out of the solution as sulfide.

Another source of sulfide ion is the sulfur trapped in oceanic basalt. Ocean ridge basalts are estimated to contain an average of about 800 ppm sulfur (Moore and Schilling, 1973) some of which can be extracted by the circulating hydrothermal solutions. Such extraction is indicated by data showing that fresh basalts from the Mid-Atlantic Ridge between 24° and 30°N contain from 900 to 1,400 ppm sulfur, while adjacent greenstones, presumably subjected to hydrothermal metamorphism, contain as little as 50 ppm (Scott and Frank, 1974). A third possible source of sulfur consists of the aforementioned volatiles of direct mantle derivation. The relative contribution of each of these sources is hard to assess quantitatively at present.

The factors determining transport, selective deposition, and fractionation of the various metal species in hydrothermal solutions are complex. These factors are in principle identical to those operating in hydrothermal systems on land and have been treated extensively in the literature (Barnes and Czamanske, 1967; Skinner and Barton, 1974).

Fractionation of metals in sub-sea-floor hydrothermal systems

We assume that Fe and Mn are initially leached from the basalt by the hydrothermal water in about the same ratio as they are contained in the basalt. This assumption is based on theoretical considerations (both elements occupy the same sites in basaltic Fe-Mg silicates and Fe-Ti oxides) and on laboratory experiments showing that, by flushing hot seawater or slightly acidic solutions through ground basalt, Fe and Mn are extracted in roughly the same ratio as they are contained in the rock (Krauskopf, 1956; 1957). In the basalt-seawater experiments of Bischoff and Dickson (1975), performed in sealed vessel at 200°C and over extended periods of time, the solution resulting from the reactions and sampled at regular intervals starting 24 hours after initiation of the runs contained Fe and Mn in ratios substantially lower than those of basalt. These results could be explained by assuming that in a closed system and with a relatively low seawater/basalt ratio, as in Bischoff and Dickson's experiments, Fe may precipitate rapidly from the initial solution, lowering its initial Fe/Mn ratio.

In the deeper portions of sub-sea-floor hydrothermal systems, where constantly circulating solutions, slightly reduced and acidic conditions, and temperatures up to a few hundred degrees centigrade might be expected, Fe as initially extracted from basalt is probably kept in solution, unless brought down by sulfides.

During the sub-bottom hydrothermal circulation the partition coefficients of the various metals between sulfide solid phases and solution are determined by a number of variables: among them are, the activity of the pertinent ions, temperature, redox conditions, pH, etc. Cu and Zn, in addition to Fe, partition strongly with sulfide phases in a range of conditions reached in hydrothermal systems, as shown by experimental work (Barnes and Czaman-

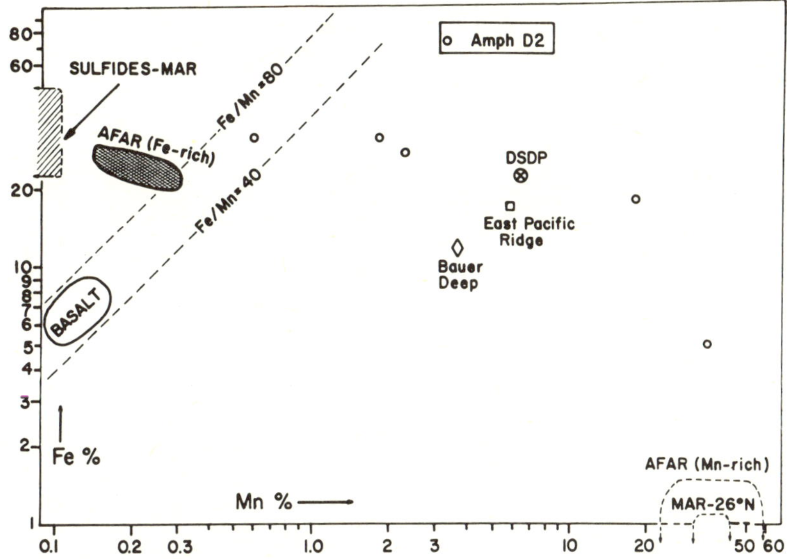

FIG. 12. Fe versus Mn ratio in a number of metalliferous sediments from spreading centers, compared with the ratio in ocean-ridge basalts and in metal sulfide mineralizations from the equatorial Mid-Atlantic Ridge. Field between dashed lines indicates the range of Fe/Mn ratios in oceanic basalts. This diagram illustrates the Fe-Mn fractionation in hydrothermal systems, with the Fe/Mn ratio higher than basalt in sulfide deposits and lower than basalt in metalliferous sediment deposits. Source of data for sulfide samples is this paper and Bonatti et al., 1976c; for metalliferous sediments is Bonatti, 1975.

ske, 1967; Skinner and Barton, 1974) and by their abundance in hydrothermal metal sulfide deposits. Other elements such as Mn are instead kept preferentially in solution during sulfide precipitation. As a result, fractionation of the metals occurs, particularly of Fe from Mn, so that in those hydrothermal systems where abundant sulfides have been deposited beneath the sea floor, the solutions at discharge may show substantially lowered Fe/Mn ratios. Additional fractionation may result from deposition of iron oxide (hematite) during the near surficial, partly oxygenated return flow of the sub-bottom circulation, while Mn is still kept in solution. Fe-Mn fractionation by deposition of Fe-sulfides and hematite has been observed in the basalt-seawater laboratory experiments (Hajash, 1975). Bonatti (1975) suggested that this type of sub-sea-floor fractionation may partly explain why most "hydrothermal" metalliferous sediments from spreading centers contain much lower Fe/Mn ratios than basalt, from which both metals were presumably extracted, as illustrated in Figure 12. Conversely, metal sulfide deposits from the oceanic crust contain Fe/Mn ratios higher than basalt (Fig. 12). A similar fractionation of Fe and Mn is suggested by the geochemistry of metal sulfide deposits and of metalliferous sediments from the Apennine ophiolite complexes, presumably formed at a Mesozoic spreading center by processes identical to those discussed in this paper (Bonatti et al., 1976c).

"Massive" sulfide deposits at modern and ancient spreading centers

The model outlined above predicts that "massive-type" sulfide deposits exist within the oceanic crust, as also suggested by Sillitoe (1973), and that they originated at spreading centers by the same processes which gave rise to the "disseminated"- and "stockwork"-type mineralizations described earlier in this paper. In fact, "disseminated"- and "stockwork"-type mineralizations are commonly found closely associated with "massive"type in land deposits. "Massive" sulfide deposits are widespread in ancient fragments of oceanic crust presently exposed on land, as in ophiolite complexes. These ophiolitic sulfide deposits are frequently associated with metalliferous sediments, similar to those formed in modern spreading centers, as seen in the Mesozoic ophiolites from Cyprus (Elderfield et al., 1972; Constantinou and Govett, 1973) and from the Apennines (Bonatti et al., 1976b) and in Paleozoic ophiolites from Newfoundland (Duke and Hutchinson, 1974).

We conclude that processes of metallogenesis similar to those occurring at modern spreading centers took place also in ancient, Paleozoic, and Mesozoic spreading centers and are responsible for the generation of the ophiolitic "massive" sulfide ore deposits.

Summary and Conclusions

(1) Stockwork- and disseminated-type Cu-Fe sulfide deposits have been found in metabasalts at the Romanche and Vema offsets of the Mid-Atlantic Ridge. It is probable that the basalts were originally emplaced at the axial zone of the Ridge.

(2) The major component of the deposits is chalcopyrite, partly altered to Fe-hydroxide, and atacamite plus a number of minor Cu oxide, chloride, and Cu-Fe sulfide phases. Pyrite and pyrrhotite are also present.

(3) We suggest that the sulfides were deposited from hydrothermal systems which operate in the oceanic crust at spreading centers. Reactions by hydrothermal water-basalt supply metals to the water; the sulfide ion is supplied also by reduction of seawater SO_4^{-2}. Minor contributions from mantle volatiles are also possible.

(4) Fractionation of metals, particularly of Fe from Mn, occurs during sulfide deposition from hydrothermal systems. The same hydrothermal systems which form sulfide deposits are responsible, upon their discharge through the sea floor, for the formation of metalliferous sediments associated with spreading centers.

(5) "Massive" sulfide deposits are probably formed at spreading centers and are present in the oceanic crust. Sulfide ore deposits from ophiolite complexes are ancient equivalents of sulfide deposits from modern spreading centers and were probably formed by the same processes.

Acknowledgments

We are grateful to W. Melson, E. Jarosevich, I. Ridley, and M. Adams for help in performing some of the electron probe analyses and to Marylou Zickl for editing and typing the manuscript. Research was supported by the National Science Foundation Grant OCE75-14536 and (IDOE) Grant GX 40428, and the Office of Naval Research Contract N00014-75-C0210. The paper is a contribution from the Rosenstiel School of Marine and Atmospheric Science, University of Miami, and is No. 2419 from the Lamont-Doherty Geological Observatory of Columbia University.

E. B.
LAMONT-DOHERTY GEOLOGICAL OBSERVATORY
COLUMBIA UNIVERSITY
PALISADES, NEW YORK 10964

B.-M. G.-H. AND J. H.
ROSENSTIEL SCHOOL OF MARINE AND ATMOSPHERIC SCIENCE
UNIVERSITY OF MIAMI
MIAMI, FLORIDA 33149
November 6, 1975; May 13, 1976

REFERENCES

Arrhenius, G., and Bonatti, E., 1965, Neptunism and volcanism in the ocean, *in* Sears, M., ed., Progress in oceanography: London, Pergamon Press, v. 3, p. 7–22.

Barnes, H. L., and Czamanske, G. K., 1967, Solubility and transport of ore minerals, *in* Barnes, H. L., ed., Geochemistry of hydrothermal ore deposits: New York, Holt, Rinehart and Winston, p. 334–441.

Baturin, G. N., 1971, Sulfide ores in the Arabic-Indian Ridge, *in* Zenkevitch, L. A., ed., The history of the world ocean: Moscow, Nauka, p. 259–265.

Bianchi, G., and Longhi, P., 1973, Copper in sea water, potential pH diagrams: Corrosion Science, v. 13, p. 8–59.

Bischoff, J. L., and Dickson, F. W., 1975, Sea water-basalt interaction at 200°C and 500 bars: Earth Planet. Sci. Letters, v. 25, p. 385–397.

Bonatti, E., 1973, Origin of the offsets of the Mid-Atlantic Ridge in fracture zones: Jour. Geology, v. 81, p. 144–156.

—— 1975, Metallogenesis at oceanic spreading centers: Ann. Rev. Earth Planet. Sci., v. 3, p. 401–431.

—— and Joensuu, O., 1966, Deep sea iron deposits from the South Pacific: Science, v. 154, p. 643–645.

Bonatti, E., and Nayudu, Y. R., 1965, Origin of manganese nodules on the ocean floor: Am. Jour. Sci., v. 263, p. 17–32.

Bonatti, E., Guerstein-Honnorez, B.-M., Honnorez, J., and Stern, C., 1976a, Hydrothermal pyrite concretions from the Romanche Trench (equatorial Atlantic): Metallogenesis in oceanic fracture zones: Earth Planet. Sci. Letters, v. 32, p. 1–10.

Bonatti, E., Zerbi, M., Kay, R., and Rydell, H., 1976b, Metalliferous deposits from the Apennine ophiolites: Mesozoic equivalent of deposits from modern spreading centers: Geol. Soc. America Bull., v. 87, p. 83–94.

Bostrom, K., 1974, The origin and fate of ferromanganoan active ridge sediments: Stockholm Contrib. Geol., v. 27, p. 129–243.

—— and Peterson, M. N. A., 1966, Precipitates from hydrothermal exhalations on the East Pacific Rise: ECON. GEOL., v. 61, p. 1258–1265.

Brewer, P. G., Spencer, D. W., and Robertson, D. E., 1972, Trace element profiles from the GEOSECS test station in the Sargasso Sea: Earth Planet. Sci. Letters, v. 16, p. 111–116.

Cabri, L. J., and Hall, S. R., 1972, Mooihoekite and haycockite, two new copper-iron sulfides and their relationship to chalcopyrite: Am. Mineralogist, v. 57, p. 689–708.

Clarke, W. B., Beg, M. A., and Craig, H., 1969, Excess ^3He in the sea: Evidence for terrestrial primordial helium: Earth Planet. Sci. Letters, v. 6, p. 213–217.

Constantinou, G., and Govett, G. J. S., 1973, Geochemistry and genesis of the Cyprus sulfide deposits: ECON. GEOL., v. 68, p. 843–858.

Corliss, J. B., 1971, The origin of metal-bearing submarine hydrothermal solutions: Jour. Geophys. Research, v. 76, p. 128–138.

Craig, H., Clarke, W. B., and Beg, M. A., 1975, Excess ^3He in deep water on the East Pacific Rise: Earth Planet. Sci. Letters, v. 26, p. 125–132.

Deffeyes, K. S., 1970, The axial valley: A steady-state feature of the terrain, *in* Johnson, H. and Smith, B. L., eds., Megatectonics of continents and oceans: New Brunswick, N. J., Rutgers Univ. Press, 194 p.

Dmitriyev, L. V., Barsukov, V. L., and Udintsev, G. B., 1970, Ocean rift zones and the problem of ore formation: Geochemija, v. 8, p. 935–944.

Duke, N. A., and Hutchinson, R. W., 1974, Geological relationships between massive sulfide bodies and ophiolitic volcanic rocks near York Harbor, Newfoundland: Canadian Jour. Earth Sci., v. 11, p. 53–69.
Elderfield, H., Gass, I. G., Hammond, A., and Bear, L. M., 1972, The origin of ferromanganese sediments associated with the Troodos massif of Cyprus: Sedimentology, v. 19, p. 1–19.
Elder, J. W., 1965, Physical processes in geothermal areas: Am. Geophys. Union Mon. 8, p. 211–239.
Gorini, M., 1976, The tectonic fabric of the equatorial Atlantic and adjoining continental margins: Unpub. Ph.D. thesis, Columbia Univ.
Hajash, A., 1975, Hydrothermal processes along mid ocean ridges: An experimental investigation: Contrib. Mineralogy Petrology, v. 55, p. 245–258.
Jenkins, W. J., Beg, M. A., Clarke, W. B., Wangersky, P. J., and Craig, H., 1972, Excess ^{3}He in the Atlantic Ocean: Earth Planet. Sci. Letters, v. 16, p. 122–126.
Lister, C. R. B., 1972, On the thermal balance of a mid-ocean ridge: Royal Astron. Soc. Geophys. Jour., v. 26, p. 515–535.
Krauskopf, K. B., 1956, Separation of manganese from iron in the formation of manganese deposits: Internat. Geol. Cong., 20th, Mexico City, 1956, Repts., p. 119–131.
—— 1957, Separation of manganese from iron in sedimentary processes: Geochim. et Cosmochim. Acta, v. 12, p. 61–84.
—— 1967, Introduction to geochemistry: New York, McGraw-Hill Book Co., p. 243–257.
Moore, S. G., and Schilling, J. G., 1973, Vescicles, water and sulfur in Reykjanes ridge basalt: Contrib. Mineralogy Petrology, v. 41, p. 105–118.
Palmason, G., 1967, On heat flow in Iceland in relation to the Mid-Atlantic Ridge, *in* Iceland and the Mid-Atlantic Ridge: Soc. Sci. Iceland, v. 38, p. 111–127.
—— and Semundson, K., 1974, Iceland in relation to the Mid-Atlantic Ridge: Ann. Rev. Earth Planet. Sci., v. 2, p. 25–46.
Ramdohr, P., 1969, The ore minerals and their intergrowth: Oxford and New York, Pergamon Press, 1174 p.
Rydell, H. S., and Bonatti, E., 1973, Uranium in submarine metalliferous deposits: Geochim. et Cosmochim. Acta, v. 37, p. 2557–2565.
Scott, R. B., and Frank, D. J., 1974, Distribution of sulfur in the oceanic crust: Geol. Soc. America Abstracts with Programs, v. 6, p. 945.
Sillitoe, R. H., 1973, Environments of formation of volcanogenic massive sulfide deposits: Econ. Geol., v. 68, p. 1321–1326.
Skinner, B. J., and Barton, P. B., 1974, Genesis of mineral deposits: Ann. Rev. Earth Planet. Sci., v. 2, p. 183–204.
Skornyakova, I. S., 1964, Dispersed iron and manganese in Pacific Ocean sediments: Internat. Geol. Rev., v. 7, p. 2167–2174.
Spooner, E. T. C., and Fyfe, W. S., 1973, Sub-sea-floor metamorphism, heat and mass transfer: Contrib. Mineralogy Petrology, v. 42, p. 287–304.
Van Andel, T. H., Corliss, J. B., and Bowen, V. T., 1967, The intersection between the Mid-Atlantic-Ridge and the Vema Fracture Zone: Jour. Marine Research, v. 25, p. 343–351.
Yoder, H. S., and Tilley, C. E., 1962, Origin of basalt magmas: Jour. Petrology, v. 3, p. 342–532.
Williams, D. L., Von Herzen, R. P., Sclater, J. G., and Anderson, R. N., 1974, Lithospheric cooling and hydrothermal circulation in the Galapagos spreading center: Roy. Astron. Soc. Geophys. Jour., v. 38, p. 587–608.

27

Reprinted from *Nature* **262**:567–569 (1976)

IMPLICATIONS OF METAL DISPERSION FROM SUBMARINE HYDROTHERMAL SYSTEMS FOR MINERAL EXPLORATION ON MID-OCEAN RIDGES AND IN ISLAND ARCS

D. S. Cronan

METAL enrichments in sediments on open ocean mid-ocean ridge segments and in island arcs have been known for some time[1–4], but, to date, no deposits have been found which are rich in copper, zinc and other heavy metal sulphides to a degree comparable to those in the Red Sea. The fact that concentrated brines were thought to have played an essential role in the formation of the Red Sea metalliferous sediments may have inhibited the search for similar deposits elsewhere, because brines of the Red Sea type are unlikely to occur in the open ocean. The discovery at Matupi Harbour, New Britain, however, that high salinity is not a prerequisite for metal enrichment in hydrothermal solutions[5], indicates that any area of submarine volcanic activity has a potential for high grade metalliferous sediment formation given suitable depositional conditions. In this report the dispersion of metals from submarine hydrothermal sources is discussed, and its usefulness in developing guidelines for undersea mineral exploration in volcanic areas is considered.

Metalliferous sediments on mid-ocean ridges and in island arcs are thought to form largely as a result of hydrothermal leaching of the oceanic crust by circulating seawater, which in turn becomes both acid and reduced[6,7]. Precipitation of the metals so derived gives rise to deposits which are characteristically rich in iron oxides and sometimes in a variety of other metals such as manganese, copper, zinc and lead; and occasionally, as in the Red Sea, include metal sulphides of localised distribution and very variable composition. It is the latter which are of potential economic value.

Evidence that mid-ocean ridge sediments vary in composition over short distances in the open ocean, and thus may locally be of high grade has been obtained on the Mid-Atlantic Ridge at 45°N (ref. 8). The chemical variations, particularly in the relative dispersion of iron and manganese, were thought to be related partly to the proximity of the samples analysed to submarine hydrothermal sources of the metals and partly to their selective precipitation on mixing with seawater. Unfortunately, the implications of these variations, in terms of geochemical exploration, for metalliferous sediments could not be evaluated because the location of the presumed hydrothermal outlets on the sea floor were not known. More recently, the selective dispersion and precipitation of iron and manganese from a known hydrothermal source has been noted in the FAMOUS area on the Mid-Atlantic Ridge (X. Le Pichon, personal communication), and hydrothermal manganese deposits have been found on the Mid-Atlantic Ridge at 26°N (ref. 9). Nevertheless, there has been no systematic study published to date of metal dispersion from a submarine hydrothermal source in the deep ocean. Fortunately, however, the likely nature of deep-sea hydrothermal metal dispersion can be deduced from studies of submarine hydrothermal systems in the Red Sea and off Santorini. On this basis a model has been constructed which attempts both to predict the most favourable environments for high grade metalliferous sediment formation on the ocean floor and to aid in the location of the sediments.

Metalliferous sediments of hydrothermal origin have been studied in both the Red Sea, and off the volcano of Santorini in the Cyclades Volcanic Arc in the eastern Mediterranean Sea. In each case there is a dispersion halo of metals around the hydrothermal source, showing, among other things, the selective dispersion of manganese from iron[10,11] (Fig. 1). A similar sequence has been observed around hydrothermal springs debouching into Matupi Harbour, New Britain[12]. These variations probably reflect the well known separation of manganese from iron with changing *p*H and *E*h (ref. 13). Poorly oxidised or reduced acid hydrothermal solutions precipitate the metals selectively on mixing with more alkaline oxidising seawater. Metals other than manganese and iron also exhibit a fractional precipitation in submarine hydrothermal systems. In the Atlantis II Deep of the Red Sea, for example, the first formed precipitates on discharge of the hydrothermal solutions are heavy metal sulphides which occupy a restricted area of the Deep adjacent to the hydrothermal vents[14] Similarly, sulphides occur underneath the sediments rich in iron oxides near the hydrothermal vents in the caldera of Santorini[15], and immediately adjacent to fumaroles off the Island of Vulcano in the Eolian Archipelago, Western Mediterranean[16]. It seems, therefore, that a model for the complete sequence of precipitates on mixing of hydrothermal solutions with seawater would include the fractional dispersion and precipitation of metal sulphides, followed by iron silicates and/or iron oxides, and finally manganese oxides.

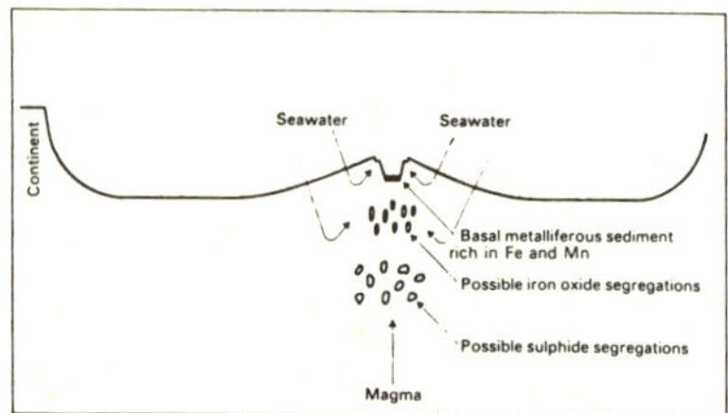

Fig. 1 Hypothetical vertical sequence of precipitates on mixing of ascending metal-bearing hydrothermal solution with seawater beneath mid-ocean ridge crests in the open ocean. This sequence occurs horizontally on the sea floor around hydrothermal centres in restricted environments like the Atlantis II Deep in the Red Sea and in shallow water volcanic areas, indicating that the nature of the environment of deposition controls the initiation of the depositional process.

This complete sequence of precipitates has not been reported to date on mid-ocean ridges in the open ocean. The iron- and manganese-rich metalliferous sediments characteristic of such ridges, however, are compositionally similar to the widely dispersed iron and manganese oxides around the Atlantis II Deep where the complete sequence is observed. This suggests that the former may be the equivalent of the sediment in the outer zone of the Atlantis II dispersion halo. If this is so, the sulphide deposits which the model predicts should be present in association with the iron and manganese oxides may occur either in restricted locations near to the hydrothermal vents or within the oceanic crust below the oxide deposits. The general physicochemical nature of seawater over mid-ocean ridge crests in the open ocean would favour the latter alternative in most cases. Oxidising seawater can be expected to penetrate fractures and fissures in the upper part of the oceanic crust leading to mixing with ascending hydrothermal solutions and the precipitation of metal sulphides before the solutions reach the surface (Fig. 1). Disseminated sulphides within rocks of the oceanic crust have been reported by several authors[17-19]. Special conditions leading to a reducing environment at elevated temperature, such as occur in the Atlantis II Deep of the Red Sea, would be required to delay sulphide precipitation until the hydrothermal solutions debouch on to the sea floor.

If sulphides are largely precipitated from submarine hydrothermal solutions within the oceanic crust, where else on the sea floor other than in the Red Sea and in small restricted embayments around volcanic islands might they be expected to form? One obvious locale is within the deep fracture zones which cut mid-ocean ridges in all the oceans and which often expose the oceanic crust to considerable depths. Deep hydrothermal circulation under these fracture zones coupled with the possibility of locally reducing conditions in narrow deep basins, could lead to the precipitation of heavy metal sulphides. Bonatti *et al.*[20] have explained high heat flow in some fracture zones on the basis of subseafloor hydrothermal circulation, and have reported the occurrence of pyrite concretions of supposed hydrothermal origin in the Romanche Fracture Zone, Equatorial Atlantic. In addition, Scott *et al.*[21] have reported the occurrence of manganese deposits of probable hydrothermal origin in the Atlantis II Fracture Zone in the North Atlantic. An additional likely locale for heavy metal sulphide precipitation is in the basins associated with island arcs. These can be both volcanically active[22] and sufficiently topographically diverse to possibly lead to a locally restricted circulation and reducing conditions favourable for sulphide deposition. Ancient marginal seas associated with island arcs are thought to have been the environment of formation of some massive sulphide deposits now exposed on land[23]. Iron- and manganese-rich sediments similar to those on mid-ocean ridges have been found in association with the island arc system in the south-western Pacific Ocean[24], and submarine sulphide deposits may also be present.

It is evident from the above that mid-ocean ridge fracture zones and island arcs warrant more attention in the search for ocean floor metalliferous sediments. The dispersion model discussed here may additionally be of use in locating such deposits. The dispersion halo of manganese around the Atlantis II Deep deposit is up to 20 km in diameter even though the sulphide deposit itself is much smaller[25]. Thus in designing a grid sampling program to locate heavy metal sulphides on the ocean floor, one could look initially for manganese dispersion haloes on a wide-spaced sampling pattern, and, if located, design a smaller grid to search the vicinity for heavy metal sulphides. Tests of the model as an exploration tool are currently being conducted around the TAG hydrothermal field on the Mid-Atlantic Ridge at 26°N, and in a fracture zone transecting the East Pacific Rise at 9°S.

This work was supported by the NERC.

1 Bostrom, K., and Peterson, M. N. A., *Econ. Geol.*, **61**, 1258 (1966).
2 Bostrom, K., Peterson, M. N. A., Joensuu, O., and Fisher, D. E., *J. geophys. Res.*, **74**, 3261 (1969).
3 Bertine, K., *Geochim. cosmochim. Acta*, **38**, 629 (1974).
4 Cronan, D. S., in *The Sea*, **5** (edit. by Goldberg, E. D.), 491 (Wiley, New York, 1974).
5 Ferguson, J., and Lambert, J. B., *Econ. Geol.*, **67**, 25 (1972).
6 Corliss, J. B., *J. geophys. Res.*, **76**, 8128 (1971).
7 Spooner, E., and Fyfe, W., *Contrib. Miner. Petrol.*, **42**, 287 (1973).
8 Cronan, D. S., *Can. J. earth Sci.*, **9**, 319 (1972).
9 Scott, R. B., Rona, P. A., McGregor, B. A., and Scott, M. R., *Nature*, **251**, 301 (1974).
10 Bignell, R., Cronan, D. S., and Tooms, J. S., *Trans. Inst. Min. Metal.* (in the press).
11 Smith, P. A., and Cronan, D. S., *Oceanology International Conference*, Brighton, 111 (BPS Exhibitions, London, 1975).
12 Ferguson, J., and Lambert, J. B., *Econ. Geol.*, **67** (1972).
13 Krauskopf, K., *Geochim. cosmochim. Acta*, **12**, 61 (1957).
14 Backer, H., and Richter, H., *Geol. Rdsch.*, **62**, 697 (1973).
15 Butuzova, G., *Dokl. Acad. Nauk. SSSR*, **168**, 1400 (1966).
16 Honnorez, J., Honnorez-Guerstein, B., Valette, J., and Wauschkuhn, A., in *Ores in Sediments* (edit. by Amstutz, G., and Bernard, A.), 139 (Springer, Berlin, 1972).
17 Baturin, G. N., in *The History of the World Ocean* (edit. by Zenkevitch, L. A.), 259 (Nauka, Moscow, 1971).
18 Dmitriyev, L. V., Barsukov, V. L., and Udintsev, G. B., *Geochemistry*, **8**, 935 (1970).
19 Bonatti, E., Honnorez-Guerstein, B., and Honnorez, J., *Econ. Geol.* (in the press).
20 Bonatti, E., Honnorez-Guerstein, M., Honnorez, J., and Stern, C., *Earth planet. Sci. Lett.* (in the press).
21 Scott, R., Rona, P. A., Butler, L. W., Nalwalk, A. J., and Scott, M. R., *Nature*, **239**, 77 (1972).
22 Weissel, J. K., and Watts, A. B., *Earth planet. Sci. Lett.*, **28**, 121 (1975).
23 Taylor, G., *Trans. Inst. Min. Metal.*, **83**, 120 (1974).
24 Bertine, K., *Geochim. cosmochim. Acta*, **38**, 629 (1974).
25 Bignell, R., Cronan, D. S., and Tooms, J. S., *Trans. Inst. Min. Metal.* (in the press).

28

Original article prepared especially for this Benchmark volume

THE ORIGIN OF FERROMANGANOAN ACTIVE RIDGE SEDIMENTS

Kurt Boström

University of Luleå, Sweden

ABSTRACT

Sediments on spreading (active) oceanic ridges are exceptionally rich in Fe, Mn, and many trace elements, and poor in Al, Ti, and Si compared to other pelagic sediments. The accumulation rates for Fe, Mn, Ba, and many trace elements are very high. Ordinary sediment-distributing processes, like eolian transport, surface currents and biological transport, cannot explain these sediments that are found exclusively on active ridges. Furthermore, the flow rates and metal contents of Pacific deep waters are too low to explain the accumulation rates on the East Pacific Rise. Pelagic sediments are mixtures of a terrigenous phase and Fe-Mn-rich phase of very definite composition that are identical to the sediments found on the crestal areas of the East Pacific Rise.

A well-defined exponential relation exists between rate of ocean floor spreading and the ratio (Fe+Mn)/Al in oceanic sediments. In areas of slow or no spreading this ratio is about 0.65, whereas in areas with spreading rates close to 10 cm/year, the ratio can be about 100--that is, spreading rates and chemistry of the ridge sediments are probably controlled by the same process. Such sediments could form by hydrothermal leaching of basaltic rock by sea water at large depths in the crust in areas of high heat flow that removes Fe, Mn, Ba, Cu, and some other elements. Shallow hot springs, conversely, rarely separate Fe and Al sufficiently well to explain the formation of active ridge sediments; nor can this process explain the distribution of several trace metals, such as U, Ti and Sc.

Volcanism is commonly associated with large-scale emission of

---

*This article is based on a paper entitled "The origin and fate of ferromanganoan active ridge sediments" that appeared in Stockholm Contr. Geology 2:149-243. Section 9 (Other Fe-Mn deposits of submarine exhalative origin and related deposits), which dealt with continental ore deposits of presumed active ridge origin and ophiolites, has been deleted because of spatial restrictions. A postscriptum has been added that summarizes some important studies after 1973.

carbon dioxide. CO_2 is a major volatile phase of deep-seated origin and is often rich in Mn, Fe, Ba, and P, and poor in Al and Ti. A carbon dioxide phase could therefore be another carrier of several constituents in the active ridge sediments.

These exhalative deposits may locally resemble ores and show similarities with several ophiolite-manganese ores-jasper occurrences that seem to occur even in Archean rocks. Many active ridge sediments probably become subducted at an island arc or at the edge of a continent. During this process volatile constituents are driven off upwards, whereas more insoluble constituents remain behind. Such a process could explain some hydrothermal ores or some secondary exhalative deposits in island arc settings.

INTRODUCTION

This paper considers the active ridge sediments, which appear to occur on all spreading oceanic ridges (figure 1), and the processes by which these sediments are formed. Emphasis is placed on those deposits that are best developed, namely those in the Indopacific area. The ferroan muds in the brine-filled depressions in the Red Sea are not discussed in detail; they have been considered elsewhere (Degens and Ross 1969).

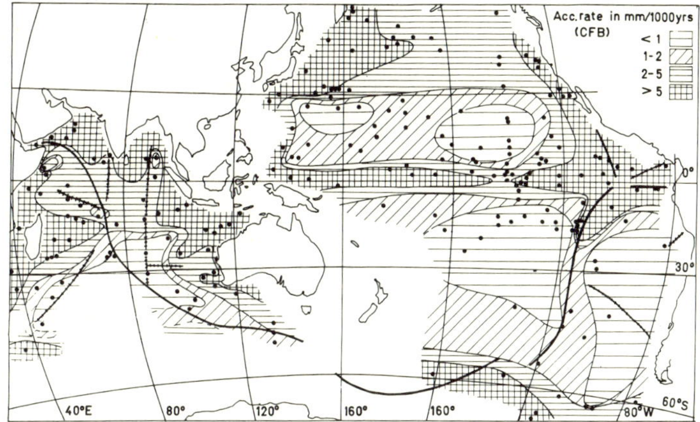

Figure 1. Heavy solid lines show the distribution of active (spreading) ridges in the Indo-Pacific area. Cross-hatched thin lines represent inactive ridges (as the Ninety-East Ridge at 90°E and the Nazca ridge at 80°W, 15°S) or ridges where the spreading center is short compared to the length of shear- and fracture zones, as the Chile ridge at 45°S in the Pacific and the South-West Indian Ocean ridge. Dashed lines represent fracture zones. The map also shows rate of sediment accumulation on a carbonate-free basis (Boström et al. 1973b). Solid dots show the points that have been used for the construction of the accumulation rate map for Mn in figure 3.

Historical Background

During the last twenty years our concepts of the Earth have changed at a pace without counterpart in modern geology. Thus, not until the late 50s and early 60s was it clear that many mid-oceanic ridges were centers of high heat flow; particularly rapid advances in this field took place between 1962 and 1965 (Revelle et al. 1952; von Herzen et al. 1966). Detailed studies of heat flow showed that much of the scatter in the values is due to topographic effects. Elder (1965), Bodvarsson et al. (1972), and Lister (1972) developed such observations into a "hydrothermal cooling model," in which oceanic water enters openings in lavas and acts as a heat conductor in addition to the thermal conduction in solid rock.

The concept of ocean floor spreading (Vine 1966) led to a new approach for studies of global tectonics. Morgan (1968) and Heirtzler et al. (1968) suggested that the Earth's crust could be treated as blocks or plates that move relative to each other. These plates fit into a global system of ridges, trenches, earthquake centers, high heat flow areas, etc., often referred to as the New Global Tectonics (Isaacs et al. 1968; Sykes et al. 1970). It is thus evident that some oceanic ridges have spreading centers (active ridges), while others do not (inactive ridges); see figure 1.

Until recently, the deep oceans were generally considered to be permanent features of the Earth. In these basins, debris accumulated for billions of years and formed a record spanning the major part of the Earth's geological history. Some sediment-filled seas, the geosynclines, were birthplaces of major orogenies with volcanism and mountain formation, but it was generally believed that no deep sea sediments took part in this cycle. A history of these concepts is given by Kuenen (1950, ch. 2).

The permanency of the ocean basins lead to many geochemical and sedimentological difficulties. Thus, the mass of these deep-sea deposits should be many times larger than the continental deposits, reaching many kilometers in average thickness (Kuenen 1946, 1950). However, geochemical balance calculations suggested that the mass of sediment derived from weathering must be considerably smaller (Goldschmidt 1954; Wickman 1954; Horn and Adams 1966). To circumvent these difficulties, it was suggested that pre-Tertiary erosion rates were considerably lower than the present ones, but the reasons for such differences are hard to defend (Gilluly 1955). Seismic profiles now show that the sediment cover in the deep ocean rarely exceeded 1 km in thickness and does not attain 100 m over large areas of the Pacific (Ewing et al. 1966; Ewing et al. 1970); the sediments are invariably underlain by basalt (data from the Deep Sea Drilling Project).

The concept of ocean floor spreading has revolutionized our thinking on marine sedimentation. In addition to the ordinary deep sea sediments (Murray and Renard 1891; Philippi 1910; Revelle 1944; Goldberg and Arrhenius 1958; Wedepohl 1960; El Wakeel and Riley 1961; and Landergren 1964), a new type of deep sea deposits, comprising active ridge deposits (or sediments) has been recognized. Up to 1964 there was a strong tendency to consider these iron-manganese-rich active ridge sediments as insignificant and local freaks of nature. However, it should be pointed out that a volcanic source for some components in deep sea sediments, such as Fe, Mn and Cu, had been suggested by several authors (von Gumbel 1878; Murray and Renard 1891; Arrhenius 1952; Wedepohl 1960).

The Nature of Active Ridge Sediments

Skornyakova (1964) was the first to demonstrate that the iron-manganese-rich sediments usually occurred on oceanic ridges in the Pacific; she suggested a volcanic origin for these sediments. Independently, Arrhenius and Bonatti (1965) pointed out the remarkable concentrations of Ba in East Pacific Rise sediments and suggested a volcanic origin for Ba. Boström and Peterson (1966) stressed the close correlation between high heat flow and the concentrations of Fe, Mn, and Cu in the sediments, suggesting an exhalative submarine volcanic origin. Concurrently with these studies East Pacific Rise sediments were analyzed for a volatile constituent, Hg, which should be a indicator of major degassing of the mantle, but the Hg results were not published until 1969 (Boström and Fisher).

It was early obvious that Mn was an element of great geochemical interest in the active ridge sediments. Geochemical balance calculations by Horn and Adams (1966) suggested that more manganese is present in the oceans than can be explained by continental weathering. Submarine weathering of basalt could hardly be the explanation, as demonstrated by various arguments collated in Boström (1967a). Thus, Goldberg et al. (1958) pointed out the absence of Al-Si-Ti-enriched residues, which should occur somewhere if Fe and Mn derived from submarine weathering. Possibly redox processes at depth in pelagic sediments or leaching of basalt in submarine volcanoes mobilized manganese (Krauskopf 1957; Boström 1967a), since the elevated temperatures should accelerate such processes. However, sediment cover is thin or absent in the central areas of the East Pacific Rise, contrary to what was earlier believed; consequently there are no thick sediment layers from which to leach manganese, etc. The active ridge sediments must therefore derive (1) from their substratum, or (2) by deposition of ordinary authigenic and diogenous matter, or (3) by deposition of some anomalous terrigenous matter.

These concepts were further developed by Boström et al. (1969b, 1969c, 1969d; 1971; 1972a), Boström (1970b), Fisher et al. (1969), and Bender et al. (1971). They demonstrated that:

(a) the sediments in areas of high heat flow and rapid crustal spreading are unusually rich in Fe, Mn, B, V, U, and As, and very poor in Al and Ti; the Si values are furthermore lower than is common in oceanic lutites. This paucity of aluminium is remarkable in view of the abundance of Al in most crustal deposits.

(b) these sediments are the first to form on newly generated oceanic crust (figure 2) and appear to form the basal sediment layer on all spreading ridges.

(c) such Fe-Mn-rich, Al-Ti-Si-poor sediments are characteristic of all spreading ridges, but not of inactive ridges such as the Ninety-East Ridge and the Nazca Ridge (Böstrom 1973) (see figure 2). They are not related to $CaCO_3$ deposition.

These results (a-c) have been verified by extensive studies of material obtained by the Deep Sea Drilling Project (von der Borch et al. 1970; Drever 1971; and Cook et al. 1971). Further studies have shown that:

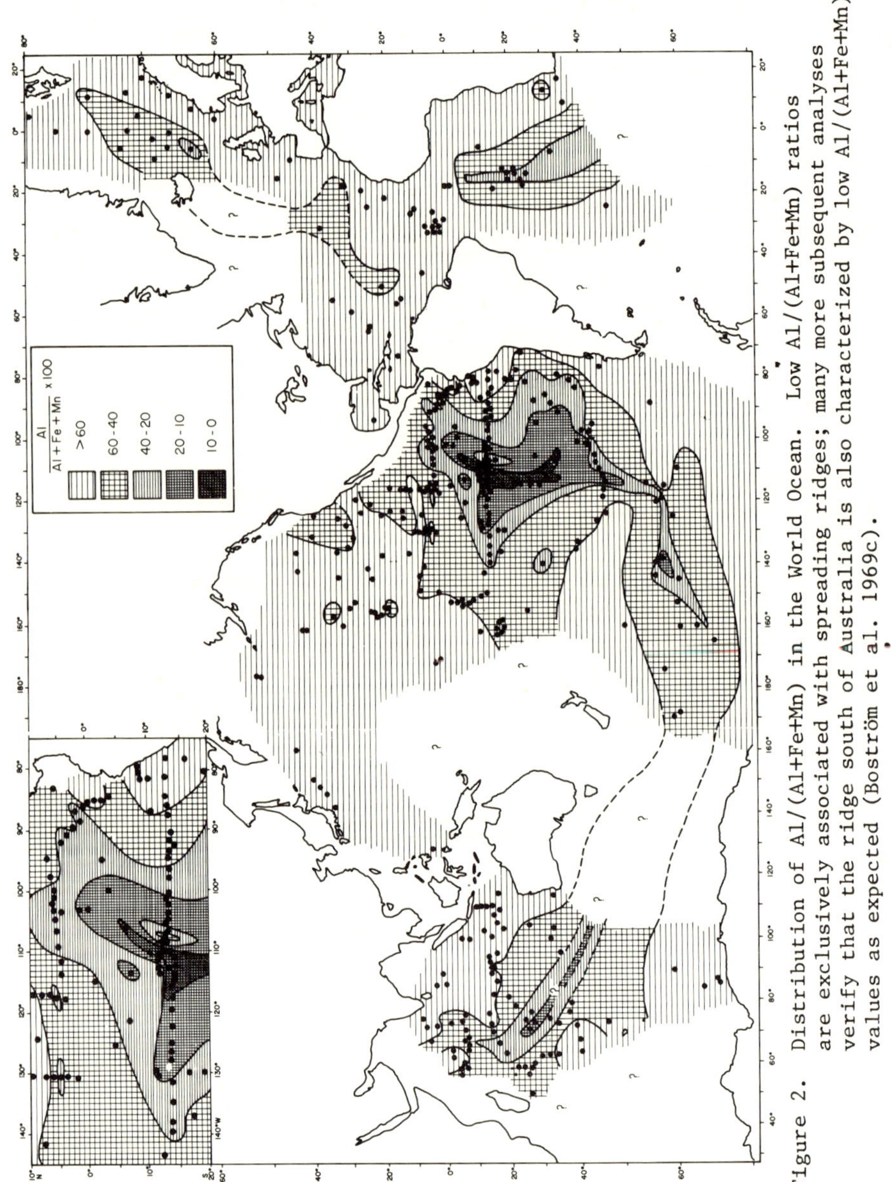

Figure 2. Distribution of Al/(Al+Fe+Mn) in the World Ocean. Low Al/(Al+Fe+Mn) ratios are exclusively associated with spreading ridges; many more subsequent analyses verify that the ridge south of Australia is also characterized by low Al/(Al+Fe+Mn) values as expected (Boström et al. 1969c).

(d) almost all Al and Ti in the oceans, even on active ridges is of continental origin, and only a small quantity derives from basaltic matter; active ridge volcanism delivers no Al and Ti of any significance (figure 2) (Boström 1970a, 1973; Boström et al. 1972a, 1972b, 1973b, 1974).
(e) the active ridge volcanism at its maximum intensity introduces iron and manganese and some other elements at rates that are unsurpassed elsewhere in the oceans except close to land (figure 3).

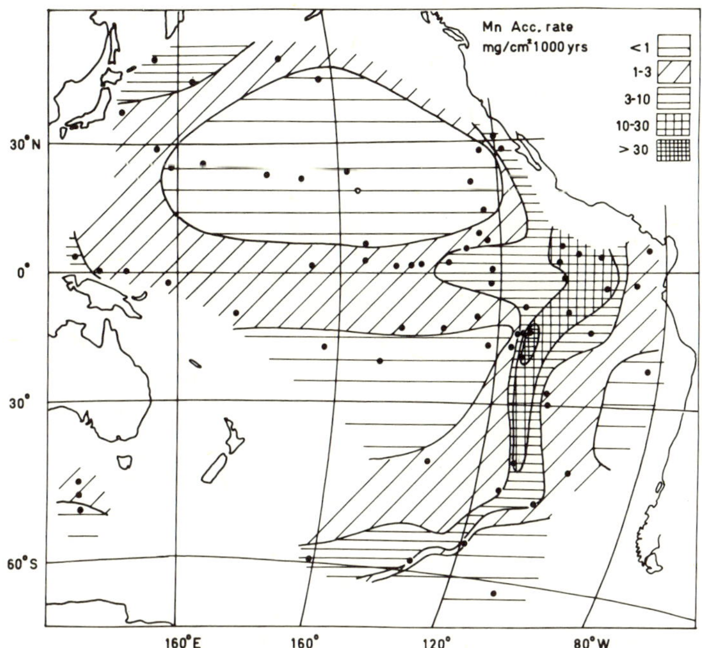

Figure 3. Accumulation rates of Mn in Pacific Ocean. Fe shows similar patterns (Boström et al. 1973b). Similar relations have also been observed for Cu and Ba except that their accumulation rates also are high in the equatorial regions.

(f) the high accumulation rates on the East Pacific Rise are not due to slumping, preferential winnowing or turbidite deposition of material into deep depressions, but are indeed indicative of a local origin, since some metalliferous sediments occur as mounds on top of isolated hills on the crest of the East Pacific Rise.

Thus (Boström et al. 1974), during cruise GS7202, much sediment was found also on some gentle slopes and hill crests, for instance at stations 33 and 35; piston coring showed that the sediment thickness at these stations exceeded 9 m. A short survey around station 35, which is located 80 km from the spreading center, revealed a well-developed sediment cap, about 80 km^2 in area, on top of a small hill that is elevated 50-200 m above surrounding valley floors (figure 4). This indicates that winnowing and turbidite flows cannot add sediments, but rather that such processes would detract from the high

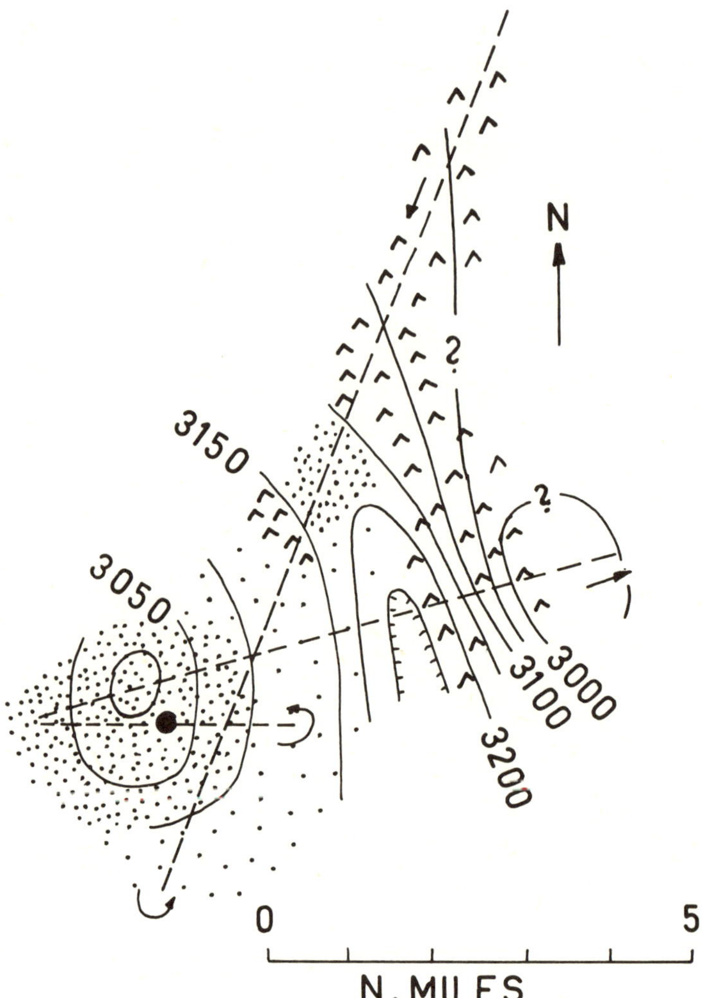

Figure 4. Physiography at Site 35. Black dot = location of core GS-7202-35 (14°47.9'S,113° 30.2'W). Dotting = sediments, heavy dotting = well developed sediment cover. Cusps = rocky terrain. Distances in nautical miles. Isobaths with 50 m intervals, except at top of the hill, where isobath corresponds to 3025 m. (Boström et al. 1974).

accumulation rates reported below. Reworking is furthermore excluded on paleontological grounds; the fossils show no indications of long transport, such as crushed tests or mixed faunas.

The recovered piston core, GS-7202-35P, is 8.50 m long and consists of dark brown ferruginous calcareous ooze; resemblance to sediments from the nearby stations 33, 34, and 36 and from the Risepac Expedition, (Boström et al. 1966, 1969b) is striking. The results of chemical analyses and accumulation rates for the core agree with previously found accumulation rates for Fe, Mn, etc. on the East Pacific Rise by Boström et al. (1969b), Boström (1970a) and Bender et al. (1971). The sediment-forming processes took place for a considerable time and were not short-lived ephemeral events (table 1).

The isotope distributions in the core are furthermore remarkable (Rydell et al. 1973b). The present activities of the U and Th isotopes suggest that at about 100,000 years B.P. the sediment was injected from below by Th-poor solutions containing much U with a $^{234}U/^{238}U$ activity ratio of 1.00. Hg, As, Fe, V, and P_2O_5 concentrations are also high, suggesting a joint deep-seated origin.

(g) The spreading rate and chemical composition of the active ridge sediments show a distinct relation (figure 5).

(h) Hemipelagic sediments are rich in little-degraded material and poor or lacking in authigenic constituents (Boström et al. 1972c), making it possible to assess the content of Al, Fe, Mn, Ba, Cu, etc. in the detrital fraction. Since most aluminium is located in detrital phases, one can then filter out the Fe, Mn, Ba, etc. in detrital phases and thus obtain an approximative but adequate estimate of how authigenic components are distributed in pelagic sediments (see table 1).

The authigenic accumulation rates (table 1), estimates of biological productivities (Bogorov 1967a, 1967b), and distances to nearest spreading center permit a quantitative study of how ridge processes and biological activities partake in the sediment forming processes. Combined effects exist for many constituents, but are particularly well pronounced for Ba, for which correlations with both volcanism and biological activity are clearly displayed. Similar two-fold trends probably exist for the P distribution, which is better correlated with volcanic sediments than with the $CaCO_3$ distribution.

The overall accumulation rates for several elements in the ocean (table 2) likewise show similar orders of magnitude as their river inputs. This is hardly surprising; the Earth has existed for about 4.5 billion years, and therefore the annual degassing can at most deliver $1/(4.5 \times 10^9)$ of the crustal abundance of a given constituent. This assumes a constant degassing rate, but probably most degassing took place during the first 1-2 billion years (Armstrong 1968). Any proof of mantle degassing as a source for active ridge sediments therefore has to be found elsewhere, as discussed below.

(i) Some element ratios of interest are Fe/Mn and (Fe + Mn)/(Co + Ni). Iron and manganese are separated to some extent during the sedimentation; much iron, mostly in detrital matter, codeposits near the continents, whereas Mn migrates to the open ocean, where well-oxidized sediments can form. Close to the continents, turbidites and much organic matter deposit, enhancing reducing conditions and thus the removal of Mn (Lynn et al. 1965). These sedimentation processes are dependant on variations in climatic conditions and sea level, and therefore marked fluctuations in content of Mn, Cu, Co, Ni, and other constituents occur in sediments close to the continents as a function of glacial-interglacial variations (Boström, 1970c; 1973; Boström et al. 1972c).

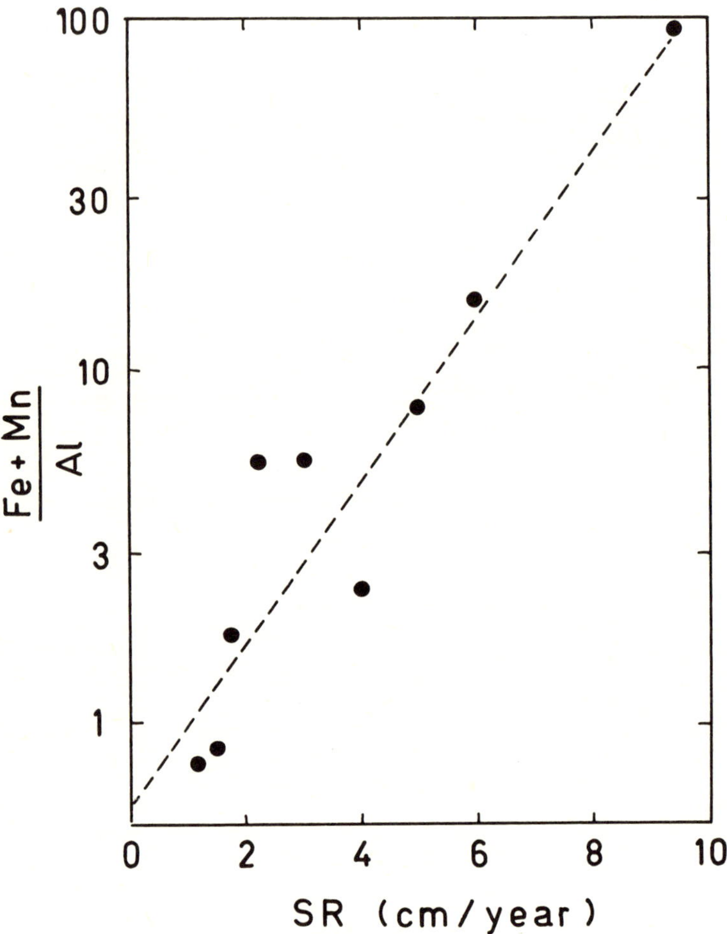

Figure 5. Relations between crustal spreading rates (SR) and the ratio (Fe+Mn)/Al in sediments (Pitman et al. 1968; Heirtzler et al. 1968; Herron 1972; Boström et al. 1969c). The correlation coefficient log $\frac{Fe+Mn}{Al}$ - SR is 0.93 and is significant at the 1% level. The regression line represents $\frac{Fe+Mn}{Al} = 0.59 \times e^{0.53SR}$. For SR = 0 $\frac{Fe+Mn}{Al}$ = 0.59; this value is close to the values 0.60-0.69 for average shale and continental crust. The highest point if from station GS 7202-35.

Co and Ni have a stronger tendency than do Fe and Mn to deposit in oxidizing environments far from the areas of hemipelagic sedimentation (figure 6), but the distribution pattern on the active ridges cannot be explained by such a mechanism. Reducing sediments, turbidites, and climatically controlled sedimentation processes are not

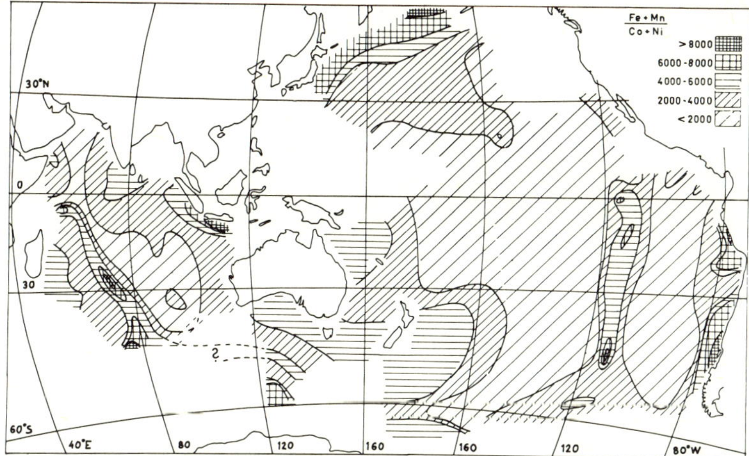

Figure 6. Relations between Fe, Mn, Co and Ni. The data suggest two major source areas for the elements; one on the continents and one on the active ridges.

known on the active ridges, nor is this distribution depth-controlled; otherwise this process should be operative on both active and inactive ridges. It thus seems that a source exists for several transition metals on the active ridges (Boström 1970a). Fe and Mn oxidize more readily than Ni and Co and should therefore precipitate first (Pourbaix et al. 1963), as figure 6 shows. The Fe and Mn distributions on the active ridges show similar trends, in which Fe deposits before Mn (Bonatti et al. 1966), but the trend is only poorly perceivable.

(j) Rare earth elements (REE) studies by Bender et al. (1971), Dymond et al. (1973) and Graham (1973) suggest a seawater origin for many elements. Similar origin may be suspected for As, Hg, and U.

(k) Isotope analyses of active ridge sediments (Veeh et al. 1971; Dasch et al. 1971; Boström et al. 1972b, 1974; Rydell et al. 1973a, 1973b) likewise suggest that a large fraction of the trace constituents in the active ridge sediments derive from seawater by adsorption. However, results by Fisher et al. 1969, Veeh et al. (1971), and Rydell et al. (1973b) also point out that there must be an additional source for some of the U in these cores. Leaching of basaltic rock is an adequate explanation in some cases, but probably cannot explain the postdepositional intrusion of U-rich solutions into the sediments at site GS-7202-35 and all anomalous $^{234}U/^{238}U$ ratios. (Rydell et al. 1973a, 1973b). Evidence for a deep-seated origin has also been furnished by lead-isotope data (Bender et al. 1971; Dasch et al. 1971; Dymond et al. 1973).

Isotope evidence for other elements, which occur abundantly in seawater, must be treated with more caution. The presence of seawater ratios for such isotopes, indicates that seawater was a diluting contaminant, but does not rule out the existence of a volcanic source for certain elements.

QUANTITIES OF BOTTOM WATERS AND BASALTIC ROCKS THAT CAN PARTAKE IN LEACHING PROCESSES

To understand how the active ridge sediments form, the quantity of water involved must be known. This can be estimated by studies of flow rates (a) through basaltic rocks and (b) over the ocean floor. The possible extent of leaching can also be evaluated from some geological evidence (c).

a) Water flowing through basaltic rocks.

a-1. Lister (1972) and Bodvarsson et al. (1972) suggested that a rapid cooling of crestal rocks occurs by open hydrothermal circulation perhaps down to depths of 5-10 km. However, their calculations do not show how much water is involved.

a-2. In Iceland, the permeability of the rocks is poor and permits permanent river runoff. Precipitation in areas with little rain (Sömme 1961) shows that less than 0.05 l/cm^2 year penetrates the basalts. Another estimate gives percolation rate of 0.004 l/cm^2 year for the hot spring area in Yellowstone National Park (data from Allen et al. 1935).

a-3. The heat flow near the East Pacific Rise can be up to 220 calories $/cm^2$ year. If all heat is carried by resurfacing heated seawaters at an average temperature of about 200^oC, the flow is 0.0011 l/cm^2 year.

b) Water flowing over the ocean floor, partaking in the formation of sediments.

Temperature and radon distributions in the East Pacific (Chung et al. 1972) indicate that the lowermost 120 m of the bottom waters are mixed. Furthermore, in several places in the Southeastern Pacific Knauss (1962) and von Herzen et al. (1972), found a temperature minimum in the bottom waters, below which point the temperatures increase adiabatically toward the bottom ($0.13^oC/km$). This temperature distribution must be expected if hot waters from below mix with cold overlying oceanic waters. The position over the bottom of this minimum varies, but for the supply rates given below it is assumed that the average value of 340 m over the bottom is representative of the water mass that participates in the sediment-forming processes.

^{14}C measurements (Knauss 1962) suggest flow rates of the bottom waters in the southeastern Pacific of about 16 km/year towards the northeast. The area with high accumulation rates of Mn, Ba, etc., extends 7-30° along this direction, varying with location of traverse and element. This means that bottom waters require 49-210 years or more to pass over this area. In a 340 m deep layer, at most 0.16-0.69 l/cm^2 year may participate in sediment-forming processes. No correction is made in this estimate for ^{14}C added from disintegrating calcite tests; thus the bottom water flows could be slower.

c) Geological evidence for the permeability and depth of penetration of water in oceanic basalts and the quantity of basaltic rock available for leaching.

It is well known that basaltic rock can be porous and rich in water conducting voids, as suggested by studies of groundwater behavior (Meinzer 1923) and submarine volcanoes (seamounts) (Harrison et al. 1959; McBirney 1971). Seventeen-hundred meter deep water conducting cracks are known in Icelandic basalts (Bjornsson et al. 1970), but large permeabilities are found along scoriaceous contacts between lava flows and interbeds. However, the tendency to develop pores and scoriaceous surfaces decreases rapidly with water pressure (McBirney 1971), as most rock cores from the deep drillings indicate (Moberly et al. 1971). Most deep sea basalts are therefore compact. Furthermore, joints tend to become sealed with increasing lithostatic pressure (Meinzer 1923). In oceanic areas with high heat flow (over 6.5 10^{-6} cal/cm^2 sec) the thermal gradients in the rocks exceed 160°C/km. At a depth of a few kilometers, the rocks become plastic under the lithostatic pressure; the presence of many extensive joints for transport of water therefore appears unlikely. Additional support for this conclusion comes from well drillings (Meinzer 1923), which show that most water-bearing strata are at depths of 0-100 m. Furthermore, many deep mines are dry; groundwater rarely starts to seep down before underground work has loosened up joints, etc.

Most hydrothermal cooling must hence be confined to shallow sections of the basalts, even if a few cracks possibly penetrate down to large depths. A few such cracks may transport much heat by conductive and convective flow, but the rock-seawater interaction is probably small. This implies that on the ridges permeability in basalt probably is low, even at the very top of the lavas, particularly if dyke injection is the main emplacement mechanism at spreading centers.

Another possibility is that rifts might open, subsequently filling by a sequence of flows (McBirney 1971), possibly forming several horizons of water-conducting voids. This type of infill should be most common on the Mid-Atlantic and Central Indian Ocean ridges, where well-developed rifts are present. The Atlantic and Indian Ocean ridges should thus be more likely places for formation of Fe-Mn-rich sediments by leaching of oceanic basalts. This is contrary to what is observed (figure 2).

Furthermore, most of the easily leached basaltic rock may not form at spreading ridges, judging from the paucity of submarine volcanoes and large abyssal hills on the East Pacific Rise; most volcanoes are consequently formed a long time after the generation of new crust. The total mass of extruded rock in identifiable seamounts and volcanoes is 2.4 x 10^7 km^3 (Menard 1964); if large abyssal hills that probably are volcanoes are included, this figure increases to 3 x 10^7 km^3. Assuming an average crustal age of 65 m.y. for the Pacific (McManus et al. 1969; Atwater et al. 1970), the mass of volcanic matter extruded in nonspreading areas thus amounts to more than 460 km^3/1000 years; lava tunnels, vesiculation, comminution, etc. make most of this matter easily accessible for leaching. The heating of the seawater will establish a chimney effect that will pump water through a seamount. That certain seamounts probably are highly permeable is well established (Harrison et al. 1959).

The length of the spreading center in the Pacific is about 13000 km, and the average opening rate is 8 cm/year; that is, 1050 km^2 of basaltic crust is generated in 1000 years. Assuming an average water penetration depth of 500 m on the ridges and realizing that even along these openings only a small fraction of the rock is available to leaching, one can then conclude that much less than 500 km^3/1000 years is available for leaching on the active ridges; similar conclusions were drawn by Sigvaldason (1966) from studies of thermal waters in Iceland. Thus, the quantity of basaltic rock available for leaching on active ridges is at most the same but probably much smaller than outside the active ridge areas. The total production rate of ferro-manganoan sediments close to seamounts should thus, if leaching of basalts is the primary source of these sediments, be as high or higher than on the active ridges. This is contrary to what chemical relations in Pacific sediments suggest. Thus ferromanganoan aluminium-poor sediments are known in the northeastern Pacific, in spite of the abundance of volcanoes; the slow accumulation of terrigenous matter there should make any volcanic sediment easy to detect (table 1, figures 2 and 3).

LEACHING AND ALTERATIONS OF BASALTIC ROCKS AS A SOURCE OF Fe, Mn-RICH Al-POOR SOLUTIONS

Fe-Mn-rich Al-poor authigenic crusts are ubiquitous on submarine volcanoes, suggesting a volcanic origin of the Fe and Mn (Zelenov 1964; Bonatti et al. 1966), possibly by leaching of rocks. The anomalous $^{234}U/^{238}U$ ratios in such oxide coatings support these explanations (Veeh et al. 1971; Scott et al. 1972; Rydell et al. 1973a). However, most crusts form very slowly (table 2); Fe and Mn accumulate about 10-100 times faster in unconsolidated sediments on the East Pacific Rise.

Hydrothermal alterations of basalt that deliver Fe and Mn for active ridge deposition must dispose of all Al and Ti and most Si. Separation of Fe from Al during precipitation is possible on a small scale, but is ruled out on the East Pacific Rise where Al-rich Fe-poor residual precipitates do not occur. This means that the separation occurs within the basalts.

The hot springs and brines at the Salton Sea and the Red Sea contain much dissolved Fe and Mn and only little Al (Degens et al. 1969; White 1965). However, many of these solutions probably originate by seepage of metoric waters through saline deposits (Craig 1966); such saline deposits do not exist near spreading centers in the open ocean. Possibly saline solutions could form when seawater under high pressure reacts with basic lavas. Cl would tend to be enriched in the residual solutions, simultaneously also forming Cl-rich minerals. This process has so far, however, not been observed.

Volcanic processes unrelated to salt deposits produce geothermal fluids (figure 7), which generally have low Fe/Al ratios and therefore are questionable sources for the active ridge sediments. Thus, in acid Icelandic springs (pH = 2.1-4.0) Barth (1950) found 80-650 ppm Fe and Al, but the Al/Fe ratio was fairly large, 1.7-1.1. Alkaline or neutral springs had only 1-7 ppm total Al and Fe. Bjornson et al. (1970) reported Fe/Ti ratios of 13-93, ratios that are lower than the ones close to 1000 in East Pacific Rise sediments.

Sulfate and silica sinters in Icelandic hot springs (Barth 1950, pp. 46-47) also show that Al and Fe were present in the same quantities in the solutions, and that most of the original Al, Fe and Ti is left in situ as new minerals in some strongly metasomatically altered lavas, which likewise suggests little differentiation of Al, Fe and Ti during hydrothermal alteration. In some instances this leads to the formation of Fe-Al-enriched residues, the total amount of Al-Fe-oxides sometimes being almost 60 percent (Barth 1950, p. 53). Furthermore, silica is always present in higher concentrations in the hot springs than are Fe and Al. In low temperature springs ($t<150^{o}$) silica content often is 75 to 125 ppm SiO_2 and is 600 ppm in hot springs. However, active

Table 1.

Compositions and accumulation rates of Pacific pelagic sediments

Average chemical compositions and accumulation rates of Pacific pelagic sediments. A = sample number, B and C are latitude and longitude in whole degrees, D = accumulation rate of carbonate-free fraction (in mm/1000 years) and E number of cores used for each average. 45 average core analyses (based on about 90 analyses) derive from the literature (Revelle 1944; Goldberg et al. 1958; El Wakeel et al. 1961; Landergren 1964; Bender et al. 1971). 93 average core-analyses were made at the Rosenstiel School, based on 350 analyses. Several hundred analyses performed at the Rosenstiel School in addition to the ones used here indicate that the analyses used are representative for the regions they come from. Concentrations are given on a carbonate-free basis and are in % except for Cu and Ni that are given in ppm. Accumulation rates were calculated in accordance with the methods discussed in Böstrom (1970a) and Böstrom et al. (1973b). Asterisks indicate that the values are for the non-detrital fraction, which was calculated from the following accumulation rate relations:

$$Fe^* = Fe - 0.55\ Al$$
$$Mn^* = Mn - 0.0118\ Al$$
$$Ba^* = Ba - 0.0044\ Al$$
$$Cu^* = Cu - 0.00056\ Al$$
$$Ni^* = Ni - 0.00042\ Al$$

These relations were derived from the element relations in hemi-pelagic sediments rich in poorly degraded continental matter and poor in authigenic and biogenic matter from the Guiana basin off the Amazon (Böstrom et al. 1972c, 1973a).

					Concentrations on a carbonate free basis								cumulation rates in mg/cm² 1000 years									
A	B	C	D	E	SiO_2	Al	Fe	Mn	Ba	Cu	Ni	Ti	SiO_2Biol	Fe*	Mn*	Ba*	Cu*	Ni*	Al	Fe	Mn	A
1	43-N	164-E	7.2	3	57	7.2	4.1	0.58	0.089	175	125	0.34	86	0	2.6	0.31	0.085	0.051	39	22	3.1	1
2	30-N	155-E	7.2	3	53	8.8	4.6	0.45	0.038	170	102	0.46	16	0	1.8	0	0.080	0.030	48	25	2.4	2
3	26-N	159-E*	0.9	2	52	9.6	5.0	0.51	0.040	185	170	0.50	−2	0	0.26	0	0.011	0.008	6.5	3.4	0.34	3
4	43-N	162-W	3	2	58	9.0	5.3	0.33	0.038	270	160	0.50	16	1.0	0.50	0	0.056	0.028	20	12	0.74	4
5	24-N	178-W	2.3	2	51	10.5	5.0	0.41	0.048	175	140	0.51	−15	−1.3	0.50	0.003	0.025	0.016	18	8.7	0.71	5
6	22-N	169-W	1.4	1	48	9.0	5.6	0.30	0.15	180	135	0.75	−3.1	0.7	0.21	0.12	0.017	0.010	9.5	5.9	0.32	6
7	25-N	157-W	1.0	1	50	11.5	5.8	0.33	0.23	400	140	0.78	−12	0	0.15	0.13	0.028	0.007	8.6	4.3	0.25	7
8	29-N	125-W	2.0	1	58	9.4	4.6	0.46	0.60	660	160	0.49	6	−0.8	0.52	0.84	0.095	0.018	14	6.9	0.69	8
9	31-N	120-W	13	1	55	8.9	5.1	0.36	0.59	350	140	0.50	39	2.1	2.7	5.4	0.32	0.10	87	50	3.7	9
10	29-N	119-W	18	1	56	10.1	5.1	0.41	0.54	410	130	0.50	−14	−6	3.9	6.7	0.51	0.12	136	69	5.5	10
11	22-N	128-W	2.2	2	51	9.0	5.0	0.60	0.55	333	230	0.49	0	0	0.18	0.84	0.050	0.032	15	8.3	0.99	11
12	16-N	125-W	2.1	2	50	9.1	5.2	0.47	0.57	600	170	0.43	−3.1	0.3	0.57	0.84	0.091	0.021	14	8.2	0.74	12
13	10-N	126-W	3	3	55	6.2	3.9	0.47	0.75	350	170	0.27	45	1.1	0.89	1.6	0.075	0.032	14	8.8	1.06	13
14	9-N	124-W	2.4	3	50	6.1	4.1	0.56	0.92	420	190	0.24	27	1.4	0.87	1.6	0.073	0.029	11	7.4	1.0	14
15	7-N	129-W	4.5	4	56	5.3	3.4	1.20	1.02	660	295	0.22	88	1.7	3.9	3.3	0.22	0.092	18	11.5	4.1	15
16	8-N	153-W	1.0	5	61	5.4	3.3	0.54	0.18	240	150	0.26	22	0.3	0.36	0.12	0.017	0.009	4.1	2.5	0.41	16
17	2-N	167-W	5.5	4	55	7.3	5.0	0.80	0.090	300	190	0.33	54	4.5	2.9	0.24	0.11	0.065	30	21	3.3	17
18	2—3-N	139-W*	3	3	56	3.8	4.1	0.82	0.80	420	130	0.19	72	4.5	1.7	1.8	0.093	0.070	8.6	9.2	1.8	18
19	4-N	132-W	5.5	1	45	2.5	7.5	1.07	1.07	410	280	0.12	124	25	4.2	4.4	0.17	0.12	10.3	31	4.3	19
20	8-N	104-W	12	3	25	3.3	6.9	2.15	0.67	315	222	0.16	54	45	19	5.9	0.27	0.19	30	62	19	20
21	5-N	101-W	12	3	32	4.6	5.9	1.3	0.49	230	360	0.23	54	30	11.5	4.2	0.20	0.30	41	53	12	21
22	0-N	104-W	7	1	—	1.0	2.2	0.25	0.30	263	80	—	—	8.7	1.2	1.6	0.14	0.040	5.3	12	1.3	22
23	2-S	93-W	12	2	63	1.8	3.5	1.4	—	—	—	0.08	477	23	12	—	—	—	16	32	12	23
24	1-S	87-W	11	3	44	4.2	2.8	0.27	0.33	145	130	0.19	330	3.7	1.8	2.5	0.11	0.14	35	23	2.2	24
25	8-S	102-W	7.3	3	25	3.2	13	4.1	1.5	1300	1040	0.15	38	62	22	7.8	0.71	0.56	17	71	23	25
26	21-S	82-W	0.4	3	30	4.6	6.7	0.64	1.2	195	105	0.35	1.2	1.2	0.17	0.35	0.005	0.0026	1.4	2.0	0.19	26
27	13-S	97-W	1.5	4	24	3.9	9.9	3.1	2.7	860	880	0.15	2.2	8.8	3.4	3.1	0.096	0.096	4.4	11.2	3.5	27
28	28-S	106-W	0.8	2	19	2.2	15	3.8	0.63	1330	760	0.37	4.2	8.0	2.2	0.37	0.080	0.080	1.3	8.7	2.3	28
29 a	18-S	113-W	5.2	1	15	0.5	30	10	0.28	—	575	—	—	116	40	1.1	—	0.22	1.8	117	40	29 a
b	17-S	116-W	0.7	1	26	3.2	20	4.0	—	—	525	—	—	10	2.1	—	—	0.028	1.7	11	2.1	b
c	16-S	122-W	0.45	1	25	1.6	12	1.9	—	—	510	—	—	3.6	0.65	—	—	0.018	0.5	3.9	0.65	c
30	7-S	113-W	1.6	1	36	0.74	8.7	2.7	0.58	380	174	0.040	38	10	3.3	0.70	0.046	0.021	0.9	10.5	3.3	30

A	B	C	D	E	SiO_2	Al	Fe	Mn	Ba	Cu	Ni	Ti	SiO_2Biol	Fe*	Mn*	Ba*	Cu*	Ni*	Al	Fe	Mn	A
31 a	13-S	111-W	14	1	13	0.25	21	6.8	0.15	740	420	0.027	126	220	71	1.6	0.78	0.44	2.6	221	71	31 a
b	13-S	112-W	5.2	1	13	0.50	23	8.8	0.25	860	520	0.020	39	87	34	0.99	0.34	0.20	2.0	88	34	b
c	13-S	113-W	4.0	1	11	0.80	22	7.9	0.25	910	480	0.030	18	66	24	0.74	0.27	0.14	2.4	67	24	c
32	12-S	144-W	0.9	3	34	5.4	7.1	2.7	0.31	1020	670	0.53	2	2.8	1.8	0.19	0.068	0.044	3.7	4.8	1.8	32
33	20-S	149-W	0.9	3	47	9.1	7.5	1.4	0.29	390	370	0.56	−3.4	1.7	0.88	0.17	0.025	0.022	6.2	5.1	0.95	33
34	17-S	163-W	0.6	7	47	9.9	6.8	1.4	0.086	680	305	0.93	−4.1	0.6	0.58	0.019	0.030	0.012	4.5	3.1	0.63	34
35	9-S	175-W	1.6	2	46	8.3	7.2	0.88	0.065	330	165	0.58	−1.2	3.1	1.0	0.035	0.038	0.016	10	8.6	1.1	35
36	1-S	158-W	4.4	2	—	—	4.2	0.45	—	—	—	—	—	—	—	—	—	—	—	14	1.5	36
37	1-N	148-W	5.4	2	—	—	6.2	0.43	—	—	—	—	—	—	—	—	—	—	—	25	1.7	37
38	1-N	141-W	6.6	2	—	—	6.4	0.29	—	—	—	—	—	—	—	—	—	—	—	32	1.4	38
39	41-S	133-W	1.4	3	49	7.5	5.2	1.0	0.87	370	290	0.34	6.3	1.2	1.0	0.37	0.037	0.027	7.9	5.5	1.1	39
40	46-S	114-W	0.6	3	16	2.2	7.8	2.5	0.88	390	180	0.077	1.8	3.0	1.1	0.40	0.018	0.0077	0.96	3.5	1.1	40
41 a	40-S	111-W	0.95	1	—	—	42	16	—	1270	220	—	—	—	—	—	—	—	—	30	11	41 a
b	47-S	104-W	2.5	1	—	—	5.4	1.6	—	850	—	—	—	—	—	—	—	—	—	10	3	b
42	41-S	98-W	0.6	4	31	5.7	14	4.7	1.0	780	700	0.32	−0.5	4.7	2.1	0.44	0.034	0.031	2.6	6.1	2.1	42
43	54-S	120-W	1.3	2	54	2.8	1.9	0.12	0.21	70	40	0.13	37	0.3	0.09	0.20	0.0062	0.0028	2.7	1.8	0.12	43
44	61-S	108-W	4.5	1	70	3.4	1.9	0.23	0.15	85	45	0.15	172	0.2	0.65	0.46	0.026	0.010	11	6.4	0.78	44
45	58-S	131-W	6.5	2	43	1.7	5.1	1.1	0.18	340	1300?	0.19	225	9.6	2.6	0.80	0.09	0.29?	7.5	14	2.7	45
46	59-S	160-W	6.8	2	67	6.1	4.2	1.3	0.36	390	670?	0.38	160	4	6.0	1.7	0.19	0.33?	31	21	6.4	46
47	52-S	134-E	0.51	1	83	1.3	3.2	0.35	0.20	115	41	0.11	29	0.9	0.13	0.074	0.0043	0.0014	0.5	1.2	0.13	47
48	50-S	134-E	2.7	1	57	2.6	9.3	1.1	0.29	290	73	0.16	84	16	2.1	0.56	0.057	0.013	5.2	19	2.2	48
49	47-S	134-E	2.1	1	50	6.6	6.4	0.83	0.45	220	117	0.39	18	4.3	1.2	0.65	0.031	0.014	10.2	9.9	1.3	49
50	48-S	154-E	26	1	60	6.7	4.2	0.52	—	—	—	0.31	430	10	8.5	—	—	—	130	82	10	50
51	47-N	173-W	7.5	1	—	—	4.6	0.32	—	—	—	—	—	—	—	—	—	—	—	26	1.8	51
52	37-N	146-E	17	4	59	7.9	3.6	0.19	0.038	82	45	0.37	200	−10	1.2	0.04	0.074	0.014	101	46	2.4	52
53	4-N	135-E	18	2	—	—	5.0	0.5	—	—	—	—	—	—	—	—	—	—	—	68	6.8	53
54	6-N	84-W	29	1	55	8.4	6.0	0.05	—	44	84	—	150	30	−1.0	—	0.049	0.10	180	130	1.1	54
55	4-N	96-W	5.3	3	42	5.3	5.9	2.7	0.58	240	395	0.20	47	11	10.5	2.2	0.089	0.07	21	23	10.7	55
56	4-N	106-W	6.5	3	20	2.8	4.8	2.9	0.59	330	340	0.13	20	16	14	2.3	0.16	0.15	13	23	14	56
57	1-N	123-W	3.5	1	45	3.2	6.3	2.2	1.07	530	350	0.20	74	12	5.6	2.8	0.14	0.089	8.3	16.5	5.7	57
58	1-S	123-W	3.0	4	25	2.8	7.4	1.9	0.79	640	410	0.12	20	13	4.1	1.8	0.14	0.089	6.3	16.7	4.2	58
59	12-S	134-W	0.5	3	34	4.8	8.4	3.3	0.32	1250	950	0.29	2.7	2.2	1.3	0.11	0.048	0.035	1.8	3.2	1.3	59
60	10-S	127-W	0.8	3	16	3.6	10.8	2.2	0.53	870	690	0.10	−3	5.3	1.3	0.31	0.051	0.040	2.2	6.5	1.3	60
61	13-S	113-W	3.8	1	14	0.32	28	10.7	0.42	1500	675	0.033	28	63	24	0.94	0.33	0.15	0.72	63	24	61

* represent (each) 2 closely spaced stations

Table 2.

Accumulation rates in Pacific sediments and river water contents

	1	2	3	4	5	6	7
Al	-	-	-	-	0.072 -0.43	4.1 -18	0.92
Ti	-	-	-	-	0.015 -0.090	0.19 - 0.88	0.094
Fe	4.06 -3.31	369 -301	52	670	0.26 -1.6	2.5 - 8.7	81
Mn	2.20 -1.90	200 -173	58	7	0.43 -2.6	0.25 - 0.99	31
Cu	0.071-0.065	6.5- 6.0	15	7	0.0088-0.098	0.012- 0.054	0.42
Ni	0.054-0.050	4.9- 4.5	19	0.3	0.015 -0.089	0.011- 0.038	0.19
Ba	0.61 -0.51	56 - 47	43-45	10	-	-	1.20

1. Average accumulation rate for Pacific sediments in mg/cm^2 1000 years.
2. Required quantity in μg/l in average world river water to support the accumulation rates in column 1.
3. Measured river water concentration (μg/l) in the U.S. (Kopp et al. 1967; Puchelt 1967).
4. River water composition (μg/l) (Turekian 1969).
5. Range in accumulation rates of some elements, depositing as concretionary products (manganese nodules, slabs, oxide coatings) (Boström et al. 1973b).
6. Range in accumulation rates in sediments that coexist with the concretionary deposits listed under A (Boström et al. 1973b).
7. Accumulation rates of elements in the East Pacific Rise Core GS-7202-35, average values for the total core (8.5 meters long) (Boström et al. 1974).

Note that if the average for total World Ocean is used, the values in columns 1 and 2 may be somewhat lower (Boström et al. 1973a). All values in columns 1 and 2 are for non-detrital fractions only. The spread in the values indicates how different weighting procedures influence the results.

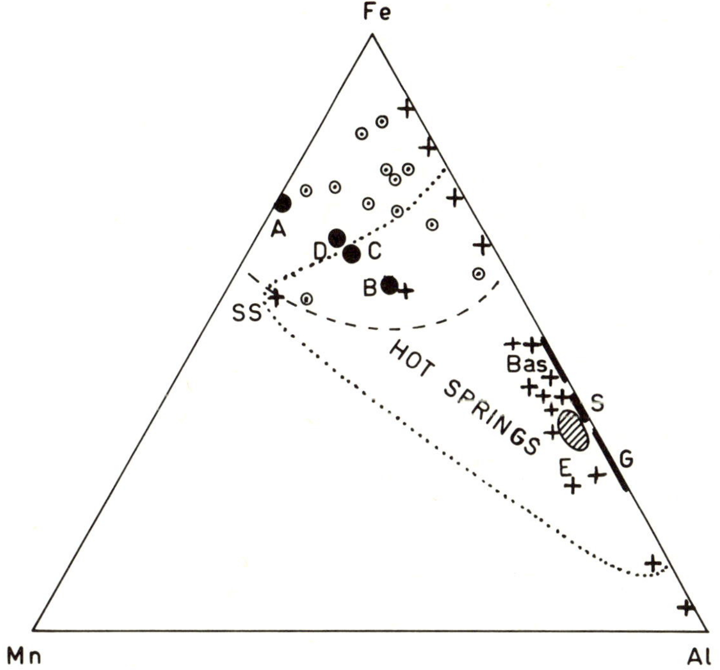

Figure 7. Ternary plot, showing the Al-Fe-Mn relations in ferro-manganoan deposits from the active rise-rift system (black dots) and in carbonatites (dot-rings) (Boström 1973). Note the close correspondence, all values occurring within the dashed boundary. Note the discrepancy relative to other Pacific sediments (shaded area at E). For comparison the relations in basalts (=Bas), shales (=S) and granites (=G) and hot springs (+) are given. Compositions of G, S, and Bas (short, heavy lines) from Krauskopf 1967 and Engel et al. 1970. Hot springs and geothermal brines have highly variable compositions. Manganese-rich waters are rare, except for the Salton Sea brine (=SS). The crosses represent data from Barth 1950; Naboko 1963; and White 1965; almost all points represent acid waters. Note the remarkable spread in the ratio Fe/(Fe+Al), from 0.90 to 0.05. In basaltic terrain most spring values are close to 0.38-0.47 -- that is, close to the ratios in the rocks. Hot springs in basaltic areas are therefore unlikely sources of active ridge sediments. The Salton Sea brine (SS) has the composition of a possible precursor of the active ridge sediments, but available isotope evidence (Craig 1966) suggests that this brine forms by leaching of saline and other sedimentary deposits, which do not occur on the active ridges in the open ocean.

ridge sediments are not particularly rich in Si. In addition, studies of palagonitization (Honnorez 1971) show that the palagonites differed little from the original rocks except for alkalies and water.

Crystal settling and transport out of the system causes only small variations in the composition of lavas as they crystallize (Peck et al. 1966). Interstitial liquid that was filter-pressed into gash fractures and drill holes shows differences from the bulk composition of the solidified crust, but the change in Al/Fe is negligible and Ti shows a considerable increase in the ooze; such oozes can not be the source of Fe-rich Al-Ti-poor sediments. Corliss (1971), however, has reached the opposite interpretation.

The lava studied by Peck et al. (1966) cooled at low pressure and exhibited extensive vesiculation and loss of volatiles. Lava dykes formed at pressures of several hundred bars on a spreading ridge will have much less vesiculation (McBirney 1971) and probably no movement of oozes.

It thus appears that alterations at low or at high temperatures (250° or less; or $\sim 900^{\circ}$ or higher) are unfavorable for differentiation of Fe and Ti. Present knowledge about spilitization reveals what alterations are possible at intermediate temperatures, in the range of 200 to 400°C. Spilites represent altered basaltic rocks in which calcic feldspar and Fe-Mg minerals are replaced by albite and chlorites; many low-temperature minerals, such as calcite and epidote are also formed. The alteration takes place at 200-400°C, as indicated by fluid inclusion thermometry (Stoiber et al. 1959; Herrmann et al. 1970).

Analyses show that Ca and Mg are lost during spilitization and Na, H_2O and CO_2 are gained, whereas the concentrations of Si, Al, Fe, Ti, Mn, and Ti appear unchanged (Boström 1973). Studies by Cann (1969) show that spilitization under oceanic conditions involves a loss of much Ca and possibly some loss of Al, while Si, Fe and Na are simultaneously gained. This conclusion is supported by analyses of minerals in spilites. Microprobe analysis (Herrmann et al. 1970) shows that the alteration products chlorite and pumpellyite can be rich in Fe and Mn. Similar evidence is summarized by Vallance (1960).

Alternatively, it is possible that spilites and Fe-Mn exhalations derive by the same process, since much of the carbon in the spilites is of magmatic origin (Schidlowski et al. 1970) and the carbonates in the altered rocks are rich in Fe and Mn, the Fe-Mn proportion being 3:1 (Lehmann 1941). The fact that whole pillows can be replaced by manganese minerals (Park 1946) is indicative of a post-pillow intrusion of hot manganiferous solutions.

Trichet (1969) studied alteration of basaltic glasses in hot solutions in experiments that lasted a few weeks to two years. Reactions with water, in the presence or absence of CO_2, altered the glasses and removed Al and Si, whereas Fe and Ti remained in residues. A considerable fractionation between Al and Fe could occur, but in the system Fe-H_2O redox reactions complicate the presentation of the solubilities; iron and sulfur are the only oxidizable elements in the basalts that are present in sufficiently large quantities to control the redox processes.

Most Fe in basalts in present in magnetite, olivines, and pyroxenes. Of these olivine and pyroxene react readily with oxygenated sea water:

$$4\ FeSiO_3 + 10\ H_2O + O_2 \rightarrow 4\ FeOOH + 4\ H_4SiO_4$$

$$4\ FeSiO_3 + 8\ H_2O + O_2 \rightarrow 2\ Fe_2O_3 + 4\ H_4SiO_4$$

The last reaction is more likely, since ferric hydroxides quickly alter to hematite even during short exposure to hot sea water (Boström et al. 1972d); it is pertinent to note that spilites contain more hematite than magnetite (Vallance 1960). It is thus likely that pH and e^o will be controlled by the relation $Fe^{2+} - Fe_2O_3 - Fe_3O_4$ (Boström 1973).

If sea water is in prolonged contact with basalt all oxygen will be consumed. The sulfate ions, which exist in large quantities in sea water, then become the main oxidizer of iron compounds (Boström 1967a, 1967b):

$$12\ FeSiO_3 + 24\ H_2O + SO_4^{2-} \rightarrow 4\ Fe_3O_4 + 12\ H_4SiO_4 + S^-$$

The protolytic reactions:

$$S^{2-} + H_2O \rightarrow HS^- + OH^- \qquad pK\ (25^oC) = 14$$

$$H_4SiO_4 \rightarrow H_3SiO_4^- + H^+ \qquad pK\ (25^oC) = 9$$

will tend to create an alkaline solution, buffered close to pH = 9, in which the iron compounds are comparatively insoluble compared to aluminium. Alkaline silica-rich solutions occur in hot springs in basaltic terrains (Sigvaldason 1966). However, sulphate reduction will hardly shift the redox potential in the system Fe - H_2O, since the boundary $S^{2-} - SO_4^{2-}$ is close to the boundary $Fe_2O_3 - Fe_3O_4$. A consequence of this is that Al and Fe differ little in solubility at high activities; that is, negligible tendency exists for them to separate from each other, whereas at low total activities Al is preferentially dissolved.

Little is known about how complexing will affect the solubilities of Fe and Al. Chloride, sulphate, and hydroxyl complexes of Fe and Al exist; data (Sillén et al. 1964; Naumov et al. 1971) suggest that the true solubilities of Fe and Al both increase about 1-2 orders of magnitude. Helz (1971) showed that complexes like $Fe_2(OH)_6^0$ increase the solubility of magnetite and hematite considerably at higher temperature (200-300°C), but only at low pH and high temperatures does the quantity of dissolved iron substantially exceed 10 ppm. No corresponding study of Al-compounds has been doen, but solubilities of feldspars increase rapidly at higher temperatures and pressures (Clark 1966). In summary, it appears that no aqueous electrolyte system exists in which a sharp segregation of Fe and Mn from Al and Ti takes place under conditions that are likely to exist on the active oceanic ridges in the open ocean.

BASALT LEACHING AND SEAWATER AS SOURCES OF ACTIVE RIDGE DEPOSITS

The quantity of seawater that may participate annually in sediment-forming processes on the East Pacific Rise is probably about 0.16-0.69 l/cm^2; of this quantity about 0.001-0.1 l/cm^2 year of seawater may have penetrated the ocean floor and participated in leaching processes (Section 3). Table 3 shows the maximum possible contributions

Table 3.

Supply rates of various elements on the East Pacific Rise

	1	2	3	4	5
Al	0.72	1	76,200	0.72	4.4
Ti	0.074	1	14,400	0.074	0.84
P	4.10	68	1,200	0.06	0.070
Fe	63	5	86,000	12.6	5.0
Mn	24	0.4	1,300	60	0.075
V	0.21	1.9	245	0.11	0.014
Cu	0.33	2.5	127	0.13	0.0074
Co	0.022	0.025	40	0.88	0.0023
Ni	0.15	2.7	85	0.056	0.0049
Zn	0.14	5	100	0.028	0.0058
Pb	0.065	0.03	2	2.2	0.00012
Sc	0.0008	0.0008	30	1.0	0.0017
Y	0.029	0.013	25	2.2	0.0015
La	0.031	0.0034	10	9.1	0.00058
As	0.11	2.6	1	0.042	0.000058
Ba	0.94	27	244	0.035	0.014
Hg	0.000056	0.15	0.080	0.00037	0.0000046
U	0.010	3.3	0.05	0.0033	0.0000029
Zr	0.032	0.026	110	1.2	0.0064

1. Accumulation rates in mg/cm^2 1000 years on the East Pacific Rise (Boström et al. 1973b, 1974, Table 1).
2. Content of seawater in µg/l; when available data for 3000 m depth have been used (Turekian 1969; Wolgemuth 1970; Spencer et al. 1970; Barnes 1957).
3. Average composition of Pacific tholeiitic basalt. Major elements after Manson 1968; traces after Prinz 1968; Fairbridge 1972; Boström et al. 1969d; Fisher, D.E., personal communication. Data by Engel et al. 1970 suggest lower values for Ba (in ppm).
4. Total quantity of seawater in l/cm^2 year needed to sustain the rates in column 1.
5. Total quantity of material added to sediments if all constituents are leached out of basalt proportionally to Fe and 5 mg Fe/cm^2 1000 years has a basaltic origin. Units: mg/cm^2 1000 years.

from this quantity of seawater to the seafloor. Seawater may easily account for many constituents, but not for some of the most important (e.g., Fe and Mn). Some of these constituents may derive from basalts by leaching.

This hydrothermal leaching may be common in seamounts since considerable quantities of basalt in seamounts are easily accessible for leaching by seawater. Upon leaving the seamount the solutions mix with well-oxygenated seawater and the easily oxidized Fe then is among the first constituents to precipitate, followed by Mn, whereas Al and Si, partly as colloids, may migrate further before deposition. Indeed, the Al/Ti ratios in unconsolidated sediments from seamount areas suggest a partly basaltic origin whenever the accumulation rates of terrigenous matter is low (Boström et al. 1973b). This may be partly due to admixture of chemically little-altered basalt debris, but some Al, Ti, and Si may derive from residual leaching products. The slowly forming Fe-Mn-crusts on seamounts could therefore be products of leaching (Bonatti et al. 1966). Furthermore, even low flow rates of leaching solutions suffice to carry enough dissolved matter to the ocean floor. A water flux of 0.001 l/cm^2 year with 1 ppm Fe will deposit 1 mg Fe/cm^2 1000 years; a flow rate as high as 0.1 l/cm^2 year and a concentration of 0.1 ppm would deposit 10 mg Fe/cm^2 1000 years. These accumulation rates resemble natural ones (table 2).

The situation on the active ridges is different. If leaching of basalt is the origin for the active ridge sediments, higher flow rates and higher concentrations of dissolved matter are required; a solution with 10 ppm Fe and flow rate of 0.01 l/cm^2 year could explain the accumulation of 100 mg Fe/cm^2 1000 years. The transport capacity of the leaching solutions is consequently not the limiting factor. More significant is that nowhere do the compositions of the sediments suggest a basaltic origin.

In a previous section it was shown that Fe and Al will generally not be well separated during leaching, except that Al will tend to remain in solution if the flow rate is high--that is, if the rock/water ratio is very low. The accumulation rates of Al should thus be about the same as for iron, but are in fact much lower in the South Pacific (Bostrom et al. 1973b) (figure 6). The accumulation rates of Al over the ridge suggest that hydrothermally leached iron accounts for less than 5 mg Fe/cm^2 1000 years, or at most 8 percent of the Fe on the East Pacific Rise. The clay mineral distribution and the Al/Ti relations in Pacific sediments display latitudinal patterns--that is, clays, Al, and Ti are largely distributed by winds and surface currents (Bostrom et al. 1972a, 1972b), whereas leaching of any intensity would have formed sediments richer in Al. Another indication of insignificant leaching is that Sc, Y, and La are all present in about equal quantities in basalts, but the ratio Sc/Y is much lower in the sediments on the crest of the East Pacific Rise. This differentiation is hardly to be expected from hydrothermal leaching of basalts, since Sc, La, and Y exhibit similar behavior in solution: (Bostrom 1973).

Important evidence against simple hydrothermal leaching is also provided by the remarkably uniform composition of the volcanic sediments. A plot of superior analyses of pelagic sediments in a Fe/Ti - Al/(Al + Fe + Mn) diagram shows that the points are distributed along a well-defined mixing curve: active ridge sediment-average continental crust debris (see figure 8). This means that on active ridges Fe, Mn, Al, and Ti are delivered in the proportions 10:3.8:0. 1:0.01. These ratios do not vary substantially with time at a given locality; thus at station GS-7202-35 the overall content of Fe, Mn, Al, and Si vary only little over 600,000 years; the only variation being in the ratio Al/Ti. This constancy is not to be expected from seawater interactions with lavas, since variations in (a) temperature, (b) rate of seawater flow due to closing or opening of fissures, and (c) chemical composition of leaching waters due to varying admixture of minor amounts of volcanic gases will promote different seawater-rock interactions and lead to the formation of hot spring solutions of variable composition, as is observed in continental hot spring

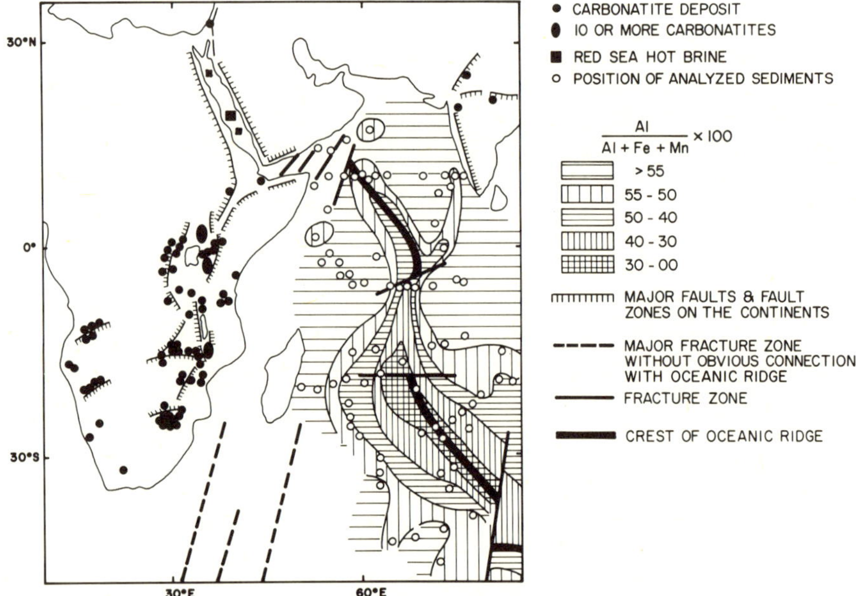

Figure 8. The relations between Fe/Ti and Al' in Pacific and Indian Ocean sediments. Al' is a transformation of Al/(Al+Fe+Mn) such that the mixing function between EPRD and TM can become a straight line. The solid lines represent the linear regressions giving the best fits to the data in the graphs, and dashed lines the standard errors of estimate for the linear regression. EPRD (=East Pacific Rise deposit) is equal in composition to the average active ridge sediment and TM (=terrigenous matter) represents average continental crust (After Löfgren et al. 1979).

areas (figure 7). Remarkably, many hot acid springs are enriched in Al, and in basaltic areas the Fe/Al ratios of the springs resembles that in the basaltic rocks.

The primary source of the sediments could hardly be seawater; it is difficult to understand why the dissolved Fe and Mn would deposit on active ridges only. The accumulation rates and water concentration of Fe and Mn indicate that 13 and 60 l of seawater/cm^2 year, respectively, are required to form such deposits. This is more water than is available in the lowermost 1000 m of the slowly moving bottom waters (2-0.48 l/cm^2 year).

However, once formed, oxides and hydroxides of Fe and Mn may act as adsorbers. Table 3, column 4, shows that Cu and V require very high adsorption efficiences in order to derive from seawater, whereas P, Ni, Ba, As, Zn and particularly Sr, U, and Hg are possible adsorbates. This conclusion is supported by studies of U- and Sr-isotopes, but other observations contradict these conclusions. Thus, some cases exist (Veeh et al. 1971; Scott et al. 1972; Rydell et al. 1973a, 1973b) in which the U-isotopes indicate sources different from seawater, probably including a deep-seated source (Rydell et al. 1973b). Similar conclusions hold for Ba, which has biological and volcanic sources (Boström et al. 1973a). Indeed, in the northwestern Pacific, adsorption of Ba from seawater on hydroxides should be proceeding undisturbed by other processes (primarily terrigenous deposition); yet in this area the accumulation rates of authigenic Ba is about zero (table 1). This observation contradicts the inference that adsorption from seawater is an important deposition process for Ba. Furthermore, even in the laboratory there can be difficulties with a quantitative removal of matter by adsorption. Low adsorption efficiences must therefore occur also in the ocean.

In conclusion, it can be stated that a deep-seated source must be considered for Fe and Mn. Probably a large fraction also of Cu, V, Ni, Ba, P, and As have the same source. It is likely that part of the Sr, Zn, U, and Hg derive from seawater; however, for U a fraction must have a deep-seated origin. The nature of this deep-seated source is unknown, but a few possibilities may be suggested:
a) Water and other volatiles liberated during the fractionation of pyrolite into basalt and residual peridotite leach material from the basalt through which they ascend.
b) Residual solutions are released during the crystallization of basaltic melts.

Such mechanisms imply a leaching from below. However, during the crystallization of basalt little water is liberated, probably 1 percent at most -- that is, 20 l water/cm^2 from a 7 km-thick basalt layer. About 100 g of Fe per cm^2 have been deposited on the East Pacific Rise at 14°S before the end of the formation of the active ridge sediments. One liter of surfacing water must therefore carry about 5000 ppm iron. This is the same value as in the Salton Sea brine. However, the Salton Sea brine is rich in halogens, which presumably keep Fe and Mn in solution as chloride complexes, whereas basaltic magmas are not rich in halogens. The lack of separation of the major constituents Fe and Al is thus the major stumbling block for any hypothesis that tries to explain the origin of the active ridge sediments by processes linked to the upper crust of the Earth.

CARBONATIC EMANATIONS AS A SOURCE OF ACTIVE RIDGE SEDIMENTS

The current opinion of the upper mantle processes under active ridges, where crustal generation takes place, is that relatively undifferentiated primitive mantle material (pyrolite) is fractionated into basalt and residual peridotite (Ringwood 1969; Green 1964; and Hess 1964). During such a process geochemically volatile matter is likely to escape upwards. Such emanations are probably very small in volume but are likely to contain constituents that do not fit the high-pressure, high-temperature region of the mantle. Such mantle rejects are probably characterized by:

a) high vapor pressures, like CO_2, H_2O, B, As, Hg, and rare gases;
b) high solubilities, like NaCl and KCl;
c) large ionic radii and low densities, like U and Ba; and
d) a tendency to be enriched in residual melts, like Fe and Mn.

Most volcanic emanations contain considerable quantities of CO_2; recalculated on water-free basis CO_2 is by far the predominating constituent. Furthermore, whereas almost no water, except perhaps 0.5-2 percent is magmatic, most evidence favors such an origin for much of the CO_2. Recent geophysical and petrological evidence indicates that much CO_2 is present in the asthenosphere, where it acts as a "lubricant" for moving crustal plates (Green 1972). According to Green's calculations the CO_2 content should reach a maximum in the asthenosphere; this explains the presence of much CO_2 in many ultrabasic and ultramafic rocks. In the zone of crustal spreading and petrogenesis on the active ridges, extensive pressure relief occurs; it is therefore likely that a large scale degassing of CO_2 takes place on the active ridges.

A CO_2-rich phase should initiate related processes, whether it ascends in oceanic or in continental mantle and crust. A study of carbonatite magmas might therefore reveal what would be enriched in a disperse oceanic carbonate phase (figure 7; Boström 1973). The data show that the ratios Fe', Al', Mn', P', and Ba' are very similar in both carbonatites (especially in their carbonate fractions) and active ridge deposits; in these expressions Me' represents the ratio Me/(Fe+Al+Mn). Furthermore, the ratio Fe/Ti in carbonatites is closer to that in active ridge sediments than in basalts -- that is, less hydrolysis of Ti in the crust has to be invoked to explain the Fe/Ti ratios observed in the active ridge sediments. It is even conceivable that some of the Al, Ti, and Sc in the active ridge sediments derive from carbonate-rich emanations. Carbonate lavas are also rich in U but poor in Th (Poole 1963), which is an excellent indication of a deep-seated origin for the U in some active ridge sediments (Rydell et al. 1973b).

Additional support for these conclusions can be derived from several relations:

1) Spreading rates and "anomality" of spreading ridge sediments are closely interrelated (see figure 5), which suggests that they are caused by the same subcrustal process.

2) Hydrothermal authigenic carbonates occur in sediments on the East Pacific Rise (Bonatti 1966) and in rocks from the Mid-Atlantic Ridge (Bonatti et al. 1971).

3) The worldwide rise-rift system is predominantly submarine, but in the Gulf of Aden the Indian Ocean Rift system joins the East African Rift system, which is subaerial. These continental tensional rifts exhibit negligible spreading (Girdler 1964; Baker et al. 1971; Bailey 1964), which has been intermittent since early Mesozoic without breakup of the African continent (Le Bas 1971; Sowerbutts 1972). This rise-rift system is closely correlated with occurrences of carbonatites and associated alkalic rocks on the continents (Heinrich 1966; Tuttle et al. 1966) and with active ridge sediments on the ocean floor (figures 2 and 9).

4) Rift systems like the East African may be precursors of spreading ridges like the Mid-Atlantic ridge, as indicated by physiographical and geological similarities between the Afar region and the Brazil-Nigeria junction (Burke et al. 1971) and between the borders of the South Atlantic Ocean, which is fringed with several occurrences of alkalic rocks and carbonatites (Le Bas 1971; Sowerbutts 1972) that were emplaced in the Pre-Atlantic rift zone. "Oceanic" carbonatites of various types are also known (Allegre et al. 1971; Allen et al.

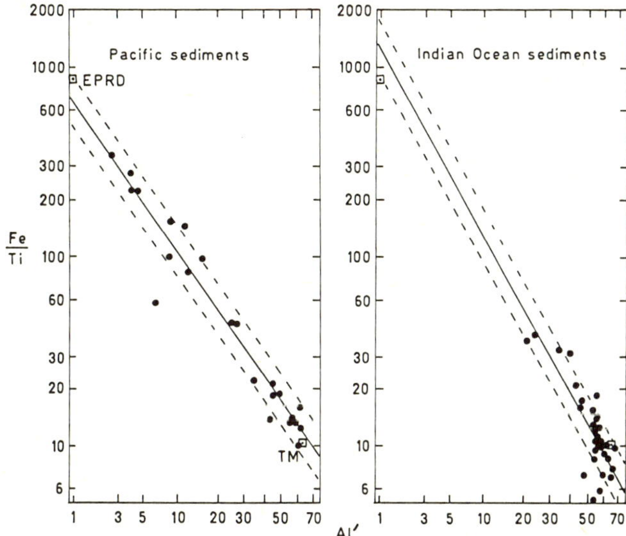

Figure 9. Distribution of active ridge sediments and carbonatites. Based on data in Boström et al. 1969c; Boström, unpublished analyses; Heinrich 1966; Heezen et al. 1963; Girdler 1964; Bailey 1964; and Tuttle et al. 1966.

1965; Bradshaw 1968; Heinrich 1966; Tuttle et al. 1966; Honnorez et al. 1970).

5) Carbonatites are not known on active ridges, but the time required for the gas phase buildups after the initial tholeiitic volcanism is considerable (Pecora 1956; Verschure 1966). On a rapidly spreading ridge, the time available may be too short for the gas buildup; furthermore, the faster the spreading and rate of pyrolite unmixing is, the more complete the loss of CO_2. Thus, the East Pacific Ocean floor, with spreading rates close to 8-10 cm/year, has only little alkalic volcanism; the Atlantic Ocean, with spreading rates of 1-2 cm/year, displays more alkalic rocks. The East African rift system, with average spreading rates of almost zero, shows spectacular examples of alkalic and carbonatitic rocks.

6) Quantitative relations show that carbonatitic volcanism may well be the source of active ridge sediments. The required deposition of Mn on the ocean floor (table 3) demands only 8,000 g of carbonatite matter with 0.53 percent Mn per cm^2 of ocean floor; in a mantle that is 75 km deep, this corresponds to only 0.03 percent. An ascending CO_2 phase with 28.5 percent CO_2 annually adds 0.009 m moles CO_2 per cm^2 to the ocean floor, which hardly changes the CO_2 concentration in the bottom waters; oxidation of debouching Fe and Mn consumes only 0.00027 moles O_2/cm^2 1000 years.

7) The capacity of CO_2-rich phases to transport Mn and Fe is suggested also by the Fe-Mn contents of carbonates in altered basic rocks, in which Lehman (1941) and Sundius (1923) reported Fe-Mn-containing calcites with 3.4 percent $FeCO_3$ and 1.4 percent $MnCO_3$. Puchelt (1973) found that much Fe, Mn, and other constituents but little Al, surface by exhalations that are rich in CO_2 and deposit iron carbonates in

shallow bays on Santorini. A CO_2-rich phase could thus transport much Fe, Mn, as well as other elements, and the ascendance of such phases on spreading ridges is very likely. It is therefore probable that most Fe, Mn, and Pb in active ridge sediments on the crest of the East Pacific Rise at about 15°S derive from such emanations, and furthermore that a considerable fraction (perhaps more than 50 percent) of the Co, P, Cu, V, Ni, Ba, and As in these sediments have this origin.

EMPLACEMENT OF THE EXHALATIVE DEPOSITS

Little is known about how the solutions come up, whether as few very intense hot springs or as slow widespread seeping through the ground. The occurrence of continuous sequences (Core Gs-7202-35; see Boström et al. 1974; Bender et al. 1971) and postdepositional injections into the sediments (Rydell et al. 1973b) suggest a quiet volcanic process with slow seeping and no explosive events. Obviously, there could be intermittent variations in the accumulation of ferric lutites. However, on a long time basis, the activity at site GS-7202-35 has remained fairly constant; the same conclusion applies to the sites described by Bender et al. (1971).

The precipitation process that forms the ferro-manganoan aluminum-poor sediments has not yet been discussed. The metal-depositing solutions are more likely reducing, and become oxidized in the sea water where they precipitate their metal content. A large-scale oxidation at depth in the basalts by seawater could not have taken place in the basalts, since the ferric hydroxides quickly form hematite at high temperatures (Boström et al. 1972d). Since x-ray diffraction analysis of East Pacific Rise samples shows no hematite, this possibility is even more remote -- that is, such oxidizing conditions could at most have a shallow extent in the basalts, and the circulating solutions only have very short residence times. However, postdepositional injections (Rydell et al. 1973b) suggest that the solutions cannot be very reducing, otherwise the high concentrations of Mn in the lower part of core GS-7202-35, would not have remained intact.

TOTAL MASS OF EXHALATIVE SEDIMENTS IN THE OCEAN

While few would deny that some volcanism and associated production of sedimentary matter takes place on the active ridges, there is still no consensus on how much of the pelagic sediments are of exhalative origin. A major contribution to these difficulties is that even if an exhalative origin is accepted for much of the sedimentary matter on active ridges these sediments will, due to spreading and subsequent admixture with terrigenous matter, become covered and inaccessible for quantitative studies by conventional oceanographic procedures. This distribution pattern of active ridge sediments was predicted by Boström and Peterson (1969b); evidence from subsequent Deep Sea Drilling Project cruises has given excellent support for this hypothesis.

The (Fe + Mn)/Al ratio is an exponential function of the spreading rate (see figure 5). The rate of accumulation of Al in midoceanic areas is low and can as a first approximation be considered constant -- that is, the rate of accumulation of Fe, Mn, etc. is largely a function of the spreading rates. Consequently, in areas of very slow spreading, the basal iron-manganese-rich sediments should form a very thin layer, but should approach considerable dimensions in areas of fast spreading. This conclusion is verified by results from the East Pacific Rise and site GS-7202-35. These results, combined with a comparison with accumulation rate maps (Boström et al. 1973b) and with spreading rates, suggest that an exponential relation exists between accumulation rate and spreading rate. An approximate relation of spreading rate (SR) to relative accumulation rate (RAR) for exhalative components appears to be:

$$\text{RAR } (\%) = 1.2e^{0.47SR}$$

where SR is in cm/year. This expression can be used to derive some general ideas of how volcanic sediments are distributed:

Half-spreading rates (cm/year)	Relative (%)	Accumulation rates: Fe Mg/cm^2 / 1000 years	Accumulation rates: Mn mg/cm^2 / 1000 years
0	1.2	1.2	0.36
2	3	3	0.90
4	8	8	2.4
6	19	19	5.7
8	50	50	15
9.5	100	100	30

Most of the World Rift-Rise system has SR = 4 cm/year of less (Heirtzler et al. 1968) and thus minimal accumulation of exhalative deposits. These relations also explain the difficulties in detecting volcanogenic accumulation rates for Fe, Mn, etc. in the Indian Ocean (Boström et al. 1973b), where SR generally is less than 3 cm/year.

Known spreading rates for various parts of the ridge system and the areas of the ocean that have been generated by a given ridge segment permit a calculation of the amount of Fe, Mn, and Ba of possible exhalative origin. Furthermore, estimates of the mass and average composition of other deep sea deposits permit a calculation of the total mass of Fe, Mn, and Ba that have accumulated in ordinary pelagic sediments. In the whole ocean the total mass of exhalative iron, Fe(exh), is 2.9×10^{22}g, Mn(exh) = 1.1×10^{22}g and Ba(exh) = 4.4×10^{20}g. This means that Fe(exh) constitutes 1.1 percent of all iron in deep-sea deposits (3 percent in the Pacific); Mn(exh) 4.7 percent of all manganese in deep-sea sediments (10 percent in the Pacific); and Ba(exh) 0.97 percent of all barium in deep-sea deposits (2.2 percent in the Pacific). Assuming that the ocean floors are rejuvenated every 200 m.y., the last 3 b.y. have involved fifteen cycles of ocean floor generation with associated sediment buildup -- that is, in this period quantities of Fe, Mn, and Ba equivalent to 17 percent of all the Fe, 70 percent of all the Mn, and 15 percent of all the Ba now in the ocean have surfaced by submarine volcanism.

REFERENCES

Allègre, C. J., F. Pineau, M. Bernat, and M. Javoy. 1971. Evidence for the occurrence of carbonatites on the Cape Verde and Canary Islands. Nature Phys. Sci. 233:103-104.

Allen, E. T., and A.L. Day. 1935. Hot springs of the Yellowstone National Park. Carnegie Inst. Washington Pub. 466, 525 pp.

Allen, J. B., and T. Deans. 1965. Ultrabasic eruptives with alnöitic-kimberlitic affinities from Malaita, Solomon Islands. Mineralog. Mag. 34:16-34.

Armstrong, R. L. 1968. A model for the evolution of strontium and lead isotopes in a dynamic Earth. Rev. Geophys. 6:175-199.

Arrhenius, G. O. S. 1952. Sediments from the East Pacific. Reports of the Swedish Deep Sea Expedition 1947-48, vol. Göteborg: Göteborgs Kungl. Vetenskaps- och Vitterhetssamhälle, 227 pp.

Arrhenius, G. O. S., and E. Bonatti. 1965. Neptunism and volcanism in the ocean. In Progress in Oceanography, vol. 3, M. Sears, ed., London, Pergamon Press, pp. 7-22.

Atwater, T., and H. W. Menard. 1970. Magnetic lineations in the northeast Pacific. Earth and Planetary Sci. Letters 7:445-450.

Bailey, D. K. 1964. Crustal warping - A possible tectonic control of alkaline magmatism. Jour. Geophys. Research 69:1103-1111.

Baker, B. H., and J. Wohleenberg. 1971. Structure and evolution of the Kenya Rift Valley. Nature 229:538-542.

Barnes, H. 1957. Nutrient elements. In Treatise on marine ecology and paleoecology, vol. 1, J. W. Hedgpeth, ed., Geol. Soc. America Mem. 67:297-344.

Barth, T. F. W. 1950. Volcanic geology, hot springs and geysers of Iceland. Carnegie Inst. Washington Pub. 587:174 pp.

Bender, M., W. Broecker, V. Gornitz, U. Middel, R. Kay, S. S. Sun, and P. Biscay. 1971. Geochemistry of three cores from the East Pacific Rise. Earth and Planetary Sci. Letters 12:425-433.

Björnsson, S., S. Arnorsson, and J. Tomasson. 1970. Exploration of the Reykjanes thermal brine area. U. N. Symposium on Geothermal Energy, Pisa, 1970 (preliminary manuscript).

Bodvarsson, G, and R. P. Lowell. 1972. Ocean-floor heat flow and the circulation of interstitial waters. Jour. Geophys. Research 77:4472-4475.

Bogorov, V. G., ed. 1967a. Biology of the Pacific Ocean; Book 1, Plankton. Moscow. Nauka , pp. 225-227.

Bogorov, V. G. 1967b. Biological transformations and exchange of energy and matter in the ocean. Okeanologia 7:649-665.

Bonatti, E. 1966. Authigenic deep-sea carbonates. Science 153:534.

Bonatti, E., J. Honnorez and G. Ferrara. 1971. Peridotite-gabbro-basalt complex from the Equatorial Mid-Atlantic Ridge. Royal Soc. London Philol. Trans. A 268:385-402.

Bonatti, E., and O. Joensuu. 1966. Deep-sea iron deposits from the South Pacific. Science 154:643-645.

Borch, C. C. von der, and R. W. Rex. 1970. Amorphous iron oxide precipitates. In Initial Reports of the Deep Sea Drilling Project, vol. 5, D. A. McManus et al., eds., Washington, D.C.: Government Printing Office, pp. 541-544.

Boström, K. 1967a. The problem of excess manganese in pelagic sediments. In Researches in Geochemistry, vol. 2, P. H. Abelson, ed., New York: John Wiley & Sons, pp. 421-452.

Boström, K. 1967b. Some pH controlling redox reactions in natural waters. In Equilibrium Concepts in Natural Water Systems, Advances in Chemistry Series No. 67, Washington, D.C.: The Am. Chemical Soc., pp. 286-311.

Boström, K. 1970a. Submarine volcanism as a source for iron. Earth and Planetary Sci. Letters 9:348-354.

Boström, K. 1970b. Geochemical evidence for ocean floor spreading in South Atlantic Ocean. Nature 227:1041.

Boström, K. 1970c. Deposition of manganese rich sediments during glacial periods. Nature 226:629-630.

Boström, K. 1973. The origin and fate of ferromanganoan active ridge sediments. Stockholm Contr. Geology 24:149-243.

Boström, K., and M.N.A. Peterson. 1966. Precipitates from hydrothermal exhalations on the East Pacific Rise. Econ. Geology 61:1258-1265.

Boström, K., and D. E. Fisher. 1969a. Distribution of mercury in East Pacific Rise sediments. Geochim. et Cosmochim. Acta 33:743-745.

Boström, K., and M.N.A. Peterson. 1969b. Origin of aluminium-poor ferro-manganoan sediments in areas of high heat flow on the East Pacific Rise. Marine Geology 7:427-447.

Boström, K., M.N.A. Peterson, O. Joensuu, and D. E. Fisher. 1969c. Aluminium-poor ferro-manganoan sediments on active oceanic ridges. Jour. Geophys. Research 74:3261-3270.

Boström, K., and S. Valdés. 1969d. Arsenic in ocean floor. Lithos 2:351-360.

Boström, K., and D. E. Fisher. 1971. Volcanogenic uranium, vanadium and iron in Indian Ocean sediments. Earth and Planetary Sci. Letters 11:95-98.

Boström, K., O. Joensuu, S. Valdés, and M. Riera. 1972a. Geochemical history of South Atlantic Ocean sediments since late Cretaceous. Marine Geology 12:85-121.

Boström, K., B. Farquharson, and W. Eyl. 1972b. Submarine hot springs as a source of active ridge sediments. Chem. Geology 10:189-203.

Boström, K., and D. E. Fisher. 1972c. Lateral fluctuations in pelagic sedimentation during the Pleistocene glaciations. Boreas 1:275-288.

Boström, K., and A. Horowitz. 1972d. Origin of pH variations and inorganic carbonates in pelagic sediments. Geol. Fören. Stockholm Förh. 94:513-533.

Boström, K., O. Joensuu, C. Moore, B. Bostrom, M. Dalziel, and A. Horowitz. 1973a. Geochemistry of Ba in pelagic sediments. Lithos 6:159-174.

Boström, K., T. Kraemer, and S. Gartner. 1973b. Provenance and accumulation rates of opaline silica, Al, Ti, Fe, Mn, Cu, Ni, and Co in pelagic sediments. Chem. Geology 11:123-148.

Boström, K., O. Joensuu, T. Kraemer, H. Rydell, S. Valdés, S. Gartner, and G. Taylor. 1974. New finds of exhalative deposits on the East Pacific Rise. Geol. Fören. Stockholm Förh. 96:53-60.

Bradshaw, N. 1968. Petrographic examination of seven ultrabasic eruptive rocks and one limestone from Malaita, British Solomon Islands. Report 78, In The British Solomon Islands Geological Record, III, 1963-1967, pp. 51-53, Honiara, Guadalcanal.

Burke, K., T.F.J. Dessauvagie, and A. J. Whiteman. 1971. Opening of the Gulf of Guinea and geological history of the Benue Depression Niger Delta. Nature Phys. Sci. 233:51-55.

Cann, J. R. 1969. Spilites from the Carlsberg Ridge, Indian Ocean. Jour. Petrology 10:1-19.

Chung, Y., and H. Craig. 1972. Excess radon and temperature profiles from the Eastern Equatorial Pacific. Earth and Planetary Sci. Letters 14:55-64.

Clark, S. P., Jr. 1966. Solubility. In Handbook of physical constants, S. P. Clark, Jr., ed., Geol. Soc. America Mem. 97:415-436.

Cook, H. E., and I. Zemmels. 1971. X-ray mineralogical studies--Leg 8. In Initial Reports of the Deep Sea Drilling Project, vol. 8, J. I. Tracey et al., eds., Washington, D.C.: Government Printing Office, pp. 901-950.

Corliss, J. B. 1971. The origin of metal-bearing submarine hydrothermal solutions. Jour. Geophys. Research 76:8128-8138.

Craig, H. 1966. Isotopic composition and origin of the Red Sea and Salton Sea geothermal brines. Science 154:1544.

Dasch, E. J., J. R. Dymond, and G. R. Heath. 1971. Isotopic analysis of metalliferous sediment from the East Pacific Rise. Earth and Planetary Sci. Letters 13:175-180.

Deep-Sea Drilling Project, Initial Reports: see Cook et al. 1971; Drever, 1971; Fischer et al., 1971; Hollister et al., 1972; Maxwell et al., 1970; Moberly et al., 1971; von der Borch et al., 1970 and Winterer et al., 1971.

Degens, E. T., and D. A. Ross. 1969. Hot Brines and Recent Heavy Metal Deposits in the Red Sea. New York: Springer-Verlag, 600 pp.

Drever, J. I. 1971. Chemical and mineralogical studies, site 66. In Initial Reports of the Deep-Sea Drilling Project, 7-2, E. L. Winterer et al., eds., Washington, D. C.: Government Printing Office, pp. 965-976.

Dymond, J., J. B. Corliss, G. Ross Heath, C. W. Field, E. J. Dasch, and H. H. Veeh. 1973. Origin of metalliferous sediments from the Pacific Ocean. Geol. Soc. America Bull. 84:3355-3372.

Elder, J. W. 1965. Physical processes in geothermal areas. In Terrestrial Heat Flow, H. K. Lee, ed., Am. Geophys. Union Geophys. Mon. 8:211-239.

El Wakeel, S. K., and J. P. Riley. 1961. Chemical and mineralogical studies of deep sea sediments. Geochim. et Cosmochim. Acta 25: 110-146.

Engel, A. E. J., and C. Engel. 1970. Mafic and ultramafic rocks. In The Sea, 4-I, A. E. Maxwell, ed., New York: Wiley-Interscience, pp. 465-519.

Ewing, J., and M. Ewing. 1970. Seismic reflections. In The Sea, 4-I, A. E. Maxwell, ed., New York: Wiley-Interscience, pp. 1-51.

Ewing, M., J. Ewing, R. F. Houtz, and R. Leyden. 1966. Sediment distribution in the Bellingshausen Basin. In Symposium on Antarctic Oceanography, R. I. Currie, ed., Santiago, Chile: Scott Polar Research Inst., pp. 89-100.

Fairbridge, R. W., ed. 1972. Encyclopedia of Geochemistry and Environmental Sciences. New York: Van Nostrand & Reynolds Co. 1321 pp.

Fisher, D. C., and K. Boström. 1969. Uranium-rich sediments on the East Pacific Rise. Nature 224:64-65.

Gilluly, J. 1955. Geologic contrasts between continents and ocean basins. In Crust of the Earth, A. Poldervaart, ed., Geol. Soc. America Spec. Paper 62:7-18.

Girdler, R. W. 1964. Geophysical studies of rift valleys. In Physics and Chemistry of the Earth, vol. 15, London, Pergamon Press, pp. 121-156.

Goldberg, E. D., and G.O.S. Arrhenius. 1958. Chemistry of Pacific pelagic sediments. Geochim. et Cosmochim. Acta 13:153-212.

Goldschmidt, V. M. 1954. Geochemistry Oxford: Clarendon Press. 730 pp.

Graham, A. 1973. Rare-earth abundances in East Pacific Rise sediments. In preparation.

Green, D. H. 1964. The petrogenesis of the high temperature peridotite intrusion in the Lizard area, Cornwall Jour. Petrol. 5:134-188.

Green II, H. W. 1972. A CO_2-charged asthenosphere. Nature Phys. Science 238:2-5.

Gümbel, G. von. 1878. Über die im Stillen Ocean auf dem Meeresgrunde vorkommenden Manganknollen. Sitz. Ber. Kgl. Bayer. Akad. Wiss., München, Math-Physik Kl. 1877, pp. 189-209.

Harrison, J. C., and W. D. Brisbin. 1959. Gravity anomalies off the west coast of North America. 1: Seamount Jasper. Geol. Soc. America Bull. 70:929-934.

Heezen, B. C., and H. W. Menard. 1963. Topography of the deep-sea floor. In The Sea, vol. 3 M. N. Hill, ed., New York: Wiley-Interscience, pp. 233-280.

Heinrich, E. W. 1966. The Geology of Carbonatites, Chicago: Rand McNally & Co, 608 pp.

Heirtzler, J. R., G. O. Dickson, E. M. Herron, W. C. Pitmann III, and X. LePichon. 1968. Marine magnetic anomalies, geomagnetic field reversals and motions of the ocean floor and continents. Jour. Geophys. Research 73:2119-2136.

Helz, G. R. 1971. Hydrothermal Solubility of Magnetite. Ph.D. diss., Pennsylvania State University, 110 pp.

Herrmann, A. G., and K. H. Wedepohl. 1970. Untersuchungen an spilitischen Gesteinen der variskischen Geosynkline in Nordwestdeutschland. Contr. Mineralogy and Petrology 29:255-274.

Herron, E. M. 1972. Seafloor spreading and Cenozoic history of the East Central Pacific. Geol. Soc. America Bull. 83:1671-1692.

Hess, H. H. 1964. The oceanic crust, the upper mantle and the Mayaguez serpentinized peridotite. In A study of the serpentinite, C. A. Burk, ed., Natl. Acad. Sci. - Natl. Research Council Pub. 1188

Honnorez, J. 1971. La palagonitisation Zürish: Vulkaninstitut Immanual Friedländer, 131 pp.

Honnorez, J., and E. Bonatti. 1970. Nepheline gabbro from the Mid-Atlantic Ridge. Nature 228:850-852.

Horn, M. K., and J.A.S. Adams. 1966. Computer-derived geochemical balances and element abundances. Geochim. et Cosmochim. Acta 30:279-297.

Isacks, B., J. Oliver, and L. R. Sykes. 1968. Seismology and the new global tectonics. Jour. Geophys. Research 73:5855-5899.

Knauss, J. A. 1962. On some aspects of the deep circulation of the Pacific. Jour. Geophys. Research 67:3943-3954.

Kopp, J. F., and R. G. Kroner. 1967. Trace Metals in Waters of the United States U. S. Dept. Interior. Div. Pollution Surveillance, 212 pp.

Krauskopf, K. B. 1957. Separation of manganese from iron in sedimentary processes. Geochim. et Cosmochim. Acta 12:61-84.

Krauskopf, K. B. 1967. Introduction to Geochemistry New York: McGraw-Hill Co. 721 pp.

Kuenen, Ph. H. 1946. Rate and mass of deep-sea sedimentation. Am. Jour. Science 244:563-372.

Kuenen, Ph. H. 1950. Marine Geology New York: John Wiley & Sons, 568 pp.

Landergren, S. 1964. On the geochemistry of deep-sea sediments. Reports of the Swedish Deep Sea Expedition, X, Spec. Investig. 5:1-154.

Le Bas, M. J. 1971. Peralkaline volcanism, crustal swelling and rifting. Nature Phys. Science 230:85-87.

Lehmann, E. 1941. Eruptivgesteine and Eisenerze im Mittelund Oberdevon der Lahnmulde. Technisch-Pädagogischer Verlag. Weitzlar. 391 pp.

Lister, C. R. B. 1972. On the thermal balance of a mid-oceanic ridge. Geophys. Royal Astron. Soc. Jour. 26:515-535.

Lynn, D. C., and E. Bonatti. 1965. Mobility of manganese in diagenesis of deep-sea sediments. Marine Geology 3:457-474.

Manson, V. 1968. Geochemistry of basaltic rocks: major elements. In Basalts, vol. I, H. H. Hess and A. Poldervaart, eds., New York: John Wiley & Sons, pp. 215-269.

McBirney, A. R. 1971. Oceanic volcanism-- a review. Revs. of Geophysics and Space Physics 9:523-556.

McManus, D. A., and R. E. Burns 1969. Scientific report on deep-sea drilling project, leg V. Ocean Industry 4:40-42.

Meinzer, O. E. 1923. The occurrence of ground water in the United States. U. S. Geol. Surv. Water Supply Paper 489, Washington, D. C.: Government Printing Office.

Menard, H. W. 1964. Marine Geology of the Pacific New York: McGraw-Hill 271 pp.

Moberly, R., and G. R. Heath. 1971. Volcanic rocks from the western and central Pacific; leg 7, Deep-sea drilling project, In Initial Reports of the Deep-Sea Drilling Project, vol. VII E. L. Winterer et al., eds. Washington, D.C.: Government Printing Office, pp. 1011-1025.

Morgan, W. J. 1968. Rises, trenches, great faults, and crustal blocks. Jour. Geophys. Research 73:1959-1982.

Murray, J., and A. F. Renard. 1891. Deep-Sea Deposits. Challenger Exped. Reports, vol. 3, London: Her Majesty's Stationary Office, 496 p.

Naboko, S. I. 1963. Gidrotermalnyi metamorfizm porod v vulkanitjeskikh oblastjakh. Izd. Akad. Nauk.,Moscow, 172 pp.

Naumov, G. B., B. N. Ryzhenko, and I. L. Khodakovskii. 1971. Spravochnik termodinamicheskikh Velichin Moscow: Atomizaat., 239 pp.

Park, C. F. Jr. 1946. The spilite and manganese problems of the Olympic Peninsula, Washington, Am. Jour. Sci. 244:305-323.

Peck, D. L., T. L. Wright, and J. G. Moore. 1966. Crystallisation of tholeiitic basalt in Aloe Lava Lake, Hawaii. Bull. Volcanol. 29:629-655.

Pecora, W. T. 1956. Carbonatites, a review. Geol. Soc. America Bull. 67:1537-1556.

Phillipi, E. 1910. Die Grundproben der Deutschen Süd-Polar-Expedition, 1901-1903. Deutsche Süd-Polar-Expedition II, vol. 6 Berlin: Georg Reimer, pp. 413-416.

Pitman, W. E., III, E. M. Herron, and J. R. Heirtzler 1968. Magnetic anomalies in the Pacific and sea floor spreading. Jour. Geophys. Research 73:2069-2085.

Poole, J. H. J. 1963. Radioactivity of sodium carbonate lava from Oldoinyo Lengai, Tanganyika, Nature 198:1291.

Pourbaix, M., N. De Zoubov, and J. Muylder, et al. 1963. Atlas d' equilibres electrochimiques Paris: Gauthier-Villars & Cie.

Printz, M. 1967. Geochemistry of basaltic rocks: Trace elements. In Basalts, vol. I, H. H. Hess, and A. Poldervaart, eds., New York: Interscience, pp. 271-323.

Puchelt, H. 1967. Zur Geochemie des Bariums im exogenen Zyklus. Sitz. Berlin Heidelberg. Akad. D. Wissensch. Mat.-Nat. Klasse, Springer-Verlag. 205 pp.

Puchelt, H. 1973. Recent iron sediment formation at the Kameni islands. Santorini (Greece.) In Ores in Sediments, G. C. Amstutz and J. Bernal, eds., Berlin: Springer-Verlag, pp. 227-245.

Revelle, R. R. 1944. Marine bottom samples collected in the Pacific by the Carnegie on its seventh cruise. Carnegie Inst. Washington Publ. 556, 180 pp.

Revelle, R. R., and A. E. Maxwell. 1952. Heat flow through the floor of the eastern North Pacific Ocean. Nature 170:199-202.

Ringwood, A. E. 1969. Composition and evolution of the upper mantle. In The Earth's Crust and Upper Mantle, Amer. Geophys. Union Geophys. Mon. 13. p. 1-17.

Rydell, H. S., and E. Bonatti. 1973a. Uranium in submarine metalliferous deposits. Geochim. et Cosmochim. Acta 37:2557-2565.

Rydell, H. S., T. Kraemer, K. Boström, and O. Joensuu. 1973b. Postdepositional injections of solutions of deep-seated origin into active ridge sediments. Later pub. in Marine Geology 17:151-164.

Schidlowski, M., W. Stahl, and G. C. Amstutz. 1970. Oxygen and carbon isotope abundances in carbonates of spilitic rocks from Glarus, Switzerland. Naturwissenschaften 57:542-543.

Scott, M. R., J. K. Osmond, and J. K. Cochran. 1972. Sedimentation rates and sediment chemistry in the South Indian Basin. In Antarctic Oceanology II, Antarctic Research Ser. 19, D. E. Hayes, ed., pp. 317-334.

Sigvaldason, G. E. 1966. Chemistry of thermal waters and gases in Iceland. Bull. Volcanol. 24:589-604.

Sillén, L. G., and A. E. Martell. 1964. Stability Constants of Metal-Ion Complexes. London: Chemical Society. 754 pp.

Skornyakova, I. S. 1964. Dispersed iron and manganese in Pacific Ocean sediments. Int. Geology Rev. 7:2161-2174.

Sömme, A. ed. 1961. A Geography of Norden. London: Heinenmann. 364 pp.

Sowerbutts, W. T. C. 1972. Rifting in Eastern Africa and the fragmentation of Gondwanaland. Nature 235:435-437.

Spencer, D. W., D. E. Robertson, K.K. Turekian, and T. Folson. 1970. Trace element calibrations and profiles at the Geosecs test station in the northeast Pacific Ocean. Jour. Geophys. Research 75:7688-7696.

Stoiber, R. E., and E. S. Davidson. 1959. Amygdule mineral zoning in the Portage Lake lava series. Econ. Geology 54:1250-1277.

Sundius, N. 1923. Grythyttefältets geologi. Sveriges Geol. Undersökning, Ser. C 312, 354 pp.

Sykes, L. R., J. Oliver, and B. Isacks. 1970. Earthquakes and tectonics. In The Sea, 4-I, A. E. Maxwell, ed., New York: Wiley-Interscience, pp. 353-420.

Trichet, M. J. 1969. Contribution a l'etude de l'alteration experimentale des verres volcaniques. Ph.D. diss., Univ. of Paris, AO-3149, May 7, 1969.

Turekian, K. K. 1969. The ocean streams and atmosphere. In Handbook of Geochemistry, 1 K. H. Wedepohl, ed., Springer-Verlag, pp. 297-323.

Tuttle, O. F., and J. Gittins, eds. 1966. Carbonatites, New York: Interscience, 591 p.

Vallance, T. G. 1960. Concerning spilites. Proceedings of the Linnean Society of New South Wales 85:8-52.

Veeh, H. H., and K. Boström. 1971. Anomalous $^{234}U/^{238}U$ on the East Pacific Rise. Earth and Planetary Sci. Letters 10:372-374.

Verschure, R. H. 1966. Possible relationships between continental and oceanic basalt and kimberlite. Nature 211:1387-1389.

Vine, F. J. 1966. Spreading of the ocean floor: new evidence. Science 154:1405-1415.

Von Herzen, R. P., and R. N. Anderson. 1972. Implications of heat flow and bottom water temperature in the Eastern Equatorial Pacific. Royal Astronom. Soc. Geophys. Jour. 26:427-458.

Von Herzen, R. P., and M. G. Langseth. 1966. Present status of oceanic heat-flow measurements. Phys. Chem. Earth 6:365-407.

Wedepohl, K. H. 1960. Spurenanalytische Untersuchungen an Tiefseetonen aus dem Atlantik. Geochim. et Cosmochim. Acta 18:200-231.

White, D. E. 1965. Metal contents of some geothermal fluids. In Problems of postmagmatic ore deposition, II; Academy of Sciences Symposium Prague: pp. 432-443.

Wickman, F. E. 1954. The total amount of sediments and the composition of the average igneous rock. Geochim. et Cosmochim. Acta 5:97-110.

Wolgemuth, K. 1970. Barium analyses from the first geosecs test cruise. Jour. Geophys. Research 75:7687.

Zelenov, K. K. 1964. Iron and manganese in exhalations from the submarine volcano Banu Vuhu (Indonesia). Akad. Nauk. SSSR Doklady 155:1317-1320.

POSTSCRIPTUM

The active ridge sediments and their origin have received increasing attention since 1971. The subjects of these advances broadly fall into the following categories, that will be treated in sections a-d below. Some other aspects are treated in Bonatti (1975).

a. Composition and Distribution of Unconsolidated (Normal) Active Ridge Sediments

Several studies have described additional occurrences of unconsolidated active ridge sediments, which often form the basal section of the deep sea deposits (Cook 1972; Cronan 1973; Dymond et al. 1973; Piper 1973; Boström et al. 1976; Dymond et al. 1976a, 1976b). The accumulation rates of Fe and Mn in such sediments are frequently very high (Boström et al. 1976; Dymond et al. 1976a). Similar sediments have been found at DSDP site 183 (Natland 1973). Furthermore, some iron-manganese-enriched sediments from DSDP-leg 35 could represent mixtures of terrigenous and volcanic matter, as has also been found in the Atlantic Ocean, see figure 9 (Bogdanov et al. 1976; Donelly et al. 1976a; Drewer 1976; and Gieskes et al. 1976). Another type of basal sediment is enriched in iron and titanium and has most likely another origin than the active ridge sediments (Jenkyns et al. 1976). The distribution of U and Th have been studied by Boström et al. (1979).

b. Composition and Distribution of Indurated (Hard-Crusts) Active Ridge Sediments

Prior to 1973 the only known hard-crust deposit at a spreading center was Amph D2 on the East Pacific Rise (Bonatti et al. 1966; Veeh et al. 1971); many more such crusts are now known. They consist of varying amounts of manganese and iron oxides, sometimes enriched in different layers, and grow exceptionally fast (Scott et al. 1974a; 1974b; Piper et al. 1975; and Moore et al. 1976). These crusts thus are strikingly different from ordinary manganese nodules in shape, composition, distribution, and rate of growth (Horn 1972; Ku 1977). Mapping of physiography and bottom water temperaturesat the TAG Hydrothermal Field and a small inactive hydrothermal deposit in the FAMOUS area of the Mid-Atlantic ridge suggest that hot spring vents exist near the crests (Rona 1976; Rona et al. 1976; Lalou et al. 1976). Another vent area with barite-opaline silica deposits occurs at the Lau spreading center (Bertine et al. 1975). Spectacular observations from research submarines of active hot spring vents have been made at the Galapagos spreading center (Corliss et al. 1977a, 1977b). These observations convincingly demonstrate the connection between hydrothermal vents and the deposition of iron and manganese oxides. Nevertheless, the number of such deposits is still small; there are hundreds of dredge sites near the spreading centers where no such crusts have been found. The unconsolidated iron-manganese-rich active ridge deposits thus probably contain 99.0-99.9 percent of the hydrothermally derived elements like iron and manganese.

c. Genesis of Metalliferous Solutions

Lister (1975, 1977) has shown that hot lavas newly emplaced at spreading centers should react rapidly with cold seawater and develop cracks that quickly spread to depths of several kilometers. In the openings formed, seawater may descend down to large depths, become heated, and resurface in very short time, from a few hours to a few days. Seawater-rock interactions may thus be an important source for metalliferous solutions. This has been emphasized also by Spooner et al. (1973).

Several experimental investigations of such processes have been made (Bischoff et al. 1975; Hajash 1975; Mottl 1976), showing that solutions with the same Fe/Mn-ratio as in the active ridge deposits and with much dissolved Fe and Mn can be produced by such processes (figure 10). Most students consider this process to be the main generator of the metalliferous solutions. However, studies of natural

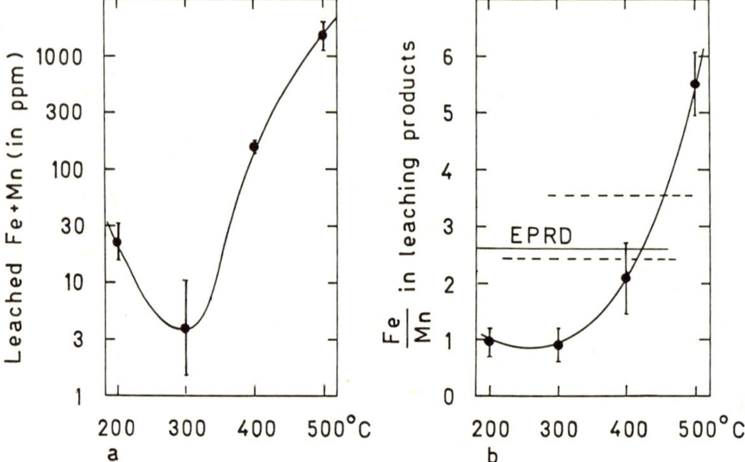

Figure 10. Results of experimental leaching of basalt with seawater between 200° and 500°C and pressures of respectively 500 to 1000 bars (data from Mottl 1976). The errors (one standard deviation) are shown as vertical bars.
a. Amount of Fe+Mn (in ppm) that is dissolved in seawater.
b. The ratio of Fe/Mn in the formed solutions. Solid line shows position of EPRD (=East Pacific Rise sediments, table 1), whereas dashed lines represent positions of error limits for the pure Fe-Mn phase in Pacific sediments.

springs in Iceland show higher CO_2 contents than can be explained by descending water-rock interactions (Mottl 1976) and studies of He distributions in bottom waters over the East Pacific Rise suggest a degassing process involving the mantle (Clarke et al. 1969). To date this information does not prove any transfer of metals from the mantle to the top of the oceanic crust, but the possible role of a carbonate-charged phase has been emphasized in the main article. Another problem that has not been studied is how sealing of cracks by hot plastic rocks will affect the life time of water-conducting cracks. This problem was pointed out in the main article, but is difficult to resolve with present theoretical tools (Lister, personal communication).

The changes in chemical composition of the altered rocks recovered to date (Flower 1975) do not show the source of iron and manganese. Iron and manganese may even increase and aluminum decrease in the rocks as alteration progresses (Aumento et al. 1976)--that is, the compositional changes are opposite to those required to form the metalliferous solutions. This suggests that several series of alterations occur in the oceanic crust. In the upper part of the crust several low-temperature processes are proceeding, as is well shown by several articles in the Initial Reports of the DSDP.

A large source depth is suggested by Löfgren et al. (1979). Their statistical study of several hundred sediment analyses from the Atlantic, Indian, and Pacific Oceans shows that the abundances of the elements Fe, Mn, Al, and Ti vary as if they were caused by simple mechanical mixtures of a well-defined active ridge component (identical to the noncarbonate fraction at site GS-7202-35) and average terrigenous matter (Löfgren et al. 1979; see figure 8). This constancy in the active ridge component can be explained if the metalliferous solutions are formed at large depths, where the conditions are uniform because of ample supplies of heat; the leaching process must proceed at 400-450°C and considerable pressures according to data shown in figure 10. Shallow leaching is ruled out since it should give metalliferous solutions of highly varying compositions because of rapid variations in redox state, ionic composition of the solutions, temperature, and pressure.

d. Sedimentation Models

In the active ridge sediments, the hydrothermal component is mixed with other components like terrigenous matter, biogenous matter, and adsorbates from sea water; a clear understanding of the origin of active ridge sediments obviously requires a good knowledge of these additional element sources. The extreme type of active ridge sediment found at 13°S and 111-113°W (see table 1) probably exhibits the main element composition of the hydrothermal precipitates, but the origin of the minor and trace elements is more difficult to ascertain. To circumvent these difficulties various procedures have been tried. One is to analyze isolated separates from deep sea sediments to get an understanding of the composition of admixed terrigenous and biological phases and subsequently deduct proportional quantities of other elements, assuming that all Al derives from terrigenous matter and all $CaCO_3$ and excess silica derive from biological sources (see, for instance, table above). The relation Al/Ti could furthermore indicate the relative abundance of detritus from the continental and oceanic (basaltic) crusts. The "leftovers" from such analyses of data should thus be the volcanic and authigenic components in the sediments. Such models have been discussed by Boström (1973, 1976) and Donelly et al. (1976a, 1976b).

Another approach is to determine the composition of the original rock and biogenous phases before they are deposited on the ocean floor, and subsequently to mix these components in arbitrary proportions and test whether the ensuing mixtures resemble pelagic sediments. Such mathematical mixing models show striking similarities with sediments far from spreading centers, but fail to explain the compositions of active ridge sediments (Boström et al. 1974, 1976, 1978a, 1978b). Indeed, successful mixing models all must assume that an Fe-Mn-Ba-V-rich phase is added to lithogenic and biogenous phases to reproduce the element patterns found in sediments near spreading centers.

These studies show that biological detritus and not adsorption from seawater accounts for most of the trace elements in deep sea sediments (Boström et al. 1978a, 1978b). The results are in excellent agreement with the concept of biological element vectors (e.g., fecal pellets, exoskeletons, and tissue detritus) discussed by Ferrante et al. (1977), Higgo et al. (1977) and Lal (1977). It should be pointed out that the success with the mixing models outlined above to a large extent depends on the relatively constant minor and trace constituent composition of biological matter (Boström et al. 1974; Moore et al. 1978). This does not exclude the possibility that adsorption of As, U, Hg, etc., may be an ongoing process near spreading centers, where the sediments are exceptionally rich in highly adsorbing amporphous ferric hydroxides.

These mixing models have also been used to trace small admixtures of hydrothermal precipitates in predominantly biogenic sediments. Thus, studies of sediments from DSDP leg VIII suggest that small but varying volcanic inputs have taken place during the whole Cenozoic in the Equatorial Pacific (Boström et al. 1978b).

These results are in excellent agreement on several points with results obtained by Dymond et al. (1976a), who could demonstrate, using factor-analysis, that only three factors explain almost all element variations in the area. One factor enriched in Fe, Mn, Cu, and Zn predominates in sediments from the East Pacific Rise; another factor enriched in Ni and Al is well associated with the Bauer deep; and a third factor with much Si is associated with the Equatorial high productivity zone.

REFERENCES TO POSTSCRIPTUM

Aumento, F., W. S. Mitchell, and M. Fratta. 1976. Interaction between sea water and oceanic layer two as a function of time and depth. 1. Field evidence. Canadian Mineralogist 14:269-290.

Bertine, K. K., and J. B. Keene. 1975. Submarine barite-opal rocks of hydrothermal origin. Science 188:150-152.

Bischoff, J. L., and F. W. Dickson. 1975. Seawater-basalt interactions at 200°C and 500 bars: Implications for origin of sea-floor heavy-metal deposits and regulation of sea water chemistry. Earth and Planetary Sci. Letters 25:385-397.

Bogdanov, Yu. A., et al. 1976. Chemical composition of sediments. In Hollister, C. D. et al., eds. Initial Reports Deep-Sea Drilling Project, Washington D. C.: Government Printing Office, pp. 447-464.

Bonatti, E. 1966. Authigenic deep-sea carbonates. Science 153:534.

Bonatti, E. 1975. Metallogenesis at oceanic spreading centers. In Donath, F., Stehli, F. and Wether III, G., eds. Annual Review of Earth and Planetary Sciences, vol. 3 Palo Alto, Calif.: Annual Reviews, Inc., pp. 401-431.

Boström, K. 1973. The origin and fate of ferromanganoan active ridge sediments. Stockholm Contr. Geology 24:149-243.

Boström, K. 1976. Particulate and dissolved matter as sources for pelagic sediments. Stockholm Contr. Geology 30:15-79.

Boström, K., O. Joensuu, and I. Brohm. 1974. Plankton - Its chemical composition and its significance as a source of pelagic sediments. Chem. Geology 14:255-271.

Boström, K., O. Joensuu, S. Valdés, W. Charm, and R. Glaccum. 1976. Geochemistry and origin of East Pacific sediments, sampled during DSDP leg 34. In Yeatz et al., eds. Initial Reports of the Deep-Sea Drilling Project, pp. 559-574.

Boström K., L. Lysén, and C. Moore. 1978a. Biological matter as a source of authigenic matter in pelagic sediments. Chem. Geology 23:11-20.

Boström, K., C. Moore, and O. Joensuu. 1979. Biological matter as a source for Cenozoic deep sea sediments in the equatorial Pacific. Ambio Special Report No. 6, pp. 11-17.

Boström, K., and H. Rydell. 1979. Geochemical behavior of U and Th during exhalative sedimentary processes. In Lalou, C., ed. Colloque Internat. du CNRS, N. 289, Sur la Genese des Nodules de Manganèse, Gif Sur Yvette, 1978. In press.

Clarke, W. B., M. A. Beg, and H. Craig. 1969. ^{3}He in the sea: evidence for terrestrial primordial helium. Earth and Planetary Sci. Letters 6:213-220.

Cook, H. E. 1972. Stratigraphy and sedimentation. In Hays, J. D., ed. Initial Reports of the Deep-Sea Drilling Project, leg IX, pp. 933-943.

Corliss, J. B., J. Dymond, M. Lyle, R. Cobler, D. Williams, R. Von Herzen, and Tj. van Andel. 1977. a. Observations of the sediment mounds of the Galapagos Rift during the Alvin diving program. (Abs.). b. Movie about hydrothermal site at the Galapagos spreading center. Geol. Soc. America Annual Meeting, Seattle.

Cronan, D. S. 1973. Basal ferruginous sediments cored during leg 16, Deep-Sea Drilling Project. In van Andel, T. H. et al., eds. Initial Reports of the Deep Sea Drilling Project XVI, pp. 601-604.

Donelly, T. W., and J. L. Wallace. 1976a. Major element chemistry of the Tertiary rocks at site 317 and the problem of the origin of the non-biogenic fraction of pelagic sediments. In Schlanger, S. O. et al., eds., Initial Reports of the Deep-Sea Drilling Project XXXIII, pp. 557-562.

Donelly, T. W., and J. L. Wallace. 1976b. Major- and minor-element chemistry of Antarctic clay-rich sediments: sites 322, 323, and 325, DSDP leg 35. In Hollister, C. D. et al., eds. Initial Reports of the Deep-Sea Drilling Project, XXXV, pp. 427-446.

Drewer, J. I. 1976. Chemical and mineralogical studies, site 323. In Hollister, C. D. et al., eds. Initial Reports of the Deep-Sea Drilling Project, XXXV, pp. 471-477.

Dymond, J., J. B. Corliss, G. R. Heath, C. W. Field, E. J. Dasch, and H. H. Veeh. 1973. Origin of metalliferous sediments from the Pacific ocean. Geol. Soc. America Bull. 84:3355-3372.

Dymond, J., J. B. Corliss, and R. Stillinger. 1976a. Chemical composition and metal accumulation rates of metalliferous sediments from sites 319, 320B, and 321. In Yeatz et al., eds. Initial Reports of the Deep-Sea Drilling Project, XXXIV, pp. 575-589.

Dymond, J., and J. B. Corliss. 1976b. The IDOE Nazca plate project: origin of metalliferous sediments. 25th Internat. Geol. Congress, Sydney, Abs. vol. 3, pp. 736.

Ferrante, J. B., and J. I. Parker. 1977. Transport of diatom frustules by copepod fecal pellets to the sediments of Lake Michigan. Limnology and Oceanography 22:92-98.

Flower, M. F. J. 1975. Low temperature alterations of sea floor basalts, DSDP leg 37. (Abs.) Internat. Conf. on Nature of Oceanic Crust, Dec. 4-6, 1975, La Jolla, Calif. Organizer: American Geophysical Union.

Geological Society of America, 1977. Annual Meetings. Seattle. Geophysics division symposium: Hydrothermal systems at oceanic spreading centers, Organizers: P. A. Rona and R. P. Lowell.

Gieskes, J. M., and J. R. Lawrence. 1976. Geochemistry, leg 35: Introduction and summary. In Hollister, C. D. et al., eds. Initial Report of the Deep-Sea Drilling Project, XXXV, pp. 403-406.

Hajash, A. 1975. Hydrothermal processes along mid-ocean ridges: an experimental investigation. Contrib. Mineralogy and Petrology 53:205-226.

Higgo, J. J. W., R. D. Cherry, M. Heyrand, and S. W. Fowler. 1977. Rapid removal of plutonium from the oceanic surface layer by zooplankton faecal pellets. Nature 226:623-624.

Horn, D. R., ed. 1972. Ferro-manganese deposits on the ocean floor. International Decade of Ocean Exploration - National Science Foundation, Washington D. C. 293 pp.

Jenkyns, H. C., and R. G. Hardy. 1976. Basal iron-titanium rich sediments from hole 315A. In Schlanger, S. C., et al., eds. Initial Reports of the Deep-Sea Drilling Project, XXXIII, pp. 833-836.

Ku, T. L. 1977. Rates of accretion. In Glasby, G., ed. Marine Manganese Deposits Amsterdam: Elsevier Scientific Publ. Co. pp. 249-268.

Lal. D. 1977. The oceanic microcosm of particles. Science 198:997-1009.

Lalou, C., E. Brichet, T. L. Ku, and C. Jehanno. 1976. The FAMOUS hydrothermal deposit, radiochemical and SEM/EDAX study. Jt. Oceanogr. Assem., Edinburgh, 1976. Late Abs., unpaginated.

Lister, C. R. B. 1975. Rapid evolution of geothermal systems in new oceanic crust predicts mineral output mainly near ridge crests (abs). EOS 56:1074.

Lister, C. R. B. 1977. Stability of estimates of water-penetration rate to theoretical assumptions. Geol. Soc. America Annu. Meet., Seattle, Abs. p. 1072

Löfgren, C., and K. Boström. 1979. The origin of the volcanic component in active ridge sediments. In The Dynamic Environment of the Ocean Floor, K. Fanning and F. T. Manheim, eds., D. C. Heath & Co., Lexington Books, in press.

Moore, C.,and K. Boström. 1978. The elemental compositions of lower marine organisms. Chem. Geology 23:1-9.

Moore, W. S., and P.R. Vogt. 1976. Hydrothermal manganese crusts from two sites near the Galapagos spreading axis. Earth and Planetary Sci. Letters 29:349-356.

Mottl, M. J. 1976. Chemical Exchange Between Seawater and Basalt During Hydrothermal Alteration of the Oceanic Crust. Ph.D. diss., Harvard University, 188 pp.

Natland, J. 1973. Basal ferromanganoan sediments at DSDP site 183 and site 192. In J. S. Creager et al., eds. Initial Reports of the Deep-Sea Drilling Project, XIX, pp. 629-636.

Piper, D. Z. 1973. Origin of metalliferous sediments from the East Pacific Rise. Earth and Planetary Sci. Letters 19:75-82.

Piper, D. Z., H. H. Veeh, W. G. Bertrand, and R. L. Chase. 1975. An iron-rich deposit from the northeast Pacific, Earth and Planetary Sci. Letters 26:114-120.

Rona, P. A. 1976. Pattern of hydrothermal mineral deposition: Mid-Atlantic ridge crest at latitude 26°N. Marine Geology 21:M59-M66.

Rona, P. A., R. N. Harbison, B. G. Bassinger, R. B. Scott, and A. Nalwalk. 1976. Tectonic fabric and hydrothermal activity of Mid-Atlantic ridge crest (lat. 26°N). Geol. Soc. America Bull. 87:661-674.

Scott, M. R., R. B. Scott, P. A. Rona, L. W. Butler, and A. J. Nakwalk. 1974a. Rapidly accumulating manganese deposit from the median valley of the Mid-Atlantic ridge. Geophys. Research Letters 1:355-358.

Scott, R. B., P. A. Rona, B. A. McGregor, and M. R. Scott. 1974b. The TAG hydrothermal field. Nature 251:301-302.

Spooner, E. T. C., and W. S. Fyfe. 1973. Sub-seafloor metamorphism, heat and mass transfer. Contr. Mineralogy and Petrology 42: 287-304.

Veeh, H. H., and K. Boström 1971. Anomalous $^{234}U/^{238}U$ on the East Pacific Rise. Earth and Planetary Sci. Letters 10:372-374.

29

Reprinted from *Contrib. Mineralogy and Petrology* **42**:287-304 (1973)

Sub-Sea-Floor Metamorphism, Heat and Mass Transfer

E.T.C. Spooner*

Dept. of Geology, The University, Manchester M13 9PL, England

W.S. Fyfe

Dept. of Geology, The University of Western Ontario, London 72, Canada

Received August 10, 1973

Abstract. The ophiolitic rocks of E. Liguria, Italy contain a „spilitic" metamorphic assemblage sequence, cross-cut by hydrothermal veins, which developed in the oceanic environment. Metamorphic parageneses indicate that temperatures as high as ~400° C were realised at depths as shallow as 300 m below the original rock/water interface. The inferred temperature interval was equivalent to a geothermal gradient of ~1300° C/km.

It is suggested that metamorphism took place in a sub-sea-floor geothermal system, and that such systems are an integral part of the sea-floor spreading process. Modern evidence is provided to support this hypothesis, and to suggest that heavy metal rich solutions discharged from such systems are responsible for the formation of a metal enriched sedimentary component. A unified model of sub-sea-floor metamorphism and mass transfer is proposed, and possible differences between sub-sea-floor and terrestial geothermal systems are discussed. In the light of the model, the origins of certain aspects of bedded cherts found associated with ophiolitic rocks, of ophiolitic massive sulphide deposits and of certain trace element patterns are considered.

Introduction

Recognition of the structural (Gass, 1968; Cann, 1970; Thayer, 1969; Coleman, 1971; Dewey and Bird, 1971; Moores and Vine, 1971; Khan *et al.*, 1972) and geochemical (Pearce and Cann, 1971) similarities which exist between ophiolite sequences and the oceanic crust and Upper Mantle has led to the wide acceptance of the hypothesis that they are genetically equivalent; the former being tectonically emplaced fragments of the latter (De Roever, 1957; Dietz, 1963; Gass and Masson-Smith, 1963; Vuagnat, 1963; Hess, 1965; Moores, 1969). The study of ophiolitic rocks, therefore, provides an opportunity for studying the nature of the sub-sea-floor processes which occur at sea-floor spreading ridges. A unified model for sub-sea-floor metamorphism, heat and mass transfer, partly derived from an examination of the metamorphic ophiolitic rocks of the E. Ligurian Apennines of N.W. Italy, is here presented.

The Ophiolitic Rocks of E. Liguria

These ophiolitic rocks (Bortolotti and Passerini, 1970) consist of a characteristic association of altered ultramafic tectonites and cumulates (Bezzi and Piccardo, 1971), coarse-grained metagabbros locally cross-cut by metadiabase dykes and a volcano-sedimentary unit containing pillow lavas and massive flows intercalated by bedded radiolarian cherts (Thurston, 1972), ophiolitic breccias and

* Present address: Dept. of Geology and Mineralogy, Parks Road, Oxford OX1 3PR, England.

turbidites, and shales. A „sheeted intrusive dyke complex" is absent, and contacts between the major lithologic units are tectonic expect for those above and below the volcano-sedimentary unit. This unit passes upwards into a sedimentary sequence, and rests stratigraphically on a base of metagabbro and serpentinite. Ophicalcite is locally present at the contact. Ophiolitic rocks constitute the stratigraphic base of the allochthonous, eugeosynclinal 2–3 km thick Jurassic-Palaeocene Vara supergroup, the tectonically uppermost thrust sheet of the E. Ligurian Apennines, and part also of the tectonically underlying 6 km thick Jurassic-Mid-Eocene Trebbia supergroup (Voltaggio sequence) (Abbate and Sagri, 1970; Gelati and Pasquarè, 1970). Fragments of ophiolitic rocks also occur tectonically associated with the two supergroups and as olistrosome constituents (Abbate *et al.*, 1970).

Fission track dating of zircons from the related Tuscan ophiolitic rocks has shown that the latter were generated throughout Jurassic time (Bigazzi *et al.*, 1972), although an Upper Jurassic age is conventionally assigned to the immediately overlying sediments (Abbate and Sagri, 1970). Tectonic emplacement, with an inferred lateral translation of the order of 300 km in the same sense as the overall N.E. directed orogenic polarity, took place in Palaeocene-Lr. Miocene time (Abbate and Sagri, 1970).

In the region studied around Bonassola and Anzo (between Genoa and La Spezia), the ophiolitic rocks have been extensively metamorphosed at low to medium grades. Because fragments of all the types of metamorphic rock occur in overlying and intercalated breccias which contain an unmetamorphosed matrix, it is clear that the metamorphism which affected these rocks occurred *before tectonic emplacement*, and is therefore of sub-sea-floor origin. This earlier phase of alteration is separable from a later, but similarly low grade, prehnite-pumpellyite/pumpellyite-actinolite facies burial metamorphism of syn- and post-tectonic emplacement age which has also locally affected the ophiolitic rocks (Galli and Cortesogno, 1970).

In this area, a prograde sequence of metamorphic assemblages occurs in the lower part of a 225 m thickness of pillow lavas and subordinate massive flows (see Table 1). The metamorphosed lavas are structurally inverted, and form the S.W. limb of a recumbent syncline. They rest (stratigraphically) on a base of altered harzburgite and dunite which is cross-cut locally by a conjugate set of aphyric and glomoporphyritic feeder dykes. The dykes are both single and composite. Parageneses have been determined by optical and X-ray diffractometric techniques, and also by use of the electron microprobe.

The significant features of the metamorphic sequence may be summarized:

a) Haematite and calcite, which are abundant in the red pillow lavas, decrease in abundance sharply at the interface with blue-green pillow lavas. This decrease corresponds with the appearance and increase of pumpellyite, which characterises the blue-green pillow lavas.

b) In the deeper levels of the red pillow lavas there is a definite replacement of smectite by chlorite within a transition zone containing mixed layer smectite-chlorite.

c) Pumpellyite, epidote, actinolite and hornblende appear sequentially and in this particular order.

Table 1. Metamorphic assemblage sequence in the ophiolitic metapillow lavas and metadolerites of the Bonassola-Anzo region of E. Liguria, Italy

Lithologic unit	Mineral appearance	Mineral disappearance	Total assemblage
Red metapillow lavas			1. Albite-smectite-haematite-calcite-sphene
	Chlorite		2. Albite-smectite-(chlorite)-haematite-calcite-sphene
	Pumpellyite		3. Albite-chlorite-(smectite)-pumpellyite-haematite-calcite-sphene
		Smectite (Haematite and calcite decrease to accessory amounts)	
Blue-green metapillow lavas			4. Albite-chlorite-pumpellyite-sphene-(haematite-calcite)
	Epidote Actinolite		
Yellow-green metapillow lavas			5. Albite-epidote-pumpellyite-actinolite-chlorite-sphene-(haematite-calcite)
	Magnetite Hornblende	Haematite	
Metadolerite feeder dykes			6. Albite-epidote-pumpellyite-actinolite-hornblende-chlorite-magnetite-sphene-(calcite)

d) A level at which magnetite replaces haematite as the stable iron oxide is identified, and occurs at the base of the yellow-green pillow lavas between the first appearances of actinolite and hornblende.

These hydrous „spilitic" assemblages replace the constituents of the pillow lavas, whose original igneous textures are faithfully preserved, and may completely replace material interstitial to pillows: originally hyaloclastite. Original glassy material is extensively replaced and coarse-grained product phases occur in cracks and vesicles. Plagioclase is topotactically replaced by albite with low T optics and by randomly oriented calcic phases, twin planes having been inherited during cation diffusion. Clinopyroxene is frequently an unaltered metastable constituent of the paragenesis, and olivine is replaced by calcite-haematite in the red pillow lavas, and by chlorite + pumpellyite + actinolite in the deeper levels. Hornblende bearing assemblages occur only in the metadiabase dykes. Where these are glomoporphyritic, the original megacrysts have acted as partially closed chemical subsystems during metamorphism. Hence, the total disequilibrium assemblage is a mosaic of equilibrium sub-assemblages defined by pseudomorph boundaries.

Over the intrusive zone (100 m) original dunite and harzburgite are replace by a metasomatically developed assemblage containing tremolite (talc-chlorite-calcite). Adjacent to dyke margins intensive metasomatism has produced mono-mineralic actinolite. Outside the zone, ultramafic material is replaced by lizardite-chrysotile-talc-chlorite-magnetite-ferritchromite. Both the metadiabase dykes and the metasomatised ultramafic material are extensively in-situ brecciated. These breccias are cross-cut by unbroken epidotic hydrothermal veins.

Similar, but thicker (up to 10–20 cms), originally steeply inclined, fracture controlled hydrothermal veins cross-cut all grades of metapillow lavas, and also intercalated breccia units. High vein densities occur in specific zones. The vein parageneses consist of a restricted number of phases from the total group (quartz-albite-epidote-pumpellyite-actinolite-chlorite-sphene-calcite), which is equivalent to assemblage 5 of the metapillow lavas. Where such veins cross-cut lower grade material wallrock alteration is extensive. In particular, selvedges of proximal pillow lavas are replaced by epidotic assemblages, and vein wallrock is replaced by higher grade material over 1–5 cm widths. The defect structure of albite from one such vein has been examined by high voltage transmission electron microscopy (Lorimer *et al.*, 1972).

Interpretation of the Metamorphic Rocks

The „spilitic" assemblages described developed by irreversible (Helgeson, 1967; Helgeson, 1968; Helgeson *et al.*, 1969) hydration, cation exchange, oxidation and carbonation reactions from an initial, essentially anhydrous basaltic parent (water content of deep sea pillow lavas = 0.06–0.42% H_2O + : Moore, 1970), in response to the introduction of extraneous hydrothermal fluid. Assemblages were produced in local reversible equilibrium both amongst themselves and with the interacting aqueous electrolyte solution at the particular physical conditions of any given level in the sequence. Since both parent (eg. clinopyroxene, and relict spinodally decomposed plagioclase feldspar found by high voltage transmission electron-microscopy within albite pseudomorphs) and product phases are present, the intrinsic reaction kinetics and fluid mass flux (Helgeson *et al.*, 1970) prevented complete reconstitution of the original basaltic material.

Comparison with the assemblage sequence developed in the Reykjanes geothermal system of Iceland (Tómasson and Kristmannsdóttir, 1972), suggests that the possibility of developing the sequence described here by *isothermal* interaction with a fluid of variable chemistry is to be discounted. Specifically, the smectite-chlorite transition and the sequential appearance of pumpellyite, epidote, actinolite and hornblende indicate that the ophiolitic metamorphic sequence developed in response to rising temperature, originally downwards.

Likewise, wall rock alteration of lower grade metapillow lavas to higher grade assemblages adjacent to hydrothermal veins clearly indicates that such veins originally contained a higher temperature fluid introduced from below, which was rising originally upwards.

Both the occurrence of extensive metasomatism in the intrusive zone, and the presence of silicates precipitated from solution in the veins, indicate that extensive solution and mass transfer of material occurred in a moving hydrothermal fluid. Further comparison with the Reykjanes and other geothermal systems, combined with experimental work, allows some estimates of the temperature distribution to be made.

a) The similarity between the smectite-chlorite transition zone of Liguria, with that of Reykjanes and Hveragerdi (Sigvaldason, 1962) suggests a temperature of 210°C ± 20°C for this level.

b) The unique finding of actinolite at 320°C in the Salton Sea geothermal system (Keith *et al.*, 1968), whereas no actinolite has yet been found in Icelandic geothermal systems where temperatures up to 300°C occur, suggests a temperature greater than 320°C for actinolite bearing assemblages.

c) The occurrence of prehnite in metagabbros related to the deep level hydrothermally altered material shows that the temperature was less than 400°C at this level (Liou, 1971); a temperature of 370°C ± 30°C is considered likely.

This data suggests, therefore, that the ophiolitic assemblage sequence developed over a maximum temperature interval of 180°C—400°C. The maximum inferred temperature of ~400°C occurred at a depth of only 300 m below the original rock/water interface. This was equivalent to a geothermal gradient of 1300°C/km. Metamorphic paragenesis/thickness relations indicate that, along strike, the original geothermal gradient was lower at ~900°C/km: an inferred temperature of 200°C occurring at a 225 m depth.

From the presented evidence that large amounts of extraneous fluid were introduced, which interacted with basalt while rising in temperature to 370°C ± 30°C, and which subsequently discharged upwards along fractures towards the water/rock interface, it is concluded that the ophiolitic metamorphic sequence developed in a sub-sea-floor geothermal system involving convective circulation of brine, which underwent a chemical and thermal evolution prior to discharge into sea water.

Origin of the Hydrothermal Fluid

The evidence that large amounts of extraneous fluid were incorporated in and fluxed through the ophiolitic sequence indicates that a volumetrically large reservoir of readily available fluid lay in close proximity.

Basaltic magma is undersaturated with respect to a water dominated gas phase under the hydrostatic load pressure of sea water overlying a spreading ridge (Moore, 1970). Hence, juvenile water cannot generally form a separate phase with an independent physical behaviour: it is retained in glass or in amphibole. In general, therefore, basic magmas may be expected to absorb water from their environment when erupted in the deep oceans. In the ophiolitic rocks, however, there is evidence of circulation of large volumes of aqueous fluid. It could not, therefore, have been juvenile. Also, the presence of oxidised haematite-calcite bearing assemblages indicates that the original fluid was oxygenated and rich in dissolved CO_3, and SO_4. Juvenile water could not have produced such alteration since it is reduced and H_2 rich. These facts, and the geometrical correlation of the isograds with the original rock/water interface indicate that the original fluid was predominantly sea water. Modified sea water, therefore, was the heat transfer medium in the geothermal system, as it is in the currently active Reykjanes system of Iceland (Tómasson and Kristmannsdóttir, 1972).

Sea water circulation, however, may constitute one stage in the transport of juvenile water from the oceanic Upper Mantle. Firstly, it travels in solution in basic magma formed by partial melting. Then, it is remobilised and exchanged by hydrothermal sea water, and discharged into the hydrosphere.

The predominance of sea water in the hydrothermal brine has been confirmed by separate mineral carbon and oxygen isotopic analysis (Spooner *et al.*, in press).

Modern Evidence for the Existence of Sub-Sea-Floor Geothermal Systems

Heavy Metal Enriched Sediments. It has been shown that the accumulation rate of heavy metal enriched sediments (Böstrom and Peterson, 1966: Böstrom *et al.*, 1969: Böstrom *et al.*, 1972) on the active spreading ridges is too large for a terrestrial source (Bender *et al.*, 1971), and that the lead isotopes indicate a volcanic origin for the heavy metal component (Bender *et al.*, 1971: Dasch *et al.*, 1971). Although the present distribution of such sediments is restricted to the high heat flow regions of the ridges, it has been shown that similar sediments form a common basal facies to the sedimentary sequences drilled far from ridges (Von der Borch and Rex, 1970: Von der Borch *et al.*, 1971).

Boström *et al.* (1972) have proposed that such sediments are the products of inorganic precipitation from heavy metal enriched submarine ,,hot springs" on interaction with sea water. According to the model presented here, such ,,hot springs" may be identified as the discharge solutions of sub-sea-floor geothermal systems.

Ocean Floor Metamorphic Rocks. Metabasic rocks, with which the metamorphosed Ligurian ophiolitic rocks show distinct similarities, and which range from ,,clay mineral" to ,,granulite" facies have been described from the mid-oceanic ridges (Melson *et al.*, 1968: Cann, 1969: Cann, 1971: Miyashiro *et al.*, 1971). The intensity of alteration is variable, and such rocks can show a well developed foliation. Since all are hydrated, an extraneous fluid was required for metamorphism. Muehlenbachs and Clayton (1972) have shown from oxygen isotope analysis that this fluid was sea water. Corliss (1971) has presented evidence for the depletion in heavy metals of ocean floor pillow lavas, and has proposed removal by heated sea water.

Oceanic Heat Flow. Large scale convective circulation of sea water has been proposed to account for the local scatter and low mean values of the heat flow at oceanic ridges (Talwani *et al.*, 1971; Anderson, 1972; Lister, 1972). Deffeyes (1970) has concluded that it would be difficult to prevent hot sea water circulating in the crack systems in the oceanic crust. Palmason (1967) has shown, for Icelandic geothermal terrain, that the Rayleigh number exceeds the threshold value of $4\pi^2$ above which free convection is possible.

Iceland. The occurrence of 17 active geothermal systems in Iceland which are restricted to the neovolcanic zone (Bödvasson, 1961) demonstrates that such systems can exist in active ridge segments. The Reykjanes system (Tómasson and Kristmannsdóttir, 1972) involves recirculated sea water, whereas the remaining 16 involve meteoric water (Craig, 1963).

The Red Sea. Ross (1972) has shown that the Atlantis II deep hot brine is supplied by discharge of a heavy metal enriched solution at a temperature of 104°C and at a rate of 3,150 litres/sec. Precipitation from this brine is responsible for the formation of the heavy metal enriched sediments (Degens and Ross, 1969). Backer and Schoell (1972) describe the presence of a further 13 brine pools distributed along the active axis of the Red Sea.

Since discharged hot brines are related directly to sub-sea-floor geothermal systems the findings from the Red Sea and Iceland combine to support the hypothesis that convective circulation of sea water constitutes an important geochemical process operative at oceanic ridges. Discharge of such brines may be help responsible for the formation of heavy metal enriched sediments.

A Unified Model for Sub-Sea-Floor Metamorphism

Linking together the modern evidence and the evidence derived from the E. Ligurian ophiolitic rocks allows the formulation of a general model for sub-sea-floor metamorphism and mass transfer. This is divisible into three types whose feature are summarised below.

Background Metamorphism. The permeability of oceanic crust allows the penetration of sea water to depth (Deffeyes, 1970), but geothermal systems are sporadic in occurrence because they require a distinct thermal anomaly for activation. Hence, it is likely that the bulk of oceanic crust, which is not involved with geothermal activity, is subjected to low intensity hydration producing prograde assemblage sequences controlled by temperature rising at the geothermal gradients which occur at ocean ridge axial zones. This model is closely similar to the occurrence of the prograde zeolitic sequence, developed in response to a background geothermal gradient of 65°C/km (Palmason, 1967), which occurs in the Icelandic basalts (Walker, 1960).

Although the average global axial conductive heat flow is 2.5 H.F.U. (Langseth and Von Herzen, 1970) which gives an average thermal gradient of 50°C/km taking a thermal conductivity of 0.005 cals/cm. sec (Clarke, 1966), the actual heat flow at ridges is highly variable: values as high as 8.3 H.F.U., corresponding to a thermal gradient of 166°C/km, have been recorded (Langseth and Von Herzen, 1970). It is likely, therefore, that the intensity and rate of increase in grade with depth of the background metamorphism will vary in a manner similar to the observed heat flow variation. The prograde metamorphic sequence described from the Lower Pillow lavas and Sheeted intrusive complex of the Troodos Massif of Cyprus by Gass and Smewing (1973), which is closely similar to the model predicted by Cann (1970), may be characteristic of this type of sub-sea-floor metamorphism.

Geothermal System Metamorphism. In Iceland 17 geothermal systems occur over 250 km and in the Red Sea at least 14 systems occur over 900 km: the latter is a minimum value since discharge points can give brine pools only under specific topographic conditions. Geothermal systems, therefore, may occur at sea-floor spreading axes as discrete entities causing intense metamorphism under the influence of geothermal gradients higher than the background value. Such systems may grow and decay as thermal perturbations change during sea-floor spreading, and may be sporadically affected by faulting (Tómasson and Kristmannsdóttir, 1972). Relicts of geothermal activity may occur in oceanic crust displaced from ridges since all oceanic crust is generated at spreading axes.

Since geothermal systems probably occur along all the ridge systems it is clear that there may be massive interaction between sea water and oceanic crust. The scale of this process may be estimated by considering the heat transfer involved. Deffeyes (1970) has estimated that about 12.5 km^3/year of basaltic material appears at the ridge systems at the present time. The bulk of this, about 7–10 km^3 is considered to be intrusive under a shallow volcanic cover. The heat produced can drive sub-sea-floor convection. When 1 km^3 of basic magma ($\sim 3 \times 10^{15}$ g) cools from 1200°C down to 300°C, heat is liberated corresponding to the latent heat of crystallisation ($\sim 3 \times 10^{17}$ cals.), and the heat capacity ($\sim 7 \times 10^{17}$ cals.); in all about 10^{18} cals. To heat 1 km^3 of sea water to 300°C requires about 3×10^{17} cals. Hence, 1 km^3 of cooled basic magma can heat 3 km^3 of sea water to 300°C[1]. This phenomenon is massive, and must constitute the major process involved in the cooling of the intrusive material of the oceanic crust. The entire present ocean volume could be heated during circulation through oceanic crust in 100 my. at the present rate of volcanism. The possible influences on the geochemical balance of sea water are impressive, since the discharged fluid is chemically quite unlike the initial input sea water.

Although the exact nature of the chemical modification of the brine is unknown, because the chemistry of basalt-sea water interaction is unknown, certain features of the solution chemistry are apparent. As indicated below, we consider that the sub-sea-floor convective process may approximate to a simple single pass system, in which water infiltrates the upper volcanic layers, descends and heats up without significant recirculation, and finally returns to the water/rock interface.

The initial sea water is an oxygenated aqueous electrolyte of ionic strength 0.7, whose anion content is dominated by Cl^-, Br^-, HCO_3^-, CO_3^{--} and SO_4^{--}. During descent it must become progressively reduced as it reacts with a basaltic mineralogy which contains the olivine-magnetite buffer system. In this process, reactions such as

$$\underset{\text{Fayalite}}{Fe_2SiO_4} + 1/2\, O_2 = \underset{\text{Haematite}}{Fe_2O_3} + 1/2\, \underset{\text{Quartz}}{(SiO_2)_{solid}} + 1/2\, (SiO_2)_{solution}$$

and

$$11\,\underset{\text{Fayalite}}{Fe_2SiO_4} + 2SO_4^{--} + 4H^+ = 7\,\underset{\text{Magnetite}}{Fe_3O_4} + \underset{\text{Pyrite}}{FeS_2} + 11\,SiO_2 + 2H_2O$$

can cause massive oxidation. The relative quantities involved indicate that sulphate is the more important oxidising agent. If 3 km^3 of sea water interacts with 1 km^3 of basalt, as the previous mass calculation suggests, then about 30% of the iron in 1 km^3 of basaltic material may be oxidised by SO_4^{--} and O_2.

Dissolved HCO_3^-, CO_3^{--} and SO_4^{--} are removed from solution by fixation in solid phases, because the solubility of calcite and anhydrite decreases rapidly with increasing temperature (Holland, 1967). The abundance of calcite and particularly anhydrite in the Reykjanes geothermal system demonstrates this process (Tómasson and Kristmannsdóttir, 1972).

1 This is a consevativer estimate, since heat is released by exothermic hydration of basic material.

With progressive removal of the oxidising species, the brine becomes reduced: a trend which increases with increasing temperature. In this state large quantities of Mn^{2+} and Fe^{2+} may pass into solution: as is seen, for example, in the Salton Sea brines. The fluid may become enriched in other transition metals (e.g. Cu, Ag, Au, Ni) by a high temperature leaching process in which they are removed into solution as chloride complexes (Helgeson, 1964: Helgeson, 1969). The stability of such complexes increases with temperature. The solution of nickel and similar elements is unaffected oxygen fugacity.

Na: K: Ca ratios may change due to temperature dependent exchange reactions, and the brine may be expected to increase (slightly) in salinity due to fixation of hydroxyl in metamorphic hydrates.

When the enriched brines return towards the surface partial precipitation and exchange reactions may occur in response to decreasing temperature. This stage in the process is ideal for the formation of hydrothermal ore deposits.

Finally, on discharge into sea water, the dissolved heavy metals may be absorbed onto precipitated colloidal „hydrous ferric oxide" and manganese dioxide. Flocculation in the electrolyte and sedimentation may then form a metal enriched sedimentary component. That such a geochemical mechanism of mass transfer is possible is shown by the composition of the brine discharged from the Reykjanes geothermal system in which hot sea water/basalt interaction occurs. This brine contains 374 ppm. SiO_2, 485 ppb Fe, 10.9 ppb Mo, 0.7 ppb V and 21610 ppm Cl^-, (Björnsson *et al.*, 1970). These figures represent enrichment factors of $\times$ 125 for SiO_2, $\times$ 243 for Fe, $\times$ 36 for Mo, $\times$ 2 for V and $\times$ 1.6 for Cl^-, relative to ambient sea water. It is clear that the chloride content of the brine is only slightly increased during the process, whereas the other elements are significantly enriched. Circulation of hot brine in geothermal systems, therefore, not only causes intense metamorphism but is also an important means of transporting a variety of elements from the oceanic crust into sea water, and, since the fluxed basalts are initially derived by mantle anatexis, a significant stage in the transport from the Upper Mantle to the hydrosphere. The process is geologically continuous since sea-floor spreading is geologically continuous.

Dynamothermal Metamorphism. The occurrence of foliated metamorphic rocks in the oceanic crust and mantle (Miyashiro *et al.*, 1971) shows that, at depth, ductile deformation by a creep process may occur in the presence of minor amounts of extraneous fluid. Sheppard and Epstein (1970) have shown that the amphibolites of St. Pauls rocks have interacted with sea water. Such a process may account for the formation of foliated amphibolites which occur associated with ophiolitic rocks in parts of Jugoslavia (Pamić *et al.*, 1973) and also the amphibolites associated with the Lizard ultramafic complex (Green, 1964).

Characteristics of Sub-Sea-Floor Geothermal Systems

Various aspects of terrestrial geothermal systems have been described. In particular, Ellis (1967) has summarised the chemistry of explored systems, Elder (1965) has described the mass flux and heat transfer characteristics of the Wairakei system and Craig (1963) has presented combined D/H and $^{18}O:^{16}O$ data which shows that the waters involved in most systems are of meteoric origin. Recently, models based on geothermal systems have been proposed to account for hydrothermal ore deposition (Henley, 1973), oxygen isotopically light hydrothermally altered high level intrusive bodies (Taylor and Forester, 1971) and combined D/H and ^{18}O:^{16}O results from hydrothermally altered material associated with porphyry copper deposits (Sheppard *et al.*, 1969; Sheppard *et al.*, 1971).

Consideration of the different conditions which apply in the sub-sea-floor environment suggest that certain aspects of sub-sea-floor convective systems may be different from their terrestrial analogues. If sea water is approximated to by a pure NaCl brine, then the critical point is displaced from 374°C/217.7 bars to 408°C/304 bars (data of Sourirajan and Kennedy, 1962). Since the elevation of sea-floor spreading ridges is in the range − 2 to − 4 km (ave. of 3.6 km for Mid-Atlantic Ridge and Mid-Indian Ocean Ridge, Heezen and Ewing, 1963; Heezen and Menard, 1963) and since pressure increases at the rate of 0.1 bar/metre, the hydrostatic pressure of overlying sea water at ridges is 200–400 bars. Therefore, water in sub-sea-floor geothermal systems is either supercritical or just sub-critical. In the latter case, it will be rare for the water involved to boil. This deduction has three significant implications:

1. The geothermal gradient is *not* constrained by the boiling point curve for water: which permits a maximum temperature of 310°C to occur at a depth of 1 km in a hydrostatic pressure gradient.

2. The severe chemical effects on the brine which occur during boiling (Browne and Ellis, 1970) cannot occur.

3. The significant mechanism of heat transfer by boiling and vapourisation (Elder, 1965) likewise cannot occur.

A second major difference, is that the air/ground boundary condition operative in terrestrial systems is replaced by a water/water-rock interface in sub-sea-floor systems. Hence, the gravitational constraint on the mass flux is removed. In terrestrial systems this boundary condition acts to retain some water in the circulatory system for lengths of time of the order of 10^4 years (Elder, 1965). The removal of this boundary condition allows the free motion of discharged brine out of the system. Hence the mass flux, isotherm distribution and residence time of water in the system may be changed:

a) Since the water is free to discharge, the amount of recirculation is reduced. Therefore sub-sea-floor systems may approximate to single-, as opposed to multi-, pass systems.

b) The freedom of motion of water out of the system will reduce the size of the mushroom shaped high temperature region caused by recirculation (Elder, 1965). In terrestrial systems this property dominates the distribution of isotherms at high levels (0–2.5 km at Wairakei). In sub-sea-floor systems it may not. This factor is especially encouraged if discharged water is focussed along fractures (Elder, 1965): which has been shown to be the case in the ophiolitic rocks of E. Liguria, Italy.

It is therefore considered that the isotherm distribution in sub-sea-floor systems is dominated not by a recirculatory zone, but by a delocalised recharge zone in which temperature rises rapidly downwards. The discharge zone is localised, and may be confined to fractures. The postulated difference may be illustrated diagrammatically:

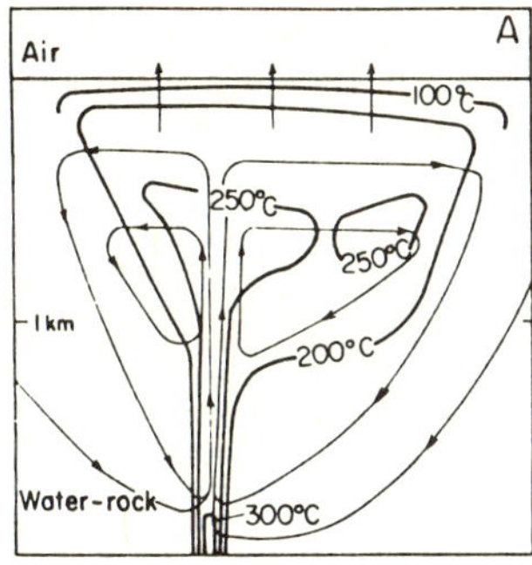

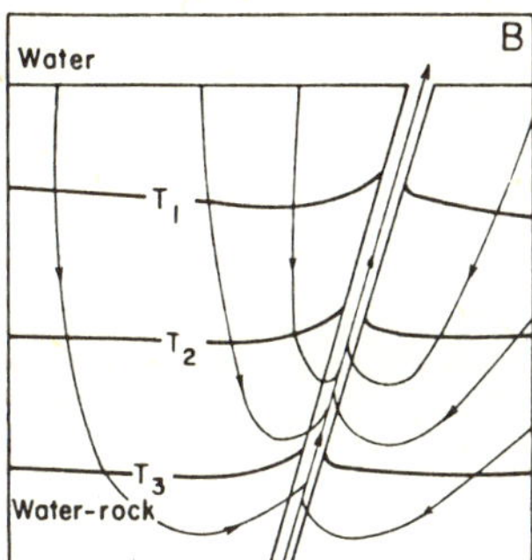

Fig. 1 A and B. Isotherm and flow line distribution in: A. The upper 2 km of the terrestrial Wairakei geothermal system (modified after Elder, 1965) B. A possible sub-sea-floor geothermal system where discharge is fracture focussed

Implications of the Model

The proposed model can be extended to provide interpretations of certain other features of ophiolitic and associated rocks. In particular, the origins of certain aspects of bedded cherts and massive sulphide deposits may be considered, together with the behaviour of certain transition elements.

Bedded Cherts Associated with Ophiolitic Rocks. Although not universal (Abbate *et al.*, 1972), the basal sedimentary facies overlying ophiolitic rocks is commonly a succession of thin bedded multi-coloured radiolarian cherts with pelitic interlayers (Thurston, 1972; Thurston, in press). This lithology constitutes one member of the „Steinmann Trinity" (Bailey and McCallien, 1960). Such sediments consist of four major components (Thurston, in press): radiolarian tests, diagenetically redeposited silica, detrital material and minute particles of haematite. The latter component is enriched in heavy metals (Thurston, personal communication). The heavy metal rich nature and morphological similarity to the „hydrous ferric oxide" particles of the heavy metal enriched sediments of oceanic ridges, together with the direct association of the bedded cherts with tectonically emplaced oceanic crust, indicate that this component is an inorganic precipitate (Thurston, 1970) which formed by interaction of hot brine, discharged from a sub-sea-floor geothermal system, with overlying deep ocean water.

The precipitate formed (Zelenov, 1964) on cooling, oxidation and increasing the pH of the iron rich discharged brine (cf. Reykjanes with 485 ppb Fe) during mixing with sea water. The resultant negatively charged sol, (Glasstone and Lewis, 1965) consisting of minute platelets (50 Å) of „hydrous ferric oxide", has a hexagonally close packed oxygen structure similar to haematite, but with interlayer hydrogen bonded molecular H_2O (Towe and Bradley, 1967). The chemistry is variable, but Towe and Bradley (1967) deduce a formula of $Fe_5O_{7.5} \cdot 4.5H_2O$ for colloidal material precipitated by hydrolysis of ferric nitrate solution. Because this sol and hydrated MnO_2 sol are strong absorbents of heavy metal cations (Krauskopf, 1956), precipitation, flocculation and sedimentation can remove other heavy metals discharged from hydrothermal solution to form a heavy metal rich sedimentary component. Such particles have no thermodynamic stability relative to haematite (Berner, 1969; Langmuir, 1971) and transform to this phase on ageing: the process being relatively fast as the slightly elevated temperatures which occur during burial. Krauskopf (1956) has shown that this removal process of heavy metals is highly efficient and that ocean water is undersaturated with respect to such elements. It is probable, therefore, that the bulk of the heavy metals removed from underlying oceanic crust and mantle by high temperature leaching and mass transfer are fixed in the overlying sediments; only a minor component being released into solution in deep ocean water. Since Robertson and Hudson (1973) have shown that the Upper Cretaceous Umber deposits of Cyprus are genetically equivalent to the heavy metal enriched sediments of the oceanic ridges, it is considered that discharged geothermal brine was also responsible for their formation.

Discharged geothermal brines are saturated with respect to quartz at the temperature of the reservoir from which they derive (Mahon, 1966). The solubility of quartz is relatively unaffected by dissolved salts (Holland, 1967), and is ten orders of magnitude greater at 200°C than at 50°C (Holland, 1967). Brines discharged on the ocean floor may therefore be 10 to 20 orders of magnitude supersaturated with respect to quartz when cooled. However, Thurston (in press) has shown that the silica of the E. Ligurian cherts is almost certainly of biogenic origin, and not an inorganic precipitate. Estimates of the depth of the ocean in which the cherts formed, which are based on pillow lava vesicularity and fluid inclusion properties, both give values of 2–3 km. Radiolaria, which grew in the top 100 m. of the oceans, were therefore separated from the sea-floor by 2 kms of sea water dominated by complex horizontal currents. Therefore, usage by the Radiolaria of silica discharged in hot brines was probably impossible. The silica rich nature of cherts associated with ophiolitic rocks, therefore, is considered to be a biogenic phenomenon and not related directly to discharge of silica rich solutions.

The absence of a detectable inorganic silica contribution may be explained by consideration of two points. Firstly, if solutions are derived from temperatures less than 150°C then on cooling they will be supersaturated with respect to quartz but not with respect to amorphous silica (Holland, 1967). And, secondly, if derived from temperatures greater than 150°C solutions supersaturated with respect to amorphous silica can exist metastably for long periods of time without precipitation (White *et al.*, 1956). Since solutions mix with sea water on discharge, dilution may reduce the supersaturation values and further inhibit precipitation.

It is possible, however, that a minor inorganic component, which was incorporated by colloidal absorption, may be present whose identity was lost during the diagenesis to which cherts are particularly prone (Thurston, in press). Bischoff (1969) has suggested that silica is incorporated in the sediments of the Atlantis II deep of the Red Sea by absorption and polymerisation on precipitated „hydrous ferric oxide" sol.

Nisbit and Price (in press) have shown that the bedded nature of the cherts, associated with the ophiolitic rocks of Othrys, E. Greece, is due to a turbiditic mode of resedimentation. The cherts of E. Liguria also contain tubiditic sedimentary structures.

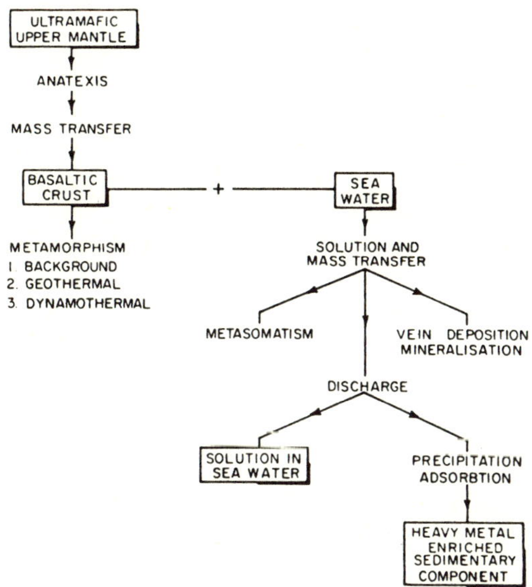

Fig. 2. Flow diagram of geochemical processes involved in sub-sea-floor metamorphism and mass transfer

Ophiolitic Massive Sulphide Deposits. The upper levels of ophiolitic complexes can contain massive sulphide deposits (Sillitoe, 1972). Descriptions of the deposits in Cyprus (Constantinou and Govett, 1972; Johnson, 1972) show them to consist of a stratiform body, whose upper part is leached, overlying a mineralized stockwork enclosed in a sheath of intensely hydrothermally altered pillow lavas containing disseminated sulphides. In the light of the model described here, the stockwork may be interpreted as the product of prolonged precipitation from hydrothermal solutions which moved upwards in the discharge zone of a sub-sea-floor geothermal system, after intense high temperature leaching at depth. The stratiform body may be identified as a heavy metal enriched sediment produced by inorganic precipitation under low oxygen fugacity conditions. Since each deposit may be identified as the discharge zone of a specific system, the model explains their sporadic occurrence; the pattern of occurrence being directly comparable to the distribution pattern of geothermal systems in the neovolcanic zone of Iceland (Pálmason, 1967).

Trace Element Geochemistry. A consequence of the model is that intensely altered metabasic material, which was fluxed by hydrothermal brine, should be stripped of the elements in which overlying sediments and associated mineralized zones are enriched.

The data presented by Govett and Pantazis (1971) for the distribution of nickel, in the Cyprus pillow lavas confirms this prediction. The bulk of the Cypriot pillow lavas contain 80–200 ppm Ni, which corresponds to the nickel content of 70–200 ppm in deep sea basalts (Nicholls and Islam, 1971), whereas intensely altered pillow lavas in class I and a mines class are depleted, and contain 20–30 ppm Ni and 50–100 ppm Ni respectively. Preliminary data on the E. Ligurian metabasic rocks described here show that they are variably depleted in Au, Ag, (Spooner and Fyfe, in press) W, Ni and Co relative to the contents of unaltered deep sea basalts.

Conclusion

On the basis of summarised modern evidence and the interpretation of the metamorphic assemblage sequence in the ophiolitic rocks of E. Liguria, Italy, it is concluded that vigorous convective circulation of brine within the oceanic crust is a significant process at axes of sea-floor spreading. The components of the overall process, which constitutes one stage in the transport of material from the oceanic mantle to the oceanic crust and the hydrosphere, are summarised in Fig. 2.

The fixation of heavy metals, which are discharged into sea water in the solutions derived from geothermal systems, by colloidal absorption is considered to be particularly important, since, if this process did not operate, then the chemistry of sea water and the chemistry of deep ocean water around spreading ridges would be significantly different from that observed. This part of the process prevents the trace element concentration of deep ocean water increasing to toxic levels.

Acknowledgements. We are most grateful to Dr. A. Bezzi, Dr. G. B. Piccardo, Dr. L. Cortesogno, Dr. E. Abbate, and Dr. V. Bortolotti for assistance during fieldwork, to Dr. D. R. Thurston, Mr. R. B. Foster and Mrs. J. Spooner for helpful discussion, and to Professor I. G. Gass, Professor J. W. Elder and Dr. A. B. Thompson for critically reading a preliminary copy of the manuscript. The first author gratefully acknowledges the receipt of an N.E.R.C. grant.

References

Abbate, E., Bortolotti, V., Passerini, P.: Olistrosomes and olistoliths. In: Development of the Northern Apennines Geosyncline, G. Sestini, ed. Sediment. Geol. **4**, 521–557 (1970)

Abbate, E., Bortolotti, V., Passerini, P.: Studies on mafic and ultramafic rocks. 2-Palaeogeographic and tectonic considerations on the ultramafic belts in the Mediterranean area. Boll. Soc. Geol. Ital. **91**, 239–282 (1972)

Abbate, E., Sagri, M.: The eugeosynclinal sequences. In: Development of the Northern Apennines Geosyncline, G. Sestini, ed. Sediment. Geol. **4**, 251–340 (1970)

Anderson, R. N.: Petrologic significance of low heat flow on the flanks of slow-spreading mid-ocean ridges. Geol. Soc. Am. Bull. **83**, 2947–2956 (1972)

Backer, H., Schoell, M.: New deeps with brines and metalliferous sediments in the Red Sea. Nature Physical Science. **240**, 153–158 (1972)

Bailey, E. B., McCallien, W. J.: Some aspects of the Steinmann trinity, mainly chemical. Quart. Jl. Geol. Soc. Lond. **116**, 365–395 (1960)

Bender, M., Broecker, W., Gornitz, V., Middel, U., Kay, R., Sun, S.: Geochemistry of three cores from the East Pacific rise. Earth Planet. Sci. Lett. **12**, 425–433 (1971)

Berner, R. A.: Goethite stability and the origin of red beds. Geochim. Cosmochim. Acta. **33**, 267–273 (1969)

Bezzi, A., Piccardo, G. B.: Structural features of the Ligurian ophiolites: petrologic evidence for the „oceanic" floor of the Northern Apennines geosyncline; a contribution to the problem of the Alpine type Gabbro-peridotite associations. Mem. Soc. Geol. Ital. **10**, 53–63 (1971)

Bigazzi, G., Bonadona, F.P., Ferrara, G., Innocenti, F.: Fission track ages of zircons and apatites from North Apennine ophiolites. Fortschr. Mineral. **50**, 51–53 (1972)

Bischoff, J.L.: Red Sea geothermal brine deposits; their mineralogy, chemistry and genesis. In: Hot brines and recent heavy metal deposits in the Red Sea, E.T. Degens, D.A. Ross, eds. p. 368–401. Berlin-Heidelberg-New York: Springer 1969

Björnsson, S., Arnórsson, S., Tómasson, J.: Exploration of the Reykjanes Brine Area. Report to the U. N. Symposium on Geothermal Energy. Pisa. p. 1–25 1970

Bödvarsson, G.: Physical characteristics of natural heat resources in Iceland. Jökull. **11**, 29–38 (1961)

Bortolotti, V., Passerini, P.: Magmatic activity. In: Development of the Northern Apennines geosyncline, G. Sestini, ed. Sediment. Geol. 4, 599–624 (1970)

Böstrom, K.: Submarine volcanism as a source of iron. Earth Planet. Sci. Lett. **9**, 348–354 (1970)

Böstrom, K., Farquarson, B., Eyl, W.: Submarine hot springs as a source of active ridge sediments. Chem. Geol. **10**, 189–203 (1972)

Böstrom, K., Peterson, M.N.A.: Precipitates from hydrothermal exhalations on the East Pacific Rise. Econ. Geol. **61**, 1258–1265 (1966)

Böstrom, K., Peterson, M.N.A., Joensuu, O., Fisher, P.E.: Aluminium-poor ferromanganoan sediments on active oceanic ridges. J. Geophys. Res. **74**, 3261–3270 (1969)

Browne, P.R.L., Ellis, A.J.: The Ohaki-Broadlands hydrothermal area, New Zealand: Mineralogy and related geochemistry. Am. J. Sci. **269**, 97–130 (1970)

Cann, J.R.: Spilites from the Carlsberg Ridge, Indian Ocean. J. Petrol. **10**, 1–19 (1969)

Cann, J.R.: A new model for the structure of the ocean crust. Nature. **226**, 928–930 (1970)

Cann, J.R.: Petrology of basement rocks from Palmer Ridge, N. E. Atlantic. Phil. Trans. Roy. Soc. Lond. Ser A. **268**, 605–617 (1971)

Clark, S.P., Jr.: Thermal conductivity. In: Handbook of physical constants. S. P. Clark, Jr. ed. Geol. Soc. Mem. **97**, 459–482 (1966)

Coleman, R.G.: Plate tectonic emplacement of Upper Mantle Peridotites along Continental edges. J. Geophys. Res. **76**, 1212–1222 (1971)

Constantinou, G., Govett, G.J.S.: Genesis of sulphide deposits, ochre and umber of Cyprus. Trans. Inst. Mining Met. (Sect. B: Appl. earth sci.). **81**, B32–46 (1972)

Corliss, J.B.: The origin of metal-bearing submarine hydrothermal solutions. J. Geophys. Res. **76**, 8128–8138 (1971)

Craig, H.: The isotopic geochemistry of water and carbon in geothermal areas, conference on nuclear geology in geothermal areas, Spoleto, Italy. p. 17–53 (1963)

Dasch, E., Ross Heath, G., Dymond, J.: Isotopic analysis of metalliferous sediments from the East Pacific Rise. Earth Planet. Sci. Lett. **13**, 175–180 (1971)

Degens, E.T., Ross, O.A., eds.: Hot brines and recent heavy metal deposits in the Red Sea. Berlin-Heidelberg-New York: Springer 1969

Deffeyes, K.S.: The axial valley: a steady state feature of the terrain. In: Megatectonics of continents and oceans, p. 194–222. New Brunswick: Rutgers Univ. Press 1970

De Roever, W.P.: Sind die Alpinotypen Peridotit-massen vielleicht tektonisch verfrachtete Bruchstücke der Peridotitschale? Geol. Rundschau. **46**, 137–146 (1956)

Dewey, J.F., Bird, J.M.: Origin and emplacement of the ophiolite suite: Appalachian ophiolites in Newfoundland. J. Geophys. Res. **26**, 3179–3206 (1971)

Dietz, R.S.: Alpine serpentinites as oceanic rind fragments. Geol. Soc. Am. Bull. **74**, 947–952 (1963)

Elder, J.W.: Physical processes in geothermal areas. In: Terrestrial heat flow. Am. Geophys. Union Geophys. Mon. **8**, 211–239 (1965)

Ellis, A.J.: The chemistry of some explored hydrothermal systems. In: Geochemistry of hydrothermal ore deposits, H. L. Barnes, ed. p. 465–514. New York: Holt, Rinehart and Winston 1967

Galli, M., Costesogno, L.: Studi petrografici sulle formazioni ofiolitiche dell'Appennino Ligure. Nota XIII- Fenomeni de Metamorfismo de baso grado in alcune rocce della formazione ofiolitica dell'Appennino Ligure. Rend. Soc. Mineral. Ital. **26**, 599–647 (1970)

Gass, I. G.: Is the Troodos Massif of Cyprus a fragment of Mesozoic ocean floor? Nature. **220**, 39–42 (1968)
Gass, I. G., Masson-Smith, D.: The geology and gravity anomalies of the Troodos Massif, Cyprus. Phil. Trans. Roy. Soc. London, Ser. A **255**, 417–467 (1963)
Gass, I. G., Smewing, J. D.: Intrusion, extrusion and metamorphism at constructive margins: evidence from the Troodos Massif, Cyprus. Nature **242**, 26–29 (1973)
Gelati, R., Pasquarè, G.: Interpretazione Geologica del limite Alpi-Appennini in Liguria. Riv. Ital. Palaeont. **76**, 513–578 (1970)
Glasstone, S., Lewis, D.: Elements of physical chemistry, p. 578. London: Macmillan and Co. Ltd. 1965
Govett, G. J. S., Pantazis, Th. M.: Distribution of Cu, Zn, Ni and Co in the Troodos Pillow lavas series, Cyprus. Trans. Inst. Mining Met. (Sect. B: Appl. earth sci.) **80**, B27–46 (1971)
Green, D. H.: The petrogenesis of the high-temperature peridotite intrusion in the Lizard Area, Cornwall. J. Petrol. **5**, 134–188 (1964)
Heezen, B. C., Ewing, M.: The mid-oceanic ridge. In: The sea, M. N. Hill, ed. vol. 3, p. 388–410. New York: Interscience 1963
Heezen, B. C., Menard, H. W. J.: Topography of the deep-sea floor. In: The sea, M. N. Hill ed. vol. 3, p. 233–280. New York: Interscience 1963
Helgeson, H. C.: Complexing and hydrothermal ore deposition. New York: Pergamon Press 1964
Helgeson, H. C.: Solution chemistry and metamorphism. In: Researches in geochemistry, P. H. Abelson, ed. vol. II, p. 362–404. New York: John Wiley 1967
Helgeson, H. C.: Evaluation of irreversible reactions in geochemical processes involving minerals and aqueous solutions-I. Thermodynamic relations. Geochim. Cosmochim. Acta **32**, 853–877 (1968)
Helgeson, H. C.: Thermodynamics of hydrothermal systems at elevated temperatures and pressures. Am. J. Sci. **267**, 729–804 (1969)
Helgeson, H. C., Brown, T. H., Nigrini, A., Jones, T. A.: Calculations of mass transfer in geochemical processes involving aqueous solutions. Geochim. Cosmochim. Acta **34**, 569–592 (1970)
Helgeson, H. C., Garrels, R. M., Mackenzie, R. T.: Evaluation of irreversible reactions in geochemical processes involving minerals and aqueous solutions-II. Applications. Geochim. Cosmochim. Acta **32**, 455–482 (1969)
Henley, R. W.: Some fluid dynamics and ore genesis. Trans. Inst. Mining Met. (Sect. B: Appl. earth sci.) **82**, B1–B8 (1973)
Hess, H. H.: Mid-ocean ridges and tectonics of the sea-floor. In: Submarine geology and geophysics, 17th Colston Research Symposium, Bristol, England, W. F. Whittard, R. Bradshaw, eds., p. 317–333. London: Butterworths 1965
Holland, H. D.: Gangue minerals in hydrothermal systems. In: Geochemistry of hydrothermal ore deposits, H. L. Barnes, ed. p. 382–436. New York: Holt, Rinehart and Winston 1967
Johnson, A. E.: Origin of Cyprus pyrite deposits. 24th. I.G.C., Sect. 4, 291–298 (1972)
Keith, T. E. C., Muffler, L. J. P., Cremer, M.: Hydrothermal epidote formed in the Salton Sea geothermal system, California. Am. Mineralogist **53**, 1635–1644 (1968)
Khan, M. A., Summers, C., Bamford, S. A. D., Chroston, P. N., Poster, C. K., Vine, F. J.: Reversed seismic refraction line on the Troodos Massif, Cyprus. Nature Physical Sciences **238**, 134–136 (1972)
Krauskopf, K. B.: Factors controlling the concentrations of thirteen rare metals in sea-water. Geochim. Cosmochim. Acta **9**, 1–32 B (1956)
Langmuir, D.: Particle size effect on the reaction goethite → haematite and water. Am. J. Sci. **271**, 147–156 (1971)
Langseth, M. G., Herzen, R. P. von: Heat flow through the floor of the world oceans. In: The sea, A. E. Maxwell, ed., vol. 4, pt. 1 p. 299–352. New York: Intersience 1970
Liou, J. G.: Synthesis and stability relations of Prehnite, $Ca_2Al_2Si_3O_{10}(OH)_2$. Am. Mineralogist **56**, 507–531 (1971)
Lister, C. R. B.: On the thermal balance of a Mid-ocean ridge. Geophys. J. R. astr. Soc. **26** 515–535 (1972)

Lorimer, G.W., Champness, P.E., Spooner, E.T.C.: Dislocation distributions in naturally deformed Omphacite and Albite. Nature Physical Science **239**, No. 94, 108–109 (1972)

Mahon, W.A.F.: Silica in hot water discharged from drillholes at Wairakei, New Zealand. New Zealand J. Sci. **9**, 135–144 (1966)

Melson, W.G., Thompson, G., Andel, Tj.H. van: Volcanism and metamorphism in the mid-Atlantic Ridge. 22°N latitude. J. Geophys Res. **73**, 5925–5941 (1968)

Miyashiro, A., Shido, F., Ewing, M.: Metamorphism in the mid-Atlantic ridge near 24°N and 30°N. Phil. Trans. Roy. Soc. London Ser. A. **268**, 589–603 (1971)

Moore, J.G.: Water content of basalt erupted on the ocean floor. Contr. Mineral. and Petrol. **28**, 272–279 (1970)

Moores, E.M.: Petrology and structure of the Vourinos ophiolite complex of northern Greece. Geol. Soc. Am. Spec. Pap. **118**, (1969)

Moores, E.M., Vine, F.J.: The Troodos Masif, Cyprus and other ophiolites as oceanic crust: evaluation and implications. Phil. Trans. Roy. Soc. London. Ser. A. **268**, 443–466 (1971)

Muehlenbachs, K., Clayton, R.N.: Oxygen isotope geochemistry of submarine greenstones. Can. J. Earth Sci. **9**, 471–478 (1972)

Nicholls, G.D., Islam, M.R.: Geochemical investigations of basalts and associated rocks from the ocean floor and their implications. Phil. Trans. Roy. Soc. London. Ser. A. **268**, 469–486 (1971)

Palmason, G.: On heat flow in Iceland in relation to the Mid-Atlantic ridge. In: Iceland and Mid-ocean ridges. S. Bjornsson, ed. Soc. Sci. Islandica **38**, 111–127 (1967)

Pamić, J., Šćavničar, S., Medjimorec, S.: Mineral assemblages of amphibolites associated with Alpine-type Ultramafics in the Dinaride Ophiolite Zone (Yugoslavia). J. Petrol. **14**, 133–157 (1973)

Pearce, J.A., Cann, J.R.: Ophiolite origin investigated by discriminant analysis using Ti, Zr and Y. Earth Planet. Sci. Lett. **12**, 339–349 (1971)

Robertson, A.H.F., Hudson, J.D.: Cyprus umbers; chemical precipitates on a Tethyan ocean ridge. Earth Planet. Sci. Lett. **18**, 93–101 (1973)

Ross, D.A.: Red Sea hot brine area: revisited. Science **175**, 1455–1457 (1972)

Sheppard, S.M.F., Epstein, S.: D/H and $^{18}O/^{16}O$ ratios of minerals of possible mantle or lower crustal origin. Earth Planet. Sci. Lett. **9**, 232–239 (1970)

Sheppard, S.M.F., Nielsen, R.L., Taylor, H.P., Jr.: Oxygen and hydrogen isotope ratios of clay minerals from porphyry copper deposits. Econ. Geol. **64**, 755–777 (1969)

Sheppard, S.M.F., Nielsen, R.L., Taylor, H.P., Jr.: Hydrogen and oxygen isotope ratios in minerals from porphyry copper deposits. Econ. Geol. **66**, 515–542 (1971)

Sigvaldason, G.E.: Epidote and related minerals in two deep geothermal drill holes, Reykjavik and Hveragerdi, Iceland. U.S. Geol. Surv. Profess. Papers **450-E**, 77–79 (1962)

Sillitoe, R.H.: Formation of certain massive sulphide deposits at sites of sea-floor spreading. Trans. Inst. Mining Met. (Sect. B: Appl. earth Sci.) **81**, B141–148 (1972)

Sourirajan, S., Kennedy, G.C.: The system H_2O-NaCl at elevated temperatures and pressures. Am. J. Sci. **260**, 115–141 (1962)

Talwani, M., Windish, C.C., Langseth, M.G., Jr.: Reykjanes ridge crest: a detailed geophysical study. J. Geophys. Res. **76**, 473–517 (1971)

Taylor, H.P., Jr., Forester, R.W.: Low-^{18}O igneous rocks from the intrusive complexes of Skye, Mull and Ardnamurchan, Western Scotland. J. Petrol. **12**, 465–497 (1971)

Thayer, T.P.: Peridotite-gabbro complexes as keys to the petrology of mid-ocean ridges. Geol. Soc. Am. Bull. **80**, 1515–1522 (1969)

Thurston, D.R.: Ph.D. Thesis, University of Manchester. 1970

Thurston, D.R.: Studies on bedded cherts. Contr. Mineral. and Petrol. **36**, 329–334 (1972)

Tómasson, J., Kristmannsdóttir, H.: High temperature alteration minerals and thermal brines, Reykjanes, Iceland. Contr. Mineral. and Petrol. **36**, 123–134 (1972)

Towe, K.M., Bradley, W.F.: Mineralogical constitution of colloidal „Hydrous ferric oxides". J. Colloid and Interface Sci. **24**, 384–392 (1967)

Von der Borch, C.C., Rex, R.W.: Amorphous iron oxide precipitates in sediments cored during Leg 5, Deep Sea Drilling Project. In: Initial reports of the Deep Sea Drilling Project, V, p. 541–544 Washington, D.C.: U.S. Govt. Printing Office 1971

Von der Borch, C.C., Nesteroff, W.D., Galehouse, J.S.: Iron-rich sediments cored during Leg 8 of the Deep Sea Drilling Project. In: Initial Reports of the Deep Sea Drilling Project, v VIII, p. 829–833. Washington, D.C.: U.S. Govt. Printing Office 1971

Vuagnat, M.: Remarques sur la trilogie serpentinites-gabbros-diabases dans le bassin de la Mediterranée occidentale. Geol. Rundschau **53**, 336–358 (1963)

Walker, G.P.L.: Zeolite zones and dyke distribution in relation to the structure of the basalts of eastern Iceland. J. Geol. **68**, 515–528 (1960)

White, D.E., Brannock, W.W., Murata, K.J.: Silica in hot-spring waters. Geochim. Cosmichim .Acta **10**, 27–59 (1956)

Zelenov, K.K.: Iron and Manganese in exhalations from the submarine volcano Banu Wuhu (Indonesia). Dokl. Akad. Nauk S.S.S.R. **155** (6), 1317–1320 [in Russian] (1964)

E.T.C. Spooner
Dept. of Geology and Mineralogy
Parks Road
Oxford OX1 3PR
England

30

Reprinted from *Jour. Geology* **84**:249-275 (1976)

HYDROTHERMAL CIRCULATION AND GEOCHEMICAL FLUX AT MID-OCEAN RIDGES[1]

THOMAS J. WOLERY AND NORMAN H. SLEEP

Department of Geological Sciences, Northwestern University, Evanston, Illinois 60201

ABSTRACT

The flow rate of sea water through sub-sea-floor hydrothermal systems at mid-ocean ridges has been estimated at 1.3–9 × 10^{17} g/yr by consideration of the rate at which circulating fluid must advect heat out of the spreading plates into the oceans. The rate of hydrothermal heat advection was obtained by computing the difference (40 ± 4 × 10^{18} cal/yr) between the theoretical heat production associated with sea-floor spreading and observed heat flow measurements. Effects of exothermic chemical reactions, direct heat loss from flows extruded on the ocean floor, and heat of crystallization of basalt were minor and yielded insignificant contributions to the estimate. The majority of dredged ocean-floor metamorphic rocks appear to represent the products of intensive hydrothermal reaction with hot sea water. The chemical trends (loss of Ca and K, gain of Mg, Na, and H_2O) and alteration mineral assemblages of these rocks closely resemble those observed in both laboratory alteration experiments and in subaerial geothermal systems where sea water circulates in hot basic rocks. Alteration of oceanic crust appears strongly dependent on the presence of permeable features, especially cracks, and may be restricted mainly within the vicinity of major fault zones. An examination of submarine weathering, volcanic effusion, and hydrothermal mass transfers between basaltic crust and sea water for silica, manganese, and iron demonstrated that the relative importance of the three processes may vary markedly from element to element. A similar examination of magnesium, calcium, oxidants-reductants, and water indicated that only about a kilometer or less of basalt actually reacts with sea water in hydrothermal systems.

INTRODUCTION

It has been proposed that massive hydrothermal circulation occurs beneath mid-ocean ridges (Lister 1972; Spooner and Fyfe 1973; Williams and Von Herzen 1974). Mid-ocean ridges form the accreting boundaries of the spreading oceanic lithosphere and hence are principal loci of contact between the exogenic (surface) and endogenic (deep) geochemical systems. Thus, the proposed large-scale hydrothermal circulation has interesting implications for chemical transfer between these two systems.

The exogenic system may be defined as the collection of all geochemical reservoirs which comprise the atmosphere, the hydrosphere, sediments (including pelagic sediments), and sedimentary rocks. These are distinguished from other geochemical reservoirs by their high concentration of volatiles. Sediments which have been subjected to high-grade metamorphism and have lost most of their volatile content may be regarded as having been returned to the endogenic system. It is convenient to define an extended exogenic system to include also that part of the upper crystalline continental crust which participates in the sedimentary rock cycle. Most geochemical cycling models (e.g., Garrels and Mackenzie 1972) have dealt strictly with the extended exogenic system and treated

[1] Manuscript received May 7, 1975; revised December 18, 1975.

it as a closed system. In general these models have been rather successful in describing the surface of the earth over Phanerozoic time.

The endogenic system may be defined as including all material underlying the exogenic system to the maximum depth penetrated by subducted oceanic lithosphere, which is at least the 700 km depth of deep-focus earthquakes (Isacks et al. 1968). It is composed mainly of material which is volatile-deficient and reduced compared with the exogenic system (Robertson and Wyllie 1971). A simple calculation shows that the current production rate of oceanic lithosphere, 2.94 km^2/yr (Chase 1972), is probably insufficient to produce a steady state between the endogenic and exogenic systems over geologic time. Assuming a 5 km thick oceanic crust, a 100 km thick oceanic lithosphere over an area of 300×10^6 km^2, and a 700 km depth of circulation, 1×10^9 yr is required to circulate material through the oceanic lithosphere, 4×10^9 yr to circulate it through the upper 25 km zone of magma segregation (Kay et al. 1970), and 20×10^9 yr to cycle material through the oceanic crust. Lead model ages of 1 to 2×10^9 yr on fresh basalts (Oversby and Gast 1970) can be interpreted as giving a cycle time for oceanic lithosphere compatible with the above calculation (Garfunkel 1975).

In this paper we summarize the main aspects of the hydrothermal circulation beneath the ridges and estimate the magnitude of the water flow. The implications of the flow rate estimate for alteration of the oceanic crust and the associated geochemical fluxes of calcium, magnesium, silica, manganese, and oxidants-reductants are then examined and discussed in terms of balances in the exogenic system.

HYDROTHERMAL ACTIVITY AT MID-OCEAN RIDGES

Hydrothermal circulation at mid-ocean ridges has been proposed to explain: (1) the presence and distribution of metalliferous pelagic sediment (Bostrom et al. 1971; Dymond et al. 1973; Piper 1973), (2) the scattered and anomalously low heat flow observed on the ridge crest and flanks (Lister 1972; R. Anderson 1972; Williams and Von Herzen 1974), (3) compositional differences between glassy pillow basalts and more slowly cooled holocrystalline basalts dredged from mid-ocean ridges (Corliss 1970, 1971), (4) isotope geochemistry and field observations of ophiolite complexes thought to represent old ocean floor (Spooner and Fyfe 1973; Spooner et al. 1974), (5) the decay in the intensity of magnetic remanence away from the ridge axis (Irving et al. 1970; Irving 1970), and (6) microearthquake swarms that occur at mid-ocean ridges (Sykes 1971).

Unequivocal direct evidence of presently active hydrothermal discharges near submerged ridge crests has yet to be reported, despite intensive searches in several ridge localities. Small thermal anomalies (up to 0.1°C) recorded in bottom waters may be of hydrothermal origin (Sott et al. 1974; Williams et al. 1974) or the result of normal hydrologic processes (Fehn et al. 1975). Nor have any chemical anomalies been detected in bottom waters (Garner and Ford 1969; Fanning et al. 1974). Thick manganese crusts recovered from the crest of the Mid-Atlantic Ridge (Scott et al. 1975; Thompson et al. 1975) seem to be the strongest evidence of local hydrothermal discharge (not necessarily still active). At present, actual observation of hydrothermally circulating sea water on a ridge is known only from subaerial geothermal fields on the Reykjanes Peninsula in Iceland (Bjornsson et al. 1972; Mottl et al. 1975).

The most conclusive single factor indicating that hydrothermal circulation is a massive process beneath the ridges is the difference between the expected and observed heat flow which appears within several tens of kilometers of the ridge axis (Talwani et al. 1971; Hyndman and Rankin 1972; Le Pichon and Langseth

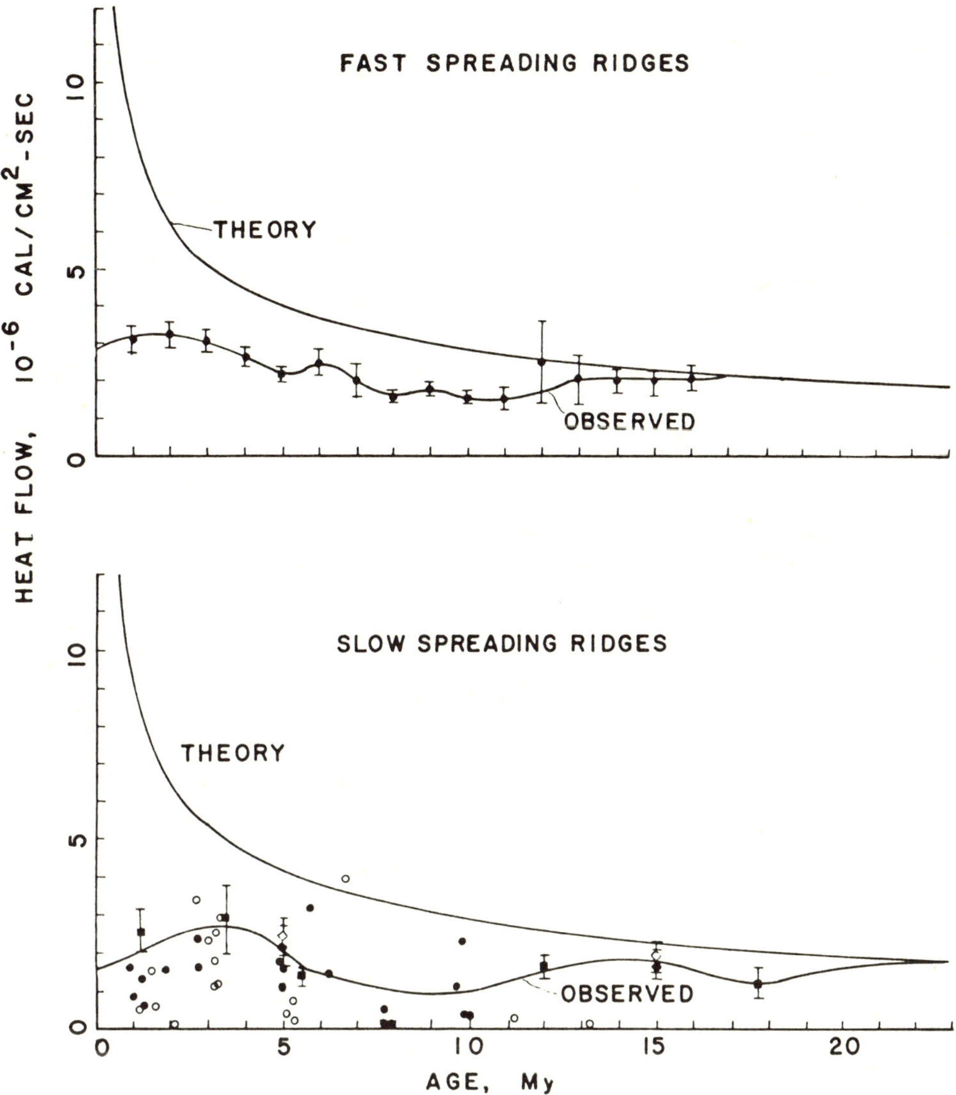

FIG. 1.—Theoretical vs. observed heat flow profiles as a function of crustal age for fast-spreading (top) and slow-spreading (bottom) mid-ocean ridges. Top: (●) East Pacific Rise (Von Herzen and Anderson 1972). Bottom: (●) Reykjanes Ridge (Talwani et al. 1971); (○) Mid-Atlantic Ridge near 45°N (Hyndman and Rankin 1972); (◆, ◇) Mid-Atlantic Ridge, South Atlantic (Sclater and Francheteau 1970); (■) Atlantic-Indian ocean mean (Le Pichon and Langseth 1969). Error bars indicate standard errors for points obtained by averaging a number of individual measurements.

1969; Williams et al. 1974). Figure 1 illustrates this anomaly for both a fast-spreading (3–6 cm/yr) ridge (the East Pacific Rise) and a composite of slow-spreading (1–2 cm/yr) ridges. (Compare fig. 7, Sclater et al. 1974, for a profile along the fast-spreading Galapagos spreading center). Observed heat flow measurements are compared with a theoretical heat flow curve derived from a thermal model for a spreading lithospheric plate (Sleep 1974; see also the Appendix). This thermal model is similar to that of McKenzie (1967), but assumes the realistic boundary condition that the plane of the ridge axis is a source of constant heat flux instead of a plane of constant temperature.

The actual heat flow measurements, taken in sediment ponds, were subject to thermal refraction effects caused by the conductivity contrast between sediment and crystalline crust. The magnitude of this effect is small, since the vertical/horizontal aspect ratios of ocean floor sediment bodies are generally very small (Lister 1972; Williams et al. 1974; Sclater et al. 1974). Thus, these measurements give a reasonably accurate measure of the heat actually conducted out of the oceanic lithosphere.

The thermal model requires the assumption of a number of parameter values. Those assumed in calculating the theoretical heat flow curves in figure 1 were: (1) thickness of lithosphere: λ = 100 km; (2) adiabatic constant: $\gamma = 1$; (3) thermal conductivity: K = 0.006 cal/cm^3-sec-°C; (4) volume specific heat: ρc = 0.91 cal/cm^3-°C; (5) temperature at the base of the lithosphere: T_λ = 1200°C. Of these, the parameter which is most poorly known is λ, the thickness of the lithosphere. It is sufficiently great that, to a fairly good approximation, the lithosphere may be taken as an infinite half-space in calculating the heat flow profile (Lister 1972). Thus, even though this parameter is not well-known, there is little corresponding error in the computed heat flow. The temperature at the base of the lithosphere, T_λ, must be sufficiently great to give rise to the basaltic melts that are extruded and intruded at the ridge axis to form the oceanic crust. The value of 1200°C assumed here may be slightly low. Sclater and Francheteau (1970) seem to prefer higher values (~1,400°–1,500°C), which result in higher computed heat flows near the ridge. An important point in the thermal spreading model is that most of the surface heat flow derives from that portion of the lithosphere which underlies the basaltic crust in which hydrothermal circulation takes place.

Estimation of heat advection due to hydrothermal circulation.—The heat production advected by hydrothermal circulation may be estimated by integrating the heat flow anomaly (fig. 1) from the time of formation of new plate material (t = 0 yr) to the time at which the theoretical and observed curves merge (t = ~17 m.y. for fast-spreading ridges, ~23 m.y. for slow-spreading ridges). The hydrothermally advected heat per unit area of new plate was found to be 11.5 × 10^8 cal/cm^2 for fast-spreading ridges, and 15.1 × 10^8 cal/cm^2 for slow-spreading ridges. The total heat predicted by the thermal model (see eq. A15 in Appendix) associated with the process of production of new plate material by sea-floor spreading is 36.4 × 10^8 cal/cm^2. Thus, approximately 32% (fast-spreading rates) to 42% (slow-spreading rates) of this heat production is hydrothermally advected. By assuming values of 2.94 km^2/yr (Chase 1972; Williams and Von Herzen 1974) for the areal rate of sea-floor spreading and a moderate mean linear spreading rate (2.74 cm/yr; Williams and Von Herzen 1974), one obtains an estimate of 40 ± 4 × 10^{18} cal/yr for present-day hydrothermal heat advection.

Upper limits on other possible causes of heat flow anomalies at ridges are negligible for purposes of this paper (table 1). However, some discussion of our method of estimating the contributions due to chemical reactions is necessary because these effects have often been presumed to be important. Also, latent heat does not increase the heat production due to sea floor spreading but rather shifts a minor part of it from the zone of melting and magma segregation of about 25 km depth directly to the crest of the growing plate (Kay et al. 1970; Sleep 1974, 1975).

Chemical reactions between the circulating hydrothermal fluid and the

TABLE 1

SUMMARY OF THERMAL EFFECTS

Effect	Heat Production (10^8 cal/cm²)	Global Effect (10^{18} cal/yr)	Remarks and References
Observed heatflow anomaly	11.5 (fast ridges)–15.1 (slow ridges)	40	This paper
Fast heat loss from extrusive flows	0.7	2	Assumes extrusion of 0.5 km of flows onto the ocean floor at 1,200°C (includes latent heat)
Latent heat	(1.5)	(5)	5 km basaltic crust; latent heat 300 cal/cm³ (Kay et al. 1970); see discussion in text
Chemical reactions:			
Hydration and Oxidation by sulfate	1	3	5 km basaltic crust; hydration reactions after Fyfe (1974)
Oxidation by dissolved O_2.....	0.01	0.02	Total reaction of O_2 in water flow of 6 × 10^{17} g/yr (see later estimates in this paper)
Oxidation by O_2 plus hydration	(6)	(18)	Reaction of 10 km basaltic crust and 5 km peridotite. Fyfe (1974)
Corrected global anomaly*		38–42	Hydrothermal heat advection

* Estimates in parentheses not included in this calculation.

basaltic crust have been proposed as a major source of heat at ridges (Fyfe 1974; see table 1). This estimate is probably excessive. First, hydrothermal circulation can be inferred not to penetrate beyond 4 to 5 km depth from petrological models of the crust (Spooner et al. 1974; Peterson et al. 1974) and from the distribution of microearthquake swarms on the exposed ridge in Iceland (Ward and Bjornsson 1971; Einarsson et al. 1973). Secondly, most of Fyfe's (1974) heat production is derived by highly exothermic oxidation of ferrous silicates by dissolved oxygen, which, however, is insufficiently present in sea water (2.8 × 10^{-4} mole/l; Garner and Ford 1969) to oxidize the bulk of the crust. Oxidation by dissolved sulfate (2.8 × 10^{-2} mole/l; Riley and Chester 1971) is much less exothermic, as the enthalpy per mole Fe_2SiO_4 destroyed equals −7.8 kcal for sulfate and −61 kcal for dissolved oxygen (at 25°C and 1 atm, data from Robie and Waldbaum 1968).

Absorption of heat by endothermic dehydration reactions due to isothermic rebound after the cessation of hydrothermal activity (R. Anderson 1972) represents only a partial reversal at the exothermic reaction discussed above, since the initial fraction of water in ocean basalt is small. Sclater et al. (1974) have developed a model which demonstrates that isothermic rebound is probably insufficient to initiate much dehydration prior to lithospheric subduction.

Calculation of the mass flow rate of circulating fluid.—The mass flow rate of the sub sea-floor geothermal systems may now be estimated from the rate at which it must advect heat out of the oceanic crust. This requires that the following parameters be known: (1) heat capacity of water: $c_w \approx 1.0$ cal/g; (2) temperature of entering fluid: $T_1 = 3°C$ (Garner and Ford 1969); (3) mean exit temperature of the hydrothermal fluid: $T_2 = 50–300°C$; (4) H = rate of hydrothermal heat advection (40×10^{18} cal/yr). The mass flow rate Q may then be estimated from the equation,

$$Q = \frac{H}{c_w(T_2 - T_1)}, \quad (1)$$

to be in the range $1.3 - 9 \times 10^{17}$ g/yr.

The mean exit temperature T_2 is the most poorly known parameter, and probably lies in the interval 150–300°C. Such temperatures are in agreement (1) subsurface (below the boiling zone) temperatures in Icelandic geothermal systems (Ward and Bjornsson 1971); (2) reaction temperatures inferred from $\delta^{18}O$ studies of dredged submarine greenstones (Muehlenbachs and Clayton 1972) and ophiolites (Spooner and Fyfe 1973; Spooner et al. 1974); and (3) magnetic blocking temperatures of fresh oceanic basalts (Irving et al. 1970). These yield a best estimate of the flow rate of $1.3–2.7 \times 10^{17}$ g/yr, which would cycle the mass of the oceans ($13{,}700 \times 10^{20}$ g; Garrels and Mackenzie 1971) in only 5–11 m.y.

The uptake of hydrothermal water by hydration of basalt is negligible compared to the flow rate, as may be shown by a simple calculation. Using previous data the volume flow rate is $1.3–9 \times 10^{17}$ cm^3/yr and the volume rate of ocean crust (5 km) production is 1.47×10^{16} cm^3/yr. These yield water-basalt ratios (v/v) of 9:1 to 58:1. The uptake of water by hydration of basalt is only 0.1 cm^3 water/cm^3 basalt (from data of R. A. Hart 1973; Miyashiro et al. 1969), or at most $\sim 1\%$ of the flow, even assuming that all 5 km of the crust are hydrated. The residence time of an individual batch of water in the hydrothermal system is about 100 to 10,000 yr for porosity of 0.1 to 1% and the above flow rate and anomaly duration.

Nature and distribution of the hydrothermal circulation.—Hydrothermal flow beneath the mid-ocean ridges requires sufficient permeability in the oceanic crust to permit circulation. Porosity and permeability are restricted to: (1) the sediment of Layer I; (2) the upper 200–300 m of Layer II which is composed of intermingled extrusive flows, pelagic sediment, basalt rubble, and intrusions (Melson et al. 1974); (3) major faults and fault zones, which are presumed to be associated with steep scarps on the bottom topography (e.g., Williams et al. 1974); and (4) occurrences of features such as cracks (Lister 1974), minor faults, amygdules, and brecciation. Studies of the oxygen isotopes of basalts drilled from Layer II during DSDP Legs 34 and 37 demonstrate that reaction with sea water has taken place extensively, but at temperatures of less than 20°C (Muehlenbachs 1975*a*, 1975*b*). It appears on the basis of this admittedly limited sampling of Layer II that (2) above is probably not a main locus of hydrothermal activity. Williams et al. (1974) have pointed out a correlation between areas of high heat flow and the presence of lineaments of sediment mounds, which they suggest are hydrothermal exit vents. The mounds in turn roughly parallel and are in proximity to fault scarps. From their figure 7, the scarps have an approximate spacing of 3 km, or about one-half of the 6 ± 1 km wavelength of the heat flow profile normal to the ridge axis. This is highly suggestive that major fault zones constitute the major part of the hydrothermal plumbing, being alternately zones of intake and discharge on a traverse normal to the ridge axis.

The theoretical (Sleep 1974; Appendix) and observed heat flow curves shown in figure 1 converge of within reasonable error at 15–20 m.y. for fast-spreading ridges and at 21–26 m.y. for slow-spreading ridges. This observation implies cessation of hydrothermal activity and resumption of purely conductive heat flow (R. Anderson 1972; R. Anderson and Hobart 1975; Sclater et al. 1974). Circulation ceases for one or any combination of the following reasons: (1) the thermal gradients which drive the convection lose their intensity (Lister 1972), (2) permeability becomes greatly reduced due to precipitation of vein minerals or increase of volume of alteration minerals over their predecessors (Fyfe 1974; Sclater et al. 1974), (3) deposition of pelagic sediment chokes off the intake and discharge zones (Lister 1972; R. Anderson 1972; Williams et al. 1974).

These ages represent upper limits to the lifetime of hydrothermal activity since a period of time is required after cessation of circulation for the rebound of lithospheric isotherms (Sclater et al. 1974). Estimates of the actual lifetimes were obtained using a simple model in which hydrothermal activity was presumed to fix the upper 5 km of oceanic crust as a cold layer for a fixed period of time, during which isotherms were depressed into the underlying lithosphere. The cold layer boundary condition was then removed (simulating cessation of circulation) and the time required for approximate restoration of normal conductive heat flow was determined. The model was evaluated using simple expressions of the form $z = C\sqrt{Kt}$, where z is the distance over which conductive heat flow is "felt" and C is an arbitrary constant whose value depends on the tolerance desired in defining z. Ages for the crust at approximate convergence of the heat flow curves in figure 1 were calculated as a function of assumed lifetime of hydrothermal activity and compared with the observed ages. In this way the lifetimes of the circulation were estimated to be 6–14 m.y. for fast-spreading ridges and 11–19 m.y. for slow-spreading ridges.

The ages calculated above correspond to the presence of hydrothermal activity at distances in excess of 200 km from the ridge axis. However, the process of hydrothermal heat advection appears to be much more intense closer to the ridge axis, and the same is most probably true for the water flow rate. Figure 2 shows the cumulative integrated heat flow anomaly for fast- and slow-spreading ridges. Half of the cumulative anomaly occurs within 2–3 m.y. However, if the effect of the latent heat shift of 5×10^{18} cal/yr is considered, it is apparent that even a larger fraction of the total heat flow anomaly occurs within lithosphere of that age and younger. The occurrence of mid-ocean ridge geothermal fields is thus by no means limited to the axial valley, but likely becomes much less frequent beyond distances on the order of 50 km or more from the ridge axis.

SEA WATER-BASALT REACTION AND ALTERED OCEAN CRUST

Massive hydrothermal circulation of sea water within the upper few kilometers of the oceanic crust appears to offer a viable explanation for the mechanism of observed metasomatic alteration of many dredged rock specimens. A large variety of metamorphic rocks, principally metabasalts and metagabbros, have been recovered from widely spaced localities along the worldwide system of mid-ocean ridges (see Melson and van Andel 1966; Aumento and Loncarevic 1969; Cann and Vine 1966; Bonatti et al. 1975). However, the rocks most commonly dredged from the sea floor are fresh or slightly altered basalts and gabbros; their metamorphic equivalents (metabasites) have been recovered mainly within the vicinity of fault scarps. Is the chemical alteration and resultant modified petrology of these rocks attributable to hydrothermal reaction with circulating sea water?

Metabasalts and metagabbros occur in the zeolite, greenschist, and amphibolite facies (Miyashiro et al. 1971), although it appears that greenschist facies metamorphism is most commonly developed in the oceanic crust (Melson et al. 1968; Cann 1969). Miyashiro et al. (1971) have classified the oceanic metabasites into two groups. Members of group I have undergone no significant compositional changes other than an increase in water content. The members of the other group (group II)

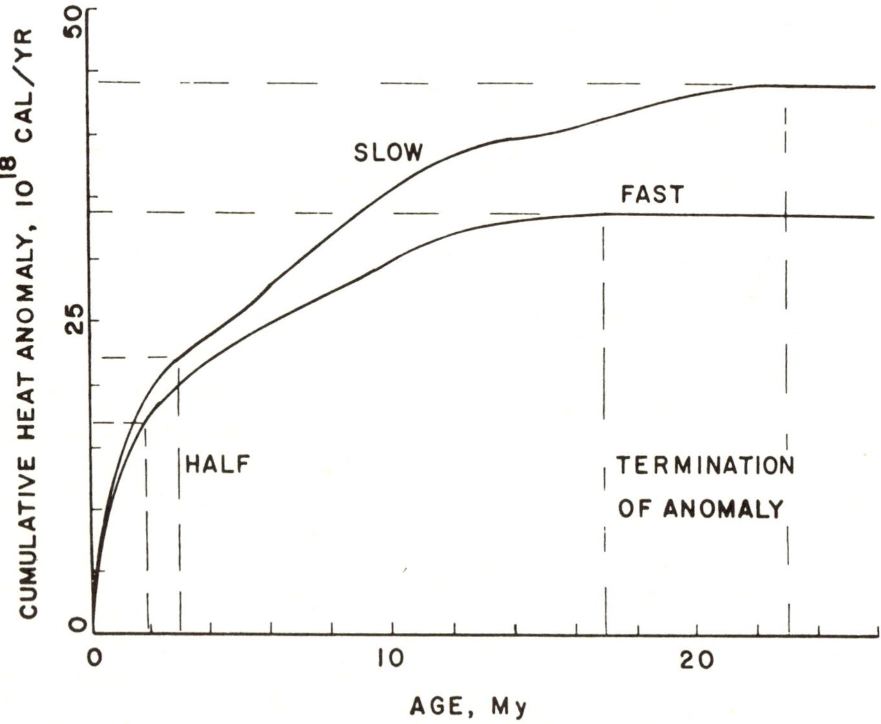

Fig. 2.—Cumulative distribution of the heat production anomaly with crustal age for a world-wide system of fast- or slow-spreading ridges; obtained by integration of data in Fig. 1 over crustal age (results in cal/cm²) followed by multiplication by the rate of production of new ocean floor (2.94 km²/yr).

have suffered intense compositional changes, which must have been accompanied by significant chemical migration. Nearly all workers who have studied the ocean-floor metabasites seem to agree that the chemical alteration was due to reaction of originally fresh basalt with hot, aqueous fluids (cf. Melson and van Andel, 1966; Cann 1971; Miyashiro et al. 1971). The available evidence suggests that the metasomatizing fluid in the case of the group II metabasites was hot sea water.

The group II metabasites have been altered in general toward a "spilitic" composition. Cann and Vine (1966) have in fact demonstrated that a complete gradation exists between fresh basalts and oceanic spilites, in which the original texture is essentially preserved. Original calcic plagioclase becomes strongly albitized, and mafic minerals are for the most part destroyed. Chlorite replaces the glassy matrix of extrusives, partially replaces plagioclase, and also forms vesicles and fractures (Cann and Vine 1966; Melson and van Andel 1966; Miyashiro et al. 1971). Other minerals which may develop include epidote, actinolite, talc, sphene, pyrite, nontronite, and quartz (see Melson and van Andel 1966). Zeolites and mixed-layer phyllosilicates of chlorite + smectite ± vermiculite may occur in zeolite facies rocks (Miyashiro et al. 1971). In terms of the bulk chemical composition, the metasomatism generally results in significant decreases in CaO, K_2O, and the Fe_2O_3/FeO ratio, a significant increase in H_2O+, and less marked increases in MgO and Na_2O (Melson and van Andel 1966; Miyashiro et al. 1971; Cann and Vine 1966).

The SiO_2 varies markedly from sample to sample, indicating that silica migrates extensively within altered oceanic crust (Miyashiro et al. 1971). The $\delta^{18}O$ of dredged greenstones indicates a metamorphic water with a $\delta^{18}O$ of -2 to $+2‰$ SMOW, consistent with reaction with sea water (Muehlenbachs and Clayton 1972).

The chemical changes inferred in the metasomatism of the group II rocks are nearly the same as (1) those observed in geothermal fields on the Reykjanes Peninsula of Iceland, where sea water circulates in basaltic rock, and (2) those observed in laboratory studies of hydrothermal reaction between sea water and basalt. The alteration minerals found at the Reykjanes geothermal field (Tomasson and Kristmannsdottir 1972) are essentially the same ones found in the group II metabasites. The sea water which circulates both in the Reykjanes system (Bjornsson et al. 1972) and in the volcanic rocks of Deception Island, Antarctica (Elderfield 1972) is enriched in calcium and potassium but depleted in magnesium; they may be slightly depleted in sodium which is sufficiently abundant in sea water that small losses are obscured. These chemical modifications of the sea water complement those inferred for the group II ocean floor rocks, thus demonstrating that the origin of the modification of these rocks is at least consistent with reaction of fresh basalt with hot sea water. Laboratory investigations of basalt-sea water interaction at sub-sea-floor pressures and hydrothermal temperatures (Hajash 1975; Mottl et al. 1974, 1975; Bischoff and Dickson 1975) have yielded essentially the same results, although small losses of sodium from sea water are more apparent. Table 2 summarizes the major chemical trends for both hydrothermal and weathering temperatures.

TABLE 2

COMPOSITIONAL CHANGES IN SEA WATER REACTING WITH BASALT AT HIGH AND LOW TEMPERATURES

Element	T = 200°–400°C	T = 0°–20°C
Ca......	+	+
Mg	−	± ?
Na	−	± ?
K	+	−
SiO_2	+	+
Al_2O_3 ...	?	?
Fe......	+	± ?
Mn	+	+ ?

NOTE.—Data compiled from Hajash 1975; Mottl et al. 1974; Bischoff and Dickson 1975; R. Hart 1973; S. R. Hart et al. 1974; and Bjornsson et al. 1972. + indicates that the concentration of the element in sea water is increased by the reaction; − indicates that the concentration of dissolved element is decreased by precipitation of the element in an alteration phase (e.g., Mg^{++} in chlorite).

The group I metabasites appear to have an origin which is not due to reaction with hot sea water. Hydration of these rocks is the only significant change in their bulk composition (Miyashiro et al. 1971; see also Bonatti et al. 1975). The original plagioclase tends to be preserved, in contrast to the strong albitization observed in the group II metabasites (Miyashiro et al. 1971), suggesting that group I rocks were not in contact with a large reservoir of hot, sodium-rich fluid. While both rock groups occur in the zeolite and greenschist facies, only group I rocks occur in the amphibolite facies (Miyashiro et al. 1971). Since amphibolite facies metamorphism requires high temperatures (about 600°C), the fluid which reacted with the group I rocks may have been derived locally by dehydration reactions associated with localized intrusions of magma.

Two alternative hypotheses for the extent of ocean-crust metasomatism have been suggested by the fact that metamorphic rocks have been recovered mainly in the vicinities of major scarps thought to have been created by normal faulting. The first is that the alteration is uniformly pervasive at depth in the oceanic crust, and that the action of faulting creates local outcropping on the ocean floor (Barrett and Aumento 1970; R. Hart 1973). The second hypothesis is that altered rocks occur only at fault zones as a result of the admittance of hydrothermal, metasomatizing fluids by the action of the shearing, crushing, and brecciation which occurred during the genesis of the fault zones (Melson et al. 1968; Cann and Vine 1966). The choice of hypothesis is critical in estimating the magnitude of geochemical flux between the oceanic crust and the overlying oceans. R. Hart (1973) has already assumed the first hypothesis of pervasive alteration to develop a model for the major element fluxes; if the second hypothesis is correct, then his estimates of the mass transfer are too high.

To assess the nature of the spatial distribution of the altered oceanic rocks, it is necessary to consider the effect of structural control (fractures, brecciation, etc.) upon the circulation of hot sea water and subsequent wall-rock alteration. Many studies (cf. Cann 1971; Sales and Meyer 1948; Carswell et al. 1974) have demonstrated that the low permeability of uncracked, massive igneous rock restricts metasomatic effects to within a few centimeters to a few meters of the cracks in which hydrothermal solutions circulate. The alteration is limited by the rates of diffusive mass transport through the growing thickness of altered material. The process is well illustrated by the model developed by Nigrini (1969) for the advancement of an alteration zone in which albite is converted to montmorillonite and quartz at 300°C. The thickness of the altered zone was found to be proportional to the square root of elapsed time: 139 yr produced a 10 cm thickness; 1 m.y., 8.5 m; and 10 m.y., 27 m.

Major fault zones, which may constitute the bulk of the hydrothermal plumbing, are spaced about a kilometer or more apart (cf. Fig. 7 of Williams et al, 1974). However, hydrothermal cooling itself may lead to further cracking and penetration of hot sea water; although this process is not well understood, it may be of great importance in controlling the amount of oceanic crust which is hydrothermally altered (Lister 1974). It is not possible at present to infer the distribution of crack spacing in the basaltic crust and thence to estimate limits for the fraction of material which is altered. In the following section we examine probable limits on selected element fluxes and their implications for the amount of altered oceanic crust.

GEOCHEMICAL FLUX AT MID-OCEAN RIDGES: SPECULATION AND CONSTRAINT

The chemical fluxes between the oceans and basaltic crust which accompany sub-sea-floor hydrothermal activity are at present difficult to estimate with confidence by direct means, since no samples of the hydrothermal solutions have yet been obtained or analyzed. It is necessary to turn instead to the compositions of solutions obtained from laboratory experiments and from "analogue" geothermal fields (especially those on the Reykjanes Peninsula). The tempting but hazardous approach is to assume that these solutions are reasonable approximations to those flowing out of submarine springs on the mid-ocean ridges. Since hydrothermal reaction between sea water and basalt is not yet well understood, this approach should be taken only with great care.

There are some significant differences in the conditions of reaction among the experiments, the "analogue" geothermal fields, and the hydrothermal fields on the mid-ocean ridges. The basalt in the oceanic crust is fairly massive, though cracked and brecciated; at Reykjanes, most of the stratigraphic section consists of tuffs (Tomasson and Kristmannsdottir 1972); and in the published experiments (Hajash 1975; Mottl et al. 1974; Bischoff and Dickson 1975) the rock had been ground to a fine powder. Thus, diffusion through a zone of altered wall-rock may be an important control on reaction beneath the ridges, but not so much at Reykjanes and probably not at all in the experiments. Pure kinetic factors may also produce differences, although the success of the experiments (less than 2 yr duration) in reproducing the alteration found at

Reykjanes (Mottl et al. 1974, 1975) suggests that these may be minor. The water:rock ratios in the experiments and at Reykjanes may be much lower than in sub-sea-floor systems. From our minimum computed flow rate (1.3×10^{17} cm^3/yr) and an assumed maximum section of 5 km of basalt which reacts hydrothermally, we calculate a minimum water:rock ratio of 9:1 (v/v) beneath the ridges. This is in the general range of the experiments (Mottl et al. 1974, 3:1 to 5:1; Hajash 1975, 3:1 to 15:1; and Bischoff and Dickson 1975, 30:1) and the Reykjanes geothermal field (3:1, Mottl et al. 1975). A direct attempt to compute hydrothermal element fluxes from these compositional data and our flow rate estimates *may* implicitly assume that most of the oceanic crust is hydrothermally altered.

Indirect estimates or constraints of the chemical fluxes at the ridges may be attempted by (1) summing the known sources and sinks of an element to and from the oceanic reservoir or (2) assumption of the growth of the exogenic reservoir of an element over a long period of earth history. For the first method to be useful, an element must be strongly leached or absorbed in sub-sea-floor hydrothermal systems so that the process makes an important contribution to assumed oceanic balance; otherwise the hydrothermal contribution would be masked by uncertainties in known source and sink fluxes. The second method requires explicit consideration of other endogenic-exogenic mass transfer processes which occur mainly near island arcs: (1) subduction, (2) obduction to high (erosional) levels, and (3) volcanism. The fate of Layer I sediment and underlying basalt at convergent plate boundaries is the subject of much dispute. Estimates range from nearly total subduction of all of about 300 m of pelagic sediment and underlying material (Scholl and Marlow 1974; Shor et al. 1970) to accumulation in arc-trench gaps of nearly all of the sediment plus a sizeable amount of the basaltic crust (Dickinson 1973; Karig and Sharman 1975). Production of volcanic material may be estimated from the volume of crust in island arcs and the time range of its formation (A. Anderson 1974, 1975). Indirect estimates suffer from cumulative uncertainties, and should be regarded as speculative unless confirmed by direct estimates.

Mass transfer between basaltic oceanic crust and the exogenic ocean reservoir may be conveniently considered for working purposes in terms of three separate processes: low-temperature weathering, volcanic "effusion," and the large-scale hydrothermal activity. Weathering encompasses reactions between fluid and rock at temperatures less than about 20°C (Compare Muehlenbachs 1975*a*, 1975*b*). Volcanic "effusion" refers to high temperature mass transfer between sea water and cooling igneous bodies, especially extrusive flows, and includes loss of both volatile components and leachable interstitial material removed by relatively small amounts of sea water circulating in cooling cracks. Physically, it includes transport by both molecular diffusion and advection by hot sea water. It is distinguished from the large-scale hydrothermal activity previously discussed in this paper in that heat is provided by the cooling volcanic material itself, whereas the main hydrothermal activity obtains most of its heat by conduction from about 100 km of underlying material which cools to form the oceanic lithosphere. Volcanic effusion is therefore a much shorter-lived process than the large-scale hydrothermal circulation.

In the remainder of this paper we discuss direct and indirect flux estimates for the above three processes for silica, manganese, iron, magnesium, calcium, oxidants-reductants, and water. The purpose of these discussions is not to develop a comprehensive model of endogenic-exogenic mass transfer, but rather to take the first steps of (1) examining the relative

TABLE 3

SUMMARY OF MASS TRANSFER DATA FOR SILICA

Flux to the Oceans	Rate (g/yr)	Remarks and References
1 Stream flux	4.27×10^{14}	Wollast (1974); dissolved load only; can be balanced within reasonable error by known sinks, implying that hydrothermal input is small
2 Hydrothermal flux	$0.3–0.9 \times 10^{14}$	Assumes flow rate of $1.3–9 \times 10^{17}$ g/yr, initial concentration of 6 ppm SiO_2 (Garrels and Mackenzie 1971, p. 100), and final concentration at chalcedony/quartz solubility of exit temperature (see text); best estimate assumes 200°C exit temperature
3 Best estimate	0.7×10^{14}	
4 Submarine weathering.	$0.3–0.8 \times 10^{14}$	Assumes degree of weathering reported by R. Hart (1973) in dredged and shallow-drilled rock specimens; weathered thickness assumed 60–200 m. (Compare thickness of basalt "rubble" zone reported by Melson et al. 1974)
5	(2.9×10^{14})	R. Hart (1973); assumes 0.66 km of weathered basalt
6	0.03×10^{14}	Harriss (1966); weathering of marine sediment only
7 Volcanic effusion	0.0003×10^{14}	Calvert (1968)
8 Total basaltic crust input . . .	$0.6–1.7 \times 10^{14}$	Sum of 2, 4, and 7

importance of the weathering, volcanic effusion, and hydrothermal processes in the total sea water-basalt interaction (for SiO_2, Mn, Fe), and (2) obtaining quantitative limits on the amount of basalt which may react with the sea water (using Mg, Ca, oxidants-reductants, H_2O). Most of the flux estimates to be presented are given as ranges which reflect the uncertainties in the calculations. Assumed values of parameters used in the calculations are given along with the estimates themselves in tables 3–7 and offer further insight into the nature of the uncertainties. For example, the weathering estimates were derived mainly from those of R. Hart (1973) by assuming weathered thicknesses on the order of those of the basalt "rubble" zone encountered on DSDP Leg 37 (up to 300 m; Melson et al. 1974). Our estimates are probably accurate to within an order of magnitude, but detailed studies to accurately determine the distribution of weathering intensity in the basaltic crust are needed to confirm and refine them.

Silica.—Silica is well-suited for a direct hydrothermal flux estimate because the factors which control its concentration in hydrothermal solutions are fairly well understood. At temperature in excess of 180°C dissolved silica is controlled by quartz solubility; below 110°C it is controlled by chalcedony solubility; and in the intervening temperature range it may vary between the two solubility curves (Arnorsson 1970, 1975). Assuming that reaction temperature is approximately the same as exit temperature, the hydrothermal silica flux to the oceans lies in the range $0.3–0.9 \times 10^{14}$ g/yr and is comparable to the input for submarine weathering ($0.3–0.8 \times 10^{14}$ g/yr; see table 3 for a summary of all silica data). The total flux to the oceans from the basaltic crust is estimated at $0.6–1.7 \times 10^{14}$ g/yr, which represents a minor input compared to the stream flux of 4.27×10^{14}

g/yr. In great contrast, R. Hart's (1973) model predicted a total flux of 6.90×10^{14} g/yr from the oceanic crust.

Manganese and iron.—A flux of manganese and iron from the oceanic crust due to hydrothermal activity beneath mid-ocean ridges has been proposed as the source for metalliferous pelagic sediments (cf. Bostrom et al. 1971; Dymond et al. 1973; Piper 1973). Leachable metals appear to occur in large part as reactive interstitial material and not bound within silicate materials (Ellis 1968; Corliss 1971). Bischoff and Dickson (1975) in a laboratory investigation of sea water-basalt reaction found that dissolved manganese and iron exhibited a rapid rise to about 35 ppm each followed by a decline to about 5 ppm each after 4752 hr (at which point they were still decreasing very slowly). These final values are in reasonably good agreement with the dissolved managanese and iron concentrations observed in Reykjanes hydrothermal fluids (Bjornsson et al. 1972; Mottl et al. 1975). It appears that the manganese content of hydrothermal sea water probably becomes fixed at 4–5 ppm and the iron content at 0.4–4 ppm by equilibria with unknown solid products (probably hydroxyoxides or sheet silicates). The estimates in table 4 show that for manganese the hydrothermal contribution ($0.5\text{–}4.5 \times 10^{12}$ g/yr) is somewhat greater than but similar in magnitude to the contributions from volcanic effusion and weathering. The total manganese flux from basalt to sea water is in remarkable agreement with indirect balance estimates (table 4). For iron, however, the hydrothermal flux is at most 3.6×10^{12} g/yr, and volcanic effusion (17×10^{12} g/yr) and weathering ($5\text{–}18 \times 10^{12}$ g/yr) dominate the total iron flux from the basalt of $22\text{–}39 \times 10^{12}$ g/yr. This total flux is sufficient to account for observed thicknesses of basal iron-rich sediment (table 4). Metalliferous pelagic sediment appears to have diverse origins in terms of the processes which extract the metals from the basaltic crust.

Magnesium.—A detailed examination of the sources and sinks of oceanic magnesium (Drever 1974) showed that the known sinks were capable of removing only half of the total input. If oceanic magnesium were building up at this rate, it would double in only 29 m.y. However, hydrothermal reactions with basaltic crust were not considered in Drever's study as being a significant sink mechanism. Studies of basalt-sea water reaction (cf. Bischoff and Dickson 1975; Mottl et al. 1974, 1975) have shown that this process removes nearly all of an original 1,300 ppm of magnesium at water:rock ratios of 3:1 to 30:1 (at higher ratios this may not be the case). A direct estimate of the hydrothermal flux of magnesium into the oceanic crust assuming the mass transfer observed in the experiments gives an estimate of $174\text{–}350 \times 10^{12}$ g/yr, whereas the flux needed for oceanic balance is only 65×10^{12} g/yr (table 5). We consider the latter estimate more reliable. Dredged greenstones (which may or may not be representative of hydrothermally altered oceanic crust) suggest that such a flux corresponds to an equivalent thickness of 420–1,300 m of reacted crust. The water: rock ratios would be 33:1 to 215:1, assuming hydrothermal exit temperatures of 150–300°C. The concentration of magnesium in the exogenic system seems well-explained by the weathering of average igneous rock (Garrels and Mackenzie 1971, p. 204–248); this strongly suggests that it is not being depleted by hydrothermal removal and subduction of oceanic crust. A balance could be achieved by obduction to high (erosional) levels of 0.1–0.3 km^3/yr of ophiolitic material followed by weathering. Corresponding thicknesses of this material are less than 120 m, or only a small fraction of the total thicknesses of oceanic crust sheared off and accumulated in arc-trench gaps in the model of Karig and Sharman (1975).

Calcium.—Most of the calcium contained in sedimentary rocks cannot be accounted for by weathering of any "average"

TABLE 4

SUMMARY OF MASS TRANSFER DATA FOR MANGANESE AND IRON

Flux to the Oceans	Rate (g/yr)	Remarks and References
Manganese:		
1 Stream flux	15.5×10^{12}	Mackenzie (1976); includes dissolved plus suspended load
2 Hydrothermal flux	$0.5–4.5 \times 10^{12}$	Flow rate of $1.3–9 \times 10^{17}$ g/yr, initial concentration of 2 ppb (Riley and Chester 1971), and final concentration of 4–5 ppm (See text); best estimate assumes 200°C exit temperature
3 Best estimate	1.4×10^{12}	
4 Submarine weathering	$0.1–0.3 \times 10^{12}$	Assumes degree of weathering reported by R. Hart (1973) over 60–200 m of basalt (compare thicknesses of the "rubble" zone reported by Melson et al. 1974)
5	0.79×10^{12}	R. Hart (1973); assumes 0.66 km of weathered basalt
6 Volcanic effusion	0.2×10^{12}	Assumes loss of 150 ppm (Corliss 1971) over 200 m of flow interiors
7 Total basaltic crust input	$0.8–4.9 \times 10^{12}$	Sum of 2, 4, and 6
8	0.9×10^{12}	T. A. Lietzke (1974, personal communication); indirect balance estimate
9	3×10^{12}	Mackenzie (1976); indirect balance estimate
Iron:		
1 Stream flux	200×10^{12}	Garrels and Perry (1974); dissolved plus suspended load
2 Hydrothermal flux	$0.07–3.6 \times 10^{12}$	Assumes above flow rate and 0.4–4 ppm increase due to hydrothermal reaction (See text); best estimate assumes 200°C exit temperature
3 Best estimate	0.8×10^{12}	
4 Submarine weathering	$5–18 \times 10^{12}$	After R. Hart (1973), assuming weathered thickness of 60–200 m of basaltic crust
5	59×10^{12}	R. Hart (1973); assumes 0.66 km of weathered basalt
6 Volcanic effusion	17×10^{12}	Assumes loss of $\approx 1\%$ Fe (Corliss 1971) from 200 m of flow interiors
7 Total basaltic crust input	$22–39 \times 10^{12}$	Sum of 2, 4, and 6
8 Equivalent thickness of iron-rich sediment	24–43 m	Equivalent of 7 as sediment with 40% Fe_2O_3 and bulk density of 1.5 g/cm^3
9 Observed thickness of basal iron-rich sediment	2–30 m	Cf. von der Borch et al. 1971; von der Borch and Rex 1970)

igneous rock (Garrels and Mackenzie 1971, p. 240–248). This calcium excess is estimated at $1,200 \times 10^{20}$ g (table 5). Garrels and Mackenzie (1971) postulated submarine weathering of ocean basalt as the source of this anomaly. The required input to the exogenic system over 4.0×10^9 yr is 30×10^{12} g/yr, and represents the excess of weathering and hydrothermal contributions over subduction of pelagic sediment. Estimates of these fluxes (table 5) suggest that submarine weathering alone could account for the anomaly if subduction of sediments is minimal—otherwise a hydrothermal contribution is required. The factors which control calcium transfer in hydrothermal reaction of basalt and sea water are incompletely known. However, assuming that observed in the experiment of Bischoff and Dickson

TABLE 5

Summary of Mass Transfer Data for Magnesium and Calcium

Flux to Oceans (Mg) or Exogenic System (Ca)	Rate (g/yr)	Remarks and References
Magnesium:		
1 Stream flux	131×10^{12}	Drever (1974); dissolved load
2 Hydrothermal flux	-65×10^{12}	Unbalanced stream flux (Drever, 1974), here assumed removed by hydrothermal activity; uncertainty is probably about $\pm 20 \times 10^{12}$ g/yr at most
3	$(-174$ to $-350 \times 10^{12})$	Assumes that exit and reaction temperatures are in the range 150°–300°C and about all of 1,300 ppm in sea water is removed (See discussion in next)
4 Submarine weathering...	$1\text{–}11 \times 10^{12}$	Assumes degree of weathering reported by R. Hart (1973) over less than 60 m of basaltic crust; the actual degree of weathering may be smaller or even of opposite sign (see S. R. Hart et al. 1974)
5	(121×10^{12})	R. Hart (1973); assumes 0.66 km of weathered basalt
6 Total basaltic crust input	-54 to -65×10^{12}	Net sum of 2 and 4
7 Equivalent greenstone thickness of 2	420–1,300 m	Assumes $\Delta MgO = 1\text{–}3\%$ due to hydrothermal reaction (Miyashiro et al. 1971; Cann 1969)
Calcium:		
1 Stream input to oceans..	480×10^{12}	Garrels and Mackenzie (1971, p. 102–103); dissolved load
2 Sediment reservoir......	$2{,}100 \times 10^{20}$ g	Assumes average 11.7% CaO (Wyllie 1971, table 7–4) and $25{,}000 \times 10^{20}$ g of sediments (Garrels and Perry 1974)
3 Contribution to 2 by continental weathering ..	900×10^{20} g	From weathering of average igneous rock (See Garrels and Mackenzie 1971, p. 240–248) to produce $25{,}000 \times 10^{20}$ g of sediments
4 Endogenic contribution..	$1{,}200 \times 10^{20}$ g	Difference of 2 and 3
5 Endogenic input	30×10^{12}	4 averaged over 4×10^{9} yr
6 Submarine weathering...	$8\text{–}57 \times 10^{12}$	Assumes 30–200 m of basalt are weathered to degree reported by R. Hart (1973)
7	(187×10^{12})	R. Hart (1973); assumes 0.66 km of weathered basalt
8 Net of hydrothermal input over subduction removal	-27 to $+22 \times 10^{12}$	Difference of 5 and 6
9 Subduction removal.....	$12\text{–}120 \times 10^{12}$	Probable limits over 4×10^{9} yr, assuming nearly all pelagic sediment is subducted (Scholl and Marlow 1974) and contains an equivalent thickness of 20–200 m of calcium carbonate sediment with 0.2 g Ca/cm^3
10 Hydrothermal flux	(22×10^{12})	Assumes no subduction; 8, but must be > 0
11	$(93\text{–}142 \times 10^{12})$	Assumes upper limit of subduction removal, which uses post-Cretaceous amounts of $CaCO_3$ in pelagic sediment; these limits should be too high for the 4 b.y. average
12	$65\text{–}135 \times 10^{12}$	Assumes flow rate of $1.3\text{–}2.7 \times 10^{17}$ g/yr (best estimates) and 500 ppm increase in dissolved calcium of the hydrothermal sea water (Bischoff and Dickson 1975); other sources (cf. Hajash 1975; Mottl et al. 1974) suggest increases of up to 1000 ppm are possible
13 Equivalent greenstone thickness of 11	170–770 m	Assumes $\Delta CaO = 3\text{–}5\%$ due to hydrothermal reaction (Miyashiro et al. 1971; Cann 1969); this range of estimates should represent an upper limit
14 Equivalent greenstone thickness of 12	210–730 m	Same sources as above; should represent probable limits

(1975), where calcium achieved a constant concentration, a direct hydrothermal flux estimate of 65–135 × 10^{12} g/yr is obtained. This estimate predicts about the same equivalent thickness as the magnesium data and also agrees well with the estimate obtained assuming significant subduction of calcium-bearing sediments (table 5).

Oxidants-reductants. — Hydrothermal and weathering reactions in the basalt of the oceanic crust may constitute an important sink for exogenic O_2. If this crust is subducted into the deep mantle, then there may exist a significant drain on the total exogenic oxidant reservoir (O_2 in the atmosphere and stored as sulfate and ferric oxide in oceans and sediments) which has not been explicitly considered in previous discussions of the exogenic oxidant cycle (cf. Cloud, 1972; Garrels and Perry, 1974). Oxidation of the ferrous iron of the basaltic crust is not limited by the inflow of dissolved O_2; in hydrothermal systems at ridges sulfate and water may act as major suppliers of oxygen. Reduction of sulfur has been noted qualitatively (by detection of H_2S odor) in experimental studies of sea water-basalt reaction (Hajash 1975; Bischoff and Dickson 1975). Occurrence of H_2 (produced when water acts as an oxygen donor) is known from high-temperature geothermal fields in Iceland (Polyak and Kononov 1974). In contrast, studies of dredged metamorphic rocks have shown that their Fe_2O_3 contents either remained unchanged (Miyashiro et al. 1971) or decreased (Melson and van Andel 1966; Cann 1971). The most plausible interpretation is that these rocks reacted with ascending limbs of reduced hydrothermal sea water and are not, at least with regard to oxidation state, representative of hydrothermally altered oceanic crust.

A useful constraint on the oxidation of oceanic basaltic crust can be developed by consideration of the Phanerozoic rock record and possible compensatory mass transfers (mainly at island arcs). Estimates of the magnitudes of the major exogenic oxidant-reductant reservoirs and pertinent chemical fluxes are given in table 6. The organic matter reservoir is inferred to have been constant to within 10% of the present magnitude over Phanerozoic time (Garrels and Perry 1974), based on (1) the invariance with time of isotopic compositions of carbonate carbon (Broecker 1970) and organic carbon (Degens 1969) and (2) the absence of a noticeable secular increase in the average organic carbon content of sediments (Ronov 1958; Gregor 1971). The significance of this is that it severely constrains the potential of the exogenic system to have produced more oxidant (O_2) by photosynthesis to compensate for losses due to oxidation of oceanic crust. For purposes of obtaining an upper bound on the latter process, let us assume that over Phanerozoic time the organic matter reservoir increased by as much as 30%. Over 600 m.y. this would yield an average 50 × 10^{10} moles O_2/yr (eq/yr) or 1 atmospheric oxygen reservoir per 80 m.y. Loss of H_2 to outer space would yield only 1.4 × 10^{10} eq/yr (Walker 1974), which is negligible for our purposes.

Further compensation could be occurring at convergent plate boundaries by subduction of highly reduced organic matter and pyrite and volcanic degassing of more oxidized carbon dioxide and sulfur dioxide. This possible role of carbon and sulfur acting as oxygen carriers is illustrated in figure 3, and could correspond to one or both of the following physical interpretations: (1) subduction of the sediments to the deep mantle and degassing of CO_2 and SO_2 from large mantle reservoirs or (2) incorporation of the sediment into the lower levels of arc-trench gaps followed by reaction with mantle-derived Fe_2O_3 during igneous activity beneath island arcs to yield CO_2 and SO_2 for subsequent degassing. Some very crude estimates (table 6) indicate that carbon and sulfur could each yield about 40 × 10^{10} eq/yr by such a process. Thus, we obtain an upper limit to the

TABLE 6

SUMMARY OF RESERVOIR AND FLUX ESTIMATES FOR OXIDANTS-REDUCTANTS

Flux or Reservoir	Moles/yr*	Eq/yr*	Remarks and References
1 Atmospheric O_2 . .	0.4×10^{20} moles	1 eq-atm	All reservoir estimates are taken from Garrels and Perry (1974)
2 "Excess" Fe_2O_3 in sediments	2.0×10^{20} moles	2.5 eq-atm	Defined as Fe_2O_3 not derived from igneous precursor (Garrels and Perry, 1974)
3 $CaSO_4$ evaporites	2.0×10^{20} moles	9.9 eq-atm	O_2 equivalent defined relative to S^{--} oxidation state
4 Marine SO_4^{--} . . .	0.4×10^{20} moles	2.2 eq-atm	As above
5 Total exogenic oxidants		15.6 eq-atm	Sum of 1 through 4
6 Organic matter . .	10.4×10^{20} moles	26.0 eq-atm	Note much greater equivalent than 5
7 Extreme upper limit on H_2 hydrothermal flux	$330\text{–}10{,}500 \times 10^{10}$	$165\text{–}5{,}250 \times 10^{10}$	Assumes flow rate of $1.3\text{–}9 \times 10^{17}$ g/yr and H_2 content controlled by QFM buffer
8 Upper limit on hydrothermal SO_4^{--} reduction .	$360\text{–}840 \times 10^{10}$	$720\text{–}1{,}700 \times 10^{10}$	Total hydrothermal inflow assuming above flow rate and 0.028 moles/l in sea water
9 Limit to net drain on 5	. . .	50×10^{10}	Assumes maximal 30% growth of CH_2O reservoir over the Phanerozoic (see text)
10 H_2 escape to space	2.8×10^{10}	1.4×10^{10}	Walker (1974)
11 O_2 hydrothermal flux	$3.5\text{–}8 \times 10^{10}$	$3.5\text{–}8 \times 10^{10}$	Assumes hydrothermal flow rate of $1.3\text{–}3 \times 10^{17}$ g/yr and reduction of all of 2.7×10^{-4} mole/l present in inflowing sea water (Garner and Ford 1969)
12 Volcanic effusion of H_2	$11\text{–}67 \times 10^{10}$	$5.5\text{–}28 \times 10^{10}$	Assumes $\Delta Fe_2O_3 = 1\%$ (compare Watkins and Haggerty 1967) over 200 m of flow interiors (lower estimate plus 1 km of shallow dikes and intrusions (upper estimate)
13 Submarine weathering.	$3.3\text{–}33 \times 10^{10}$	$1.6\text{–}16 \times 10^{10}$	Assumes $\Delta Fe_2O_3 = 1\text{–}3\%$ over 60–200 m of basalt
14 SO_2 outgassing at island arcs	$15\text{–}50 \times 10^{10}$	$20\text{–}60 \times 10^{10}$	See summary by Cadle (1975); O_2 equivalent defined relative to sulfur in pyrite
15 Subduction of pyrite (FeS_2)	$8\text{–}16 \times 10^{10}$	$20\text{–}40 \times 10^{10}$	Assumes subduction of 300 m of sediment (Scholl and Marlow 1974; Shor et al. 1970) containing 0.5–1.0% pyrite

TABLE 6—*Continued*

Flux or Reservoir	Moles/yr*	Eq/yr*	Remarks and References
16 CO_2 outgassing at island arcs	2,000 × 10^{10}	2,000 × 10^{10}	Buddemeier and Pucetti (1974); must be about an order of magnitude too high, otherwise reasonable estimates for subduction of exogenic carbon (organic matter, 17; $CaCO_3$, table 5, 9) are insufficient to prevent a massive build-up of exogenic carbon (which is not observed)
17 Subduction of organic matter ..	40 × 10^{10}	40 × 10^{10}	Assumes subduction of 300 m ot sediment and 0.2% organic carbon (cf. Gealy 1971; Pimm 1971)
18 Maximum limit to drain on exogenic oxidants	...	130 × 10^{10}	9 plus 40 × 10^{10} eq/yr from each of 14–15 and 16–17
19 Upper limit to H_2 flux from basaltic crust	260 × 10^{10}	130 × 10^{10}	18 expressed as H_2; note that this is much smaller than 7
20 Maximum hydrothermal SO_4^{--} reduction	65 × 10^{10}	130 × 10^{10}	18 expressed as sulfate reduction (SO_4^{--} to S^{--}); note that this is much smaller than 8
21 Equivalent thickness of reacted rock	106–1060 m		Assumes ΔFe_2O_3 = 1%–10% due to hydrothermal reaction (see discussion in text of Fe_2O_3 in dredged greenstones); represents upper limits to amount of rock which may be oxidized (some reacted rock is probably reduced on ascending limbs of the hydrothermal flow—see text)

* Reservoirs (1–6) are given in moles and the equivalent number of oxygenated atmospheres (eq-atm)

oxidation of oceanic basalt of 130 × 10^{10} eq/yr. If the oxidation were proceeding at this rate and not compensated by any of the above mechanisms, it would remove the total of the exogenic oxidant reservoirs in only 480 m.y. and the atmospheric oxygen reservoir alone in 30 m.y.

Submarine weathering and consumption of dissolved O_2 in hydrothermally circulating sea water only account for relatively small oxygen losses (1.6–16 × 10^{10} eq/yr and 3.5–8 × 10^{10} eq/yr, respectively). Volcanic effusion of H_2 is probably more significant (5.5–28 × 10^{10} eq/yr), although it can be estimated only crudely. The oxidation state within cooling igneous bodies of basaltic composition may be close to that predicted assuming control by a quartz-fayalite-magnetite (QFM) buffer (Carmichael and Nicholls 1967; Eugster 1972). Figure 4 shows the hydrogen content of aqueous fluid in equilibrium with this buffer assemblage as a function of temperature at sub-sea-floor pressures; these data suggest that strong H_2 fugacity gradients may develop within cooling flows, dikes, and sills of the oceanic crust. Sato and Wright (1966) suggested that

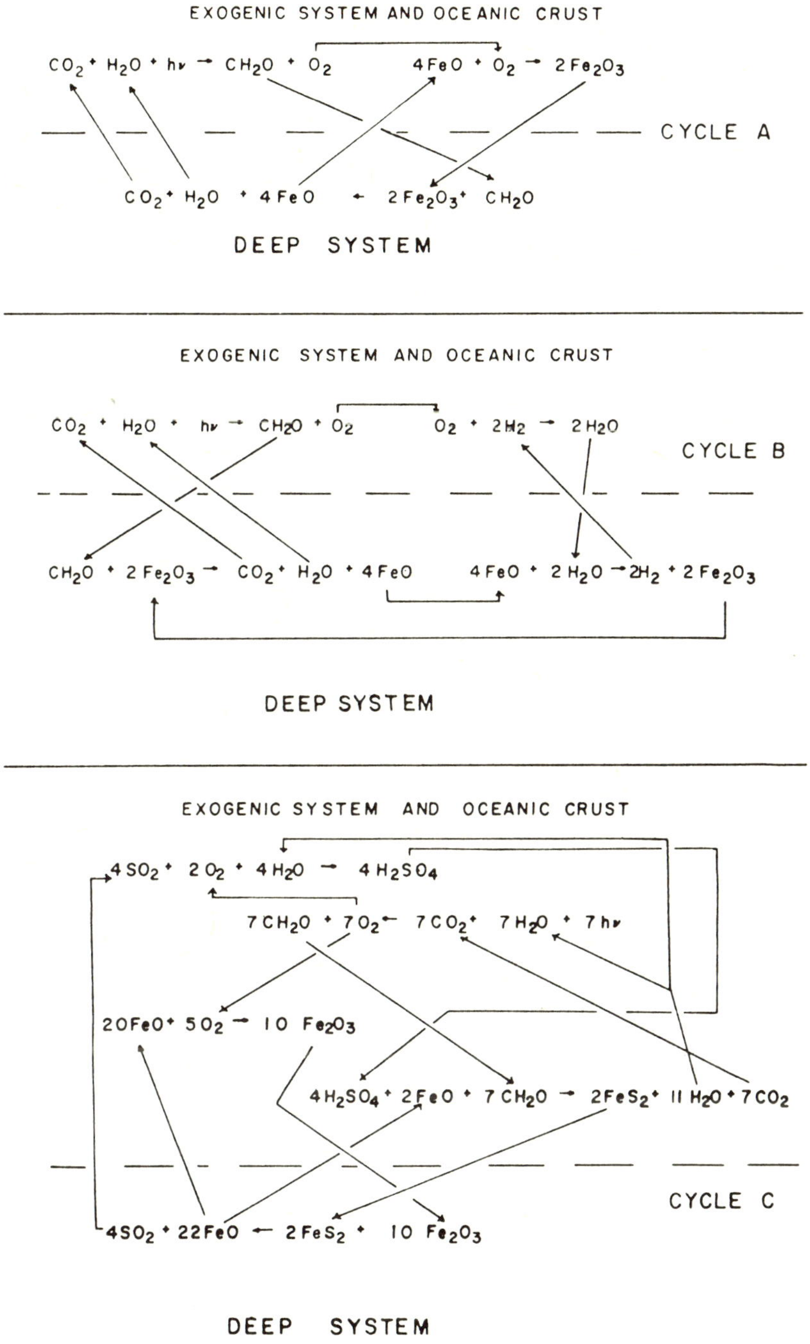

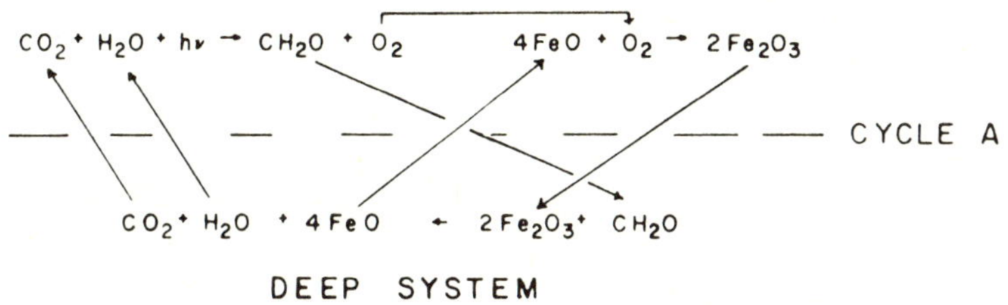

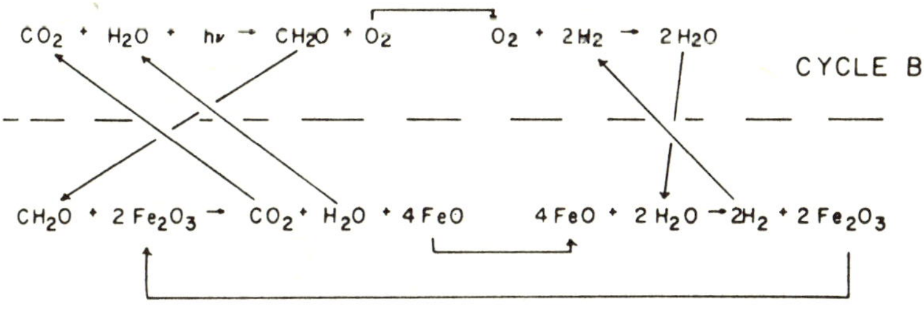

EXOGENIC SYSTEM AND OCEANIC CRUST
$CO_2 + H_2O + h\nu \rightarrow CH_2O + O_2$
$4FeO + O_2 \rightarrow 2Fe_2O_3$
CYCLE A
$CO_2 + H_2O + 4FeO \leftarrow 2Fe_2O_3 + CH_2O$
DEEP SYSTEM
EXOGENIC SYSTEM AND OCEANIC CRUST
$CO_2 + H_2O + h\nu \rightarrow CH_2O + O_2$
$O_2 + 2H_2 \rightarrow 2H_2O$
CYCLE B
$CH_2O + 2Fe_2O_3 \rightarrow CO_2 + H_2O + 4FeO$
$4FeO + 2H_2O \rightarrow 2H_2 + 2Fe_2O_3$
DEEP SYSTEM
EXOGENIC SYSTEM AND OCEANIC CRUST
$4SO_2 + 2O_2 + 4H_2O \rightarrow 4H_2SO_4$
$7CH_2O + 7O_2 \leftarrow 7CO_2 + 7H_2O + 7h\nu$
$20FeO + 5O_2 \rightarrow 10Fe_2O_3$
$4H_2SO_4 + 2FeO + 7CH_2O \rightarrow 2FeS_2 + 11H_2O + 7CO_2$
CYCLE C
$4SO_2 + 22FeO \leftarrow 2FeS_2 + 10Fe_2O_3$
DEEP SYSTEM

FIG. 3.—Three simplified models of exogenic-endogenic oxidant-reductant cycling. Cycle A (after Eugster, 1972) depicts compensation of subducted oxygen (Fe_2O_3) by subduction of organic matter (CH_2O) and outgassing of CO_2 and H_2O; cycle B shows possible role of H_2 generation in producing Fe_2O_3 in the oceanic crust instead of direct oxidation by O_2 (cycle A); cycle C depicts a possible role of sulfur which is analogous to that of carbon in cycles A and B. Exogenic-endogenic cycles may or may not be close to steady state (see remarks in introduction).

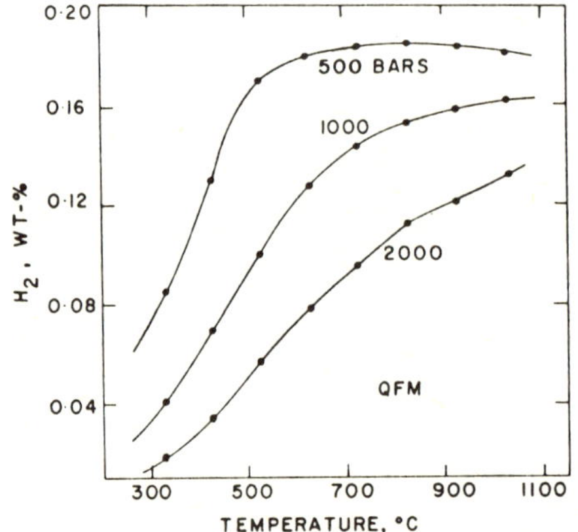

FIG. 4.—Hydrogen content of aqueous fluid in equilibrium with a quartz-fayalite-magnetite (QFM) buffer, computed from fugacity data of Eugster and Skippen (1967) and hydrogen fugacity coefficient data of Shaw and Wones (1964).

hydrogen would preferentially diffuse out of such bodies, leaving the interiors more oxidized than the margins. Their work and that of Watkins and Haggerty (1967) appear to confirm the importance of this process, but it is difficult to infer just how extensive it is within the oceanic crust; our estimate is of necessity a crude one. Production of H_2 and/or reduction of dissolved sulfate in the large-scale hydrothermal circulation may be the most important means of oxidizing oceanic basalt since the other processes noted above appear to consume at most less than half of the 130 × 10^{10} eq/yr limit (table 6). If those processes consume minimum amounts, then nearly all of that limit could be applied to H_2 generation and/or sulfate reduction. Applying it to each separately, we find that H_2 generation is limited to at most 260 × 10^{10} moles/yr (equivalent to a concentration in exiting fluid of 6–40 ppm) and that sulfate reduction is limited to 65 × 10^{10} moles/yr (8–18% of the influx of dissolved SO_4^{--}). It should be noted that acid-balanced reactions involving sulfate reduction can occur without any exotic components, for example:

$$5Mg^{++} + CaAl_2Si_2O_8 + 22Fe_2SiO_4 + 4H_2O + 4SO_4^{--}$$
$$\rightarrow Mg_5Al_2Si_3O_{10}(OH)_8 + 14Fe_3O_4 + 2FeS_2 + 21SiO_2 + Ca^{++}$$

magnesium + anorthite + fayalite + water + sulfate

→ chlorite + magnetite + pyrite + quartz + calcium.

Assuming that reacted rock gains a net average of 1–10% Fe_2O_3, the equivalent thickness of reacted rock is 106–1,060 m, in surprising agreement with the estimates based on reactions of magnesium and calcium.

A minimum exogenic residence time of 820 m.y. for sulfur is implied by the above exogenic-endogenic flux estimates. This is sufficiently great not to obscure the historical record of sharp variations of $\delta^{34}S$ of seawater in response to marine evaporite depositions in the Phanerozoic (Garrels and Perry 1974). Li (1972) successfully used $\delta^{34}S$ observed in sedimentary rocks to estimate the relative proportions of oxidized (sulfate) and reduced (pyrite) sulfur assuming a closed exogenic system. His results therefore imply a very small turnover of exogenic sulfur.

Water.—Water may be cycled in a manner similar to that postulated for CO_2 and SO_2. Although the parameters controlling exogenic-endogenic water transfer are known only very crudely, a constraint on the net imbalance is that sea level has been inferred to have changed very little over Phanerozoic time (Sloss 1963; Wise 1972). This corresponds to a net imbalance of 1.4 × 10^{14} g/yr (sign uncertain; see table 7 for a summary of the water data). Hydration of basalt by weathering and hydrothermal reactions may lead to subduction of 1.5–6 × 10^{14} g/yr of water assuming thicknesses of altered rock similar to those inferred

TABLE 7

SUMMARY OF RESERVOIR AND FLUX ESTIMATES FOR WATER

Reservoir or Flux	Magnitude	Residence Time*	Remarks and References
1 Exogenic water reservoir.........	17,200 × 10^{20} g	...	Garrels and Mackenzie (1971, p. 100)
2 Continental run-off	360 × 10^{17} g/yr	48,000 yr	Garrels and Mackenzie (1971; p. 100); includes flux of ground water; note much larger value than sub-sea floor hydrothermal circulation
3 Limit on loss or gain of exogenic water	±1.4 × 10^{14} g/yr	12 × 10^9 yr	Sea level is restricted to a 300 m net change over the Phanerozoic (Sloss 1963; Wise 1972); see Wise (1972) for models with variable continental area
4 Hydration of basalt	1.5–6 × 10^{14} g/yr	3–12 × 10^9 yr	Assumed uptake of ~0.1 g H_2O/cm^3 rock by weathering (R. Hart, 1973) and hydrothermal reaction (Miyashiro et al. 1971; R. Hart 1973) over equivalent thickness of 0.5–2 km (range suggested by previous discussion in this paper)
5 Limit by subduction of sediment	0.03–0.3 × 10^{14} g/yr	57–570 × 10^9 yr	Compacted thickness assumed 300 m (cf. Shor et al. 1970) and 0.3%–3% H_2O by volume (Rieke and Chilingarian 1974, p. 40–83; Burst 1969); includes both chemically bound and fluid water
6 Volcanic outgassing	1–5 × 10^{14} g/yr	3.4–17 × 10^9 yr	Upper value from A. Anderson (1974) assuming 10% H_2O in magma; lower value assumes 2% H_2O in magma (Chayes 1969); note close agreement of 6 with 4

* Time required to circulate a mass equivalent to the exogenic water reservoir (1).

earlier in this paper. In contrast, intense compaction (cf. Rieke and Chilingarian 1974, p. 40–83) of sediments probably expels all but a small fraction of both chemically bound and pore water, so that subduction in sediments is probably much less significant than basalt hydration (table 7). The flux of water from volcanos at island arcs (A. Anderson 1975) is estimated at 1–5 × 10^{14} g/yr (about the same as for hydration of basalt) and seems to confirm a probable balance for H_2O. Note that exogenic-endogenic exchange of water is miniscule compared to continental runoff and that the residence time of water in the exogenic system is about the age of the earth or greater (A. Anderson 1975). Volcanism releases chlorine and water in their exogenic (oceanic) ratios (A. Anderson 1975), which is not expected if water is mainly subducted in a chemically bound state to the source region of the volcanics and degassed. If chlorine is not absorbed from sea water in hydrothermal metamorphism, this would seem to imply that subducted water (and other volatiles) are returned to the deep mantle and replaced by volatiles from the large mantle reservoir. For components other than water, geochemical evidence for input of subducted material into island-arc volcanics is lacking (Oversby and Ewart 1972; but compare Armstrong 1971).

CONCLUDING REMARKS

Although sea water is inferred to hydrothermally circulate at rates sufficient to cycle the oceanic mass in only 5–11 m.y., the mass transfer considerations examined in this paper require only about 1 km or less of oceanic crust to react with this fluid. Models of the oceanic crust which include primarily hydrous and oxidized material such as serpentinite lie outside the limits of our geochemical balance estimates and can be excluded. The extent of permitted alteration is large enough, however, that physical properties largely dependent on cracks (seismic velocity and electrical conductivity) may be extensively modified. The bulk composition of the oceanic crust appears to be only slightly modified.

More extensive drilling of the basaltic ocean floor is needed to confirm and refine the admittedly crude estimates made in this paper, both to obtain samples of hydrothermal fluid and to achieve a better sample of altered rock. Mass transfer by all three processes—weathering, volcanic effusion, and hydrothermal reaction—is but crudely known on a quantitative basis. More precise examination of the cycling of elements of different geochemical behavior and of their isotopes will hopefully lead to a quantitative understanding of such problems as the fate of material at convergent plate boundaries and the possible growth of the continents.

ACKNOWLEDGMENTS.—We thank A. T. Anderson and T. A. Lietzke for helpful discussions and R. M. Garrels, F. T. Mackenzie, and the journal reviewers for critically reading the manuscript and suggesting improvements. A. T. Anderson, J. L. Bischoff, F. W. Dickson, R. Anderson, and R. Wollast kindly provided advance copies of their papers. Michael Mottl (Harvard) has developed a model for magnesium which in general agrees with our conclusion that the amount of reaction is small compared to the potential allowed by the circulation. E. T. C. Spooner (Oxford) has independently developed some of the concepts presented in this paper, including an estimate of the hydrothermal flow rate. This research was supported by the Earth Sciences Section, National Science Foundation, NSF Grant DES-74-22338.

APPENDIX

CONDUCTIVE HEAT FLOW MODEL FOR A SPREADING LITHOSPHERE

The constant heat-flux model for the spreading oceanic lithosphere (Sleep 1974) requires solution of the heat equation,

$$\frac{u\,\partial T}{\partial x} = \kappa \nabla^2 T, \qquad \kappa = \frac{K}{\rho c}, \tag{A1}$$

using the following three boundary conditions: (1) The temperature is zero at the rock-water interface,

$$T = 0 \qquad \text{at } z = 0; \tag{A2}$$

(2) the temperature at some depth λ is constant (T_λ),

$$T = T_\lambda \qquad \text{at } z = \lambda; \tag{A3}$$

(3) heat brought up in dikes at the plane of the ridge axis balances the heat carried away by conduction and convection,

$$S \equiv \frac{-K\,\partial T}{\partial x} + u\rho c\left(T - \frac{T_\lambda z}{\lambda}\right)$$

$$= \frac{\gamma T_\lambda(\lambda - z)u\rho c}{\lambda} \qquad \text{at } x = 0. \tag{A4}$$

The third boundary condition replaces the less realistic boundary condition of McKenzie (1967) which required the plane of the ridge axis to be a plane of constant temperature:

$$T = T_\lambda \qquad \text{at } x = 0. \tag{A5}$$

The solution which satisfies the three boundary conditions (A2–A4) (Sleep 1974) is:

$$T = \frac{T_\lambda z}{\lambda} + \left(\frac{2T_\lambda \gamma}{\pi}\right) \sum_{m=1}^{\infty} A_m \sin\left(\frac{m\pi z}{\lambda}\right) \times \frac{\exp(a_m x)}{m}, \quad \text{(A6)}$$

where

$$A_m \equiv 2[1 + (1 + R^2m^2)^{1/2}]^{-1}, \quad \text{(A7)}$$

$$a_m \equiv \frac{u}{2\kappa}[1 - (1 + R^2m^2)^{1/2}], \quad \text{(A8)}$$

$$R \equiv \frac{2\kappa\pi}{u\lambda}. \quad \text{(A9)}$$

The surface heat flow is obtained by substituting equation (A6) into the conductive heat equation,

$$q = \frac{K\,\partial T}{\partial z}, \quad \text{(A10)}$$

and evaluating the resulting expression at $z = 0$.

$$q = \frac{KT_\lambda}{\lambda} + \left(\frac{2\gamma KT_\lambda}{\lambda}\right) \sum_{m=1}^{\infty} A_m \exp(a_m x) = q_\infty + q'. \quad \text{(A11)}$$

Here q_∞ is the constant term which the heat flow q approaches as x approaches infinity. It is also identical with the heat flow which occurs if $u = 0$, that is, when there is no sea-floor spreading. The x-dependent term q' is the heat flow component which occurs because of the thermal processes associated with the spreading action.

It is desired to find the integral of the heat flow over the time interval from $t = 0$ to $t = t_c$, where t_c is some age of the oceanic crust, such as that when the theoretical and observed heat flow curves have converged within the error associated with the heat flow measurements. Thus,

$$\int_0^{t_c} dt\, q = \int_0^{t_c} dt\, q_\infty + \int_0^{t_c} dt\, q' = t_c q_\infty + \int_0^{t_c} dt \left[\frac{2\gamma KT_\lambda}{\lambda} \sum_{m=1}^{\infty} A_m \exp(a_m ut)\right]. \quad \text{(A12)}$$

The remaining integral in eq. (A12) may be integrated term-by-term to yield the expression:

$$\int_0^{t_c} dt\, q' = \frac{8\gamma q_\infty \kappa}{(u^2R^2)} \times \left[\frac{\pi^2}{6} - \sum_{m=1}^{\infty} \frac{\exp(a_m ut)}{m^2}\right], \quad \text{(A13)}$$

where use has been made of the relation $\sum_{m=1}^{\infty} 1/m^2 = \pi^2/6$. Eq. (A13) may be substituted back into eq. (A12) to yield:

$$\int_0^{t_c} dt\, q = \frac{KT_\lambda t_c}{\lambda} + (2\gamma\, KT_\lambda \rho c\lambda) \times \left[\frac{1}{6} - \frac{1}{\pi^2} \sum_{m=1}^{\infty} \frac{\exp(a_m ut_c)}{m^2}\right], \quad \text{(A14)}$$

where R and q_∞ have been replaced by the expressions which define them for simplification.

Two simple closed expressions may be obtained from the above analysis. First, the total excess heat per unit area of new oceanic crust is given by:

$$\int_0^{\infty} dt\, q' = \frac{\gamma T_\lambda \rho c\lambda}{3}. \quad \text{(A15)}$$

A similar expression gives the excess heat per unit length of mid-ocean ridge:

$$2\int_0^{\infty} dx\, q' = \frac{2\gamma T_\lambda \rho cu\lambda}{3}. \quad \text{(A16)}$$

REFERENCES CITED

Anderson, A. T., 1974, Chlorine, sulfur, and water in magmas and oceans: Geol. Soc. America Bull., v. 85, p. 1485–1492.

——— 1975, Some basaltic and andesitic gases: Rev. Geophysics and Space Physics, v. 13, p. 37–56.

Anderson, R. N., 1972, Petrologic significance of low heat flow on the flanks of slow-spreading mid-ocean ridges: Geol. Soc. America Bull., v. 83, p. 2947–2956.

———, and Hobart, M. A., 1976, The relation between heat flow and age in the southeastern Pacific: Jour. Geophys. Research (in press).

Armstrong, R., 1971, Isotopic and chemical constraints on models of magma genesis in volcanic arcs: Earth and Planetary Sci. Letters, v. 12, p. 137–142.

Arnorsson, S., 1970, Underground temperatures in hydrothermal areas in Iceland as deduced from the silica content of the thermal water: Geothermics (spec. issue 2), v. 2, pt. 1, p. 536–541.

——— 1975, Application of the silica geothermometer in low temperature hydrothermal areas in Iceland: Am. Jour. Sci., v. 275, p. 763–784.

Aumento, F., and Loncarevic, B. D., 1969, The Mid-Atlantic Ridge near 45°N. III. Bald Mountain: Canadian Jour. Earth Sci., v. 6, p. 11–23.

Barrett, D. C., and Aumento, F., 1970, The Mid-Atlantic Ridge near 45°N. XI. Seismic velocity, density, and layering of the crust: Canadian Jour. Earth Sci., v. 7, p. 1117–1124.

Bischoff, J. L., and Dickson, F. W., 1975, Sea water-basalt interaction at 200°C and 500 bars: implications for origin of sea-floor heavy-metal deposits and regulation of sea water chemistry: Earth and Planetary Sci. Letters, v. 25, p. 385–397.

Bjornsson, S.; Arnorsson, S.; and Tomasson, J., 1972, Economic evaluation of Reykjanes Thermal Brine Area, Iceland: Am. Assoc. Petroleum Geologists Bull., v. 56, p. 2380–2391.

Bonatti, E.; Honnorez, J.; Kirst, P.; and Radicati, F., 1975, Metagabbros from the Mid-Atlantic Ridge at 06°N: contact-hydrothermal-dynamic metamorphism beneath the axial valley: Jour. Geology, v. 83, p. 61–78.

Bostrom, K.; Farquharson, B.; and Eyl, W., 1971, Submarine hot springs as a source of active ridge sediments: Chem. Geology, v. 10, p. 189–203.

Broecker, W. S., 1970, A boundary condition on the evolution of atmospheric oxygen: Jour. Geophys. Research, v. 75, p. 3553–3557.

Buddemeier, R. W., and Pucetti, A., 1974, C-14 dilution estimates of volcanic CO_2 emission rates (abs.): Am. Geophys. Union, Trans., v. 55, p. 488.

Burst, J. F., 1969, Diagenesis of Gulf Coast clayey sediments and its possible relation to petroleum migration: Am. Assoc. Petroleum Geologists Bull., v. 53, p. 73–93.

Cadle, R. D., 1975, Volcanic emissions of halides and sulfur compounds to the troposphere and stratosphere: Jour. Geophys. Research, v. 80, p. 1650–1652.

Calvert, S. E., 1968, Silica balance in the ocean and diagenesis: Nature, v. 219, p. 919–920.

Cann, J. R., 1969, Spilites from the Carlsberg Ridge, Indian Ocean: Jour. Petrology, v. 10, p. 1–19.

——— 1971, Petrology of basement rocks from Palmer Ridge, NE Atlantic: Royal Soc. (London) Philos. Trans., ser. A, v. 268, p. 605–617.

———, and Vine, F. J., 1966, An area on the crest of the Carlsberg Ridge: petrology and magnetic survey: Royal Soc. (London) Philos. Trans., ser. A, v. 259, p. 198–217.

Carmichael, I. S. E., and Nicholls, J., 1967, Iron-titanium oxides and oxygen fugacities in volcanic rocks: Jour. Geophys. Research, v. 72, p. 4665–4687.

Carswell, D. A.; Curtis, C. D.; and Kanaris-Sotirious, R., 1974, Vein metasomatism in peridotite at Kalskaret near Tafjord, South Norway: Jour. Petrology, v. 15, p. 383–402.

Chase, C. G., 1972, The N plate problem of plate tectonics: Geophys. Jour. Royal Astron. Soc., v. 29, p. 117–122.

Chayes, F., 1969, The chemical composition of Cenozoic andesite, *in* McBirney, A., ed., Andesite Conf., Proc., Bull. 65, Oregon Dept. Geology and Mineral Industries, p. 1–9.

Cloud, P., 1972, A working model of the primitive earth: Am. Jour. Sci., v. 272, p. 537–548.

Corliss, J. G., 1970, Mid-ocean ridge basalts: 1. The origin of submarine hydrothermal solutions. 2. Regional diversity along the Mid-Atlantic Ridge: Unpub. Ph.D. dissert., University of California, San Diego.

——— 1971, The origin of metal-bearing submarine hydrothermal solution: Jour. Geophys. Research, v. 76, p. 8128–8138.

Degens, E. T., 1969, Biogeochemistry of stable carbon isotopes, *in* Eglinton, G., and Murphy, M. T. J., eds., Organic geochemistry: New York, Springer-Verlag, p. 304–328.

Dickinson, W. R., 1973, Widths of modern arc-trench gaps proportional to past duration of igneous activity in associated magmatic arcs: Jour. Geophys. Research, v. 78, p. 3376–3389.

Drever, J. I., 1974, The magnesium problem, *in* Goldberg, E. D., ed., The sea: New York, Wiley, v. 5, p. 337–357.

Dymond, J.; Corliss, J. B.; Heath, G. R.; Field, G. W.; Dasch, E. J.; and Veeh, H. H., 1973, Origin of metalliferous sediments from the Pacific Ocean: Geol. Soc. America Bull., v. 84, p. 3355–3372.

Einarsson, P.; Klein, F.; and Wyss, M. 1973, Microearthquakes on the Mid-Atlantic plate boundary in the Reykjanes Peninsula, Iceland (Abs.): Am. Geophys. Union, Trans., v. 54, p. 324.

Elderfield, H., 1972, Effects of volcanism on water chemistry, Deception Island, Antarctica: Marine Geology, v. 13, p. M1–M6.

ELLIS, A. J., 1968, Natural hydrothermal systems and experimental hot-water/rock interaction: reactions with NaCl solutions and trace metal extraction: Geochim. et Cosmochim. Acta, v. 32, p. 1356–1363.

EUGSTER, H. P., 1972, Reduction and oxidation in metamorphism (II): Internat. Geol. Congress, 24th, sec. 10, Gardenvale, Quebec, Harper's Press Cooperative, p. 3–11.

———, and SKIPPEN, G. B., 1967, Igneous and metamorphic reactions involving gas equilibria, *in* ABELSON, Ph., ed., Researches in geochemistry: New York, Wiley, v. 2, p. 492–520.

FANNING, K. A.; BETZER, P. R.; BOLGER, G. W.; MILLER, G. R.; McGREGOR, B. A.; and RONA, P. A., 1974, Silica and fluoride over the TAG hydrothermal field (Abs.): Am. Geophys. Union, Trans., v. 55, p. 293.

FEHN, U.; SIEGEL, M. D.; WILLIAMS, D. L.; and HOLLAND, H. D., 1975, Water temperatures in the FAMOUS area (Abs.): Am. Geophys. Union, Trans., v. 56, p. 376.

FYFE, W. S., 1974, Heats of chemical reactions and submarine heat production: Geophys. Jour. Royal Astron. Soc., v. 37, p. 213–215.

GARFUNKEL, Z., 1975, Growth, shrinking, and long-term evolution of plates, and their implications for the flow pattern in the mantle: Jour. Geophys. Research, v. 80, p. 4425–4432.

GARNER, D. M., and FORD, W. M., 1969, The Mid-Atlantic Ridge near 45°N. IV. Water properties in the median valley: Canadian Jour. Earth Sci., v. 6, p. 1359–1363.

GARRELS, R. M., and MACKENZIE, F. T., 1971, Evolution of sedimentary rocks: New York, Norton, p. 397.

———, — ——— 1972, A quantitative model for the sedimentary rock cycle: Marine Chemistry, v. 1, p. 27–41.

———, and PERRY, E. A., JR., 1974, Cycling of carbon, sulfur, and oxygen through geologic time, *in* GOLDBERG, E. D., ed., The sea: New York, Wiley-Interscience, v. 5, p. 303–336.

GEALY, E. L., 1971, Carbon-carbonate content of sediments from the western equatorial Pacific: Leg 7, *in* WINTERER, E. L. et al., Initial reports of the Deep-Sea Drilling Project: Washington, U.S. Govt. Printing Office, v. 7, p. 845–862.

Gregor, B., 1971, Carbon and atmospheric oxygen: Science, v. 174, p. 316–317.

HAJASH, A., JR., 1975, Hydrothermal processes along mid-ocean ridges: an experimental investigation: Unpub. Ph.D. dissert., Texas A & M University, College Station, Texas.

HARRISS, R. C., 1966, Biological buffering of oceanic silica: Nature, v. 212, p. 275–276.

HART, R., 1973, A model for chemical exchange in the basalt-seawater system of oceanic layer II: Canadian Jour. Earth Sci., v. 10, p. 799–816.

HART, S. R.; ERLANK, A. J.; and KABLE, E. J. D., 1974, Sea floor basalt alteration: some chemical and strontium isotope effects: Contr. Mineralogy and Petrology, v. 44, p. 219–230.

HYNDMAN, R. D., and RANKIN, S. J., 1972, The Mid-Atlantic Ridge near 45°N. XVII. Heat flow measurements: Canadian Jour. Earth Sci., v. 9, p. 664–670.

IRVING, E., 1970, The Mid-Atlantic Ridge at 45°N. XIV. Oxidation and magnetic properties of basalt; review and discussion: Canadian Jour. Earth Sci., v. 7, p. 1528–1538.

———, PARK, J. K., HAGGERTY, S. E., AUMENTO, F., and LONCAREVIC, B., 1970, Magnetism and opaque mineralogy of basalts from the Mid-Atlantic Ridge at 45°N: Nature, v. 228, p. 974–976.

ISACKS, B.; OLIVER, J.; and SYKES, L. R., 1968, Seismology and the new global tectonics: Jour. Geophys. Research, v. 73, p. 5585–5899.

KARIG, D. E., and SHARMAN, G. F., III, 1975, Subduction and accretion in trenches: Geol. Soc. America Bull., v. 86, p. 377–389.

KAY, R.; HUBBARD, N.; and GAST, P., 1970, Chemical characteristics and the origin of oceanic ridge volcanic rocks: Jour. Geophys. Research, v. 75, p. 1585–1613.

LE PICHON, X., and LANGSETH, M. G., 1969, Heat flow from the mid-ocean ridges and sea floor spreading: Tectonophysics, v. 8, p. 319–344.

LI, Y.-H., 1972, Geochemical mass balance among lithosphere, hydrosphere, and atmosphere: Am. Jour. Sci., v. 272, p. 119–137.

LISTER, C. R. B., 1972, On the thermal balance of a mid-ocean ridge: Geophys. Jour. Royal Astron. Soc., v. 26, p. 515–535.

———, 1974, On the penetration of water into hot rock: Geophys. Jour. Royal Astron. Soc., v. 39, p. 465–509.

MACKENZIE, F. T., 1976, Sedimentary cycling of global processes, *in* GOLDBERG, E. D., ed., The sea: New York, Wiley-Interscience, v. 6, (in press).

McKENZIE, D. P., 1967, Some remarks on heat flow and gravity anomalies: Jour. Geophys. Research, v. 72, p. 6261–6273.

MELSON, W. G., and VAN ANDEL, TJ. H., 1966, Metamorphism in the Mid-Atlantic Ridge, 22°N latitude: Marine Geology, v. 4, p. 165–186.

———; THOMPSON, G.; and VAN ANDEL, TJ. H., 1968, Volcanism and metamorphism in the Mid-Atlantic Ridge, 22°N latitude: Jour. Geophys. Research, v. 73, p. 5925–5941.

———, and scientific staff, 1974, Deep Sea Drilling Project, Leg 37, the volcanic layer: Geotimes, v. 19, p. 16–18.

MIYASHIRO, A.; FUMIKO, S.; and EWING, M., 1969, Diversity and origin of abyssal tholeiite from the Mid-Atlantic Ridge near 24° and 30°N latitude: Contr. Mineralogy and Petrology, v. 23, p. 38–52.

———; SHIDO, F.; and EWING, M., 1971, Metamorphism in the Mid-Atlantic Ridge near 24° and 30°N: Royal Soc. (London) Philos. Trans., ser. A: v. 268, p. 589–603.

MOTTL, M. J.; CORR, R. F.; and HOLLAND, H. D., 1974, Chemical exchange between sea water and mid-ocean ridge basalt during hydrothermal alteration: an experimental study (Abs.): Geol. Soc. America Ann. Mtgs., p. 879–880.

———; ———; — ———, 1975, Trace element content of the Reykjanes and Svartsengi thermal brines, Iceland (Abs.): Geol. Soc. America Ann. Mtgs., p. 1206–1207.

MUEHLENBACHS, K., 1975*a*, Oxygen isotope geochemistry of DSDP Leg 34 basalts; *in* Initial reports of the Deep-Sea Drilling Project, v. 34 (in press).

———, 1975*b*, Oxygen isotope geochemistry of DSDP Leg 37 rocks, *in* Initial reports of the Deep-Sea Drilling Project, v. 37 (in press).

———, and CLAYTON, R. N., 1972, oxygen isotope geochemistry of submarine greenstones: Canadian Jour. Earth Sci., v. 9, p. 471–478.

NIGRINI, A., 1969, Prediction of ionic fluxes in rock alteration processes at elevated temperatures: Unpub. Ph.D. dissert., Northwestern University, Evanston, Ill.

OVERSBY, V., and GAST, P., 1970, Isotopic composition of lead from oceanic islands: Jour. Geophys. Research, v. 75, p. 2097–2114.

———, and EWART, A., 1972, Lead isotopic compositions of Tonga-Kermadec volcanics and their petrogenetic significance: Contr. Mineralogy Petrology, v. 37, p. 181–210.

PETERSON, J. J.; FOX, P. J.; and SCHREIBER, E., 1974, Newfoundland ophiolites and the geology of the oceanic layer: Nature, v. 247, p. 194–196.

PIMM, A. C., 1971, Carbon carbonate results: *in* FISCHER, A. G., et al., Initial reports of the Deep-Sea Drilling Project, U.S. Govt. Printing Office, Washington, v. 6, p. 739–752.

PIPER, D. Z., 1973, Origin of metalliferous sediments from the East Pacific Rise: Earth and Planetary Sci. Letters, v. 19, p. 75–82.

POLYAK, B. G., and KONONOV, V. I., 1974, Geoenergetic zoning of hydrothermal activity in Iceland: Doklady Akademii Nauk SSSR, v. 216, p. 1364–1367.

RIEKE, H. H., III, and CHILINGARIAN, G. V., 1974, Compaction of argillaceous sediments: Elsevier Scientific Publ. Co., Amsterdam, 424 p.

RILEY, J. P., and CHESTER, R., 1971, Introduction to marine chemistry: Academic Press, London and New York, 465 p.

ROBERTSON, J. K., and WYLLIE, P. J., 1971, Rock-water systems, with special reference to the water-deficient region: Am. Jour. Sci., v. 271, p. 252–277.

ROBIE, R. A., and WALDBAUM, D. R., 1968, Thermodynamic properties of minerals and related substances: U.S. Geol. Survey Bull. No. 1259.

RONOV, A. B., 1958, Organic carbon in sedimentary rocks (in relation to the presence of petroleum): Geochemistry, p. 510–536.

SALES, R. H., and MEYER, C., 1948, Wall rock alteration at Butte, Montana: Am. Inst. Mining Metall., Engineers Trans. v. 178, p. 9–35.

SATO, M., and WRIGHT, T. L., 1966, Oxygen fugacities directly measured in magmatic gases: Science, v. 153, p. 1103–1105.

SCHOLL, D. W., and MARLOW, M. S., 1974, Sedimentary sequence in modern Pacific trenches and the deformed circum-Pacific eugeosyncline: *in* Modern and Ancient Geosynclinal Sedimentation, Soc. Econ. Paleontologists Mineralogists Spec. Publ. 19, p. 193–211.

SCLATER, J. G., and FRANCHETEAU, J., 1970, The implications of terrestrial heat flow observations on current tectonic and geochemical models of the crust and upper mantle of the earth: Geophysics J. R. Astr. Soc., v. 20, p. 509–542.

SCLATER, J. G.; VON HERZEN, R. P.; WILLIAMS, D. L.; ANDERSON, R. N., and KLITGORD, K., 1974, The Galapagos spreading center: heat flow on the north flank: Geophysics Jour. Royal Astron. Soc., v. 38, p. 609–626.

SCOTT, R. B.; RONA, P. A.; MCGREGOR, B. A., and SCOTT, M. R., 1974, The TAG hydrothermal field: Nature, v. 251, p. 301–302.

———; MALPAS, J.; UNDINTSEV, G.; and RONA, P. A., 1975, Submarine hydrothermal activity and sea-floor spreading at 26°N, MAR (Abs.): Geol. Soc. America Ann. Mtgs., p. 1263.

SHAW, H. R., and WONES, D. R., 1964, Fugacity coefficients for hydrogen gas between 0° and 1000°C for pressures to 3000 atmospheres: Am. Jour. Sci., v. 262, p. 918–929.

SHOR, G. G., JR.; MENARD, H. W.; and RAITT, R. W., 1970, Structure of the Pacific basin: *in* MAXWELL, A. E., ed., The sea, Wiley-Interscience, New York, v. 4, part II, p. 3–27.

SLEEP, N. H. 1974, Segregation of magma from a mostly crystalline mush: Geol. Soc. America Bull., v. 85, p. 1225–1232.

——— 1975, Formation of ocean crust: some thermal constraints: Jour. Geophys. Research, v. 80, p. 4037–4042.

SLOSS, L. L., 1963, Sequences in the cratonic interior of North America: Geol. Soc. America Bull., v. 74, p. 93–114.

SPOONER, E. T. C., and FYFE, W. S., 1973, Sub-sea-floor metamorphism, heat and mass transfer: Contr. Mineralogy and Petrology, v. 42, p. 287–304.

———; BECKINSALE, R. D.; FYFE, W. S.; and SMEWING, J. D., 1974, 0^{18} enriched ophiolitic metabasic rocks from E. Liguria (Italy), Pindos (Greece), and Troodos (Cyprus): Contr. Mineralogy and Petrology, v. 47, p. 41–62.

SYKES, L. R., 1971, Earthquake swarms and sea-floor spreading: Jour. Geophys. Research v. 75, p. 6598–6611.

TALWANI, M.; WINDISH, C. C.; and LANGSETH, M. G., JR., 1971, Reykjanes Ridge crest: a detailed geophysical study: Jour. Geophys. Research v. 76, p. 473–517.

THOMPSON, G.; WOO, C. C., and SUNG, W., 1975, Metalliferous deposits on the Mid-Atlantic Ridge (Abs.): Geol. Soc. America Ann. Mtgs., p. 1297.

TOMASSON, J., and KRISTMANNSDOTTIR, H., 1972, High temperature alteration minerals and thermal brines, Reykjanes, Iceland: Contr. Mineralogy and Petrology v. 36, p. 123–134.

VON DER BORCH, C. C.; NESTEROFF, W. D.; and GALEHOUSE, J. S., 1971, Iron-rich sediments cored during Leg 8 of the Deep Sea Drilling Project: *in* Initial Reports of the Deep Sea Drilling Project, v. 8, p. 829–836.

———, and REX, R. W., 1970, Amorphous iron oxide precipitates in sediments cored during Leg 5, Deep Sea Drilling Project: *in* Initial Reports of the Deep Sea Drilling Project, v. 5, p. 541–544.

VON HERZEN, R. P., and ANDERSON, R. N., 1972, Implications of heat flow and bottom water temperature in the eastern equatorial Pacific: Geophys. Jour. Royal Astron. Soc., v. 26, p. 427–458.

WALKER, J. C. G., 1974, Stability of atmospheric oxygen: Am. Jour. Sci., v. 274, p. 193–214.

WARD, P. L., and BJORNSSON, S., 1971, Microearthquakes, swarms, and the geothermal areas of Iceland: Jour Geophys. Research v. 76, p. 3953–3982.

WATKINS, N. D., and HAGGERTY, S. E., 1967, Primary oxidation variation and petrogenesis in a single lava: Contr. Mineralogy and Petrology, v. 15, p. 251–271.

WILLIAMS, D. L., and VON HERZEN, R. P., 1974, Heat loss from the earth: new estimate: Geology, v. 2, p. 327–328.

———; VON HERZEN, R. P.; SCLATER, J. G.; and ANDERSON, R. N., 1974, The Galapagos spreading center: lithospheric cooling and hydrothermal circulation: Geophys. Jour. Royal Astron. Soc., v. 38, p. 587–608.

WISE, D. U., 1972, Freeboard of continents through time: Geol. Soc. America Memoir 132, p. 87–100.

WOLLAST, R., 1974, The silica problem: *in* GOLDBERG, E. D., ed. The sea, New York, v. 5, p. 359–392.

Wyllie, P. J., 1971, The dynamic earth: John Wiley and Sons, New York.

31

Reprinted from *Econ. Geology* 73:135-160 (1978)

Criteria for Recognition of Hydrothermal Mineral Deposits in Oceanic Crust

PETER A. RONA

Abstract

Hydrothermal mineral deposits in oceanic crust include metalliferous sediments and encrustations and massive sulfides. The occurrence of these deposits is explained by the hypothesis that they are concentrated at the discharge zones of high intensity subsea-floor hydrothermal convection systems involving the circulation of sea water through oceanic crust and upper mantle at oceanic spreading centers.

Criteria that have proven most useful for recognition of hydrothermal deposits in oceanic crust based on known deposits at oceanic spreading centers include petrology of the deposits and surrounding rocks; structural conditions that create exceptionally high permeability and thermal gradients; seismicity in the form of microearthquakes and earthquake swarms; geochemical properties of hydrothermal discharge (^{3}He, ^{222}Rn, ferric hydroxides, silica); contrasts in acoustic impedence between normal sea water and hydrothermal solutions; anomalous gravity as an indicator of geologic structure; electrical properties of the hydrothermal solutions and deposits; an associated low in residual magnetic intensity attributed to hydrothermal alteration of the magnetic mineral component of basalt; patterns of deposition in early rift and advanced oceanic ridge stages of opening of an ocean basin; and distribution of hydrothermal deposits both parallel and perpendicular to an oceanic spreading center.

The seismic, geochemical, acoustic, thermal, and certain electrical criteria are applicable to recognition of active discharge zones of subsea-floor hydrothermal convection systems. The petrologic, structural, gravity, magnetic, and other electrical criteria are applicable to recognition of hydrothermal deposits in all of oceanic crust which underlies ocean basins covering two-thirds of the Earth and is emplaced on land as ophiolites.

Introduction

OCEANIC crust is the most widespread crustal type on Earth, underlying the ocean basins that cover two-thirds of the Earth and also tectonically emplaced in certain islands and continents as ophiolites (Fig. 1). The metallic mineral potential of oceanic crust other than in manganese nodules is poorly known, and its evaluation will require more exploration at sea and on land (Rona, 1977). This paper endeavors to develop criteria to guide exploration for hydrothermal mineral deposits in oceanic crust based on recent work by both oceanographers and students of ore deposits (Rona, 1976a). The deposits under consideration comprise metalliferous sediments, metalliferous encrustations, and massive sulfide bodies that share a common origin as precipitates from hydrothermal solutions in regions of volcanism and high heat flux at oceanic spreading centers. The hydrothermal deposits are distinguished from hydrogenous deposits which include metalliferous sediments, encrustations, and nodules that were precipitated from normal sea water over wide areas of ocean basins (Table 1).

The exploration criteria are based on geological, geochemical, and geophysical characteristics of hydrothermal mineral deposits where these are actually forming at oceanic spreading centers. Oceanic spreading centers are the loci of generation of all oceanic crust. The characteristics of oceanic spreading centers are related to their tectonic settings which, in turn, influence the occurrence of associated hydrothermal deposits. One evolutionary sequence of tec-

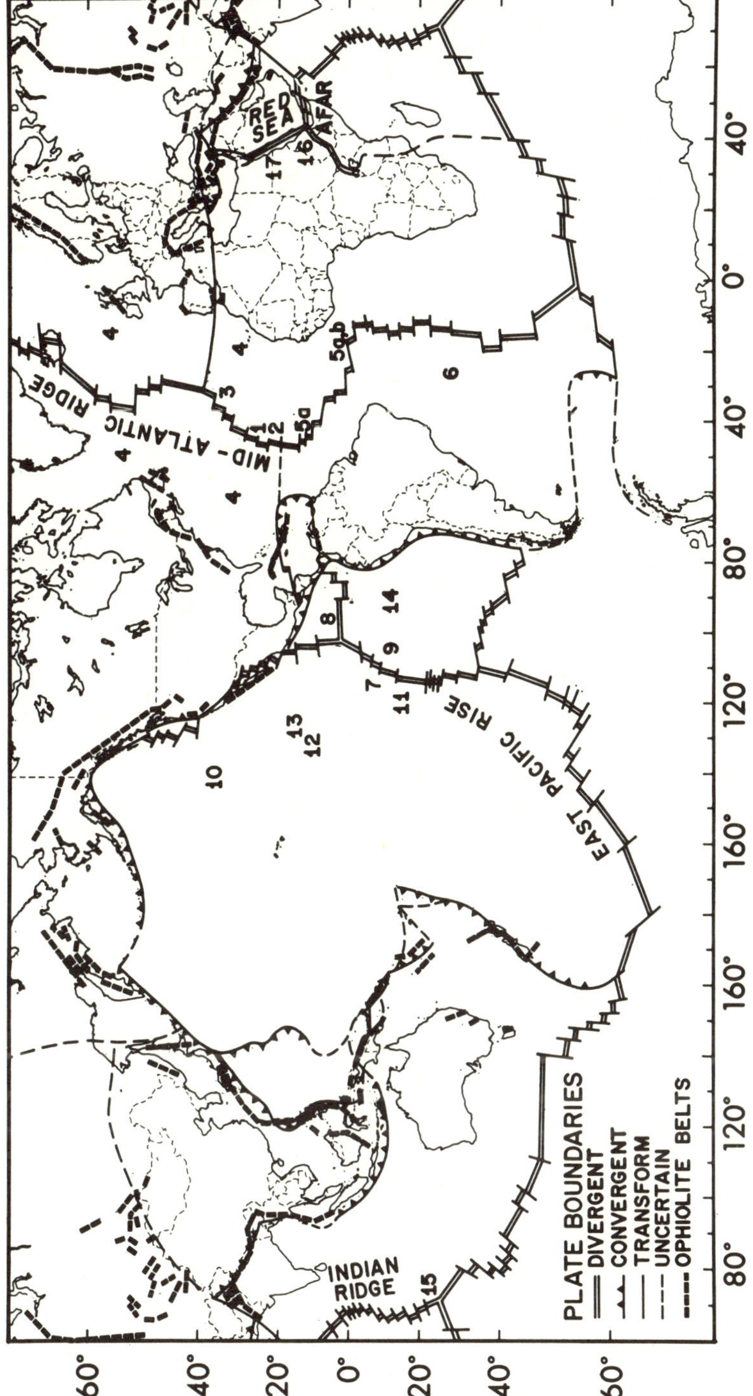

FIG. 1. World map showing lithospheric plate boundaries and location of known hydrothermal deposits in ocean basins. Numbers refer to locations described in Table 2. Ophiolite belts are also shown (after Coleman, 1971, 1977).

TABLE 1. Comparison of Hydrothermal and Hydrogenous Metalliferous Deposits (Bonatti et al., 1972b; Table 2)

	Hydrothermal		Hydrogenous		
	Sediment	Encrustation	Sediment	Encrustation	Nodule
Stratigraphic position	Overlies basalt	Overlies basalt	Present throughout sedimentary column	Overlies bedrock	Overlies sediment
Fe/Mn ratio	<0.1 and >1	Fe-rich: 12–237 Mn-rich: 0.002–0.14	0.5 to 5	≃1	≃1
Trace metal content	Low (see Fig. 4; Table 2)	Low (see Fig. 4; Table 2)	High (see Fig. 4)	High (see Fig. 4)	High (see Fig. 4)
U/Th content	>1	>1	<1	<1	<1
Accumulation rate	Fast	100 to 1,000 mm per 10^6 yr	Slow	1 to 10 mm per 10^6 yr	1 to 10 mm per 10^6 yr

tonic settings involves stages in the development of an ocean basin about a spreading center from a rift zone, exemplified by the axial trough of the Red Sea, to an oceanic ridge, exemplified by the Mid-Atlantic Ridge. Another evolutionary sequence involves the opening of a basin marginal to a continent about a spreading center associated with a volcanic island-arc system, exemplified by the marginal basins of the western Pacific (Karig, 1971).

Metalliferous sediments and encrustations of hydrothermal origin are known from a number of locations at oceanic spreading centers (Fig. 1, Table 2). The metalliferous sediments occur at the base of the sedimentary layer (layer 1) and the encrustations at the top of the basaltic layer (layer 2) of oceanic crust (Figs. 2, 3). Disseminated sulfides, particularly pyrite and chalcopyrite, are not uncommon in altered basalt of layer 2 (Figs. 2, 3; Table 2, location 5a; Dmitriev et al., 1970; Moore and Calk, 1971; Bonatti et al., 1976a; Czamanske and Moore, 1977). Massive sulfide deposits have not yet been found at active spreading centers, but their occurrence is predicted based on geochemical considerations and their existence in ophiolites (Pereira and Dixon, 1971; Sillitoe, 1972, 1973; Spooner and Fyfe, 1973; Bonatti, 1975). The massive sulfide deposits occur within tholeiitic pillow lava or altered basalt corresponding to oceanic crustal layer 2 of certain ophiolites (Figs. 2, 3); gold and silver may be associated with the sulfides (Keays and Scott, 1976). The basalt of these ophiolites may be overlain by metalliferous encrustations and sediments, like those present at oceanic spreading centers. Examples include the ophiolites of the Troodos Massif, Cyprus (Wilson and Ingham, 1959; Moores and Vine, 1971; Searle, 1972; Constantinou and Govett, 1973; Miyashiro, 1973; Spooner et al., 1974); the Semail region of Oman (Bailey and Coleman, 1975); the northern Apennines (Spooner et al., 1974; Ferrara et al., 1976; Bonatti et al., 1976c); and Newfoundland

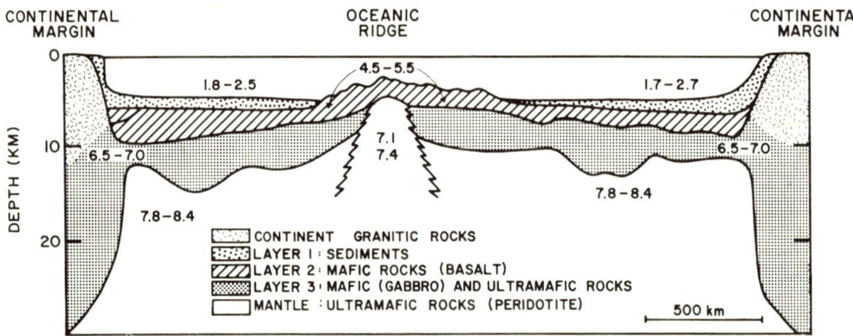

FIG. 2. A schematic section across an opening ocean basin like the Atlantic to show layers 1, 2, 3 of oceanic crust and mantle labeled with characteristic ranges of compressional wave seismic velocities in kilometers per second (Ewing, 1969). The actual structure and lithology of oceanic crust are considerably more complex (Peterson et al., 1974; Deep Sea Drilling Staff, 1977).

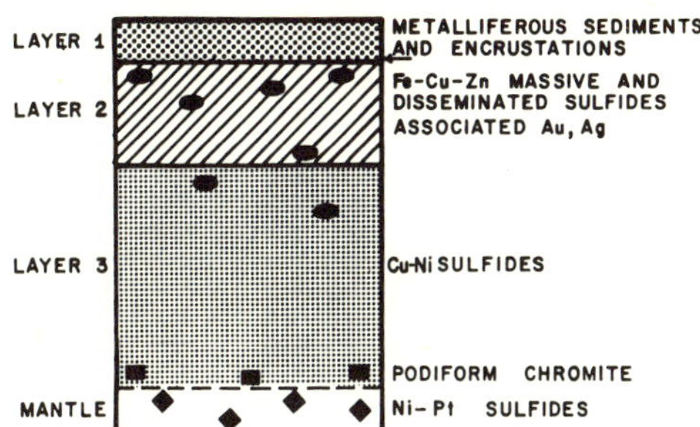

Fig. 3. Schematic section of oceanic crust and upper mantle showing potential distribution of metal deposits.

(Upadhyay and Strong, 1973; Strong, 1974a; Duke and Hutchinson, 1974). Other examples of hydrothermal ore deposits that may have formed at oceanic spreading centers are considered by Sillitoe (1972), Hutchinson (1973), Mitchell and Bell (1973), Mitchell and Garson (1976), and Coleman (1977).

The oceanic crust and incorporated hydrothermal deposits in ophiolites originated at spreading centers located at rift zones, oceanic ridges, and marginal basins associated with island arcs. Systematic compositional (Engel et al., 1965; Miyashiro, 1975) and structural (Dewey and Bird, 1971) differences in the volcanic rock series of the ophiolites may indicate the tectonic setting of their generation. Metallic deposits in addition to those under consideration are known from ophiolites, including podiform chromite (Woodli, 1964; Peters and Kramers, 1974), as well as nickel and platinum sulfide that occur near the base of layer 3 and in the upper mantle (Figs. 2, 3). The deposits of layer 3 and the upper mantle are outside the scope of this paper.

Subsea-Floor Hydrothermal Convection Systems

The effectiveness of criteria employed to search for mineral deposits depends on the validity of the hypothetical basis for the criteria. For example, the anticlinal hypothesis of oil entrapment was the basis for exploration criteria that resulted in discovery of 237 of the 266 known giant oil fields of the world (Halbouty et al., 1970). Analogously, the effectiveness of criteria to locate hydrothermal mineral deposits in oceanic crust depends on the hypothesis employed to account for concentration of the deposits.

Various lines of evidence support the hypothesis that subsea-floor convection systems concentrate hydrothermal minerals in oceanic crust (Spooner and Fyfe, 1973; Hutchinson, 1973; Sillitoe, 1973; Bonatti, 1975). A hydrothermal convection system is composed of a heat source, a recharge system, a circulation system, and a surface discharge system (Elder, 1965). Emplacement of basaltic magma at oceanic spreading centers provides a supply of heat and metals. Cooling of emplaced materials and tectonic processes operating at oceanic spreading centers result in fracture and uplift of rocks in configurations conducive to development of subsea-floor hydrothermal convection (Deffayes, 1970; Lister, 1972, 1974b; Rona et al., 1976).

A hypothetical high-intensity, subsea-floor hydrothermal system involves the following stages (Krauskopf, 1956, 1957; Helgeson, 1964, 1969; Boström and Peterson, 1966; Corliss, 1971; Boström, 1973; Hart, 1973): downward penetration of cold, dense sea water primarily through fractures in oceanic crustal layer 2 and the upper portion of layer 3, as deep as about 5 km beneath the ocean bottom (Fig. 2); progressive warming and reduction as the sea water reacts with a basaltic mineralogy recently cooled from a magmatic state; production of a sequence of metamorphic phases increasing in grade with increasing temperature and depth; high-temperature leaching and mass transfer of transition metals, Si, Ca, K, H, and S, primarily from oceanic crust, and loss of Mg, Na, and O from sea water to oceanic crust; transportation of metals and reduced S species in solution as chloride complexes; possible addition of certain elements transported in volatile phases from the mantle (He^3, F, Hg, S, B, Ba) and injected via the hydrothermal system, either directly or indirectly, by dissolution from basalt; ascent of warm, less dense, enriched solutions through gradients of decreasing temperature and pressure; partial precipitation of disseminated and massive metallic sulfides under reducing conditions in response to decreasing temperature within basalt of layer 2, involving sulfur from crustal (average sulfur content 800 ppm in tholeiitic basalt; Moore and Schilling, 1973), seawater (reduction of SO_4^{2-}), and upper mantle sources; precipitation of metallic oxide encrustations under oxidizing conditions at the basalt-sea water interface; precipitation of amorphous ferric hydroxide particles on cooling, oxidation, and increasing pH of the discharged solution during mixing with oxygenated sea water or, alternatively, precipitation of base metal sulfides on cooling of the discharged solution during mixing with oxygen-deficient sea water; scavenging of heavy metals remaining in solution by colloidal absorption on the ferric hydroxides; and finally, flocculation in the electrolyte and settling as metalliferous sediments.

It is important to distinguish between criteria for low- and high-intensity hydrothermal activity. Low-intensity hydrothermal activity is ubiquitous at oceanic spreading centers. It involves low-intensity hydration which produces prograde metamorphic assemblages within oceanic crustal layer 2; this is controlled by temperature rising at the geothermal

TABLE 2. Hydrothermal Deposits

Location number (Fig. 1)	Location	Half-rate of sea-floor spreading (cm per yr)	Structure	Type of deposit	Mineralogy
1	TAG Hydrothermal Field on Mid-Atlantic Ridge crest at latitude 26°N (Scott et al., 1974, tables 1b, 2; Rona et al., 1976)	1.3	Block-faulted topographic high	Encrustation	Birnessite, todorokite
2	Mid-Atlantic Ridge at latitude 23°N (Thompson et al., 1975)	1.4	Block-faulted topographic high	Encrustation	Todorokite
3	Mid-Atlantic Ridge crest at latitude 36°N (ARCYANA, 1975)	1.1	Fracture zone (transform fault)	Encrustation	Birnessite, todorokite, Fe- and Mn-rich smectite
4	North Atlantic Ocean basin between latitudes 20°N and 60°N (Horowitz and Cronan, 1976, table 1; DSDP holes 9A, 10, 112, 114, 117A, 118, 136, 137, 138, 141)	1.1–1.5	Basal sediment (106 m.y. ago to Recent) directly overlying basalt of oceanic crustal layer 2	Sediment[1]	Goethite, iron-rich montmorillonite (smectite), manganese hydroxyoxides
5a	Equatorial Mid-Atlantic Ridge at latitudes 11°N and 0° (Bonatti et al., 1976a; Kirst, 1976)	2.0	Vema and Romanche fracture zones (transform faults)	Disseminated grains and veins in metabasalt (chloritic)	Chalcopyrite, pyrite, pyrrotite
5b	Equatorial Mid-Atlantic Ridge at latitude 0° (Bonatti et al., 1976b, table 1)	2.0	Romanche fracture zone (transform fault)	Concretion	Pyrite, goethite
6	South Atlantic Ocean basin between latitudes 28°S and 31°S (Boström et al., 1972, table IV; DSDP holes 14, 15, 16, 19, 20, 21, 22)	2.0	Basal sediment (Late Cretaceous to Pleistocene age) overlying basalt of oceanic crustal layer 2 on west flank of Mid-Atlantic Ridge	Sediment[1]	Siderite; other mineralogical determinations incomplete
7a	East Pacific Rise at latitude 8°S, Fe-rich fraction (Bonatti and Joensuu, 1966, table 1, fraction a; Veeh and Boström, 1971, table 1)	7.5	Volcanic seamount	Encrustation	Goethite
7b	East Pacific Rise at latitude 8°S, Mn-rich fraction (Bonatti and Joensuu, 1966, table 1, fraction b)	7.5	Volcanic seamount	Encrustation	Not specified
8a	Galapagos spreading center (Moore and Vogt, 1976, table 1)	3.2	Block-faulted topographic high	Encrustation	Birnessite, todorokite
8b	Mounds Hydrothermal Field 20 to 30 km south of axis of Galapagos spreading center (Corliss et al., 1976, 1977; Lonsdale, 1977b)	3.2	Block-faulted topography	Encrustation	Manganese hydroxyoxides
9	Bauer Deep (Sayles and Bischoff, 1973, table 3; Scientific Staff, 1974; Lyle et al., 1977)	—	Oceanic basin	Sediment[1]	Smectite, goethite, todorokite

at Oceanic Spreading Centers

Composition (plotted on Fig. 4)											
Weight percent (range)					ppm (range)						
Si	Al	Fe	Mn	Fe/Mn	Ba	Co	Cu	Ni	Zn	U	Th
—	—	0.01–0.11	38–39	0.0003–0.0027	—	14–25	11–119	50–790	—	9–16	2–5
—	—	Low	High	0.0007	—	—	>1,000	—	—	—	—
—	—	High	High	0.2–237	—	<15	<15	<15	—	—	—
—	4.7–16.1 (Al_2O_3)	2.3–10.5	0.03–0.7	5–130	—		25–167	20–370	62–267	—	—
1.8–5.2	—	32–41	<0.1	>80	—	—	5–50%	—	—	—	—
12–38 (SiO_2)	1.6–7.6 (Al_2O_3)	41–73 (Fe_2O_3)	0.07–0.25 (MnO)	166–884	35–220	27–75	35–590	90–270	82–220	—	—
—	1.6–12.1	1.1–16	0.02–7.3	1–50	—	—	—	—	—	—	—
12–18 (SiO_2)	<1–1 (Al_2O_3)	29–33	0.6–2.4	12–52	100–115	32–120	60–120	90–460	—	1.3	<0.002
8.1 (SiO_2)	0.4 (Al_2O_3)	5.5	38.7	0.14	1,700	290	>500	4,500	—	—	—
—	—	Low	47–58	0.0002–0.0230	—	15–86	9–203	58–500	85–4,860	4–6	0.01–0.4
—	—	—	—	—	—	—	—	—	—	—	—
18–28	1–4	9–18	2–7	2–7	8,000–28,000	90–330	710–1,200	410–1,700	110–680	—	—

TABLE 2—(*Continued*)

Location number (Fig. 1)	Location	Half-rate of sea-floor spreading (cm per yr)	Structure	Type of deposit	Mineralogy
10	Central North Pacific Ocean basin between latitudes 32°N and 41°N (Dymond et al., 1973, tables 2, 6; DSDP sites 37, 38, 39)	—	Basal sediment (32–60 m.y.) overlying basalt of oceanic crustal layer 2	Sediment[1]	Goethite, iron-rich montmorillonite (smectite), manganese hydroxyoxides
11	East Pacific Rise crest between latitudes 12°S and 14°S (Boström and Peterson, 1969, table III; Veeh and Boström, 1971, table 1)	8.0	Oceanic ridge	Sediment[1]	Goethite, iron-rich montmorillonite (smectite), manganese hydroxyoxides
12	East Pacific Rise between latitudes 10°N and 15°N (Cronan, 1973, table 1; DSDP sites 159, 160, 161, 162, 163)	4.5	Basal sediment (Early Campanian to Late Oligocene) overlying basalt of oceanic crustal layer 2 on west flank of oceanic ridge	Sediment[1]	Goethite, iron-rich montmorillonite (smectite), manganese hydroxyoxides
13	Northeast Pacific nodule area (Bischoff and Rosenbauer, 1977, table 2); latitude 15°12.2'N, longitude 126°58.6'W; depth 4,295 m below sea level	—	Abyssal hills between Clarion and Clipperton fracture zones. Metalliferous sediment may be relict	Sediment[1]	Globules averaging 100 μm diameter composed of semi-opaque reddish yellow aggregates and detrital clay minerals of low refractive index, along with minor amounts of radiolarian debris, volcanic glass, and micronodules
14	East Pacific Rise between latitudes 9°S and 13°S (Boström et al., 1976, table 5; DSDP holes 319, 320, 321)	7.5	Basal sediment (Late Oligocene to Quaternary) overlying basalt of oceanic crustal layer 2 on east flank of oceanic ridge	Sediment[1]	Goethite, iron-rich montmorillonite (smectite), manganese hydroxyoxides
15	Indian Ocean Ridge between latitudes 16°S and 40°S (Boström et al., 1969, table 2)	4.0	Ocean ridge	Sediment	Not specified
16a	Afar Rift, Fe-rich deposits (Bonatti et al., 1972a, tables 2, 3)	—	Axial trough at rift zone of incipient oceanic spreading center	Encrustation	Goethite, nontronite (smectite)
16b	Afar Rift, Mn-rich deposits (Bonatti et al., 1972a, tables 2, 3)	—	Axial trough at rift zone of incipient oceanic spreading center	Encrustation	Pyrolusite, birnessite, todorokite, strontiobarite, rhodocrosite, opal, gypsum
17	Red Sea (Degens and Ross, 1969; Bischoff, 1969; Hendricks et al., 1969, table 8; Stephens and Wittkopp, 1969; Garson and Krs, 1976)	1.0	Basins at intersections with fracture zones in the axial trough at rift zone of an incipient oceanic spreading center	Sediment	Iron-montmorillonite (smectite), goethite, hematite, lepidocrosite, barite, sulfides (sphalerite, marcosite, chalcopyrite, pyrite, marmatite), manganosiderite, anhydrite, manganite (birnessite), todorokite

[1] Bulk chemical analysis of sediment on a $CaCO_3$-free basis.

Composition (plotted on Fig. 4)											
Weight percent (range)					ppm (range)						
Si	Al	Fe	Mn	Fe/Mn	Ba	Co	Cu	Ni	Zn	U	Th
4.3–8.9	1.5–3.4	21.1–27.8	5.7–7.6	3–4	1,500–15,000	55–110	800–1,490	570–800	480–690	2.0–5.8	—
9–38 (SiO_2)	1.5–16 (Al_2O_3)	5.7–22.6	0.6–8.8	3–9	—	75–280	510–1,800	250–1,200	190–530	0.2–12	0.2–3
—	—	2.1–30.6	0.4–9.6	3–16	—	10–183	220–1,814	62–1,680	33–518	—	—
54.92 (SiO_2)	0	23.34 (Fe_2O_3)	12.28 (MnO)	2			2,900 (CuO)				
—	1.3–8.4	4.7–23.3	0.2–9.5	2–80	1,700-14,000	18–290	270–1,700	270–805	220–1,050	—	—
—	0.2–7.4	0.03–7.0	0.1–2.0	3–12	—	—	—	—	—	—	—
14.0–20.8	3.7–5.8	22–29	0.15–0.30	70–200	135–140	17–22	11–60	<5–22	—	0.4–0.5	1.2–1.4
<0.2–4.1	0–1.2	0.03–1.64	31–54	0.0005–0.0700	135–62,500	<5–17	<5–25	<5–20	—	2–4	0.7–1.2
—	0.05–3.0	0.4–47.5	0.03–13.4	1–200	300–20,000	12–198	50–33,400	9–90	400–200,000	—	—

gradients that occur at oceanic spreading centers (Spooner and Fyfe, 1973). An example is the prograde zeolitic sequence developed in response to a background geothermal gradient of 65°C/km (Palmason, 1967) in Icelandic basalts (Walker, 1960). Evidence for the ubiquitous nature of low-level hydrothermal activity at oceanic spreading centers includes anomalously low values of conductive heat flow, implying convective removal of heat (Palmason, 1967; Deffayes, 1970; Talwani et al., 1971; Lister, 1972; Anderson, 1972; Williams et al., 1974; Sclater et al., 1974); hydrous metamorphosed oceanic crust requiring a large source of water (Miyashiro et al., 1971; Christensen, 1972); and isotopic compositions of the hydrated rocks requiring a low $\delta^{18}O$ isotopic source, such as sea water (Muehlenbachs and Clayton, 1972; Spooner et al., 1974).

In contrast to the ubiquitous nature of low-intensity hydrothermal activity at oceanic spreading centers, high-intensity hydrothermal activity is extremely localized. The high-intensity hydrothermal activity concentrates mineral deposits associated with metamorphic assemblages produced at geothermal gradients higher than background, involving temperatures and gradients as high as 400°C and 1,300°C/km within the upper 1 km beneath the ocean bottom (Spooner and Fyfe, 1973). This activity may involve thermal powers up to 15 kWm^{-2} (Lister, 1974a). Hydrothermal mineral deposits in oceanic crust are related to the special structural and thermal conditions that increase hydrothermal activity to an intensity sufficient to concentrate the deposits at discrete sites along oceanic spreading centers. Criteria for recognition of these deposits are qualified as preliminary, because subsea-floor hydrothermal convection systems are incompletely understood, and few studies exist of hydrothermal deposits at spreading centers. Emphasis is necessarily placed on those hydrothermal deposits at oceanic spreading centers that have received the most comprehensive interdisciplinary study and on those criteria that have proven most useful in recognizing the deposits. Considerable potential exists for scientific and technological transfer between studies of geothermal energy systems on land and hydrothermal mineral deposits at oceanic spreading centers (Kappelmeyer and Haenel, 1974; United Nations, 1976).

Petrologic Criteria

Metalliferous sediments

Metalliferous sediments of hydrothermal origin are recognized by their distinctive mineralogical, compositional, and isotopic characteristics, as well as their stratigraphic association with volcanic rocks.

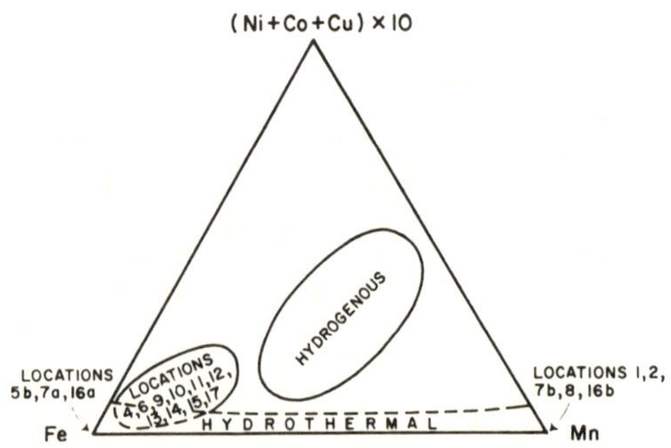

FIG. 4. Ratio Fe, Mn, (Cu + Ni + Co) of hydrothermal mineral deposits from oceanic spreading centers (modified from Bonatti, 1975). Numbers refer to locations shown in Figure 1 and described in Table 2.

A sediment layer several meters thick containing varying concentrations of metals directly overlies basalt of oceanic crustal layer 2 both at and away from spreading centers in all major ocean basins (Table 2, locations 4, 6, 9, 10, 11, 12, 13, 14, 15; Fig. 1). Metalliferous sediments of hydrothermal origin may mix with and/or exhibit transitional characteristics with metalliferous sediments of hydrogenous origin. Characteristics of metalliferous sediments considered to be predominantly hydrothermal include (Table 1; Dymond et al., 1973): high contents of transition metals (particularly Fe and Mn), low aluminum contents, high Si/Al ratios, and Fe/Mn ratios between about 1 and 200 (Table 2; Fig. 4); goethite, iron-rich montmorillonite, and manganese hydroxyoxides as the dominant minerals present on oceanic ridges, where oxidizing conditions prevail (Table 2, locations 4, 6, 9, 10, 11, 12, 13, 14, 15; Fig. 1), and in contrast, base metal sulfides which predominate during the rift stage, when near-bottom reducing conditions prevail, as in certain axial troughs of the Red Sea (Table 2, location 17; Fig. 1); sulfur, uranium, and oxygen isotopic compositions usually consistent with formation from sea water at low temperatures, but in several cases anomalous $^{234}U/^{238}U$ ratios were found that suggest leaching from oceanic crust (Veeh and Boström, 1971; Rydell and Bonatti, 1973); strontium isotopic compositions suggestive of equilibrium with modern sea water; lead isotopic compositions that resemble those of oceanic ridge tholeiitic basalts, rather than sea water; rare earth element patterns like sea water; higher U and lower Th contents than in normal pelagic sediments, attributed to more rapid growth of the metalliferous sediments that inhibits incorporation of Th from sea water (Bonatti et al., 1972b); and metal accumulation rates one to two orders of magnitude faster than pelagic sediments (approx 10 to 110 mg/cm²/10³ yr; Boström, 1970; Bender et al., 1971; Scott, 1976, 1977).

Metalliferous encrustations

Metalliferous encrustations of hydrothermal origin form on basalt of oceanic crustal layer 2 at discrete sites along oceanic spreading centers (Table 2, locations 1, 2, 3, 7a, 7b, 8a, 8b, 16a, 16b; Fig. 1). The hydrothermal encrustations are recognized by their composition, rate of accumulation, and mineralogy as follows.

Extreme chemical fractionation of Fe from Mn is of two types, an Fe-rich encrustation with Fe/Mn ratios between 12 and 237 (Table 2, locations 3, 7a, 7b, 16a, 16b; Fig. 4), and an Mn-rich encrustation with Fe/Mn ratios between 0.002 and 0.14 (Table 2, locations 1, 2, 7a, 8a, 8b, 16a, 16b; Fig. 4). This extreme fractionation distinguishes encrustations of hydrothermal origin from ferromanganese encrustations and nodules of hydrogenous origin (Table 1). The latter both generally exhibit Fe/Mn ratios of the order of 1 and high trace metal content (Ni, Co, Cu, Zn) relative to Fe and Mn (Fig. 4).

Low trace metal content (Table 2; Fig. 4) occurs as a consequence of extreme chemical fractionation and relatively rapid rate of deposition which limits scavenging from sea water.

Goethite and nontronite (Fe-smectite) are the dominant minerals in the Fe-rich encrustations (Table 1, locations 7a, 16a); birnessite, todorokite, and pyrolusite predominate in the Mn-rich encrustations (Table 2, locations 1, 2, 8a, 8b, 16b).

Minerals incorporating elements derived from hydrothermal sources are present. A prime hydrothermal indicator is barite (Table 2; Arrhenius and Bonatti, 1965; Hanor, 1966; Miller et al., 1966; Craig, 1969; Makharadze and Ikoshvili, 1970; Bonatti et al., 1972b; Bertine and Keene, 1975).

Accumulation rates of hydrothermal encrustations determined radiogenically (approx 100 to 1,000 mm per 10^6 yr; Scott et al., 1974) are one to three orders of magnitude greater than those of hydrogenous ferromanganese encrustations and nodules (approx 1 to 10 mm per 10^6 yr).

Encrustations occur as laminated layers with observed thicknesses up to 2 m (Table 2, location 16) directly overlying basalt and as matrix material in breccia of altered basalt fragments (Fig. 4; Table 2, location 1).

The petrologic characteristics described pertain to hydrothermal mineral deposits. There are two other petrologic characteristics pertaining to hydrothermal alteration and depletion of the volume of rock surrounding the deposits.

Alteration of basaltic rocks under the physical and chemical conditions prevalent at the discharge zone of a subsea-floor hydrothermal convection system produces characteristic low- to intermediate-grade metamorphic mineral assemblages in horizontal and vertical zones, such as at the Reykjanes hydrothermal field of Iceland on the Mid-Atlantic Ridge (Table 3). The mineral assemblages and associated conditions have been determined by the coordination of field observations (Sigvaldason, 1962; Steiner, 1967; Browne and Ellis, 1970; Miyashiro and Shido, 1970; Ade-Hall et al., 1971; Tómasson and Kristmannsdóttir, 1972; Spooner and Fyfe, 1973) and experimental studies (Hawkins and Roy, 1963; Helgeson, 1967, 1968; Helgeson et al., 1969; Mottl et al., 1974; Hajash, 1975; Bischoff and Dickson, 1975).

Manganese-rich hydrothermal solutions may alter sedimentary montmorillonite-group clays to palygorskite (Bonatti and Joensuu, 1968).

A halo of depletion of transition metals may occur in the volume of altered basalt or greenstone that surrounds a hydrothermal mineral deposit. The depletion halo represents the volume of rock fluxed by the hydrothermal system to produce the deposit. Depletion halos have been documented by a number of studies, although the volume of rock fluxed to produce a given deposit is poorly known. For example, holocrystalline basalts from the Mid-Atlantic Ridge at latitude 22°N are depleted in Fe, Mn, Co, Cu, Pb, and rare earth elements relative to associated glassy pillow fragments (Corliss, 1971). Most of the Troodos pillow lava series of Cyprus contains 80 to 200 ppm Ni, corresponding to a Ni content of 70 to 200 ppm in oceanic basalts (Nichols and Islam, 1971); in contrast intensely altered Troodos pillow lavas associated with hydrothermal deposits contain only 20 to 100 ppm Ni, evidencing depletion by hydrothermal processes (Govett and Pantazis, 1971). A decrease in Cu content away from the Troodos Massif may also represent hydrothermal depletion

TABLE 3. Vertical Zonation of Mineral Assemblages at Reykjanes Hydrothermal Field, Iceland (Tómasson and Kristmannsdóttir, 1972; Bjornsson et al., 1972)

Relative depth	Temperature (°C)	Mineral assemblage
Shallow	<200	Montmorillonite, calcite, zeolite, opal (≤300 m), pyrite, hematite (≤150 m), quartz (≥100 m), potassium feldspar, albite, anhydrite
Intermediate	200–280	Mixed-layer montmorillonite, chlorite, prehnite (≥200 m), pyrite, quartz, albite
Deep	>200–300	Chlorite, epidote (≥450 m), albite, pyrite, quartz

(Govett and Pantazis, 1971). A halo zone, characterized by a concentration increase of Mg and Fe and decrease of Ca and of Na/K ratio, occurs within 500 m below and to either side of a hydrothermal massive sulfide deposit ($\geq 82 \times 10^6$ tons containing 9.69 percent Zn, 3.77 percent Pb, 0.29 percent Cu, and 2.46 percent Ag per ton) in felsic volcanic rocks of the Bathhurst region of New Brunswick, Canada (volume of halo approx $125 \times 10^6 m^3$; Goodfellow, 1975).

Geochemical Criteria

Some geochemical criteria for hydrothermal mineral deposits in oceanic crust based on the deposits and surrounding rocks were considered under petrologic criteria. Another category of geochemical criteria pertains to the hydrothermal discharge that may be associated with deposits at oceanic spreading centers, as in the cases of the TAG Hydrothermal Field (Table 2, location 1), and the Galapagos spreading center (Table 2, location 8). The extent of subsurface mixing of normal sea water with hydrothermal solutions influences the chemical properties of the discharge. High dilution favors precipitation of metallic sulfides and oxides, as at the TAG Hydrothermal Field; the hydrothermal solutions are depleted in the elements precipitated. Low dilution favors direct discharge of hot, acid solutions enriched in metals, as at the Galapagos spreading center (Edmond et al., 1977).

^3He: Excess of the rare helium stable isotope ^3He above oceanic background concentrations is a sensitive indicator of sites where mantle-derived volatiles are being injected by voluminous high-intensity hydrothermal discharge at oceanic spreading centers. Multiple sources contribute to helium in sea water, and include air injection: the $^3He/^4He$ ratio in the atmosphere is 1.4×10^{-6} (Clarke et al., 1969); radiogenic ^3He: in situ decay of nuclear-era tritium in near-surface water (Jenkins and Clarke, 1976); radiogenic ^3He and ^4He: decay of U and Th to ^4He with ^3He production by reactions on Li in crustal rocks with the resulting helium characterized by $^3He/^4He \simeq 10^{-7}$, one-tenth of the atmospheric ratio (Lupton et al., 1977a); and flux of primordial helium from the mantle at oceanic spreading centers: the $^3He/^4He$ ratio of the mantle component determined in deep water on the East Pacific Rise (Craig et al., 1975) and in helium trapped in submarine basalt glasses from different oceanic spreading centers (Craig and Lupton, 1976) which approximately equals 10^{-5}, about 10 times the atmospheric ratio and 100 times the ratio in crustal helium. Measurements of excess helium in near-bottom water at different spreading centers evidencing a primordial mantle source are presented in Table 4. The measured concentrations of the isotopes and the measurement sensitivity (0.5% for $^3He/^4He$ and ^4He) indicate that detection limits are dilutions of 1.5×10^6 for ^3He and 7×10^4 for ^4He (Lupton et al., 1977b).

^{222}Rn: Enrichment of the unstable radon isotope ^{222}Rn (half-life 3.8 days) relative to oceanic background concentration may indicate discharge sites of voluminous hydrothermal convection systems where waters have circulated through the sea floor at oceanic spreading centers (Table 4; Broecker, 1965; Broecker et al., 1967; Chung and Craig, 1972). The ^{222}Rn is produced by radiogenic decay of uranium series isotopes in the thoeliitic basalt through which the hydrothermal convection system flows. Steady-state radon production in tholeiitic basalt that retains uranium series isotopes should be of the order of 7.4×10^{-2} atoms per minute per gram, corresponding to a steady-state activity of 7.4×10^{-2} disintegrations per minute per gram in the rock. The activity in pure hydrothermal discharge should be this value divided by the mass ratio of water to rock in the hydrothermal system.

TABLE 4. Measurements of Excess Helium in Sea Water

Location	$\delta(^3He)$[1] (%)	$^3He/^4He$ ($\times 10^{-6}$)	Concentration per atmospheric solubility[2] (^3He)	(^4He)	Excess ^{222}Rn[3]	Reference
East Pacific Rise deep water (latitudes 8°N and 6°S)	20–32	—	—	—	—	Clarke et al. (1969); Craig et al. (1975)
Galapagos spreading center hydrothermal plume	36–99	2.8	—	—	80–229	Lupton et al. (1977b)
Western Atlantic Ocean deep water	0–13	—	—	—	—	Jenkins and Clarke (1976)
Red Sea brine from Atlantis II Deep		12.1–12.2	3,260–3,470	370–390	—	Lupton et al. (1977a)

[1] $\delta(^3He)\% = 100\,[(R/R_A) - 1]$ where $R = {^3He}/{^4He}$ and R_A, the atmospheric ratio $= 1.4 \times 10^{-6}$.
[2] Supersaturation of ^3He and ^4He relative to atmospheric solubility equilibrium in sea water.
[3] Disintegrations per minute per 100 kilograms.

Dissolved metals: Anomalous maxima in the concentrations of Zn, Fe, and Cu have been observed together with excess ^{3}He in deep-water masses of the North Atlantic and North Pacific (Brewer et al., 1972). Analysis of hydrothermal effluent including both dissolved and particulate matter from the Galapagos spreading center, where reducing conditions and relatively high flow rates prevail, reveals an increase of Si, Ba, Mg, and Fe and a decrease of Cu, Ni, and Ca with increasing temperature (Edmond et al., 1977).

Ferric hydroxides: Amorphous ferric hydroxides precipitated from hydrothermal solutions discharging into sea water exhibit a halo of suspended weakly acid soluble particulate matter centered at the discharge zone. The ferric hydroxides effectively scavenge metals from hydrothermal solutions. As a consequence of this scavenging action, suspended particulate matter may constitute a more sensitive indicator of hydrothermal discharge than metals in solution. Nephelometer measurements reveal a dense layer of suspended particulate matter interpreted as ferric hydroxides precipitated at the interface between normal Red Sea water (oxidizing, alkaline) and metal-enriched thermal brine (reducing, acidic) in the Atlantis II Deep of the Red Sea (Ryan et al., 1969). Water samples from the central North Atlantic exhibit a decrease in suspended articulate matter below 1,500 m to nearly constant values of 2 $\mu g/l$, with marked increases in iron (10×) and manganese (10%) over the Mid-Atlantic Ridge crest, relative to the eastern and western basins (Betzer et al., 1974; Bolger, 1976). Distinguishing between suspended particulate matter directly emanating from hydrothermal discharge and that resuspended from oceanic ridge sediments may be difficult.

Silica: Discharged hydrothermal solutions are saturated with silica when leaving the high-temperature source (Mahon, 1966). The solubility of silica is relatively unaffected by dissolved salts and is ten orders of magnitude greater at 200°C than at 50°C (Holland, 1967). When hydrothermal solutions cool upon discharge from the ocean bottom, they may be 10 to 20 orders of magnitude supersaturated with respect to silica (Spooner and Fyfe, 1973), thereby providing two potential indicators of hydrothermal discharge. Silica saturation itself may indicate hydrothermal discharge, if the concentration clearly exceeds the high background level of silica in sea water. Silica derived from hydrothermal discharge may support proliferation of silica-bearing organisms, especially benthic forms, in the vicinity of the discharge. For example, sponges are observed to proliferate in areas of the rift valley of the Mid-Atlantic Ridge including the TAG Hydrothermal Field (Fig. 5; McGregor and Rona, 1975) although it is unclear whether the control is a concentrated supply of dissolved silica or simply the hard rock substrate. Radiolarian chert beds are characteristically associated with ophiolites, but the relation between planktonic organisms and a supply of hydrothermal silica is even more tenuous. Deep-sea camera and submersible observations of the hydrothermal vents at the Galapagos spreading center (Fig. 1; Table 2, location 8) reveal dense populations of giant clams (20 cm wide), mussels, crabs, sea anemones, limpets, and chitons whose presence is related to proliferation of bacteria in hydrogen sulfide-rich solutions emanating from the vents (Corliss and Ballard, 1977; Corliss et al., 1977; Lonsdale, 1977a, b). Invertebrates and bacteria are absent in the hot brines of the Red Sea deeps (Fig. 11; Watson and Waterbury, 1969).

Structural Criteria

The relation between hydrothermal mineral deposits and structure of oceanic crust is influenced by the rate of sea-floor spreading. Slow spreading at half-rates (rate to one side of a spreading center) up to about 4 cm per year is characterized by the presence of a rift valley and rough topography (Menard, 1967), exemplified by the Mid-Atlantic Ridge and the Galapagos spreading center (Table 2; Figs. 1, 6). Fast spreading at half-rates greater than about 4 cm per year is characterized by the absence of a rift valley and the presence of smooth topography, exemplified by the East Pacific Rise (Table 2). Hydrothermal mineral deposits are present at both slow and fast spreading ridges (Fig. 1, Table 2). However, the occurrence of the deposits may be influenced by differences in crustal structure, in particular permeability and the distribution of heat sources, related to the rate of sea-floor spreading.

Hydrothermal deposits have been found at four types of geologic structures at oceanic spreading centers: block-faulted topographic highs formed adjacent to the rift valley of a slowly spreading oceanic ridge (Table 2, locations 1, 2, 8; Figs. 1, 6); volcanic seamounts on oceanic ridges (Table 2, location 7; Fig. 1); fracture zones that transect oceanic spreading centers (Table 2, locations 3, 5, 17; Fig. 1) where hydrothermal deposits may form in the ridge-ridge offset (transform fault) segment of fracture zones, at the intersection of fracture zones with other ocean basin structures such as those localizing volcanoes, along active oceanic fracture zones in ocean basins (Bischoff and Rosenbauer, 1977), and at their intersections with rejuvenated fault systems at continental margins (Kutina, 1974); and basins along the axial trough present during the rift stage in the

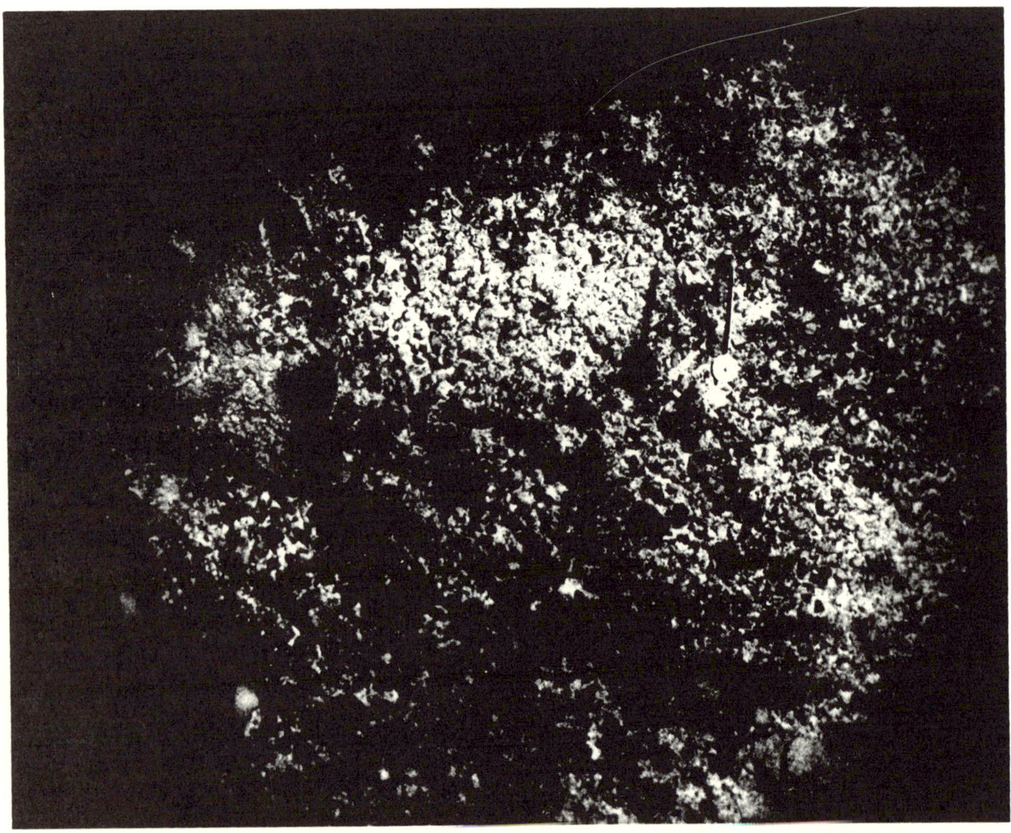

FIG. 5. Bottom photograph (field of view approximately 3 by 4 m) from the TAG Hydrothermal Field showing hydrothermal manganese oxide as a crust on (upper left) and matrix in breccia of basalt fragments (center) at 2,600 m below sea level on the east wall of the rift valley of the Mid-Atlantic Ridge (location 1 in Fig. 1 and Table 2; Rona, 1976b). A sponge is present at the lower left.

opening of an ocean basin (Table 2, location 17; Fig. 1).

The structure of block-faulted topographic highs, which form ridges trending tranverse to the rift valley of slowly spreading oceanic ridges (Fig. 6), acts to focus hydrothermal discharge. The faults and associated fractures provide conduits for hydrothermal flow, and the topographic highs create a geometric effect that forces sea water in a convection cell to flow downward beneath intervening valleys and to flow upward beneath the highs (Fig. 7; Lister, 1972, 1974b). Maximum ease of forcing should occur if the topography has a wavelength twice the cell width, which generally approximately equals the thickness of the permeable layer (Elder, 1965). As a locus of hydrothermal discharge, block-faulted topographic highs are favored localities for the occurrence of hydrothermal deposits.

Hydrothermal deposits have so far been found on only a few block-faulted topographic highs (Table 2, locations 1, 2, 8), although such structures are present adjacent to the rift valley of all slowly spreading oceanic ridges. Additional special structural conditions are required to increase the hydrothermal activity to an intensity sufficient to concentrate hydrothermal deposits. In the case of the TAG Hydrothermal Field, located on a block-faulted topographic high adjacent to the rift valley of the Mid-Atlantic Ridge (Figs. 6, 7), two additional special structural conditions are present (Rona et al., 1976).

An exceptional number of faults and faulted valleys transect the topographic high on which the TAG Hydrothermal Field is located (horizontal spacing tens of meters), compared with similar topographic highs in the region (horizontal spacing hundreds to thousands of meters; Fig. 6). The faults and as-

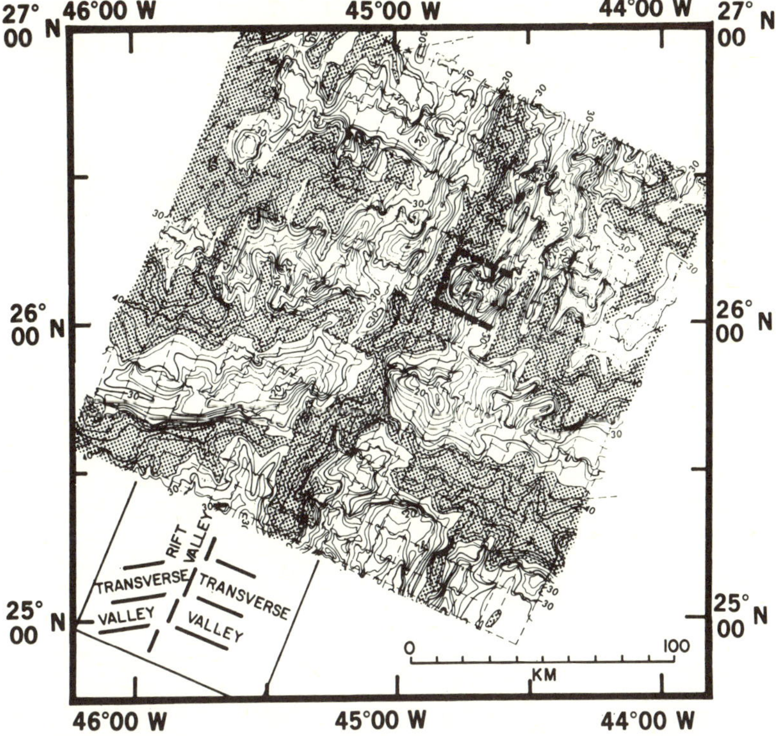

FIG. 6. Bathymetric map (modified from Rona et al., 1976) of an area of the Mid-Atlantic Ridge crest centered on the TAG Hydrothermal Field (Fig. 1, location 1) showing the relation between hydrothermal mineral deposits and topography at a slowly spreading oceanic ridge. The inset, corresponding to the area of the bathymetric map, shows the rift valley and trends of ridges (block-faulted topographic highs) and intervening valleys that strike transverse to the axial valley (shaded on the bathymetric map). The TAG Hydrothermal Field (heavy dashed lines) is located on one of the transverse ridges inferred to be the discharge zone of a subseafloor hydrothermal convection system that recharges with sea water at the intervening valleys (Fig. 7).

sociated fractures (fracture spacing millimeters to meters) increase the permeability of the rocks, thereby facilitating circulation of voluminous hydrothermal fluids.

The segment of the wall of the rift valley formed by the topographic high on which the TAG Hydrothermal Field is located projects as a salient over the floor of the rift valley and is associated with an anomalous narrowing of the valley from about 15 km to 5 km (Fig. 6). The salient is directly over intrusive heat sources beneath the rift valley, causing an exceptionally high thermal gradient that vigorously drives the upwelling limb of the hydrothermal convection system which discharges through faults in the wall of the valley at the TAG Hydrothermal Field.

Exceptionally high permeability and thermal gradients are structurally controlled conditions necessary for the development of high-intensity hydrothermal activity at all the types of structures (block-faulted topographic highs, volcanic seamounts, fracture zones, and axial troughs) in rift zones, ocean basins, and marginal basins. The position of exceptionally permeable structures directly over intrusive heat sources may make the difference between low- and high-intensity hydrothermal activity.

Thermal Criteria

Hydrothermal convection accounts for an estimated 80 percent of the heat loss from lithospheric cooling at oceanic spreading centers and 20 percent of the Earth's total heat loss of 10.2×10^{12} cal/sec (Williams and Von Herzen, 1974). Conductive heat transfer accounts for the remaining 20 percent of heat loss at oceanic spreading centers. Hydrothermal mineral deposits accumulating at oceanic

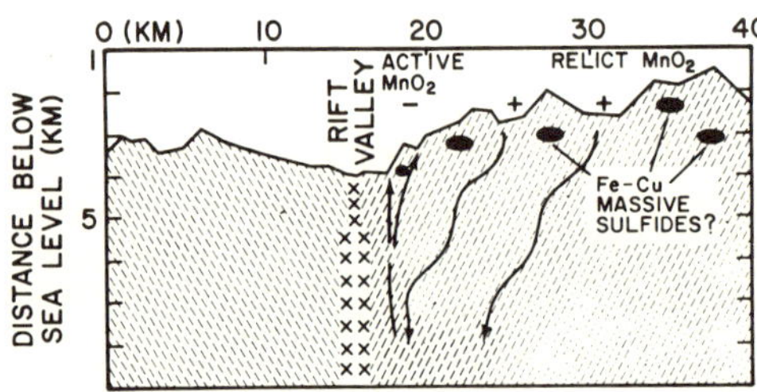

FIG. 7. Diagrammatic cross section of the Mid-Atlantic Ridge at the TAG Hydrothermal Field (vertical exaggeration × 2) showing a subsea-floor hydrothermal convection system like that hypothesized to exist (modified from Rona et al., 1976). Arrows indicate directions of hydrothermal circulation. Slant lines indicate directions of maximum permeability controlled by structural grain including fractures, faults, and dikes. Hydrothermal deposits are actively forming adjacent to the rift valley, and relict deposits are present away from the rift valley, as a consequence of sea-floor spreading. Actually, relict deposits may be covered by off-axis volcanism. Symbols: + = zone of recharge; − = zone of discharge; X = zone of igneous intrusion.

spreading centers are associated with characteristic patterns of convective and conductive heat flow. Highest values of convective heat flow occur at the discharge zones of hydrothermal convection systems, the loci of mineral deposition. The convective heat flow is manifested in the overlying water column in different ways depending on the degree of oceanic circulation. Under conditions of restricted oceanic circulation encountered in the early rift stage, the temperature of the near-bottom water in the vicinity of a hydrothermal discharge zone may increase by the order of 10°C. For example, in certain of the axial troughs of the Red Sea (Atlantis II Deep), hydrothermal solutions with an estimated discharge temperature of 100°C heat the near-bottom water to temperatures of 56°C (normal Red Sea water 22°C), associated with gradients of 1°C per meter warming downward, where stabilized by high salinity and absence of oceanic currents (Ericson and Simmons, 1969; Ross, 1972).

Near-bottom water thermal anomalies attributed to convective hydrothermal discharge under conditions of open oceanic circulation encountered on oceanic ridges are only of the order of 0.1°C due to rapid dissipation of heat by oceanic currents. For example, increases in potential temperature (temperature of a parcel of water raised to surface without any exchange of heat) of about 0.1°C along a horizontal distance between 100 and 300 m at a height of about 10 m above the ocean bottom were measured with thermistor arrays towed across hydrothermal discharge zones at the TAG Hydrothermal Field (Fig. 8; Rona et al., 1975) and at the Galapagos spreading center (Williams et al., 1974; Detrick et al., 1974; Weiss et al., 1977). A near-bottom thermal anomaly caused by hydrothermal discharge may be distinguished from downward mixing of warmer water masses by the presence in the former of a temperature gradient warming downward (0.015°C/m measured at the TAG Hydrothermal Field; Rona et al., 1975) and a potential temperature-salinity relation distinct from that of surrounding water masses (Weiss et al., 1977). Given a measurement precision of 0.004°C, the detection limit of a potential temperature anomaly from a hydrothermal solution discharged at 100°C is a dilution of 2.5×10^4; this detection limit is about two orders of magnitude lower than that cited for ^3He (1.5×10^6). Hydrothermal discharge is inferred to be episodic based on consideration of the magnitude of convective heat flux at oceanic ridges, so that convective water temperature anomalies are probably intermittent rather than steady state (Lowell and Rona, 1976).

Values of conductive heat flow, measured by thermoprobes in sediments near the axis of oceanic spreading centers at both the rift and oceanic ridge stages of the opening of an ocean basin, exhibit a large variability between adjacent values (0 to 30 Heat Flow Units) and a high average flux (approx 2.5 HFU; Erickson and Simmons, 1969; Langseth and

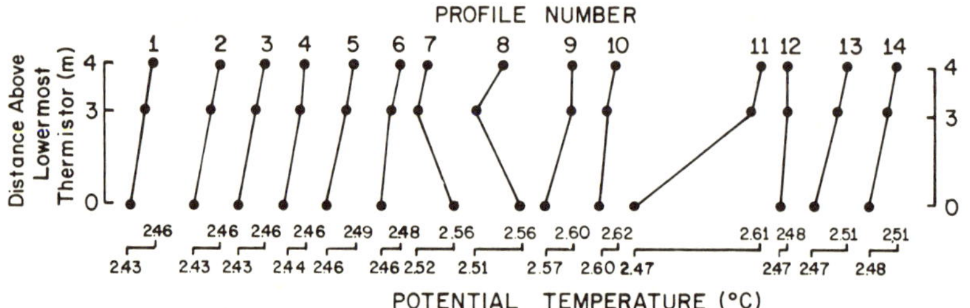

FIG. 8. Successive water temperature profiles, each about 100 m apart, made by a vertical array of three thermistors, with the lowermost thermistor about 10 m above the ocean bottom, across a portion of the TAG Hydrothermal Field (Rona et al., 1975; Lowell and Rona, 1976). An abrupt water-temperature anomaly attributed to hydrothermal discharge is evidenced by the temperature increase and inversion of thermal gradient in profiles 7 to 11.

Von Herzen, 1971; Girdler, 1970; Williams et al., 1974). Theoretical considerations suggest that conductive heat flow may systematically vary as a result of control of hydrothermal circulation by ridge crest topography (Lister, 1972, 1974b), as noted under structural criteria. Low values of conductive heat flow would occur in valleys, where subsea-floor hydrothermal convection systems are inferred to recharge with sea water. Valleys tend to depress the isotherms and therefore are places of preferential separation for the cold columns. High values of conductive heat flow would occur in localized zones on block-faulted topographic highs, where the convection systems are inferred to discharge. Limited conductive heat flow measurements from the Juan de Fuca Ridge (Lister, 1972; Davis and Lister, 1977), the Galapagos spreading center (Williams et al., 1974; Sclater et al., 1974, fig. 9; Von Herzen et al., 1977), and the Mid-Atlantic Ridge at the TAG Hydrothermal Field (Rona et al., 1976) appear consistent with the distribution predicted by theory, although variations in permeability may distort the pattern. If the proposed relation between hydrothermal circulation and topography at slow-spreading oceanic ridges proves to be a general phenomenon, then maps of bathymetry and conductive heat flow can be used to delineate localized zones of hydrothermal discharge that may be associated with mineral deposits.

Acoustic Criteria

A reflecting interface may exist between normal sea water and hydrothermal solutions when a sufficient contrast in acoustic impedence is present due to sound velocity and density differences between the fluids. Conditions of restricted oceanic circulation at the early rift stage in the opening of an ocean basin favor development of a sufficient contrast in acoustic impedance between normal and hydrothermal fluids to produce a reflecting interface. Conditions of open oceanic circulation at the advanced oceanic ridge stage inhibit development of such reflecting interfaces. For example, standard echo sounders (approx 10 kHz frequency) record a reflecting interface between normal Red Sea water (41 per mil salinity, 22°C, sonic velocity 1,550 m/sec, low turbidity) and underlying thermal brine (135 per mil salinity, 44°C, sonic velocity 1,910 m/sec, high turbidity) in the Atlantis II Deep (Dietrich and Krause, 1969; Swallow, 1969; Ostapoff, 1969; Ryan et al., 1969; Pugh, 1969). A reflecting interface is also present between layers within the thermal brine (131 per mil salinity, 44°C temperature, overlying 255 per mil, 56°C). The high content of salt and suspended particulate matter (ferric hydroxides) compensates for the temperature increase of the brine and serves to stabilize interfaces (Turner, 1969) and to establish contrasts in acoustic impedence.

Seismic Criteria

Microearthquake activity may indicate the presence of high-intensity hydrothermal convection systems. An association has been demonstrated between microearthquake activity and hydrothermal systems on continents. For example, 20 to 30 shallow-focus (< 15 km) microearthquakes per day of magnitude less than 2.0 are recorded at hydrothermal areas in the Imperial Valley (Combs and Hadley, 1977; Jarzbeck and Combs, 1976), a landward structural extension of the oceanic spreading center active in the Gulf of California. Hydrothermal areas in the Rio Grande Rift zone of the southwestern United States exhibit 5 to 70 microearthquakes per day, mostly of magnitudes less than 1.5 (Quillan and Combs, 1976; Johnson and Combs, 1976).

Measurements of seismicity in the volcanic zone of southwest Iceland on the Mid-Atlantic Ridge (Ward et al., 1969; Ward and Björnsson, 1971; Klein et al., 1973) indicate that the most active microearthquake zones often coincide with high-temperature hydrothermal systems (> 200°C in upper 1 km). Focal depths of the microearthquakes lie in the upper part of oceanic crustal layer 3 within a subsurface depth range of 2 to 6 km (Ward and Björnsson, 1971).

Earthquake swarms, each consisting of a distinctive sequence of earthquakes closely grouped in time and space with no one outstanding main shock, have been reported from 17 locations along the Mid-Atlantic Ridge crest, both north and south of Iceland. Similar earthquake swarms have been reported from sites along the axes of other slow-spreading centers including the Red Sea rift, the Indian Ridge, the Gorda Ridge, the Galapagos spreading center, and the Gulf of California rift (Francis, 1968; Sykes, 1970). The magnitude of the earthquakes in swarms observed at oceanic spreading centers is generally less than 5.0; focal depths are considered to lie within oceanic crust (< 10 km; Sykes, 1970). It is uncertain whether earthquake swarms at oceanic spreading centers are related to localized sources of high fluid pressures due to magmas or to hydrothermal solutions that may lower the effective strength of rocks and act as concentrated sources of stress (Sykes, 1970).

Gravity Criteria

No direct relation is known between gravity and hydrothermal mineral deposits. Although hydrothermal minerals generally are less dense than primary minerals, the effect on gravity is negligible. Gravity may be indirectly related to hydrothermal mineral deposits through geologic structure. For example, the central trough of the Red Sea exhibits relatively low free-air gravity values caused by terrain effect and high Bouguer values (110 to 135 mgals), consistent with dense underlying material interpreted as intrusive bodies (Phillips et al., 1969). A free-air gravity high caused by terrain effect occurs over the block-faulted topographic high on which the TAG Hydrothermal Field is located (Fig. 6; Rona et al., 1976).

Subaerial hydrothermal systems exhibit both positive (Wairakei and Niland, New Zealand, Beck and Robertson, 1955; Mesa, Imperial Valley, Biehler, 1971) and negative gravity anomalies (Larderello and Mt. Amiata, Italy, Marchesini et al., 1962). The positive and negative gravity anomalies of subaerial hydrothermal systems are controlled by local geologic structure.

Electrical Criteria

Electrical criteria for recognition of hydrothermal deposits in oceanic crust pertain to the electrical properties of metallic deposits and of hydrothermal solutions in rocks. Electromagnetic methods have been extensively applied on land to delineate hydrothermal systems (McNitt, 1965) as well as porphyry and stratabound sulfide deposits (SEG, 1966). Electromagnetic methods are at an early stage of application to the marine environment, where voltage is measured in highly conductive sea water that acts as a short circuit for electric currents.

Self-potentials involving a gradient of redox potential that determines the electromagnetic force of an oxidation-reduction electrochemical cell are associated with metalliferous sediments and massive sulfide bodies of the ocean bottom (Sato and Mooney, 1960). Redox potential gradients as great as 200 mV were measured in the Red Sea metalliferous sediments (Brooks et al., 1969). Calculation of the short-circuit effect in sea water (Brewitt-Taylor, 1975) shows that the voltage is reduced by a factor of the same order as the ratio of conductivities of sea water and the metalliferous sediment, resulting in a signal of tens of millivolts. Self-potentials have been measured in the deep ocean using a towed electrode array (Brewitt-Taylor, 1975) and on the continental shelf at a known sulfide orebody (Corwin et al., 1970).

A towed resistivity array and a towed active source audiomagnetotelluric system have detected conductivity anomalies associated with known vein copper deposits in the Great Lakes (Nebrija et al., 1976). Direct current resistivity measurements of the subaerial Reykjanes hydrothremal field in Iceland reveal values of 1 to 3 ohm meters where basalt is saturated with thermal brines, and 1,000 to 5,000 ohm meters where saturated with fresh water (Björnsson et al., 1972). A start has been made in the application of well logging techniques to the measurement of sea-floor resistivity (Bannister, 1968; Patella and Schiavone, 1974; Whiteley, 1974).

Magnetic Criteria

A growing body of field evidence supported by laboratory studies indicates that hydrothermal activity may affect the intensity of magnetization of basalt, producing a characteristic magnetic signature that may be associated with sites of hydrothermal mineral deposits. The following field evidence favors the existence of a magnetic signature indicative of hydrothermal deposits (Rona, 1978).

A distinct low in residual magnetic intensity (approx 200 gammas) coincides with the TAG Hy-

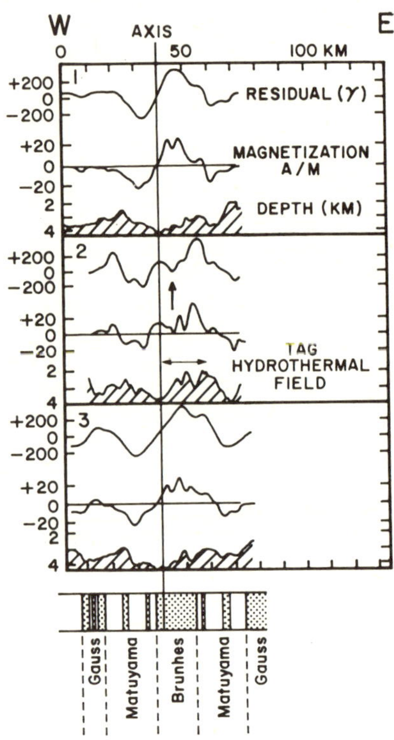

FIG. 9. Profiles of residual magnetic intensity (γ), magnetization of the basalt layer computed by the inverse method (A/M = amperes/meter = 10^{-3} emu/cm^3), and depth (KM), across the rift valley of the Mid-Atlantic Ridge at latitude 26° N (modified from McGregor et al., 1977). A distinct low in residual magnetic intensity within the positive axial magnetic anomaly (Brunhes magnetic polarity epoch) and in computed magnetization (arrow in profile 2) coincides with the TAG Hydrothermal Field. The magnetic low is absent in profiles north (profile 1) and south (profile 3) of the field. The locations of the profiles is shown in Figure 10.

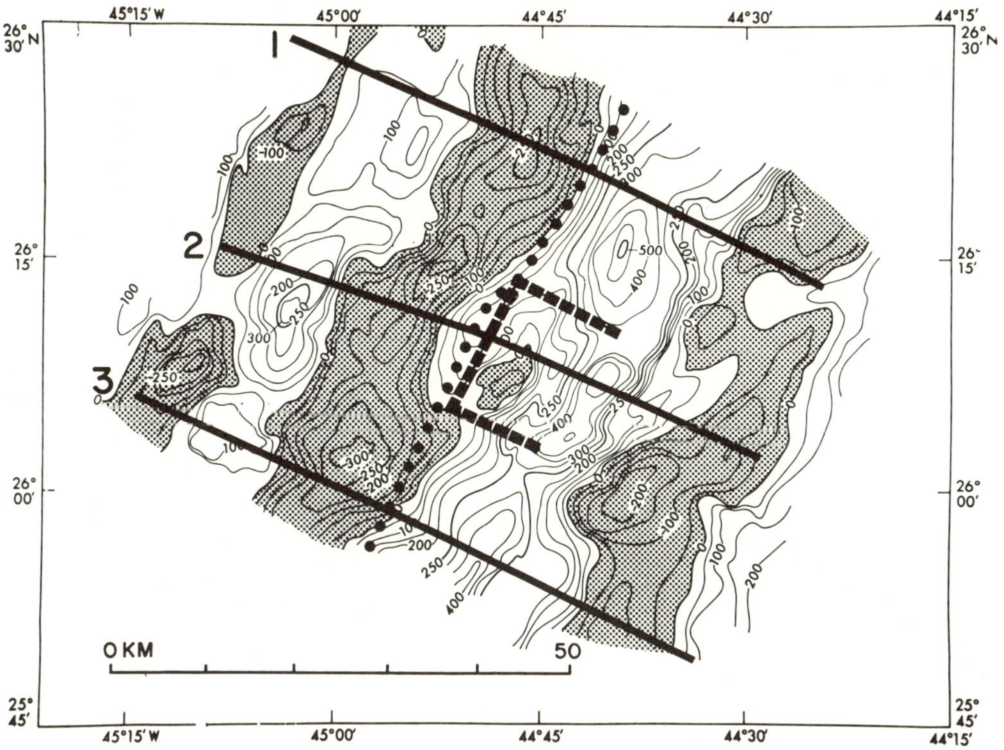

FIG. 10. Residual magnetic map (50 gamma contour interval) of the Mid-Atlantic Ridge crest at latitude 26° N (modified from McGregor et al., 1977). Shaded areas are negative. The location of profiles 1–3 in Figure 9 (solid lines), the TAG Hydrothermal Field (dashed lines), and the axis of the rift valley (dotted line) are shown. Low to negative values of residual magnetic intensity are present within the positive axial magnetic anomaly at the TAG Hydrothermal Field.

drothermal Field at the crest of the Mid-Atlantic Ridge and is absent both to the north and south of the TAG Hydrothermal Field (Figs. 9, 10; McGregor and Rona, 1975). The magnetic low occurs within the positive magnetic anomaly of the Brunhes magnetic polarity epoch that is typically associated with the axis of oceanic spreading centers.

A distinct low in residual magnetic intensity occurs at the Atlantis II Deep (approx 650 gammas) in the axial trough of the Red Sea, although the significance of this low was not noted previously (Fig. 11; Hunt et al., 1967; Phillips et al., 1969). The magnetic low is absent at the adjacent Chain Deep and Discovery Deep, which are associated with magnetic highs (approx 350 gammas) typical of the axis of oceanic spreading centers (Fig. 12). The thermal brine in these three deeps is inferred to discharge at a minimum temperature of 100°C from a single zone in the Atlantis II Deep based on distribution of metalliferous sediments, chemistry of the brine and interstitial water, and brine temperature relations (Brewer et al., 1969; Brooks et al., 1969; Ross et al., 1969; Ross, 1972). These observations indicate that the magnetic low at the Atlantis II Deep is related to the presence of hydrothermal discharge; the magnetic highs at the other two deeps occur where hydrothermal discharge is absent.

A distinct low in residual magnetic anomaly (approx 200 gammas) coincides with the subaerial Reykjanes hydrothermal field in Iceland at the axis of the Mid-Atlantic Ridge, where the thermal brine attains temperatures of 300°C (Björnsson et al., 1972).

Distinct lows in residual magnetic intensity occur in certain other subaerial hydrothermal fields, including Mesa in the Salton Trough of the southwestern United States, where measured temperatures attain 216°C (Koenig, 1967; Black, 1975, table 1); Wairakei of New Zealand, where measured temperatures attain 250°C (Studt, 1959; Banwell et al., 1957); and Rabaul, New Britain (Studt, 1961).

Laboratory studies of basalt alteration explain the cause of the lows in residual magnetic intensity observed at both submarine and certain subaerial hydrothermal fields.

Unaltered tholeiitic basalt has high magnetic susceptibility (typical values 6 to 30 emu/cm^3 × 10^4) and remanent magnetization (typical values 36 to 113 emu/cm^3 × 10^4) because of its content of fresh iron-titanium oxide, especially titanomagnetite (Luyendyk and Melson, 1967; Irving et al., 1970a, b; Watkins and Paster, 1971; Ade-Hall et al., 1971; Marshall and Cox, 1972; Fox and Opdyke, 1973).

Unaltered basalt is metamorphosed to greenstone by hydrothermal alteration within the range of temperatures and pressures encountered in hydrothermal discharge zones (Helgeson, 1968; Helgeson et al., 1969; Melson et al., 1968; Miyashiro et al., 1971; Spooner and Fyfe, 1973).

Greenstone, in which iron-titanium oxide has been replaced by sphene and other minerals, has low values of susceptibility (typical values 0.4 to 0.6 emu/cm^3 × 10^4) and remanent magnetization (typical values 0.02 to 0.11 emu/cm^3 × 10^4) (Luyendyk and Melson, 1967; Irving et al., 1970a, b; Watkins and Paster, 1971; Ade-Hall et al., 1971; Marshall and Cox, 1972; Fox and Opdyke, 1973).

The presence of altered basalts and greenstones has been confirmed where rocks have been recovered from hydrothrmal fields associated with magnetic lows (Studt, 1959, 1961; Björnsson et al., 1972; Rona et al., 1976).

Computer modeling of the magnetic low at the TAG Hydrothermal Field (Fig. 9) is consistent with attribution of the low to reduction in remanent magnetization within the basalt of oceanic crustal layer 2 (McGregor et al., 1977).

Development of magnetic lows by alteration of the magnetic mineral component of basalts in high temperature discharge zones of subsea-floor systems like those described may provide a rapid reconnaissance criterion for recognizing potential sites of hydrothermal mineral deposits in oceanic crust (Rona, 1978). This does not rule out the possibility of a magnetic high controlled by geologic structure at a hydrothermal field, as observed at certain subaerial fields such as Niland (Kelley and Soske, 1936) and Casa Diablo (Henderson et al., 1963) in the southwestern United States. The hydrothermal alteration involved specifically pertains to basalt, because alteration of other rock types may produce different magnetic effects. For example, fresh peridotites lacking magnetite alter to rocks rich in serpentine and magnetite with increase in magnetic susceptibility (Luyendyk and Melson, 1967; Fox and Opdyke, 1973).

Patterns of Deposition

Patterns of deposition are considered insofar as they are useful in locating hydrothermal mineral deposits in oceanic crust. Depositional patterns differ in the rift and oceanic ridge stages of opening of an ocean basin about an oceanic spreading center due primarily to differences in tectonic setting. At the rift stage, oceanic circulation is restricted by the high length/width raito of the opening, by the axial

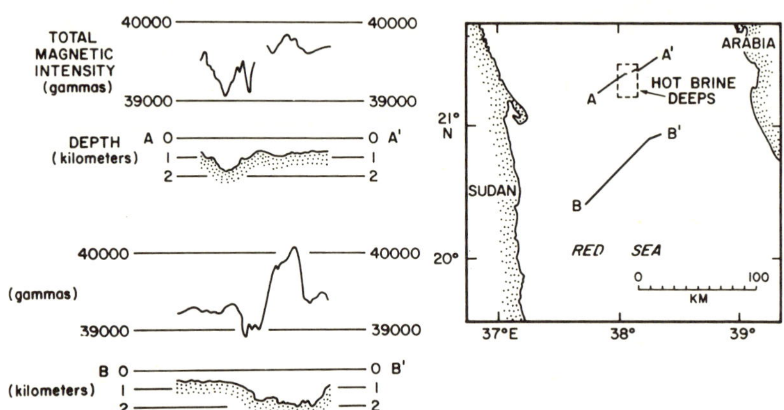

FIG. 11. Magnetic and bathymetric profiles across the axial trough of the Red Sea and an index map showing an area of deeps containing metalliferous sediments and hot brines (modified from Phillips et al., 1969). A distinct low in observed magnetic intensity occurs over the Atlantis II Deep (profile A-A'), the locus of hydrothermal discharge. A high in magnetic intensity, characteristic of the axis of oceanic spreading centers, occurs over the axial trough outside the area of the hot-brine deeps (profile B-B').

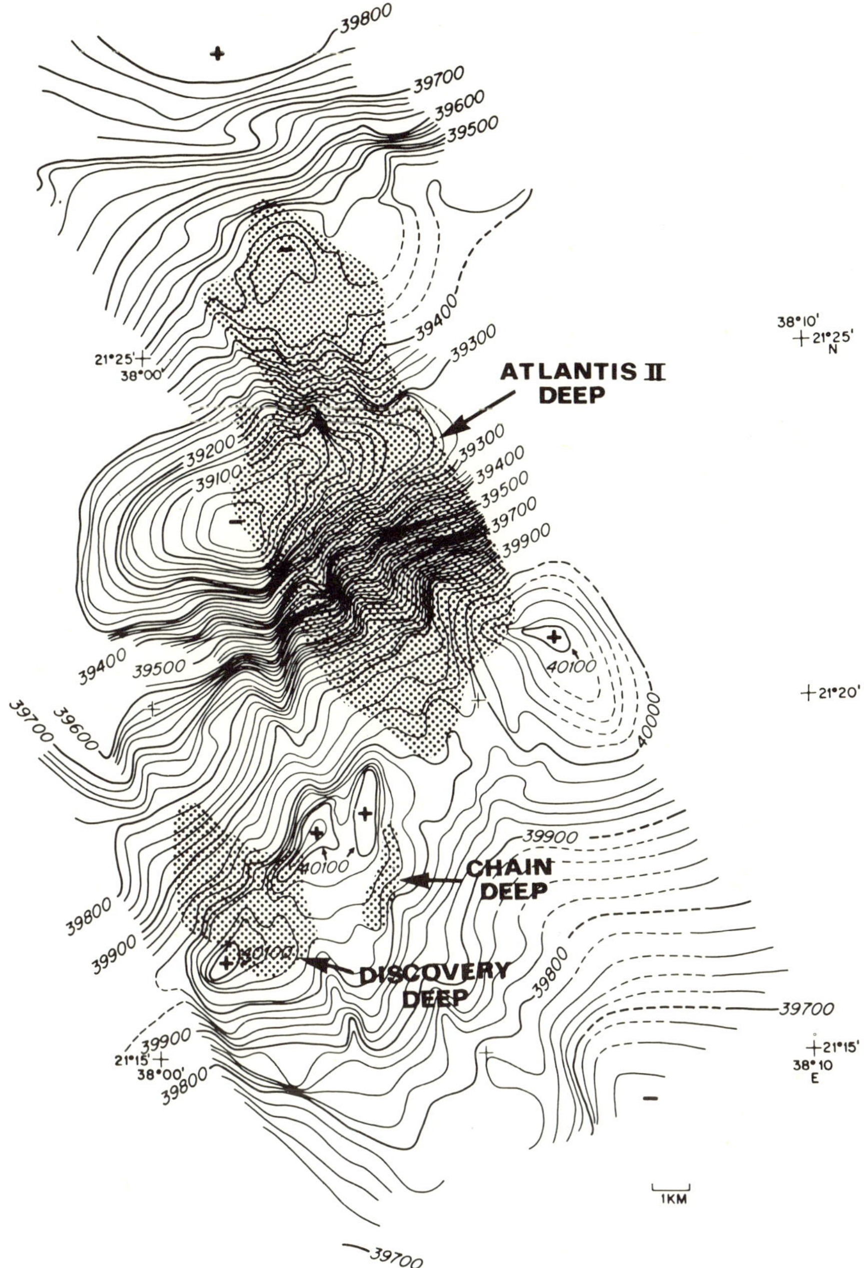

FIG. 12. Total magnetic intensity contour map (20 gamma contour interval) of the hot-brine area of the Red Sea outlined in Figure 11 (modified from Phillips et al., 1969). Each of the three deeps present is shaded. A magnetic low occurs at the Atlantis II Deep; magnetic highs occur at the Discovery Deep and Chain Deep. A hydrothermal discharge zone in the Atlantis II Deep is the source of the thermal brines and metalliferous sediments in the three deeps.

troughs that contain the hydrothermal products, and by the high salinity derived from flow through evaporites that stabilizes the hydrothermal solutions. The presence of evaporites is not simply an artifact of the Red Sea but is a regular feature of the rift stage (Rona, 1969, 1970). As a consequence of these factors, metalliferous sediments, encrustations, and massive sulfides, if present, are deposited at or near the discharge zones of subsea-floor hydrothermal convection systems and exhibit relatively pure hydrothermal compositions. For example, a dispersion halo of manganese-rich sediment about 20 km in diameter is present around the Atlantis II Deep, which is about 5 km wide by 15 km long (Fig. 12; Cronan, 1976; Bignell et al., 1976). Thermal and geochemical anomalies persist in the stably stratified water immediately overlying the deposits.

In contrast, at the oceanic ridge stage, open ocean circulation disperses metalliferous sediments, accounting for their nearly ubiquitous distribution (Fig. 1) and compositional variation (Table 2) related to dilution by biogenic, detrital, and hydrogenous components. Determination of gradients of metal content in metalliferous sediments dispersed on oceanic ridges may lead back to concentrated deposits at hydrothermal discharge zones. For example, contours of transition metal and barium contents of sea-floor sediments exhibit concentration gradients that effectively delineate oceanic spreading centers in the Atlantic, Pacific, and Indian Oceans (Boström et al., 1969, fig. 1; Boström, 1973, figs. 2, 5, 9, 10, 16). Thermal and geochemical anomalies derived from hydrothermal discharge into the near-bottom water are likewise quickly dissipated by oceanic currents and other mixing processes. Metallifeous encrustations and massive sulfides, if present, remain in situ at the hydrothermal discharge zone.

Two principal depositional patterns can be discerned from the limited data available on the distribution of hydrothermal mineral deposits perpendicular to oceanic spreading centers (Rona, 1976b).

The first is controlled by physical and chemical processes within the hydrothermal field. A major process is inferred to be sealing of talus by deposition of hydrothermal minerals from solutions discharged through faults at and adjacent to the wall of the axial valley of an oceanic ridge. In the case of the TAG Hydrothermal Field at the Mid-Atlantic Ridge crest (Table 2, location 1; Figs. 1, 6), the sealing of a given volume of talus (Fig. 5) is inferred to occur during a period of the order of 1×10^5 yr, causing successive migrations of the zone of discharge (Rona, 1976b). The resultant pattern of hydrothermal mineral deposition within the hydrothermal field is a mosaic of patchy deposits, overlapping in time and space, with a predominantly fault-controlled trend subparallel to the axis of the axial valley.

A second pattern is controlled by sea-floor spreading encompassing the entire hydrothermal field. A linear zone of hydrothermal deposits will extend from an active depositional locality at the axial valley of an oceanic ridge along the direction of sea-floor spreading, depending both on the continuity of sea-floor spreading and the persistence in time of the special structural and thermal conditions that concentrate the hydrothermal activity. In the case of the TAG Hydrothermal Field (Figs. 6, 7), the special structural and thermal conditions that have concentrated hydrothermal activity have persisted during sea-floor spreading for at least 1.4×10^6 yr (Rona, 1976b). As a relict hydrothermal deposit is carried away from the axis of an oceanic spreading center by sea-floor spreading, it may become covered by off-axis volcanism (Fig. 7). A criterion to distinguish between axial volcanics that may contain hydrothermal deposits and off-axis volcanics that may cover deposits is the higher K_2O content in the latter (Strong, 1974b).

The distribution of hydrothermal mineral deposits along oceanic spreading centers can presently only be conjectured from the known distribution of 17 active hydrothermal systems over a distance of 250 km in the neovolcanic zone of Iceland on the Mid-Atlantic Ridge (Bödvarsson, 1961) and 17 basins (at least one presently active) containing hydrothermal brine and/or sediment over a distance of 900 km in the Red Sea (Degens and Ross, 1969; Bäcker and Schoell, 1972; Bäcker, 1975). The difficulty of finding hydrothermal mineral deposits in ophiolites is comprehensible in light of their possible covering by volcanics away from oceanic spreading centers and their sporadic occurrence along oceanic spreading centers.

Conclusions

The various criteria presented can be used to guide exploration for hydrothermal mineral deposits in oceanic crust. The seismic, geochemical, acoustic, thermal, and certain electrical criteria apply to recognition of active discharge zones of subsea-floor hydrothermal convection systems, the loci of hydrothermal mineral deposition at oceanic spreading centers including rift zones, oceanic ridges, and marginal basins associated with island arcs. The petrologic, structural, gravity, magnetic, and other electrical criteria also apply to active discharge zones, as well as to recognition of hydrothermal mineral deposits in oceanic crust away from spreading centers at sea and on land (Fig. 1).

Acknowledgments

I would like to thank G. J. S. Govett of the University of New Brunswick for a helpful review and the Office of Marine Minerals of the National Oceanic and Atmospheric Administration (NOAA) for financial support. This work was performed as part of the NOAA Metallogenesis at Dynamic Plate Boundaries Project.

NATIONAL OCEANIC AND ATMOSPHERIC ADMINISTRATION
ATLANTIC OCEANOGRAPHIC AND METEROLOGICAL LABORATORIES
15 RICKENBACKER CAUSEWAY
MIAMI, FLORIDA 33149
February 23, August 29, 1977

REFERENCES

Ade-Hall, J. M., Palmer, H. C., and Hubbard, T. P., 1971, The magnetic and opaque petrological response of basalts to regional hydrothermal alteration: Royal Astron. Soc. Geophys. Jour., v. 24, p. 137–174.

Anderson, R. N., 1972, Petrologic significance of low heat flow on the flanks of slow spreading mid-ocean ridges: Geol. Soc. America Bull., v. 83, p. 2947–2956.

ARCYANA, 1975, Transform fault and rift valley from bathyscaph and diving saucer: Science, v. 190, p. 108–116.

Arrhenius, G., and Bonatti, E., 1965, Neptunism and volcanism in the ocean, *in* Sears, M., ed., Progress in oceanography: London, Pergamon Press, v. 3, p. 7–22.

Bäcker, H., 1975, Exploration of the Red Sea and Gulf of Aden during the M.S. VALDIVIA cruises "Erzschlämme A" and "Erzschlämme B": Hannover, Geol. Jahrb., D13, p. 3–78.

Bäcker, H., and Schoell, M., 1972, New deeps with brines and metalliferous sediments in the Red Sea: Nature Phys. Sci., v. 240, p. 153–158.

Bailey, E. H., and Coleman, R. G., 1975, Mineral deposits in the Semail ophiolite of northern Oman [abs.]: Geol. Soc. America, Abstracts with Programs, v. 7, p. 293.

Bannister, P. R., 1968, Determination of the electrical conductivity of the sea bed in shallow waters: Geophysics, v. 30, p. 995–1003.

Banwell, C. J., Cooper, E. R., Thompson, G. E. K., and McCree, K. J., 1957, Physics of the New Zealand thermal area: New Zealand Dept. Sci. Indus. Research Bull. 123, 109 p.

Beck, A. C., and Robertson, E. I., 1955, Geology and geophysics, *in* Grange, L.I., ed., Geothermal steam for power: New Zealand Dept. Sci. Indus. Research Bull. 117, p. 15–19.

Bender, M., Broecker, W., Gornitz, V., Middel, U., Kay, R., and Sun, S., 1971, Geochemistry of three cores from the East Pacific Rise: Earth Planet. Sci. Letters, v. 12, p. 425–433.

Bertine, K. K., and Keene, J. B., 1975, Submarine barite-opal rocks of hydrothermal origin: Science, v. 188, p. 150–152.

Betzer, P. R., Bolger, G. W., McGregor, B. A., and Rona, P. A., 1974, The Mid-Atlantic Ridge and its effect on the composition of particulate matter in the deep ocean: Am. Geophys. Union Trans., v. 55, p. 293.

Biehler, S., 1971, Gravity studies in the Imperial Valley, *in* Cooperative geological-geophysical-geochemical investigations of geothermal resources in the Imperial Valley of California: Univ. California Riverside-IGPP Rept., p. 29–42.

Bignell, R. D., Cronan, D. S., and Tooms, J. S., 1976, Metal dispersion in the Red Sea as an aid to marine geochemical exploration: Inst. Mining Metallurgy Trans., v. 85, sec. B, p. B273–B278.

Bischoff, J. L., 1969, Red Sea geothermal brine deposits, *in* Degens, E. T., and Ross, D. A., eds., Hot brines and recent heavy metal deposits of the Red Sea: New York, Springer-Verlag, p. 348–401.

Bischoff, J. L., and Dickson, F. W., 1975, Seawater-basalt interaction at 200°C and 500 bars: Implications for origin of sea floor heavy mineral deposits and regulation of seawater chemistry: Earth Planet. Sci. Letters, v. 25, p. 385–397.

Bischoff, J. L., and Rosenbauer, R. J., 1977, Recent metalliferous sediment in the North Pacific manganese nodule area: Earth Planet. Sci. Letters, v. 33, p. 379–388.

Björnsson, S., Arnórsson, S., and Tómasson, J., 1972, Economic evaluation of Reykjanes thermal brine area, Iceland: Am. Assoc. Petroleum Geologists Bull., v. 56, p. 2380–2391.

Black, H. T., 1975, A subsurface study of the Mesa geothermal anomaly, Imperial Valley, California: Washington, Natl. Sci. Found. Rept. NSF-RA-N-75-126, 58 p.

Bödvarsson, G., 1961, Physical characteristics of natural heat resources in Iceland: Jökull, v. 11, p. 29–38.

Bolger, G. W., 1976, Chemical evidence that the Mid-Atlantic Ridge is a source of hydrothermally derived suspended particulate material to the deep Atlantic: Unpub. Masters thesis, Univ. of South Florida, St. Petersburg, Office of Naval Research Contract N00014-72-A-0363-0001, 145 p.

Bonatti, E., 1975, Metallogenesis at oceanic spreading centers, *in* Annual Reviews Earth Planet. Sci., v. 3: Palo Alto, Calif., Annual Reviews, Inc., p. 401–431.

Bonatti, E., and Joensuu, O., 1966, Deep-sea iron deposit from the South Pacific: Science, v. 154, p. 643–645.

—— 1968, Palygorskite from Atlantic deep sea sediments: Am. Mineralogist, v. 53, p. 975–983.

Bonatti, E., Fisher, D. E., Joensuu, O., Rydell, H. S., and Beyth, M., 1972a, Iron-managanese-barium deposit from the Northern Afar Rift (Ethiopia): ECON. GEOL., v. 67, p. 717–730.

Bonatti, E., Kraemer, T., and Rydell, H., 1972b, Classification and genesis of submarine iron-manganese deposits, *in* Horn, D., ed., Ferromanganese deposits on the ocean floor: Washington, Natl. Sci. Found., p. 149–165.

Bonatti, E., Guerstein-Honnorez, B.-M., and Honnorez, J., 1976a, Copper-iron sulfide mineralizations from the equatorial Mid-Atlantic Ridge: ECON. GEOL., v. 71, p. 1515–1525.

Bonatti, E., Guerstein-Honnorez, B.-M., Honnorez, J., and Stern, C., 1976b, Hydrothermal pyrite concretions from the Romanche Trench (equatorial Atlantic): Metallogenesis in oceanic fracture zones: Earth Planet. Sci. Letters, v. 32, p. 1–10.

Bonatti, E., Zerbi, M., Kay, R., and Rydell, H. 1976c, Metalliferous deposits from the Apennine ophiolites: Mesozoic equivalent of modern deposits from oceanic spreading centers: Geol. Soc. America Bull., v. 87, p. 83–94.

Boström, K., 1970, Submarine volcanism as a source of iron: Earth Planet. Sci. Letters, v. 9, p. 348–354.

—— 1973, The origin and fate of ferromanganoan active ridge sediments: Stockholm Contr. Geology, v. 24, p. 149–243.

Boström, K., and Peterson, M. N. A., 1966, Precipitates from hydrothermal exhalations on the East Pacific Rise: ECON. GEOL., v. 61, p. 1258–1265.

—— 1969, The origin of aluminum-poor ferromanganoan sediments in areas of high heat flow on the East Pacific Rise: Marine Geology, v. 7, p. 427–447.

Boström, K., Peterson, M. N. A., Joensuu, O., and Fisher, D. E., 1969, Aluminum-poor ferromanganoan sediments on active oceanic ridges: Jour. Geophys. Research, v. 74, p. 3261–3270.

Boström, K., Joensuu, O., Valdés, S., and Riera, M., 1972, Geochemical history of South Atlantic Ocean sediments since late Cretaceous: Marine Geology, v. 12, p. 85–121.

Boström, K., Joensuu, O., Valdés, S., Charm, W., and Glaccum, R., 1976, Geochemistry and origin of East Pacific sediments sampled during DSDP Leg 34: Washington, U. S. Govt. Printing Office, Deep Sea Drilling Proj. Initial Repts., v. 34, p. 559–574.

Brewer, P. G., Densmore, C. D., Munns, R., and Stanley,

R. J., 1969, Hydrography of Red Sea brines, *in* Degens, E. T., and Ross, D. A., eds., Hot brines and recent heavy metal deposits in the Red Sea: New York, Springer-Verlag, p. 138–152.
Brewer, P. G., Spencer, D. W., and Robertson, D. E., 1972, Trace element profiles from the GEOSECS II test station in the Sargasso Sea: Earth Planet. Sci. Letters, v. 16, p. 111–116.
Brewitt-Taylor, C. R., 1975, Self-potential prospecting in the deep oceans: Geology, v. 3, p. 541–542.
Broecker, W. S., 1965, An application of natural radon to problems in ocean circulation: Symposium on Diffusion in Oceans and Fresh Waters, Lamont-Doherty Geol. Observatory, p. 116–145.
Broecker, W. S., Li, Y. H., and Cromwell, J., 1967, Radium 226 and radon 222: Concentration in Atlantic and Pacific oceans: Science, v. 158, p. 1307–1310.
Brooks, R. R., Kaplan, I. R., and Peterson, M. N. A., 1969, Trace element composition of Red Sea geothermal brine and interstitial water, *in* Degens, E. T., and Ross, D. A., eds., Hot brines and recent heavy metal deposits in the Red Sea: New York, Springer-Verlag, p. 180–203.
Browne, P. R. L., and Ellis, A. J., 1970, The Ohaki-Broadlands hydrothermal area, New Zealand: Mineralogy and related chemistry: Am. Jour. Sci., v. 296, p. 97–131.
Christensen, N. I., 1972, The abundance of serpentinites in the oceanic crust: Jour. Geology, v. 80, p. 709–719.
Chung, Y., and Craig, H., 1972, Excess-randon and temperature profiles from the eastern Equatorial Pacific: Earth Planet. Sci. Letters, v. 14, p. 55–64.
Clarke, W. B., Beg, M. A., and Craig, H., 1969, Excess ^{3}He in the sea: Evidence for terrestrial primordial helium: Earth Planet. Sci. Letters, v. 6, p. 213–220.
Coleman, R. G., 1971, Plate tectonic emplacement of upper mantle peridotites along continental edges: Jour. Geophys. Research, v. 76, p. 1212–1222.
—— 1977, Ophiolites: New York, Springer-Verlag, 229 p.
Combs, J., and Hadley, D. M., 1977, Microearthquake investigation of the Mesa geothermal anomaly, Imperial Valley, California: Geophysics, v. 42, p. 17–33.
Constantinou, G., and Govett, G. J. S., 1973, Geology, geochemistry, and genesis of Cyprus sulfide deposits: Econ. Geol., v. 68, p. 843–858.
Corliss, J. B., 1971, The origin of metal-bearing submarine hydrothermal solutions: Jour. Geophys. Research, v. 76, p. 8128–8138.
Corliss, J. B., and Ballard, R. D., 1977, Oases of life in the cold abyss: Natl. Geographic, v. 152, no. 4, p. 441–453.
Corliss, J. B., Dymond, J., Lyle, M., Doerge, T., Crane, K., Lonsdale, P., Von Herzen, R. P., and Williams, D., 1976, Sediment mound ridges of hydrothermal (?) origin along the Galapagos Rift [abs.]: Am. Geophys. Union Trans., v. 57, p. 935.
Corliss, J. B., Dymond, J. Lyle, M., Cobler, R., Williams, D., Von Herzen, R., and van Andel, T., 1977 Observations of the sediment mounds of the Galapagos Rift during the Alvin diving program [abs.]: Geol. Soc. America, Abstracts with Programs, v. 9, p. 937.
Corwin, R. F., Ebersole, W. C., and Wilde, P., 1970, A self potential detection system for the marine environment: Offshore Technology Conf. Houston, Texas 1970, OTC Paper 1258.
Craig, H., 1969, Geochemistry and origin of the Red Sea brines, *in* Degens, E. T., and Ross, D. A., eds., Hot brines and recent heavy metal deposits in the Red Sea: New York, Springer-Verlag, p. 208–242.
Craig, H., and Lupton, J. E., 1976, Primordial neon, helium and hydrogen in oceanic basalts: Earth Planet. Sci. Letters, v. 31, p. 369–385.
Craig, H., Clarke, W. B., and Beg, M. A., 1975, Excess ^{3}He in the deep water on the East Pacific Rise: Earth Planet. Sci. Letters, v. 26, p. 125–132.
Cronan, D. S., 1973, Basal ferruginous sediments cored during Leg 16, Deep Sea Drilling Project: Washington, U. S. Govt. Printing Office, Deep Sea Drilling Proj. Initial Repts., v. 16, p. 601–604.
—— 1976, Implications of metal dispersion from hydrothermal systems for mineral exploration on mid-ocean ridges and in island arcs: Nature, v. 262, p. 567–569.
Czamanske, G. K., and Moore, J. G., 1977, Composition and phase chemistry of sulfide globules in basalt from the Mid-Atlantic Ridge rift valley: Geol. Soc. America Bull., v. 88, p. 587–599.
Davis, E. E., and Lister, C. R. B., 1977, Heat flow measured over the Juan de Fuca Ridge; Evidence for widespread hydrothermal circulation in a highly heat transportive crust: Jour. Geophys. Research, v. 82, p. 4845–4860.
Deep Sea Drilling Staff, 1977, Deep Sea Drilling Proj. Initial Repts., v. 37: Washington, U. S., Govt. Printing Office, 1008 p.
Deffayes, K. S., 1970, The axial valley: A steady-state feature of the terrain, *in* Johnson, H., and Smith, B. L., eds., The megatectonics of continents and oceans: Camden, N. J., Rutgers Univ. Press, p. 194–222.
Degens, E. T., and Ross, D. A., 1969, eds., Hot brines and recent heavy metal deposits in the Red Sea: New York, Springer-Verlag, 600 p.
Detrick, R. S., Williams, D. L., Mudie, J. D., and Sclater, J. G., 1974, The Galapagos spreading center: Bottom water temperatures and the significance of geothermal heating: Royal. Astron. Soc. Geophys. Jour., v. 38, p. 627–637.
Dewey, J. F., and Bird, J. M., 1971, Origin and emplacement of the ophiolite suite: Appalachian ophiolites in Newfoundland: Jour. Geophys. Research, v.. 76, p. 3179–3206.
Dietrich, G., and Krause, G., 1969, The observations of the vertical structure of hot salty water by R. V. Meteor, *in* Degens, E. T., and Ross, D. A., eds., Hot brines and recent heavy metal deposits in the Red Sea: New York, Springer-Verlag, p. 10–17.
Dmitriev, L. V., Barsukov, V. L., and Udintsev, G. B., 1970, Rift zone in the oceans and the problem of ore formation: Geokhimiya, v. 4, p. 937.
Duke, N. A., and Hutchinson, R. W., 1974, Geological relationships between massive sulfide bodies and ophiolitic volcanic rocks near York Harbour, Newfoundland: Canadian Jour. Earth Sci., v. 11, p. 53–69.
Dymond, J., Corliss, J. B., Heath, J. R., Field, C. W., Dasch, E. J., and Veeh, H. H., 1973, Origin of metalliferous sediments from the Pacific Ocean: Geol. Soc. America Bull. v. 84, p. 3355 3372.
Edmond, J. M., Gordon, L. I., and Corliss, J. B., 1977, Chemistry of the hot springs on the Galapagos ridge axis [abs.]: Am. Geophys. Union Trans., v. 58, p. 1176.
Elder, J. W., 1965, Physical processes in geothermal areas, *in* Lee, W. H. K., ed., Terrestrial heat flow: Am. Geophys. Union, Geophys. Mon. Ser., v. 8, p. 211–239.
Engel, A. E. J., Engel, C. G., and Havens, R. G., 1965, Chemical characteristics of oceanic basalts and the upper mantle: Geol. Soc. America Bull., v. 76, p. 719–734.
Erickson, A. J., and Simmons, G., 1969, Thermal measurements in the Red Sea hot brine pools, *in* Degens, E. T., and Ross, D. A., eds., Hot brines and recent heavy metal deposits in the Red Sea: New York, Springer-Verlag, p. 114–121.
Ewing, J., 1969, Seismic model of the Atlantic Ocean, *in* Hart, P., ed., The Earth's crust and upper mantle: Washington, Am. Geophys. Union, Upper Mantle Proj. Sci. Rept. 21, p. 220–225.
Ferrara, G., Innocenti, F., Ricci, C. A., and Serri, G., 1976, Ocean-floor affinity of basalts from North Apennine ophiolites: Geochemical evidence: Chem. Geology, v. 17, p. 101–111.
Fox, P. J., and Opdyke, N. D., 1973, Geology of the oceanic crust: Magnetic properties of oceanic rocks: Jour. Geophys. Research, v. 78, p. 5139–5154.
Francis, T. J. G., 1968, The detailed seismicity of mid-oceanic ridges: Earth Planet. Sci. Letters, v. 4, p. 39–46.
Garson, M. S., and Krs, M., 1976, Geophysical and geological evidence of the relationship of Red Sea transverse tectonics to ancient fractures: Geol. Soc. America Bull., v. 87, p. 169–181.
Girdler, R. W., 1970, A review of Red Sea heat flow: Royal Soc. [London] Philos. Trans., A, v. 267, p. 191–203.

Goodfellow, W. D., 1975, Major and minor elements halos in volcanic rocks at Brunswick No. 12 sulfide deposit, N.B., Canada, *in* Elliott, I. L., and Fletcher, W. K. eds., Geochemical exploration 1974, 5th Internat. Geochem. Exploration Symposium: Amsterdam, Elsevier Sci. Pub. Co., p. 279–295.

Govett, G. J. S., and Pantazis, Th. M., 1971, Distribution of Cu, Zn, Ni and Co in the Troodos pillow lavas series, Cyprus: Inst. Mining Metallurgy Trans., v. 80, sec. B, p. B27–46.

Hajash, A., 1975, Hydrothermal processes along mid-ocean ridges: An experimental investigation: Contrib. Mineralogy Petrology, v. 53, p. 205–226.

Halbouty, M. T., Meyerhoff, A. A., King, R. E., Dott, R. H., Sr., Klemme, H. D., and Shabad, T., 1970, World's giant oil and gas fields, geologic factors affecting their formation, and basin classification, *in* Halbouty, M.T., ed., Geology of giant petroleum fields: Am. Assoc. Petroleum Geologists Mem. 14, p. 502–555.

Hanor, J. S., 1966, The origin of barite: Unpub. Ph.D. thesis, Harvard Univ.

Hart, R. A., 1973, A model for chemical exchange in the basalt-seawater system of oceanic layer II: Canadian Jour. Earth Sci., v. 10, p. 799–816.

Hawkins, D. B., and Roy, R., 1963, Experimental hydrothermal studies on rock alteration and clay mineral formation: Geochim. et Cosmochim. Acta, v. 27, p. 1047–1054.

Helgeson, H. C., 1964, Complexing and hydrothermal ore deposition: New York, Pergamon Press, 128 p.

—— 1967, Solution chemistry and metamorphism, *in* Abelson, P. H., ed., Researches in geochemistry: New York, John Wiley and Sons, p. 362–404.

—— 1968, Evaluation of irreversible reactions in geochemical processes involving minerals and aqueous solutions—I. Thermodynamic relations: Geochim. et Cosmochim. Acta, v. 32, p. 853–877.

—— 1969, Thermodynamics of hydrothermal systems at elevated temperatures and pressures: Am. Jour. Sci., v. 267, p. 729–804.

Helgeson, H. C., Garrels, R. M., and MacKenzie, R. T., 1969, Evaluation of irreversible reactions in geochemical processes involving minerals and aqueous solutions—II. Applications: Geochim. et Cosmochim. Acta, v. 33, p. 455–482.

Henderson, J. R., White, B. L., and others, 1963, Aeromagnetic map of Long Valley and northern Owens Valley, California: U. S. Geol. Survey Geophys. Inv. Map GP-329.

Hendricks, R. L., Reisbeck, F. B., Mahaffey, E. J., Roberts, D. B., and Peterson, M. N. A., 1969, Chemical composition of sediments and interstitial brines from the Atlantis II, Discovery, and Chain Deeps, *in* Degens, E. T., and Ross, D. A., eds., Hot brines and recent heavy metal deposits of the Red Sea: New York, Springer-Verlag, p. 407–440.

Holland, H. D., 1967, Gangue minerals in hydrothermal systems, *in* Barnes, H. L., ed., Geochemistry of hydrothermal ore deposits: New York, Holt, Rinehart and Winston, p. 382–436.

Horowitz, A., and Cronan, D. S., 1976, The geochemistry of basal sediments from the North Atlantic Ocean: Marine Geology, v. 20, p. 205–228.

Hunt, J. M., Hayes, E. E., Degens, E. T., and Ross, D. A., 1967, Red Sea: Detailed survey of hot-brine areas: Science, v. 156, p. 514–516.

Hutchinson, R. W., 1973, Volcanogenic sulfide deposits and their metallogenic significance: Econ. Geol., v. 68, p. 1223–1246

Irving, E., Park, J. K., Haggerty, S. E., Aumento, F., and Loncarevic, B., 1970a, Magnetism and opaque mineralogy of basalts from the Mid-Atlantic Ridge at 45°N: Nature, v. 228, p. 974–976.

Irving, E., Robertson, W. A., and Aumento, F., 1970b, The Mid-Atlantic Ridge near 45°N, 6, Remanent intensity, susceptibility, and iron content of dredged samples: Canadian Jour. Earth Sci., v. 7, p. 1–13.

Jarzabek, D., and Combs, J., 1976, Microearthquake survey of the Dunes KGRA, Imperial Valley, Southern California [abs.]: Geol. Soc. America, Abstracts with Programs, v. 8, p. 939.

Jenkins, W. J., and Clarke, W B., 1976, The distribution of ^{3}He in the western Atlantic Ocean: Deep-Sea Research, v. 23, p. 481–494.

Jenkins, W. J. Beg, M. A., Clarke, W. B., Wangersky, P. J., and Gray, H., 1972, Excess ^{3}He in the Atlantic Ocean: Earth Planet. Sci. Letters, v. 16, p. 122-126.

Johnson, D. M., and Combs, J., 1976, Microearthquake survey of the Kilbourne Hole KGRA, south central New Mexico [abs.]: Geol. Soc. America, Abstracts with Programs, v. 8, p. 942.

Kappelmeyer, O., and Haenel, R., 1974, Geothermics with special reference to application: Berlin, Gebruder Borntraeger, Geoexploration Monographs, ser. 1, no. 4, 238 p.

Karig, D. E., 1971, Origin and development of marginal basins in the western Pacific: Jour. Geophys. Research, v. 76, p. 2542–2561.

Keays, R. R., and Scott, R. B., 1976, Precious metals in ocean-ridge basalts: Implications for basalt as source rocks for gold mineralization: Econ. Geol., v. 71, p. 705–720.

Kelley, V. C., and Soske, J. L., 1936, Origin of the Salton volcanic domes, Salton Sea, California: Jour. Geology, v. 44, p. 496–509.

Kirst, P. W., 1976, Petrology of metamorphic rocks from the Mid-Atlantic Ridge and fracture zones: Unpub. Ph.D. thesis, Univ. Miami.

Klein, F. W., Einarsson, P., and Wyss, M., 1973, Microearthquakes on the Mid-Atlantic plate boundary on the Reykjanes Peninsula in Iceland: Jour. Geophys. Research, v. 78, p. 5084–5099.

Koenig, J. B., 1967, The Salton-Mexicali geothermal province: California Div. Mines Geology, Mineral Inf. Service, v. 20, p. 75–81.

Krauskopf, K. B., 1956, Factors controlling the concentrations of thirteen rare metals in seawater: Geochim. et Cosmochim. Acta, v. 9, p. 1-32B.

—— 1957, Separation of maganese from iron in sedimentary processes: Geochim. et Cosmochim. Acta, v. 12, p. 61–84.

Kutina, J., 1974, Structural control of volcanic ore deposits in the context of global tectonics: Bull. volcanol., v. 38, p. 1039–1069.

Langseth, M. G., and Von Herzen, R. P., 1971, Heat flow through the floor of the world oceans, *in* Maxwell, A. E., ed., The sea,: New York, Wiley-Interscience, v. 4, part 1, p. 299–352.

Lister, C. R. B., 1972, On the thermal balance of an oceanic ridge: Royal Astron. Soc. Geophys. Jour., v. 26, p. 515–535.

—— 1974a, On the penetration of water into hot rocks: Royal Astron. Soc. Geophys. Jour., v. 39, p. 465–509.

—— 1974b, Water percolation in the oceanic crust: Am. Geophys. Union Trans., v. 55, p. 740–742.

Lonsdale, P., 1977a, Clustering of suspension-feeding macrobenthos near abyssal hydrothermal vents at oceanic spreading centers: Deep-Sea Research, v. 24, p. 857–863.

—— 1977b, Deep-tow observations at the Mounds Abyssal Hydrothermal Field, Galapagos Rift: Earth Planet. Sci. Letters, v. 36, p. 92–110.

Lowell, R. P., and Rona, P. A., 1976, On the interpretation of near-bottom water temperature anomalies: Earth Planet. Sci. Letters, v. 32, p. 18–24.

Lupton, J. E., Weiss, R. F., and Craig, H., 1977a, Mantle helium in the Red Sea brines: Nature, v. 266, p. 244–246.

—— 1977b, Mantle helium in hydrothermal plumes in the Galapagos Rift: Nature, v. 267, p. 603–604.

Luyendyk, B. P., and Melson, W. G., 1967, Magnetic properties and petrology of rocks near the crest of the Mid-Atlantic Ridge: Nature, v. 215, p. 147–149.

Lyle, M., Dymond, J., and Heath, G. R., 1977, Copper-nickel-enriched ferromanganese nodules and associated crusts from the Bauer Basin, northwest Nazcas plate: Earth Planet. Sci. Letters, v. 35, p. 55–64.

McGregor, B. A., and Rona, P. A., 1975, Crest of Mid-Atlantic Ridge at 26°N: Jour. Geophys. Research, v. 80, p. 3307–3314.

McGregor, B. A., Harrison, C. G. A., Lavelle, J. W., and Rona, P. A., 1977, Magnetic anomaly pattern on Mid-Atlantic Ridge crest at 26°N: Jour. Geophys. Research, v. 82, p. 231–238.

McNitt, J. R., 1965, Review of geothermal resources, *in* Lee, H. K., ed., Terrestrial heat flow: Am. Geophys. Union, Geophys. Mon. Ser. 8, p. 240–266.

Mahon, W. A. F., 1966, Silica in hot water discharged from drill holes at Wairakei, New Zealand: New Zealand Jour. Sci., v. 9, p. 135–144.

Makharadze, A. I., and Ikoshvili, D. V., 1970, On barite from the Chiatur manganese deposit: Akad. Nauk SSSR Doklady, v. 190, p. 1204–1206 (in Russian).

Marchesini, E., Pistolesi, A., and Bolognini, M., 1962, Fracture patterns of the natural steam area of Larderello, Italy, from air photographs: Symposium on Photo Interpretation, Delft 1962, p. 524–532.

Marshall, M., and Cox, A., 1972, Magnetic changes in pillow basalt due to sea floor weathering: Jour. Geophys. Research, v. 77, p. 6459–6469.

Melson, W. G., Thompson, G., and van Andel, T. H., 1968, Volcanism and metamorphism in the Mid-Atlantic Ridge, 22° N: Jour. Geophys. Research, v. 73, p. 5925–5941.

Menard, H. W., 1967, Sea floor spreading, topography, and the second layer: Science, v. 157, p. 923–924.

Miller, A. R., Densmore, C. D., Degens, E. T., Hathaway, J. C., Manheim, F. T., McFarlin, P. F., Pocklington, H., and Jokela, A., 1966, Hot brines and recent iron deposits in deep of the Red Sea: Geochim. et Cosmochim. Acta, v. 30, p. 341–359.

Mitchell, A. H., and Bell, J. D., 1973, Island-arc evolution and related mineral deposits: Jour. Geology, v. 81, p. 381–405.

Mitchell, A. H., and Grason, M. S., 1976, Mineralization at plate boundaries: Minerals Sci. Eng., v. 8, no. 2, p. 129–169.

Miyashiro, A., 1973, The Troodos ophiolitic complex was probably formed in an island arc: Earth Planet. Sci. Letters, v. 19, p. 218–224.

—— 1975, Classification, characteristics, and origin of ophiolites: Jour. Geology, v. 83, p. 249–281.

Miyashiro, A., and Shido, F., 1970, Progressive metamorphism in zeolite assemblages: Lithos, v. 3, p. 251–260.

Miyashiro, A., Shido, F., and Ewing, M., 1971, Metamorphism in the Mid-Atlantic Ridge near 24° and 30° N: Royal Soc. [London] Philos. Trans., v. 268, p. 589–603.

Moore, J. G., and Calk, L., 1971, Sulfide spherules in vesicles of dredged pillow basalt: Am. Mineralogist, v. 56, p. 476–488.

Moore, J. G., and Schilling, J. G., 1973, Vesicles water and sulfur in Reykjanes Ridge basalt: Contr. Mineralogy Petrology, v. 41, p. 105–118.

Moore, W. S., and Vogt, P. G., 1976, Hydrothermal manganese crusts from two sites near the Galapagos spreading axis: Earth Planet. Sci. Letters, v. 29, p. 349–356.

Moores, E. M., and Vine, F. J., 1971, The Troodos Massif, Cyprus and other ophiolites as oceanic crust: Evaluation and implications: Royal Soc. [London] Philos. Trans., A, v. 268, p. 443–466.

Mottl, M. J., Corr, R. F., and Holland, H. D., 1974, Chemical exchange between sea water and mid-ocean ridge basalt during hydrothermal alteration: An experimental study [abs.]: Geol. Soc. America, Abstracts with Programs, v. 6, p. 879–880.

Muehlenbachs, K., and Clayton, R. H., 1972, Oxygen isotope geochemistry of submarine greenstones: Canadian Jour. Earth Sci., v. 9, p. 471–478.

Nebrija, E. L., Young, C. T., Meyer, R. P., and Moore, J. R., 1976, Electrical prospecting for copper veins in shallow water: Offshore Technology Conf. Houston 1976, Paper OTC 2454, p. 319–327.

Nicholls, G. D., and Islam, M. R., 1971, Geochemical investigations of basalts and associated rocks from the ocean floor and their implications: Royal Soc. [London] Philos. Trans., A, v. 268, p. 469–486.

Ostapoff, F., 1969, A fourth brine hole in the Red Sea?, *in* Degens, E. T., and Ross, D. A., eds., Hot brines and recent heavy metal deposits in the Red Sea: New York, Springer-Verlag, p. 18–24.

Palmason, G., 1967, On heat flow in Iceland in relation to the Mid-Atlantic Ridge, *in* Björnsson, S., ed., Iceland and mid-ocean ridges: Soc. Sci. Islandica, v. 38, p. 111–127.

Patella, D., and Schiavone, D., 1974, A theoretical study of the kernel function for resistivity prospecting with a Schlumberger apparatus in water: Annali di Geofisica, v. 28, p. 295–313.

Pereira, J., and Dixon, C. J., 1971, Mineralization and plate tectonics: Mineralium Deposita, v. 6, p. 404–405.

Peters, T., and Kramers, J. D., 1974, Chromite deposits in the ophiolite complex of Northern Oman: Mineralium Deposita, v. 9, p. 253–259.

Peterson, J. J., Fox, P. J., and Schreiber, E., 1974, Newfoundland ophiolites and the geology of the oceanic layer: Nature, v. 247, p. 194–196.

Phillips, J. D., Woodside, J., and Bowin, C. O., 1969, Magnetic and gravity anomalies in the central Red Sea, *in* Degens, E. T., and Ross, D. A., eds., Hot brines and recent heavy metal deposits in the Red Sea: New York, Springer-Verlag, p. 98–113.

Pugh, D. T., 1969, Temperature measurements in the bottom layers of the Red Sea brines, *in* Degens, E. T., and Ross, D. A., eds., Hot brines and recent heavy metal deposits in the Red Sea: New York, Springer-Verlag, p. 158–163.

Quillan, R., and Combs, J., 1976, Microearthquake survey of the Radium Springs KGRA, south central New Mexico [abs.]: Geol. Soc. America, Abstracts with Programs, v. 8, p. 1055.

Rona, P. A., 1969, Possible salt domes in the deep Atlantic off northwest Africa: Nature, v. 224, p. 141–143.

—— 1970, Comparison of continental margins of eastern North America at Cape Hatteras and northwestern Africa at Cap Blanc: Am. Assoc. Petroleum Geologists Bull., v. 54, p. 129–157.

—— 1976a, Criteria for recognition of sea floor hydrothermal mineral deposits: Am. Geophys. Union Trans., v. 57, p. 593.

—— 1976b, Pattern of hydrothermal mineral deposition: Mid-Atlantic Ridge crest at latitude 26°N: Marine Geology, v. 21, p. M59–M66.

—— 1977, Plate tectonics, energy and mineral resources: Basic research leading to payoff: Am. Geophys. Union Trans., v. 58, p. 629–639.

—— 1978, Magnetic signatures of hydrothermal alteration and volcanogenic mineral deposits in oceanic crust: Jour. Volcanology Geothermal Research, v. 3 (in press).

Rona, P. A., McGregor, B. A., Betzer, P. R., Bolger, G. W., and Krause, D. C., 1975, Anomalous water temperatures over the Mid-Atlantic Ridge crest at 26°N latitude: Deep-Sea Research, v. 22, p. 611–618.

Rona, P. A., Harbison, R. N., Bassinger, B. G., Scott, R. B., and Nalwalk, A. J., 1976, Tectonic fabric and hydrothermal activity of Mid-Atlantic Ridge crest (lat. 26°N): Geol. Soc. America Bull., v. 87, p. 661–674.

Ross, D. A., 1972, Red Sea hot brine area: Revisited: Science, v. 175, p. 1455–1457.

Ross, D. A., Hays, E. E., and Allstrom, F. C., 1969, Bathymetry and continuous seismic profiles of the hot brine region of the Red Sea, *in* Degens, E. T., and Ross, D. A., eds., Hot brines and recent heavy metal deposits in the Red Sea: New York, Springer-Verlag, p. 82–97.

Ryan, W. B. F., Thorndike, E. M., Ewing, M., and Ross, D. A., 1969, Suspended matter in the Red Sea brines and its detection by light scattering, *in* Degens, E. T., and Ross, D. A., eds., Hot brines and recent heavy mineral deposits of the Red Sea: New York, Springer-Verlag, p. 153–157.

Rydell, H. S., and Bonatti, E., 1973, Uranium in submarine metalliferous deposits: Geochim. et Cosmochim Acta, v. 37, p. 2557–2565.

Sato, M., and Mooney, H. M., 1960, The electrochemical mechanism of sulfide self-potentials: Geophysics, v. 25, p. 226–249.

Sayles, F. L., and Bischoff, J. L., 1973, Ferromanganoan sediments in the equatorial East Pacific: Earth Planet. Sci. Letters, v. 19, p. 330–336.
Scientific Staff, 1974, Leg 34 Oceanic basalt and the Nazca Plate: Geotimes, v. 19, p. 20–24.
Sclater, J. G., Von Herzen, R. P., Williams, D. L., Anderson, R. N., and Klitgord, K., 1974, The Galapagos spreading center: Heat-flow low on the north flank: Royal Astron. Soc. Geophys. Jour., v. 38, p. 609–626.
Scott, M. R., 1976, Accumulation rates of metals along active ocean ridges [abs.]: Joint Oceanographic Assembly, Edinburgh, Scotland 1976, Abstracts, p. 94.
—— 1977, Metal accumulation rates in sediments from the FAMOUS area on the Mid-Atlantic Ridge [abs.]: Am. Geophys. Union Trans., v. 58, p. 420.
Scott, M. R., Scott, R. B., Rona, P. A., Butler, L. W., and Nalwalk, A. J., 1974, Rapidly accumulating manganese deposit from the median valley of the Mid-Atlantic Ridge: Geophys. Research Letters, v. 1, p. 355–358.
Searle, D. L., 1972, Mode of occurrence of the cupiferous pyrite deposits of Cyprus: Inst. Mining Metallurgy Trans., v. 81, sec. B, p. B189–B197.
SEG, 1966, Mining geophysics, v. 1, case histories: Tulsa, Oklahoma, Soc. Exploration Geophysicists, 492 p.
Sigvaldason, G. E., 1962, Epidote and related minerals in two deep geothermal drill holes, Reykjavík and Hoveragerdi, Iceland: U. S. Geol. Survey Prof. Paper 450-E, p. 77–79.
Sillitoe, R. H., 1972, Formation of certain massive sulphide deposits at sites of sea-floor spreading: Inst. Mining Metallurgy Trans., sec. B, v. 81, p. B141–148.
—— 1973, Environments of formation of volcanogenic massive sulfide deposits: Econ. Geol., v. 68, p. 1321–1325.
Spooner, E. T. C., and Fyfe, W. S., 1973, Subsea-floor metamorphism, heat and mass transfer: Contr. Mineralogy Petrology, v. 42, p. 287–304.
Spooner, E. T. C., Beckinsale, R.. D., Fyfe, W. S., and Smewing, J. D., 1974, O^{18} enriched ophiolitic metabasic rocks from E. Liguria (Italy), Pindos (Greece), and Troodos (Cyprus): Contr. Mineralogy Petrology, v. 47, p.. 41–62.
Steiner, A., 1967, Clay minerals in hydrothermally altered rocks at Wairakei, New Zealand: Clays Clay Minerals, v.. 16, p. 193–213.
Stephens, J. D., and Wittkopp, R. W., 1969, Microscopic and electron beam microscopic study of sulfide minerals in Red Sea mud samples, *in* Degens, E. T., and Ross, D. A., eds., Hot brines and recent heavy metal deposits of the Red Sea: New York, Springer-Verlag, p. 441–447.
Strong, D. F., 1974a, Plate tectonic setting of Newfoundland mineral occurrences, *in* Strong, D. F., ed., Metallogeny and plate tectonics, a guidebook to Newfoundland mineral deposits: St. John's, Newfoundland, NATO Adv. Studies Inst., p. 3–27.
—— 1974b, An "off-axis" alkali volcanic suite associated with the Bay of Islands ophiolites, Newfoundland: Earth Planet. Sci. Letters, v. 21, p. 301–309.
Studt, F. E., 1959, Magnetic survey of the Wairakei hydrothermal field: New Zealand Jour. Geology Geophysics, v. 2, p. 746–754.
—— 1961, Preliminary survey of the hydrothermal field at Rabaul, New Britain: New Zealand Jour. Geology Geophysics, v. 4, p. 274–282.
Swallow, J. C., 1969, History of the exploration of the hot brine area of the Red Sea—DISCOVERY account, *in* Degens, E. T., and Ross, D. A., eds., Hot brines and recent heavy metal deposits in the Red Sea: New York, Springer-Verlag, p. 3–9.
Sykes, L. R., 1970, Earthquake swarms and sea-floor spreading: Jour. Geophys. Research, v. 75, p. 6598–6611.
Talwani, M., Windish, C. C., and Langseth, M. G., Jr., 1971, Reykjanes Ridge crest: A detailed geophysical study: Jour. Geophys. Research, v. 76, p. 473–517.
Thompson, G., Woo, C. C., and Sung, W., 1975, Metalliferous deposits on the Mid-Atlantic Ridge [abs.]: Geol. Soc. America, Abstracts with Programs, v. 7, p. 1297–1298.
Tómasson, J., and Kristmannsdóttir, H., 1972, High temperature alteration minerals and thermal brines, Reykjanes, Iceland: Contr. Mineralogy Petrology, v. 36, p. 123–134.
Turner, J. S., 1969, A physical interpretation of the observations of hot brine layers in the Red Sea, *in* Degens, E. T., and Ross, D. A., eds., Hot brines and recent heavy metal deposits in the Red Sea: New York, Springer-Verlag, p. 164–173.
United Nations, 1976, United Nations Symposium on the Development and Use of Geothermal Resources, 2nd, San Francisco, May 1975: Washington, U. S. Govt. Printing Office, Proceedings, 3 vols., p. 1–2457.
Upadhyay, H. D., and Strong, D. F., 1973, Geological setting of the Betts Cove copper deposits, Newfoundland: An example of ophiolite sulfide mineralization: Econ. Geol., v. 68, p. 161–167.
Veeh, H. H., and Boström, K., 1971, Anomalous $^{234}U/^{238}U$ on the East Pacific Rise: Earth Planet. Sci. Letters, v. 10, p. 372–374.
Von Herzen, R., Green, K. E., and Williams, D., 1977, Hydrothermal circulation at the Galapagos spreading center [abs.]: Geol. Soc. America, Abstracts with Programs, v. 9, p. 1212–1213.
Walker, G. P. L., 1960, Zeolite zones and dike distribution in relation to the structure of the basalts of eastern Iceland: Jour. Geology, v. 68, p. 515–528.
Ward, P. L., and Björnsson, S., 1971, Microearthquakes, swarms, and the geothermal areas of Iceland: Jour. Geophys. Research, v. 76, p. 3953–3982.
Ward, P. L., Pálmason, G., and Drake, C., 1969, Microearthquake survey and the Mid-Atlantic Ridge in Iceland: Jour. Geophys. Research, v. 74, p. 665–684.
Watkins, N. D., and Paster, T. P., 1971, The magnetic properties of igneous rocks from the ocean floor: Royal Soc. [London] Philos. Trans., A., v. 268, p. 507–550.
Watson, S. W., and Waterbury, J. B., 1969, The sterile hot brines of the Red Sea, *in* Degens, E. T., and Ross, D. A., eds., Hot brines and recent heavy metal deposits in the Red Sea: New York, Springer-Verlag, p. 272–281.
Weiss, R. F., Lonsdale, P., Lupton, J. E., Bainbridge, A. E., and Craig, H., 1977, Hydrothermal plumes in the Galapagos Rift: Nature, v. 267, p. 600–603.
Whiteley, R. G., 1974, Design and preliminary testing of a continuous offshore resistivity method: Australian Soc. Explor. Geophysicists Bull., v. 5, p. 9–13.
Williams, D. L., and Von Herzen, R. P., 1974, Heat loss from the Earth: New estimate: Geology, v. 2, p. 327–328.
Williams, D. L., Von Herzen, R. P., Sclater, J. G., and Anderson, R. N., 1974, The Galapagos spreading center: Lithospheric cooling and hydrothermal circulation: Royal. Astron. Soc. Geophys. Jour., v. 38, p. 587–608.
Wilson, R. A. M., and Ingham, F. T., 1959, The geology of the Xeros-Troodos area, with an account of the mineral resources: Cyprus Geol. Survey Dept. Mem. 1, p. 1–177.
Woodli, R., ed., 1964, Methods of prospecting for chromite: Paris, Organization Econ. Coop. Development, 200 p.

REFERENCES

Anderson, R. N., and M. A. Hobart. 1976. The relation between heat flow, sediment thickness, and age in the eastern Pacific. *Jour. Geophys. Research* **81**: 2968-2989.

Anderson, R. N., M. G. Langseth, and J. G. Sclater. 1977. The mechanisms of heat transfer through the floor of the Indian Ocean. *Jour. Geophys. Research* **82**: 3391-3409.

Anderson, R. N., M. A. Hobart, and M. G. Langseth, Jr. 1978. Convective heat transfer in the oceanic crust and sediment on the flanks of mid-ocean ridges in the Indian Ocean. *EOS* **59**:384.

Anderson, R. N., M. A. Hobart, and M. G. Langseth, Jr. 1979. Geothermal convection through oceanic crust and sediments in the Indian Ocean. *Science* **204**:828-832.

Arrhenius, G., and E. Bonatti. 1965. Neptunism and volcanism in the ocean. In *Progress in Oceanography*, vol. 3, M. Sears, ed. London: Pergamon Press, pp. 7-22.

Aumento, F., and K. D. Sullivan. 1974. Deep drill investigations of the oceanic crust in the North Atlantic. In *Geodynamics of Iceland and the North Atlantic Area*, L. Kristjansson, ed., NATO Adv. Study Inst., Ser. C, Math. Phys. Sci., **11**:83-103.

Bäcker, H. 1975. Exploration of the Red Sea and Gulf of Aden during the *M. S. Valdivia* cruises "Erzschlamme A" and "Erzschlamme B." *Geol. Jahrb.* **D13**:3-78.

Bender, M., W. Broecker, V. Gornitz, V. Middel, R. Kay, S. Sun, and P. Biscaye. 1971. Geochemistry of three cores from the East Pacific Rise. *Earth and Planetary Sci. Letters* **12**:425-433.

Betzer, P. R., G. W. Bolger, B. A. McGregor, and P. A. Rona. 1974. The Mid-Atlantic Ridge and its effect on the composition of particulate matter in the deep ocean (abs.). *EOS* **55**:293.

Betzer, P. R., D. W. Eggimann, G. W. Bolger, S. B. Betzer, and K. L. Carder. 1980. Iron and manganese concentrations in particulate matter over the Mid-Atlantic Ridge at 11°N: Evidence for an input of materials from the ridge to the deep ocean. In press.

References

Bignell, R. D., D. S. Cronan, and J. S. Tooms. 1976a. Metalliferous brine precipitates. In *Metallogeny and plate tectonics,* D. F. Strong, ed., *Geol. Assoc. Canada Spec. Paper 14,* pp. 147-184.

Bignell, R. D., D. S. Cronan, and J. S. Tooms. 1976b. Metal dispersion in the Red Sea as an aid to marine geochemical exploration. *Inst. Mining and Metallurgy Trans., Sect. B, Applied Earth Science* **85**:B274-278.

Bischoff, J. L. 1980. Geothermal system at 21°N, East Pacific Rise: Physical limits on geothermal fluid and role of adiabatic expansion. *Science* **207**:1465-1469.

Bischoff, J. L., and F. W. Dickson. 1975. Seawater-basalt interaction at 200°C and 500 bars: Implications for origin of seafloor heavy metal deposits and regulation of seawater chemistry. *Earth and Planetary Sci. Letters* **25**:385-397.

Bischoff, J. L., and W. E. Seyfried. 1978. Hydrothermal chemistry of seawater from 25°C to 350°C. *Am. Jour. Sci.* **278**:838-860.

Björnsson, S., S. Arnórsson, and J. Tómasson. 1972. Economic evaluation of Reykjanes thermal brine area, Iceland. *Amer. Assoc. Petroleum Geologists Bull.* **56**: 2380-2391.

Bodvarsson, G., and R. P. Lowell. 1972. Ocean floor heat flow and the circulation of interstitial waters. *Jour. Geophys. Research* **77**:4472-4475.

Bolger, G. W. 1976. *Chemical evidence that the Mid-Atlantic Ridge is a source of hydrothermally derived suspended particulate material to the deep Atlantic.* M.S. thesis, University of South Florida, St. Petersburg, 145 pp. (Office of Naval Research Contract N00014-72-A-0363-0001.)

Bolger, G. W., P. R. Betzer, and V. V. Gordeev. 1978. Hydrothermally derived manganese suspended over the Galapagos spreading center. *Deep-Sea Research* **25**: 721-733.

Bonatti, E. 1975. Metallogenesis at oceanic spreading centers. *Annual Review of Earth and Planetary Sciences,* vol. 3, Palo Alto, Cal.: Annual Reviews, Inc., pp. 401-431.

Bonatti, E., T. Kraemer, and H. Rydell. 1972. Classification and genesis of submarine iron-manganese deposits. In *Ferromanganese Deposits on the Ocean Floor,* D. Horn, ed. Washington, D.C.: National Science Foundation, pp. 149-165.

Boström, K. 1973. The origin and fate of ferromanganese active ridge sediments. *Stockholm Contr. Geology* **24**:149-243.

Browne, P. R. L. 1978. Hydrothermal alteration in active geothermal fields. *Annual Review of Earth and Planetary Sciences,* vol. 16, Palo Alto, Ca: Annual Reviews, Inc. pp. 229-250.

Bruneau, L., N. G. Jerlov, and F. Koczy. 1953. Physical and chemical methods. *Swedish Deep-Sea Exped. Report,* vol. III, Physics and Chemistry no. 4, Appendix, Table 1, Physical and chemical data, p. XXIX, pp. 101-112.

Cann, J. R., C. K. Winter, and R. G. Pritchard. 1977. A hydrothermal deposit from the floor of the Gulf of Aden. *Mineralog. Ma.* **41**:193-199.

Christensen, N. I. 1972. The abundance of serpentinites in the oceanic crust. *Jour. Geology* **80**:709-719.

Clarke, W. B., M. A. Beg, and H. Craig. 1969. Excess ^{3}He in the sea: Evidence for terrestrial primordial helium. *Earth and Planetary Sci. Letters* **6**:213-220.

Corliss, J. B., M. Lyle, J. Dymond, and K. Crane. 1978. The chemistry of hydrothermal mounds near the Galapagos Rift. *Earth and Planetary Sci. Letters* **40**: 12-24.

Corliss, J. B., J. Dymond, L. I. Gordon, J. M. Edmond, R. P. Von Herzen, R. D. Ballard, K. Green, D. Williams, A. Bainbridge, K. Crane, and T. H. van Andel. 1979. Submarine thermal springs on the Galapagos Rift. *Science* **203**:1073-1083.

Crane, K. 1979. Hydrothermal stress drops and convective patterns at three mid-ocean spreading centers. *Tectonophysics* **55**:215-238.

Crane, K., and W. R. Normark. 1977. Hydrothermal activity and crustal structure of the East Pacific Rise at 21°N. *Jour. Geophys. Research* **82**:5336-5348.

Cronan, D. S. 1976. Basal metalliferous sediments from the eastern Pacific. *Geol. Soc. America Bull.* **87**:928-934.

Deffeyes, K. S. 1970. The axial valley: A steady state feature of the terrain. In *The Megatectonics of Continents and Oceans,* H. Johnson, and B. L. Smith, eds. Camden, N.J.: Rutgers University Press, pp. 194-222.

Degens, E. T., and D. A. Ross, eds. 1969. *Hot Brines and Recent Heavy Metal Deposits in the Red Sea.* New York: Springer-Verlag, 600 pp.

Dickson, F. W., C. W. Blount, and Tunell, G. 1963. Use of hydrothermal solution equipment to determine the solubility of anhydrite in water from 100°C to 275°C and from 1 bar to 1000 bars. *Am. Jour. Sci.* **261**:61-78.

Dmitriev, L. V., V. L. Barsukov, and G. B. Udintsev. 1970. Rift zone in the oceans and the problem of ore formulation. *Geokhimya* **4**:937.

Drever, J. I., ed. 1977. *Seawater: Cycles of the Major Elements.* Stroudsburg, Pa.: Dowden, Hutchinson & Ross, 344 pp.

Drever, J. I. 1974. The magnesium problem. In *The Sea,* vol. 5, E. D. Goldberg, ed. New York: John Wiley & Sons, pp. 337-357.

Duke, N. A., and R. W. Hutchinson. 1974. Geological relationships between massive sulfide bodies and ophiolitic volcanic rocks near York Harbour, Newfoundland. *Canadian Jour. Earth Sci.* **11**:53-69.

Dymond, J., J. B. Corliss, J. R. Heath, C. W. Field, E. J. Dash, and H. H. Veeh. 1973. Origin of metalliferous sediments from the Pacific Ocean. *Geol. Soc. America Bull.* **84**:3355-3372.

Edmond, J. M., 1980. Ridge crest hot springs: The story so far. *EOS* **61**:129-131.

Elder, J. W. 1965. Physical processes in geothermal areas. In *Terrestrial Heat Flow,* W. H. K. Lee, ed., *Am. Geophys. Union Geophys. Mon.* **8**:211-239.

Elderfield, H., E. Gunnlaugsson, S. J. Wakefield, and P. T. Williams. 1977. The geochemistry of basalt-seawater interactions: Evidence from Deception Island, Antarctica, and Reykjanes, Iceland. *Mineralog. Mag.* **41**:217-226.

Fehn, U., and L. Cathles. 1978. Hydrothermal convection through oceanic crust between 0 and 70 m.y. old. *EOS* **59**:384.

Fehn, U., and L. Cathles. 1979. Hydrothermal convection at slow-spreading mid-ocean ridges. *Tectonophysics* **55**:239-250.

Ferguson, J., and I. B. Lambert. 1972. Volcanic exhalations and metal enrichments at Matupai Harbour, New Britain, T. P. N. G.: *Econ. Geology* **67**:25-37.

Francheteau, J., H. D. Needham, P. Choukroune, T. Juteau, M. Seguret, R. D. Ballard, P. J. Fox, W. Normark, A. Carranza, D. Cordoba, J. Guerrero, C. Rangin, H. Bougault, P. Cambon, and R. Hekinian. 1979. Massive deep-sea sulfide ore deposits discovered on the East Pacific Rise. *Nature* **277**:523-528.

Hajash, A. 1977. Experimental seawater/basalt interactions: Effects of water/rock ratio and temperature gradient. *Geol. Soc. America Abs. with Programs* **9**:1002.

Hall, J. M., and P. T. Robinson. 1979. Deep crustal drilling in the North Atlantic Ocean. *Science* **204**:573-586.

Hart, R. A. 1973. A model for chemical exchange in the basalt-seawater system of of oceanic layer II. *Canadian Jour. Earth Sci.* **10**:799-816.

Hart, S. R., and H. Staudigel. 1978. Oceanic crust: Age of hydrothermal alteration. *Geophys. Research Letters* **5**:1009-1012.

Hartline, B. K., C. R. B. Lister, and J. R. Booker. 1975. Topographically forced low Rayleigh number convection in a porous medium. *EOS* **56**:1065.

Hartline, B. K., C. R. B. Lister, and J. R. Booker. 1978. Topographically forced "subcritical" thermal convection in porous media. *EOS* **59**:384.

Heath, G. R., and J. Dymond. 1977. Genesis and transformation of metalliferous sediments from the East Pacific Rise, northwest Nazca plate. *Geol. Soc. America Bull.* **88**:723-733.

Hekinian, R., B. R. Rosendahl, D. S. Cronan, Y. Dmitriev, R. V. Foder, R. M. Goll, M. Hoffert, S. E. Humphris, D. P. Mattey, J. Natland, N. Petersen, W. Roggenthen, E. L. Schrader, R. K. Srivastava, and N. Warren. 1978. Hydrothermal deposits and associated basement rocks from the Galapagos spreading center. *Oceanology Acta* **1**:473-482.

Hekinian, R., M. Fevrier, J. L. Bischoff, P. Picot, and W. C. Shanks. 1980. Sulfide deposits from the East Pacific Rise near 21°N. *Science* **207**:1433-1444.

Hoffert, M., A. Perseil, R. Hekinian, P. Choukroune, H. D. Needham, J. Francheteau, and X. LePichon. 1978. Hydrothermal deposits sampled by diving saucer in Transform Fault "A" near 37°N on the Mid-Atlantic Ridge, Famous area. *Oceanology Acta* **1**:73-86.

Honnorez, J. 1969. La formation actuelle d'un gisement sousmarin de sulfures fumerolliens à Vulcano (Mer Tyrrhenienne). *Mineralium Deposita* **4**:114-131.

Honnorez, J., B. M. Honnorez-Guerstein, J. Valette, and A. Wanschukuhn. 1973. Present day formation of an exhalative sulfide deposit at Vulcano (Tyrrhenian Sea), Part II: Active crystallization of fumarolic sulfides in the volcanic sediments of the Baia di Levante. In G. C. Armstatz and A. J. Bernard, eds., *Ores in Sediments,* Int. Un. Geol. Sci., Series A., no. 3, Berlin-Heidelberg: Springer-Verlag, pp. 139-166.

Horowitz, A., and D. S. Cronan. 1976. The geochemistry of basal sediments from the North Atlantic Ocean. *Marine Geology* **20**:205-228.

Humphris, S. E., and G. Thompson. 1978. Trace element mobility during hydrothermal alteration of oceanic basalts. *Geochim. et Cosmochim. Acta* **42**:127-136.

Hunt, J. M., E. E. Hayes, E. T. Degens, and D. A. Ross. 1967. Red Sea: Detailed survey of hot brine areas. *Science* **156**:514-516.

Hutchinson, R. W. 1973. Volcanogenic sulfide deposits and their metallogenic significance. *Ecol. Geology* **68**:1223-1246.

Hyndman, R. D., R. P. Von Herzen, A. J. Erickson, and J. Jolivet. 1976. Heat flow measurements in deep crustal holes on the Mid-Atlantic Ridge. *Jour. Geophys. Research* **81**:4053-4060.

Jenkins, W. J., and W. B. Clarke. 1976. The distribution of ^{3}He in the western Atlantic Ocean. *Deep-Sea Research* **23**:481-494.

Jenkins, W. J., M. A. Beg, W. B. Clarke, P. J. Wangersky, and H. Gray. 1972. Excess ^{3}He in the Atlantic Ocean. *Earth and Planetary Sci. Letters* **16**:122-126.

Jenkins, W. J., P. A. Rona, and J. M. Edmond. 1979. Excess ^{3}He in the deep water over the Mid-Atlantic Ridge at 26°N: Evidence of hydrothermal activity. *Earth and Planetary Sci. Letters* **49**:39-44.

Karl, D. M., C. O. Wirsen, and H. W. Jannasch. 1980. Deep-sea primary production at the Galapagos hydrothermal vents. *Science* **207**:1345-1347.

Lalou, C., and E. Brichet. 1976. On some relationships between the oxide layers and the cores of deep sea manganese nodules. *Mineralium Deposita* **11**:267-277.

Lalou, C., E. Brichet, G. Poupeau, P. Romary, and C. Jehanno. 1979. Growth rates and possible age of a North Pacific manganese nodule. In *Marine Geology and*

Oceanography of the Pacific Manganese Nodule Province, J. L. Bischoff and D. Z. Piper, eds. New York: Plenum Press, 815-834.

Lapwood, E. R. 1948. Convection of a fluid in a porous medium. *Cambridge Philos. Soc. Proc.* **44**:508-521.

Langseth, M. G., Jr., and R. P. Von Herzen. 1970. Heat flow through the floor of the world oceans. In *The Sea,* vol. 4, Part 1, A. E. Maxwell, ed. New York: Wiley-Interscience, pp. 299-352.

Leinen, M., and D. Stakes. 1979. Metal accumulation rates in the central equatorial Pacific during Cenozoic time. *Geol. Soc. America Bull.* **90**:357-375.

LePichon, X., and M. G. Langseth, Jr. 1969. Heat-flow from the mid-ocean ridges and sea-floor spreading. *Tectonophysics* **8**:319-344.

Lister, C. R. B. 1974. On the penetration of water in hot rock. *Royal Astron. Soc. Geophys. Jour.* **39**:465-509.

Lister, C. R. B. 1980. "Active" and "passive" hydrothermal systems in the oceanic crust: Predicted physical conditions. In *The Dynamic Environment of the Ocean Floor,* F. T. Manheim and K. Fanning, eds. In press.

Lonsdale, P. 1977. Deep-tow observations at the mounds abyssal hydrothermal field, Galapagos Rift. *Earth and Planetary Sci. Letters* **6**:92-110.

Lonsdale, P. 1979. A deep-sea hydrothermal site on a strike-slip fault. *Nature* **281**: 531-534.

Lonsdale, P. F., J. L. Bischoff, V. M. Burns, M. Kastner and R. E. Sweeney. 1980. A high-temperature hydrothermal deposit on the seabed at a Gulf of California spreading center. *Earth and Planetary Sci. Letters* **49**:8-20.

Lowell, R. P. 1979. Hydrothermal circulation on mid-ocean ridge crests. Final Technical Report, NSF Grant #OCE-76-81876. In preparation.

Lowell, R. P. 1980. Topographically driven subcritical hydrothermal convection in the oceanic crust. *Earth and Planetary Sci. Letters* **49**:21-28.

Lowell, R. P., and P. A. Rona. 1976. On the interpretation of near-bottom water temperature anomalies. *Earth and Planetary Sci. Letters* **32**:18-24.

Lupton, J. R., R. F. Weiss, and H. Craig. 1977. Mantle helium in Red Sea brines. *Nature* **266**:244-246.

Macdonald, K. C., K. Becker, F. N. Spiess and R. D. Ballard. 1980. Hydrothermal heat flux of the "black smoker" vents on the East Pacific Rise. *Earth and Planetary Sci. Letters* **48**:1-7.

Miller, A. R., C. D. Densmore, E. T. Degens, F. C. Hathaway, F. T. Manheim, P. F. McFarlen, H. Pocklington, and A. Jokela. 1966. Hot brines and recent iron deposits in deeps of the Red Sea. *Geochim. et Cosmochim. Acta* **30**:341-359.

Miyashiro, A., F. Shido, and M. Ewing. 1971. Metamorphism in the Mid-Atlantic Ridge near 24°N and 30°N. *Royal Soc. London Philos. Trans. Ser. A* **268**:589-603.

Moody, J. B., ed. *Ophiolites and Oceanic Crust.* Stroudsburg, Pa.: Dowden, Hutchinson & Ross. In preparation.

Moore, W. S., and P. G. Vogt. 1976. Hydrothermal manganese crusts from two sites near the Galapagos spreading axis. *Earth and Planetary Sci. Letters* **29**:349-356.

Mottl, M. J., and H. D. Holland. 1978. Chemical exchange during hydrothermal alteration of basalt by seawater-I. Experimental results for major and minor components of seawater. *Geochim. et Cosmochim. Acta* **42**:1103-1116.

Muehlenbachs, K., and R. Clayton. 1976. Oxygen isotope composition of the oceanic crust and its bearing on seawater. *Jour. Geophys. Research* **8**:4365-4369.

Murray, J., and A. F. Renard. 1891. Report on deep sea deposits based on specimens collected during the voyage of the H. M. S. *Challenger* in the years 1872 to 1876. In *Report of the Scientific Results of the Voyage of H. M. S. Challenger During the Years 1872-1876,* C. W. Thompson, and J. Murray, eds. London: Her Majesty's Stationary Office, pp. 292-320.

Natland, J. H., B. Rosendahl, R. Hekinian, Y. Dmitriev, R. V. Fodor, R. M. Goll, M. Hoffert, S. E. Humphris, D. P. Mattey, N. Petersen, W. Roggenthen, E. L. Schrader, R. K. Srivastava, and N. Warren. 1979. Galapagos hydrothermal mounds: Stratigraphy and chemistry revealed by deep-sea drilling. *Science* **204**: 613-616.

Olafsson, J., and J. P. Riley. 1978. Geochemical studies on the thermal brine from Reykjanes (Iceland). *Chem. Geology* **21**:219-237.

Palmason, G. 1967. On heat flow in Iceland in relation to the Mid-Atlantic Ridge. In *Iceland and mid-ocean ridges,* S. Björnsson, ed. *Soc. Sci. Islandica* **38**:111-117.

Patterson, P. L., and R. P. Lowell. 1980. Numerical models of hydrothermal circulation for the intrusion zone at an ocean ridge axis. In *The Dynamic Environment of the Ocean Floor,* F. T. Manheim, and K. Fanning, eds. In press.

Picot, P., and M. Fevrier. 1980 Étude minéralogique d'échantillons du Golfe de Californie (campagne CYAMEX). Orleans, France: Documents du BRGM 20, 50 pp.

Piper, D. Z., H. H. Veeh, W. G. Bertrand, and R. L. Chase. 1975. An iron-rich deposit from the Northeast Pacific. *Earth and Planetary Sci. Letters* **26**:114-120.

Revelle, R. 1944. Marine bottom samples collected in the Pacific Ocean by the *Carnegie* on its seventh cruise. *Carnegie Inst. Washington Pub. 556,* Part 1, pp. 1-180.

Ribando, R. J., K. E. Torrence, and D. L. Turcotte. 1976. Numerical models for hydrothermal circulation in the oceanic crust. *Jour. Geophys. Research* **81**: 3007-3012.

Rona, P. A. 1977. Plate tectonics, energy and mineral resources: Basic research leading to payoff. *EOS* **58**:629-639.

Rona, P. A. 1978. Near-bottom water temperature anomalies: Mid-Atlantic Ridge crest at latitude 26°N. *Geophys. Research Letters* **5**:993-996.

Rona, P. A. 1980. TAG Hydrothermal field: Mid-Atlantic Ridge crest at latitude 26°N. *Jour. Geol. Soc. London:* in press.

Rona, P. A., and R. P. Lowell. 1978. Symposium report: Hydrothermal systems at oceanic spreading centers. *Geology* **6**:299-300.

Rona, P. A., and R. B. Scott, convenors. 1974. Symposium: Axial processes of the Mid-Atlantic Ridge. *EOS* **55**:292-295.

Rona, P. A., R. H. Harbison, B. G. Bassinger, R. B. Scott, and A. J. Nalwalk. 1976. Tectonic fabric and hydrothermal activity of Mid-Atlantic Ridge crest (lat. 26°N). *Geol. Soc. America Bull.* **87**:661-674.

Rona, P. A., B. A. McGregor, P. R. Betzer, G. W. Bolger, and D. C. Krause. 1975. Anomalous water temperature over the Mid-Atlantic Ridge crest at 26°N latitude. *Deep-Sea Research* **22**:611-618.

Rozanova, T. V., and G. N. Baturin. 1971. Hydrothermal ore shows on the floor of the Indian Ocean. *USSR Academy of Sciences, Oceanology* **11**:874-879. (English translation published by the American Geophysical Union.)

Schoell, M. 1976. Heating and convection within the Atlantis II deep geothermal system of the Red Sea. In *United Nations Symposium on the Development and Use of Geothermal Resources, 2nd, San Francisco, May 1975, Proceedings,* 3 vols. Washington, D.C.: Government Printing Office, pp. 583-590.

Schrader, E. L., B. R. Rosendahl, W. J. Furbish, and D. P. Mattey. 1980. Mineralogy and geochemistry of hydrothermal and pelagic sediments from the Mounds Hydrothermal Field, Galapagos spreading center: DSDP Leg 54. *Jour. Sedimentary Petrology* **50**:917-928.

Scott, M. R., R. B. Scott, J. W. Morse, P. R. Betzer, L. W. Butler, and P. A. Rona. 1978. Metal enriched sediments from the TAG Hydrothermal Field. *Nature* **276**:811-813.

Sclater, J. G., and J. Francheteau. 1970. The implications of terrestrial heat-flow observations on current tectonic and geochemical models of the crust and upper mantle of the earth. *Geophys. Jour.* **20**:509-542.

Seyfried, W. E., and J. L. Bischoff. 1977. Hydrothermal transport of heavy metals by seawater: The role of the seawater/basalt ratio. *Earth and Planetary Sci. Letters* **34**:71-77.

Seyfried, W. E., and M. J. Mottl. 1977. Origin of submarine metal-rich hydrothermal solutions: Experimental basalt-seawater interaction in a seawater-dominated system at 300°C, 500 bars. *Proc. of the Second Int. Symposium on Water-Rock Interaction,* I. A. G. C., Strasbourg, France, pp. IV173-IV180.

Shanks, W. C., III, and J. L. Bischoff. 1977. Ore transport and deposition in the Red Sea geothermal system: A geochemical model. *Geochim. et Cosmochim. Acta* **41**:1507-1519.

Shanks, W. C., III, and J. L. Bischoff. 1980. Geochemistry, sulfur isotope composition, and accumulation rates of Red Sea geothermal deposits. *Econ. Geology* **75**:445-459.

Sillén, L. G. 1967. Gibbs phase rule and marine sediments. In *Equilibrium concepts in natural water systems.* Washington, D.C.: American Chemical Society, pp. 57-69.

Sillitoe, R. H. 1972. Formation of certain massive sulfide deposits at sites of seafloor spreading. *Inst. Mining and Metallurgy Trans. Sect. B: Appl. Earth Sci.* **81**: B141-B148.

Sillitoe, R. H. 1973. Environments of formation of volcanogenic marine sulfide deposits. *Econ. Geology* **68**:1321-1336.

Skilbeck, J. N., and R. N. Anderson. 1979. Heat flow and convection in a two-layer porous medium: The oceanic crust and overlying sediments. *Jour. Geophys. Research.* In press.

Skornyakova, N. S., 1979. Zonal regularities in occurrence, morphology, and chemistry of manganese nodules of the Pacific Ocean. In *Marine Geology and Oceanography of the Pacific Manganese Nodules Province,* J. L. Bischoff, and D. Z. Piper, eds. New York: Plenum Press, p. 699-728.

Sleep, N. H., and T. J. Wolery. 1978. Egress of hot water from mid-ocean ridge hydrothermal systems—some thermal constraints. *Jour. Geophys. Research* **83**: 5913-5922.

Spiess, F. N., K. C. Macdonald, T. Atwater, R. Ballard, A. Carranza, D. Cordoba, C. Cox, V. M. Diaz Garcia, J. Francheteau, J. Guerrero, J. Hawkins, R. Haymon, R. Hessler, T. Juteau, M. Kastner, R. Larson, B. Luyendyk, J. D. Macdougall, S. Miller, W. Normark, J. Orcutt, and C. Rangin (RISE Project Group, 1980. East Pacific Rise: Hot springs and geophysical experiments. *Science* **207**:1421-1444.

Spooner, E. T. C. 1974. Sub-seafloor metamorphism, heat and mass transfer: An additional comment. *Contr. Mineralogy and Petrology* **45**:169-173.

Spooner, E. T. C. 1976. The strontium isotopic composition of seawater and seawater-oceanic crust interaction. *Earth and Planetary Sci. Letters* **31**:167-174.

Spooner, E. T. C., and C. J. Bray. 1977. Hydrothermal fluids of seawater salinity in ophiolitic sulfide ore deposits in Cyprus. *Nature* **266**:808-812.

Spooner, E. T. C., R. D. Beckinsale, W. S. Fyfe, and J. D. Smewing. 1974. 0^{18} enriched ophiolitic metabasic rocks from E. Ligurea (Italy), Pindos (Greece), and Troodos (Cyprus). *Contr. Mineralogy and Petrology* **47**:41-62.

Stieltjes, L. 1976. Research for a geothermal field in a zone of oceanic spreading: Example of the Asal Rift (French territory) of the Afars and the Issas, Afar depression, East Africa. In *United Nations Symposium on the Development and Use of Geothermal Resources, 2nd, San Francisco, May 1975, Proceedings,* 3 vols. Washington D.C.: Government Printing Office, pp. 613-623.

Stone, C., and P. Fan, 1978. Hydrothermal alteration of basalts from Hawaii Geothermal Project Well-A, Kilauea, Hawaii. *Geology* **6**:401-404.

Strong, D. F., ed. 1976. Metallogeny and plate tectonics. *Geol. Assoc. Canada Spec. Paper 14,* 660 pp.

Talwani, M., C. C. Windish, and M. G. Langseth Jr. 1971. Reykjanes Ridge crest: A detailed geophysical study. *Jour. Geophys. Research* **76**:473-517.

Temple, D. G., R. B. Scott, and P. A. Rona. 1979. Geology of a submarine hydrothermal field, Mid-Atlantic Ridge, 26°N lat. *Jour. Geophys. Research.* **84**:7453-7466.

Toth, J. R. 1980. Deposition of submarine crusts rich in manganese and iron. *Geol. Soc. America Bull.* **91**:44-54.

Walker, W., ed. 1976. *Metallogeny and Global Tectonics.* Stroudsburg, Pa.: Dowden, Hutchinson & Ross, 413 pp.

Welhan, J. A., and H. Craig. 1979. Methane and hydrogen in East Pacific Rise hydrothermal fluids. *Geophys. Res. Letters* **6**:829-831.

White, D. E. 1968. Environment of generation of some rare-metal ore deposits. *Econ. Geology* **63**:301-335.

Whitmarsh, R. B., O. E. Weser, D. A. Ross, et al. 1974. *Initial Reports of the Deep-Sea Drilling Project,* vol. 23. Washington, D. C.: Government Printing Office, 1180 pp.

Williams, D. L., K. Green, T. H. van Andel, R. P. Von Herzen, J. R. Dymond, and K. Crane. 1979. The hydrothermal mounds of the Galapagos Rift: Observations with DSRV ALVIN and detailed heat flow studies. *Jour. Geophys. Research* **84**:7467-7484.

Wright, J. B., ed. 1977. *Mineral Deposits, Continental Drift and Plate Tectonics.* Stroudsburg, Pa.: Dowden, Hutchinson & Ross, 417 pp.

Vidal, V. M. V., F. V. Vidal, and J. D. Isaacs. 1978. Coastal submarine hydrothermal activity off northern Baja, California. *Jour. Geophys. Research* **83**:1757-1777.

Zelenov, K. K. 1964. Iron and manganese in exhalations of the submarine Banu Wulu volcano (Indonesia). *Akad. Nauk. SSSR Doklady* **155**:1317-1320.

AUTHOR CITATION INDEX

SUBJECT INDEX

About the Editors

PETER A. RONA is senior research geophysicist with the Atlantic Oceanographic and Meteorological Laboratories of the National Oceanic and Atmospheric Administration (NOAA) in Miami, Florida, and adjunct professor at the University of Miami. He earned a Ph.D. degree in marine geology and geophysics from Yale University in 1967. Dr. Rona initiated and directs the NOAA Trans-Atlantic Geotraverse (TAG) Project, an interdisciplinary, multi-institutional, international cooperative investigation of the seafloor along a corridor across the central North Atlantic. He initiated a new NOAA project in 1975, Metallogenesis at Dynamic Plate Boundaries, which seeks to determine how metals are concentrated in oceanic crust and what this implies for world metal resources and pollution of the ocean environment. Dr. Rona is a consultant on seafloor resources to national and international organizations.

ROBERT P. LOWELL is an associate professor of geophysics at Georgia Institute of Technology. He holds a B.S. and M.S. in physics from Loyola University (Chicago) and Oregon State University, respectively, and a Ph.D. in geophysics from Oregon State University. Dr. Lowell's main research efforts have been in theoretical thermal geophysics, with emphasis on thermal convection processes in geothermal systems. In particular, he has developed models of convection in permeable media in which the permeability is controlled by discrete fault and/or fracture zones. His research on ocean ridge hydrothermal convection problems has been supported by grants from the National Science Foundation, and his work on continental geothermal problems has been supported by the U.S. Geological Survey. Dr. Lowell has also been involved in research on thermal models of continental margin sedimentary basins and in coastal oceanography. He is a member of the American Geophysical Union and Sigma Xi.